This book investigates the various causes of thermodynamic instability in metallic microstructures. Materials theoretically well designed for a particular application may prove inefficient or even useless unless stable under normal working conditions. The authors examine current experimental and theoretical understanding of the kinetics behind structural change. The different changes are analysed on the basis of the different driving forces for structural change: chemical, strain, interfacial, gradient etc. In each case the currently accepted mechanisms are described, analysed kinetically and then the analysis is tested against experimental evidence.

The entire text of the first edition of this popular textbook has been updated in this second edition, and a completely new chapter on highly metastable alloys has been added by a new coauthor. There are practically no thermodynamically stable metallic materials of importance to materials scientists, and the degree to which the kinetic stability of the material outweighs the thermodynamic instability is therefore very important. The useful working life of the material depends on this balance, and changes that increase total entropy or decrease free energy are nearly always possible. If, therefore, the structure is initially produced to an optimum, such changes will degrade the properties of the material.

A comprehensive and well-illustrated text, accompanied by ample references, this volume will allow final year undergraduates, graduate students and research workers in related fields to investigate in detail the stability of microstructure in metallic systems.

# Cambridge Solid State Science Series

**Stability of microstructure in metallic systems**

Second edition

TITLES IN PRINT IN THIS SERIES

**S.W.S. McKeever**
Thermoluminescence of Solids

**D.P. Woodruff and T.A. Delchar**
Modern Techniques of Surface Science – Second Edition

**P.L. Rossiter**
The Electrical Resistivity of Metals and Alloys

**D.I. Bower and W.F. Maddams**
The Vibrational Spectroscopy of Polymers

**S. Suresh**
Fatigue of Materials

**J. Zarzycki**
Glasses and the Vitreous State

**R.A. Street**
Hydrogenated Amorphous Silicon

**T-W. Chou**
Microstructural Design of Fiber Composites

**A.M. Donald and A.H. Windle**
Liquid Crystalline Polymers

**B.R. Lawn**
Fracture of Brittle Solids – Second Edition

**T.W. Clyne and P.J. Withers**
An Introduction to Metal Matrix Composites

**V.J. McBrierty and K.J. Packer**
Nuclear Magnetic Resonance in Solid Polymers

**R.H. Boyd and P.J. Phillips**
The Science of Polymer Molecules

**D. Hull and T.W. Clyne**
An Introduction to Composite Materials – Second Edition

**J.W. Martin, R.H. Doherty and B. Cantor**
Stability of Microstructure in Metallic Systems – Second Edition

**J.S. Dugdale**
The Electrical Properties of Disordered Metals

**M. Nastasi, J.W. Mayer and J.K. Hirvonen**
Ion–Solid Interactions

**T.G. Nieh, J. Wadsworth and O.D. Sherby**
Superplasticity in Metals and Ceramics

# Stability of microstructure in metallic systems

## Second edition

J.W. Martin
*Department of Materials, University of Oxford*

R.D. Doherty
*Materials Engineering, Drexel University*

B. Cantor
*Department of Materials, University of Oxford*

PUBLISHED BY THE PRESS SYNDICATE OF THE UNIVERSITY OF CAMBRIDGE
The Pitt Building, Trumpington Street, Cambridge CB2 1RP, United Kingdom

CAMBRIDGE UNIVERSITY PRESS
The Edinburgh Building, Cambridge CB2 2RU, United Kingdom
40 West 20th Street, New York, NY 10011-4211, USA
10 Stamford Road, Oakleigh, Melbourne 3166, Australia



First published 1976
First paperback edition 1980
Second edition 1997

Typeset in 10¼ on 13½pt Monotype Times

*A catalogue record for this book is available from the British Library*

*Library of Congress cataloguing in publication data*

Martin, J.W. (John Wilson), 1926–
Stability of microstructure in metallic systems / J.W. Martin,
R.D. Doherty, B. Cantor – 2nd ed.
p. cm.
Includes bibliographical references and index.
ISBN 0-521-41160-2 (hb). – ISBN 0-521-42316-3 (pb)
1. Physical metallurgy. 2. Stability. Metal crystals –
Defects. I. Doherty, R.D. (Roger Davidge), 1939–
II. Cantor, B. III. Title.
TN690.M2664 1997
620'.94–dc20 96-11097 CIP

ISBN 0 521 41160 2 hardback
ISBN 0 521 42316 3 paperback

Transferred to digital printing 2005

# Contents

# Preface to the first edition

Man's use of materials, both as a craft and more recently as a science, depends on his ability to produce a particular microstructure with desirable properties in the material when it has been fabricated into a useful object. Such a microstructure occurs, for example, in a steel crankshaft heat-treated for maximum strength, a glass lens heat-treated for fracture resistance or a small crystal of silicon containing non-uniform distributions of solute acting as a complex electronic circuit. Such microstructures are almost always thermodynamically unstable. This situation arises since for any alloy there is only one completely stable structure and there is an infinite number of possible unstable microstructures. The one with the best properties is therefore almost always one of the unstable ones. The desired structure is usually produced by some combination of heat-treatment, solute diffusion and deformation, in the course of which the transformation is arrested, normally by cooling to room temperature, at the right time to obtain the optimum structure. The success of these processes, many of which were derived from craft skills, to give materials with good strength, toughness, electrical properties, etc., is an essential part of current technology. There is, however, a price to be paid in that all these structures are potentially unstable, so that the structures can, and frequently do, transform with time into less desirable forms, especially if used at elevated temperatures.

The science of phase transformations is concerned with transformations to the desired microstructure and equally with the instability of this microstructure. The science of phase transformation has greatly advanced in the last thirty years, both with improved metallographic techniques and the development of successful quantitative theories of the mechanisms and kinetics of structural change. These subjects have been discussed in many excellent textbooks and review articles which have enabled the student to keep well up to date with advances in the subject. However, the emphasis in almost all these publications has been on the processes occurring during heat-treatments that are intended to produce the particular microstructures with desirable properties. The correlation of these properties, usually mechanical but sometimes magnetic or electrical, with microstructure has also been reviewed frequently. The stability of the microstructure produced by heat-treatment has, however, received no systematic consideration in textbooks. Microstructural stability is a topic that has been the subject of considerable research activity and in many cases the state of knowledge is already very extensive and frequently the agreement between theory and experiment is remarkably good. Despite these scientific successes and the importance and interest of the subject, there has been only limited incorporation of the state of knowledge into general textbooks of physical metallurgy and no book, as far as is known to the authors, has been published with microstructural stability as its main concern.

The present volume was written to try to correct this shortcoming by providing a book that would summarise the main elements of the current understanding of the stability of microstructure. We have attempted to describe the state of knowledge as it appears to us in the years 1972–4, although limitations of space, time and expertise precluded every topic being described exhaustively, so that we are only too aware that there are likely to be important contributions that we have not included. These omissions should be seen as criticisms of ourselves and not of the scientists whose work we have not included; a major deliberate omission is that of the effect of surface or interface reactions (e.g. corrosion) on microstructure. In particular this volume reflects the regrettable but nearly universal tendency of English and American scientists to read mainly English language papers.

The approach adopted in the book has been to consider the instabilities that occur as classified by the form of potential free energy decrease that makes the existing microstructure unstable. This approach is discussed in some detail in the first chapter, and the subsequent chapters then deal in turn with each of the various types of free energy decrease.

The authors would like to acknowledge the important contributions made by numerous colleagues in Oxford, at Sussex, and elsewhere in helping us to formulate the ideas described in this book. We would like to give particular acknowledgement to Professor G.W. Greenwood who, in addition to his published work in this area, first suggested the form of approach adopted in this book, and to Professor R.W. Cahn for his initial stimulation of the project and his continued interest and encouragement. We are grateful to the many scientists who have given us permission to use their micrographs and in many cases have provided copies for publication and to the many publishers of scientific journals from which we have quoted for permission to reproduce figures and micrographs.

July 1975

# Preface to the second edition

For the second edition, the objectives and the approach previously used have been maintained. In many areas the science base of the subject has shown little change since the first edition and here the text has only been modified by improved examples where available. In other areas the subject has advanced significantly and the text has been updated with the insights. Topics previously covered incompletely, notably the highly unstable microstructure produced initially by rapid solidification but subsequently by other processing routes, have been greatly expanded. Other significant developments that have taken place include the detailed experimental studies of homogeneous nucleation, the growth of Widmanstätten precipitates and precipitate coarsening, and the new insights into the nucleation of recrystallisation, and grain growth and its inhibition by second-phase particles. In other areas, despite the importance of the subject, progress has been disappointingly slow. As in the first edition, we have tried to indicate where there are unsolved problems. The first edition provided the authors with a rich supply of fruitful research topics and we hope that this was also true for our readers and will be equally true for the second edition. Microstructural stability of metallic (and other industrially important) materials remains a field of research with many scientific and potential engineering applications.

The authors are again grateful to Professor Robert Cahn FRS for his efforts to get this volume completed and his much-appreciated enthusiasm. We thank our many colleagues whose work and enthusiasm have stimulated our own efforts – these are fully referenced here. We are again grateful to many scientists and to the publishers of their research and scholarship for permission to use their work and to reproduce their figures. We apologize to these people whose work we have overlooked or, worse, misunderstood, and would appreciate hearing of these and other mistakes. We look forward to seeing our ideas challenged by new research. If any studies are initiated by questions or misunderstandings in this volume we shall have succeeded in one of our tasks. We would be delighted to hear of any engineering successes helped by the ideas in this volume, even if the source was the original literature rather than our review. Our electronic and postal addresses are given below to facilitate any responses from readers of this book.

1997 J.W. Martin, R.D. Doherty, B. Cantor

John W. Martin
Oxford Centre for Advanced Materials and Composites,
Department of Materials, Oxford University, Parks Road,
Oxford, OX1 3PH, UK
tel (44) 1865-273700
fax (44) 1865-283333
e-mail: john.martin@materials.ox.ac.uk

Roger D. Doherty
Department of Materials Engineering,
Drexel University, Philadelphia PA, USA 19104
tel (1) 215-895-2330
fax (1) 215-895-6760
e-mail: dohertrd@duvm.ocs.drexel.edu

Brian Cantor
Department of Materials, Oxford University, Parks Road,
Oxford, OX1 3PH, UK
tel (44) 1865-273737
fax (44) 1865-273789
e-mail: brian.cantor@materials.ox.ac.uk

# Chapter 1

# The general problem of the stability of microstructure

## 1.1 Introduction

Materials science in general and metallurgy in particular are concerned with understanding both the structure of useful materials, and also the relationship between that structure and the properties of the material. On the basis of this understanding, together with a large element of empirical development, considerable improvements in useful properties have been achieved, mainly by changes in the microstructure of the material. The term *microstructure* as normally used covers structural features in the size range from atoms (0.3 nm) up to the external shape of the specimen at a size of millimetres or metres. These structural features include the composition, the crystal structure, the grain size, the size and distribution of additional phases and so on, all of which are controlled by the normal methods of alloying, fabrication and heat treatment.

The materials scientist, having achieved some sort of optimum microstructure for a particular property or application, has not completed the task. The important area of the *stability* of the microstructure remains to be considered. This concern arises since almost none of the useful structures in materials science are *thermodynamically* stable: changes that will increase the total entropy or decrease the material's free energy are almost always possible. So if the original structure was an optimum one then such changes will degrade the material's structure and properties. This idea of inherent instability, particularly of metallic materials, is widely appreciated in the context of corrosion, where a metal, having been won from its ore, continually attempts to return to a chemically more stable form, normally that of an oxide. Although a discussion of corrosion does not concern us here, the characteristic that allows a metal to remain a metal is the same as that which maintains unstable microstructures – the *slow kinetics* of the degradation. The rates of structural and chemical change must be sufficiently slow to allow the material a useful lifetime.

In subsequent chapters of this book we shall look at the stability of metallic microstructure with respect to the various examples of thermodynamic instability. In each example, the form and amount of potential free energy decrease will be established first, then the mechanisms by which the decrease of free energy can occur will be examined and finally, using these ideas, the current experimental and theoretical understanding of the *kinetics* of the structural change will be discussed. It is not usually profitable to divide microstructural transformations into separate classes – those giving desirable microstructures and those in which desirable microstructures degrade. An obvious illustration of this is the process of grain growth (§5.8). For strength and toughness at room temperature a small grain size is usually needed. However, at temperatures above about half of the absolute melting temperature ($0.5T_m$), deformation by grain boundary sliding and diffusion creep makes a fine-grain-sized material weaker than the same material with a large grain size. As a consequence of the difficulty of determining which changes are desirable and which not, we shall discuss the structural changes irrespective of their desirability from an engineering standpoint. The distinction between different types of transformation will be made solely on the basis of the type of thermodynamic driving force involved in the transformation.

There are two different types of structural instability, the first of which is a genuine *instability* and the second is normally described as showing *metastability*. In the case of metastability, before a system can achieve the more stable state with lower energy, it must pass through an intermediate transition state having *higher* energy. This requirement of passage through a higher energy state then acts as a barrier to the transformation. Unless the necessary *activation energy* can be supplied to the system the reaction cannot take place and the system will spend extended times in the initial metastable state. For an instability, however, no such barrier exists. A clear mechanical analogy has been provided by Cahn (1968) and is shown in fig. 1.1. Here the metastable situation is provided by a rectangular block balanced on its narrow end while the unstable situation is a wedge balanced on its end.

The distinction between metastable and unstable structural transformations in materials was first made by Gibbs (1878), the founder of the subject. In a Gibbs Type I transformation, the energy barrier exists in the need to nucleate a critical-sized nucleus of the new structure before the lower energy transformed structure can grow. This process is usually described as one of *nucleation and growth.* It involves a transformation described as being large in the magnitude of the structural change but localised in a small part of the sample. That is, the partially trans-

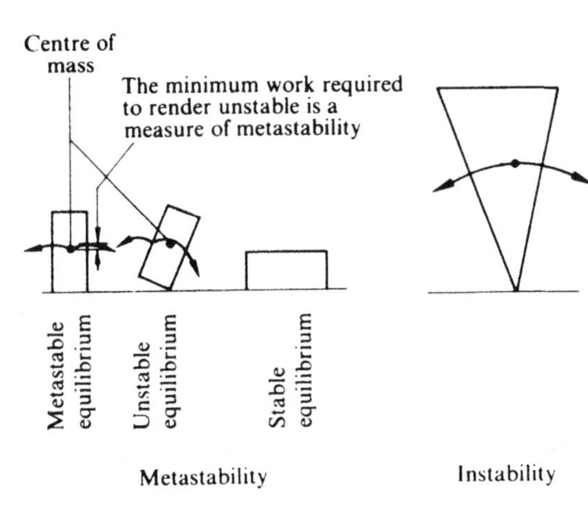


**Figure 1.1** Difference between metastability and instability. (After Cahn 1968, courtesy of AIME.)

formed structure is *heterogeneous* with small regions of the new structure embedded in the untransformed initial structure. The energy barrier to this type of transformation arises mainly from the energy of the interface between the two different structures. In the alternative mode of transformation, Gibbs Type II, the unstable structure is able to transform into the more stable form by passing through a continuous series of intermediate stages, each of which has a steadily reducing energy. Gibbs Type II transformations are ones that, at least initially, have a structural change that is small in magnitude but is unlocalised – that is, there is a small change in structure but one taking place *homogeneously* throughout the sample. The best-known example of this type of transformation is that provided by the 'spinodal decomposition' of a solid solution (§2.2.1). In spinodal decomposition a supersaturated solution decomposes into two phases with different compositions but each having the same crystal structure and orientation.

Gibbs' distinction between metastable and unstable transformations is well established. It should be noted, however, that both types of transformation involve atomic mobility and for this mobility thermal activation is necessary to overcome the energy barrier to move an atom from one site to another. As a result, the real distinction between the two types of change becomes that in a metastable change *many* atoms are involved in the activation step while in unstable changes only single atom activation is necessary. As a consequence, in none of the changes discussed in this book is the *inertia* of atoms a significant factor in the kinetics of the change. The necessity of providing thermal atomic activation means that at temperatures low compared with the absolute melting point, most of the changes discussed here will occur at a negligible rate. Significant rates of change will occur only at temperatures somewhat greater than about $0.3T_m$. For changes requiring atomic movement over more than a few atomic spacings the temperature for the onset of a significant rate of reaction is normally $0.5T_m$. That is, the instabilities do not concern most metals used at room temperature. Low melting point lead alloys are, however, exceptions to this restriction.

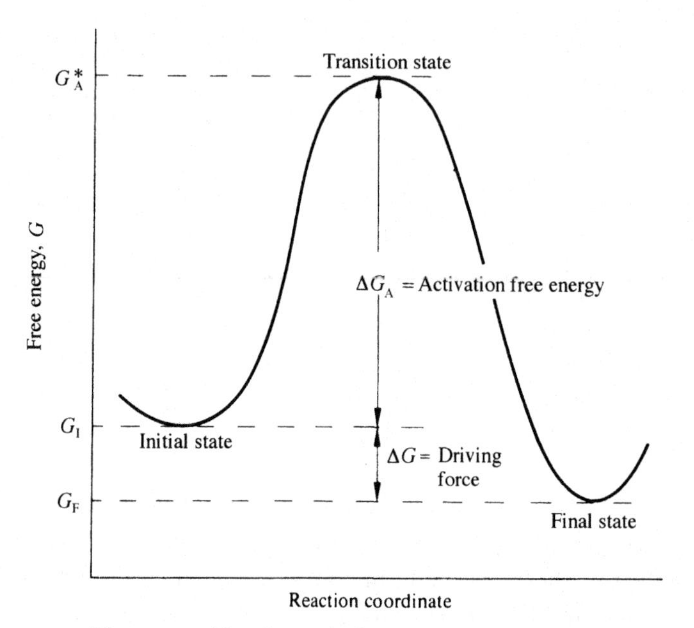


**Figure 1.2** The change in free energy of an atom as it takes part in a transition. The 'reaction coordinate' is any variable defining the progress along the reaction path.

The parameter driving all solid state structural transformations at constant temperature, $T$, and pressure, $P$, is the reduction of Gibbs free energy, $G$ of the material:

$$G = H - TS = U + PV - TS \tag{1.1}$$

Here $H$ is the enthalpy, $S$ the entropy, $U$ the internal energy and $V$ the volume of the material. The alternative form of free energy is the Helmholtz free energy, $F$, eq. (1.2), which seeks a minimum value for reactions at constant temperature and constant volume:

$$F = U - TS \tag{1.2}$$

Since constant pressure, at 1 atm, is the usual condition that most engineering materials experience we shall use the Gibbs free energy, $G$, which we will for the rest of this volume refer to merely as the free energy. In chapter 5, some consideration is given of microstructural changes that can take place in a *temperature gradient* so then it is incorrect to use the condition of a minimisation of $G$. The thermodynamic parameter that must be used is the maximisation of total entropy. The total entropy includes both the entropy change of the material and the entropy changes of the surroundings. The value of using the concept of free energy minimisation, at constant temperature, is that we need only consider changes in the material and can ignore changes in the surroundings.

The change in free energy of an atom as it takes part in a transition may be represented by fig. 1.2, in which the 'reaction coordinate' is any variable defining the progress along the reaction path. $G_I$ is the mean free energy of an atom in the initial configuration and $G_F$ is that after transformation. $\Delta G=(G_F-G_I)$ is negative and provides the driving 'force' for the transformation. The movement of the atom from the initial state to the final state is opposed by the free energy barrier $\Delta G_A$, so until the atom can temporarily acquire the necessary extra energy to carry it over this barrier, it will remain in the initial (metastable) state. The smallest energy increment ($\Delta G_A$) which will allow the atom to go over the barrier is the *activation free energy* of the reaction. An atom having the maximum free energy, $G_A^*$, in the 'transition' or 'activated' state is unstable and will either fall forward into the transformed state or back into the initial state.

The additional free energy needed for the atom to undergo the transformation is supplied by thermal energy, so it is to be expected that the reaction velocities will depend upon the magnitude of $\Delta G_A$ and upon the form of the energy distribution resulting from random thermal motion. In most problems of practical interest $\Delta G_A$ is much greater than the average thermal energy $kT$, so the rate of surmounting the energy barrier is then very small and an atom held up by the barrier spends most of its time near $G_I$.

By treating the activated state as a quasi-equilibrium state, with the number of particles per unit volume in the transition state $C_A$ and in the initial state $C_I$, then, assuming equilibrium exists between the two configurations, an equilibrium constant, $K^*$, may be written as:

$$K^*=C_A/C_I \tag{1.3}$$

$K^*$ is given by the thermodynamic relation:

$$K^*=\exp(-\Delta G_A/kT) \tag{1.4}$$

Since the rate of transformation will be proportional to $C_A$ we can write the rate as given by:

$$\text{Rate}=\text{constant}\cdot C_I\cdot \exp(-\Delta G_A/kT) \tag{1.5}$$

The constant in eq. (1.5) involves the vibration frequency $\nu\,(s^{-1})$ of atoms in the activated state. This frequency is usually approximated as the atomic vibration frequency of atoms in the initial state.

It is often more convenient to discuss the thermodynamics in terms of an enthalpy of activation $\Delta H_A$ and an entropy of activation $\Delta S_A$ which are related to

$\Delta G_A$ by the standard equation:

$$\Delta G_A = \Delta H_A - T\Delta S_A \tag{1.1a}$$

Eq. (1.5) may then be written:

$$\text{rate} = A\exp(-\Delta H_A/kT) \tag{1.6}$$

That is, the entropy component, $\Delta S_A$, is included within the pre-exponential frequency factor, $A$. Eq. (1.6) is the well-known Arrhenius equation and applies to many problems in atomic kinetics. To fit it to experimental observations it is written in the logarithmic form:

$$\ln \text{rate} = \text{constant} - \Delta H_A/kT \tag{1.7}$$

Eq. (1.7) shows that the logarithm of the rate of change should vary linearly with the reciprocal of the absolute temperature. A vast number of physical and chemical reactions, both homogeneous and heterogeneous, behave in this way, indicating that the standard Arrhenius model for the thermally activated reaction rate is valid.

The great effect of temperature on reaction rates is due to the exponential energy factor. Taking a typical value for metallurgical processes, for example, of $\Delta H_A$ of 2 eV per atom, with Boltzmann's constant $k$ of $8.6\times10^{-5}$ eV K$^{-1}$, we obtain at 1000 K, $\exp(-\Delta H_A/kT)=10^{-10}$, and at 300 K, $\exp(-\Delta H_A/kT)=10^{-33}$. This implies that a reaction taking 1 s at 1000 K will take about $3\times10^{14}$ y at room temperature. Such changes in rates are the basis of the usual heat-treatment processes to produce structures that, after cooling, are metastable at room temperature for very long times. The larger the value of $\Delta H_A$, the more rapid is the variation of reaction rate with temperature, and the easier it becomes to preserve by quenching to room temperature a structure which exists in equilibrium at high temperature but which would decompose during slow equilibrium cooling.

## 1.2 Driving forces for microstructural change – reduction of free energy

The term driving 'force' is not formally appropriate for these changes since the free energy is usually not a simple function of distance, and so the idea of a force, the differential of free energy with respect to distance, is not really appropriate. However, in the qualitative sense of causing the transformation we think of the magnitude of the free energy changes as indicating the relative 'driving forces' for the different changes to be discussed. In addition, many changes are driven

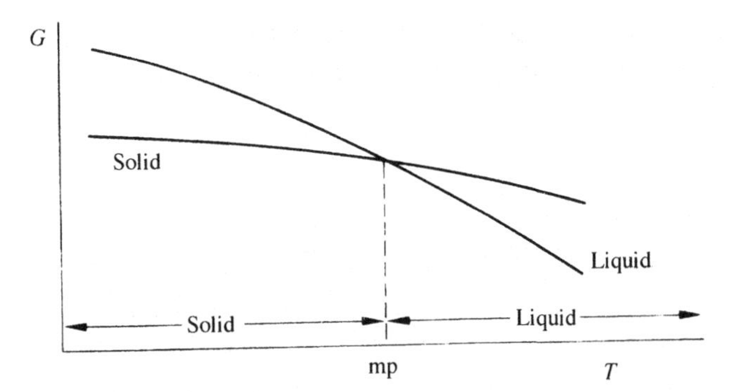


**Figure 1.3** Variation of free energy with temperature for a pure metal.

by a constant free energy change *per unit volume* in units of J m$^{-3}$, and for all such changes there is a constant pressure, a force per unit area of interface at the interface between parent and daughter phases. This pressure has the same magnitude as the free energy per unit volume but its units are N m$^{-2}$.

The driving forces for the changes to be discussed in chapters 2, 4 and 5 are the changes in *chemical, strain* and *interfacial* free energies respectively. What is meant by chemical free energy is the free energy of strain-free regions of a phase, by strain energy is the increase of free energy caused by short and long range elastic strains, and by interfacial free energy the extra free energy of atoms at interfaces between phases and at grain boundaries. There cannot be a complete distinction between these concepts since, for example, the chemical free energy of a non-uniform solute distribution (§2.1) will have a strain energy contribution if the volume per atom varies with the solute content. A change in lattice parameter requires either elastic distortions (a strain) or an array of dislocations (an interface) to accommodate the change of volume. Similarly, the presence of a curved interface between two different phases will alter the chemical composition of the two phases (§5.5.2). Despite these qualifications it is usually possible and useful to separate different microstructural changes on the basis of the principal component of the free energy driving the change.

In order to consider a particularly straightforward change in chemical free energy we can start with the solidification of a pure metal. Fig. 1.3 illustrates schematically how the free energy of a pure metal changes with temperature. At the melting point, $T_m$, the free energies of liquid and solid phases are equal. Below the melting point the solid has the lower free energy and is therefore the stable phase while above the melting point the reverse is true. The divergence of the free energy curves of the two phases as the temperature falls below the melting point produces a steadily increasing driving force for crystallisation of the supercooled liquid metal. The magnitude of the free energy change on crystallisation can be easily determined using this figure. Making the simplifying assumption that the enthalpy and entropy changes of crystallisation, $\Delta H$ and $\Delta S$, do not change significantly with temperature:

$$\Delta S = \Delta H / T_m \quad (1.8)$$

$$\Delta G = \Delta H (T_m - T)/T_m = \Delta H \Delta T / T_m \quad (1.9)$$

The assumptions, that $\Delta H$ and $\Delta S$ do not change with temperature, are valid if the specific heats of the two phases are the same, as shown, for example, by Gaskell (1983). The negative slope of either phase, $-dG/dT$ at constant pressure, is the *entropy* of that phase. The increase of entropy, $\Delta S$, on melting leads to the *discontinuous* change of $dG/dT$ at the melting point seen in fig. 1.3. The discontinuity in the *first differential* of $G$ leads to Ehrenfest's descriptions of such equilibrium structural changes as being '*first-order*' transformations. The value of $G$ itself changes continuously through the melting range. A few metallurgically important reactions, such as some, but not all, disordering transformations, involve a discontinuity in the second differential of $G$ with respect to temperature and are thus 'second-order' transformations.

The value of the latent heat of fusion, $\Delta H$, for a typical metal such as copper is 13 kJ mole$^{-1}$, so the change of free energy of crystallisation will increase from zero at $T_m$ (1356 K) to a maximum value of $-13$ kJ mole$^{-1}$ as the temperature falls towards absolute zero (crystallisation of amorphous alloys at low temperatures is discussed in §3.4.4).

Solid state polymorphic changes such as that in iron at 1183 K and in tin at 291 K have lower heats of transformation: 0.9 and 2.2 kJ mole$^{-1}$, respectively. As a result the driving forces in solid state changes are smaller than those for solidification of the same metal at similar undercoolings. This difference is, however, offset to some extent by the larger undercoolings, $\Delta T$, often encountered in solid state phase transformations. Jones and Chadwick (1971), using the standard methods of thermodynamics, have shown how the simple expression for the driving forces for crystallisation or solid state polymorphic changes is changed if the assumption of constant specific heat is relaxed. Due to compensating changes in the variation of both $\Delta H$ and $\Delta S$ with temperature, the effect of non-identical specific heats on the resulting $\Delta G$ values is usually rather insignificant, however.

Other examples of changes in chemical free energy are those occurring during diffusion in single-phase materials, for example, the homogenisation of 'cored' solid solutions, and in the precipitation of a second phase from a supersaturated solid solution. These changes can usually only be discussed semi-quantitatively since the exact forms of the relevant free energy–composition diagrams, see fig. 1.4, are not normally known. Approximate values of the free energy changes can be obtained by assuming, in the homogenisation case, that the solid solution is

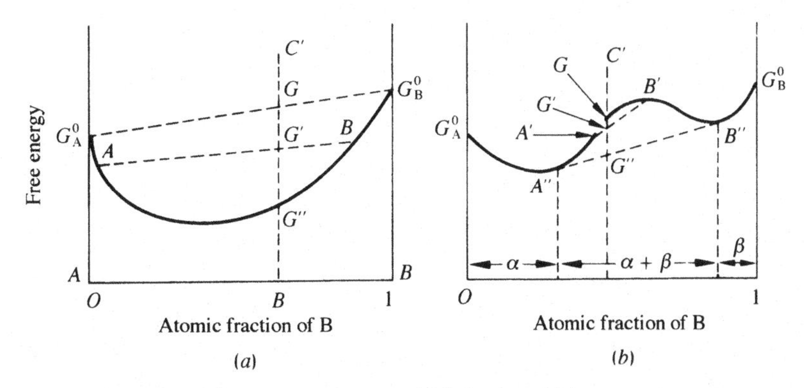


**Figure 1.4** Free energy curves which lead to (*a*) homogeneous, (*b*) heterogeneous equilibrium.

ideal, and, for precipitation, that the activity coefficient of the solute remains constant in the composition range between the supersaturated solid solution and the equilibrium solubility. A constant activity coefficient is found for 'dilute' solutions but how dilute a solution must be for this to be valid is, again, not usually known.

For an ideal solid solution the free energy of mixing, $\Delta G_m$, is given by:

$$\Delta G_m = -RT[(1-N_B)\ln(1-N_B)+N_B\ln N_B] \qquad (1.10)$$

$N_B$ is the atom fraction of component $B$ in the binary solution. An extreme value of the change would be that achieved by the interdiffusion of two pure elements forming an equiatomic solid solution ($N_B=0.5$). The value of $\Delta G_m$ in these circumstances is $-0.7RT$. At 1000 K this energy change is $-5.7$ kJ mole$^{-1}$. In the less extreme case of a cored solid solution, where the change of composition is less, a smaller value of driving force is found. For example, Doherty and Feest (1968) reported for cast copper–nickel ingots that the range of composition seen was from 48 to 77.5 wt% copper. An estimate of the free energy change on full homogenisation can be made by making the crude approximation that 50% of the alloy was at the average composition, 60% copper, 30% had the minimum composition and 20% had the maximum copper content. These assumptions yielded the result that the free energy change was $=-1.9$ kJ mole$^{-1}$ at 1000 K.

The free energy changes associated with precipitation are illustrated in fig. 1.5. Here the $\alpha$ matrix at an alloy composition of $N_0$ (given as the atomic fraction of B) is supersaturated with respect to the equilibrium solubility $N_\alpha$ while the stable $\beta$ phase has a composition $N_\beta$. The reduction of free energy on precipitation of the $\beta$ phase is $\Delta G_1$ per mole of *alloy*. However the reduction of free energy per mole of *precipitate* is the much larger value, $\Delta G_2=IJ$. This result is readily demonstrated using solution thermodynamics and fig. 1.5. To

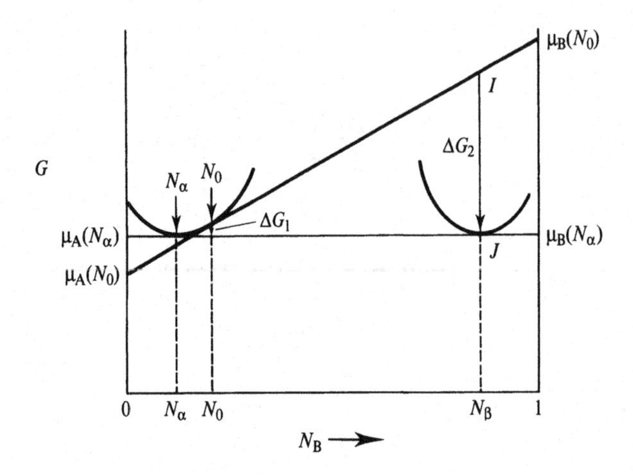


**Figure 1.5** Free energy changes on precipitation from supersaturated solid solution.

make 1 mole of the $\beta$ phase we must first remove $N_\beta$ of a mole of B atoms and $(1-N_\beta)$ of a mole of A atoms from the $\alpha$ phase. The B and A atoms have partial molar free energies, chemical potentials, of $\mu_B(N_0)$ and $\mu_A(N_0)$. One mole of such material, in the $\alpha$ phase but at the required composition of the $\beta$ precipitate, has the free energy $I$ in fig. 1.5. After transformation to the $\beta$ phase, where the two components have the chemical potentials of $\mu_B(N_\alpha)$ and $\mu_A(N_\alpha)$, the free energy per mole of the precipitate is then $J$. That is, the change of free energy $\Delta G$ per mole of precipitate, $\Delta G_2$, is $IJ$. From the similar triangles, $N_\alpha$–$I$–$J$ and $N_\alpha$–$\mu_B(N_0)$–$\mu_B(N_\alpha)$, we can relate this value of $\Delta G_2$ to the chemical potential change of the B component by eq. (1.11). (The geometry of this equation is not perfect but the errors introduced by the approximation are usually negligible.)

$$\Delta G_2=J-I=[\mu_B(N_\alpha)-\mu_B(N_0)](N_\beta-N_\alpha)/(1-N_\alpha) \quad (1.11)$$

The chemical potential of B is given by the activity, $a_B$, and thus by the product of the atomic fraction and activity coefficient, $\gamma_B$, of B atoms in the $\alpha$ phase, see eq. (1.12). $\mu_B^\circ$ is the chemical potential of 1 mole of B atoms in the 'standard state' of B where the activity of B, $a_B$, is defined as $a_B=1$. $R$ is the gas constant, 8.31 J mole$^{-1}$ K$^{-1}$.

$$\begin{aligned}\mu_B(N_\alpha)&=\mu_B^\circ+RT\ln a_B=\mu_B^\circ+RT\ln(N_\alpha\gamma_B)\\ \mu_B(N_0)&=\mu_B^\circ+RT\ln a_B=\mu_B^\circ+RT\ln(N_0\gamma_B)\end{aligned} \quad (1.12)$$

When the $\alpha$ matrix is a dilute solution, the usual case in most precipitation reactions, the activity coefficient, $\gamma_B$, is constant and so eq. (1.11) can be simplified to eq. (1.11a):

$$\Delta G_2 = RT\,\{(N_\beta - N_\alpha)/(1-N_\alpha)\}\ln(N_\alpha/N_0) \qquad (1.11a)$$

For successful precipitation hardening, see chapter 2, the driving force $\Delta G_2$, needs to have as large a negative value as possible. This is readily achieved by making the matrix phase solubility, $N_\alpha$, as small as possible. Such low solubility is usually accomplished by precipitation at low temperatures where $N_\alpha \to 0$. For example, if $N_\alpha/N_0$ is as small as $10^{-3}$ with $T$ at 300 K, and with the $\beta$ phase nearly pure B, $N_\beta \to 1$, $\Delta G_2$ from eq. (1.11a) is found to be $-17$ kJ mole$^{-1}$. The volume fraction of the precipitate phase $f_\beta$ is given by the lever rule:

$$f_\beta = (N_0 - N_\alpha)/(N_\beta - N_\alpha) \qquad (1.13)$$

In most precipitation hardening reactions $f_\beta \leq 0.1$. At $f_\beta = 0.1$:

$$\Delta G_1 = f_\beta \Delta G_2 = -1.7 \text{ kJ mole}^{-1}$$

The calculated driving pressures are greatly reduced if the precipitate and the matrix are closer in composition. For example, if the precipitate is an intermetallic phase such as $Al_2Cu$, the composition term $(N_\beta - N_\alpha)/(1-N_\alpha)$ falls from the value of about 1, when $N_\beta$ is nearly 1, to about 1/3 for $A_2B$ when $N_\beta$ is about 1/3, so the values of $\Delta G_1$ and $\Delta G_2$ both fall by a factor of about 3. For a phase such as $Al_6Mn$ in Al–Mn, where $N_\beta$ is 1/7, $\Delta G$ falls by a factor of 7.

A similar calculation to that above but for *matrix* phase changes, such as the transformation of austenitic $\gamma$ iron to ferritic $\alpha$ iron in Fe–C steels, gives the expression for the free energy reduction per mole of product phase as eq. (1.11b), Doherty (1983).

$$\Delta G_2 = [(N_\gamma - N_\alpha)/N_\gamma]\, RT \ln[(1-N_\gamma)/(1-N_0)] \qquad (1.11b)$$

Near the eutectoid temperature, at $T = 1000$ K, where $N_\alpha$ is about 0.001 and $N_\gamma$ is about 0.04 for a typical alloy composition, $N_0$, of 0.02 (0.4 wt% C), $\Delta G_2$ is then only $-170$ J mole$^{-1}$. This very small value of the driving force of the important proeutectoid ferrite reaction is due to the very small value of $\ln[(1-N_\gamma)/(1-N_0)]$. As a result of this reduced driving 'force', nucleation of the $\alpha$ phase within the $\gamma$ matrix grains is very difficult and only grain boundary nucleation is observed. Such nucleation at grain boundaries is by the process heterogeneous nucleation, see chapter 2, which uses the grain boundary energy to reduce the energy barrier to the reaction.

## 1.3 Mechanisms of microstructural change – rate controlling step

All structural changes in the solid state require atomic movement. This movement can occur over two length scales: short range and long range. Short range atomic movement occurs over distances of the order of an atomic spacing. Structural change by short range atomic movement occurs within a crystal in ordering reactions such as those that occur in bcc $\beta$ brass (Cu–Zn) or in fcc $Cu_3Au$. Short range atomic movement is very frequently seen at crystal interfaces, such as grain and phase boundaries. Grain boundary motion, which occurs by short range atomic transport across the interface, is seen in transformations such as the recrystallisation of deformed metals (§4.3) and grain growth (§5.8). The migration of phase boundaries in structural transformations such as matrix phase changes and precipitation reactions (where matrix and precipitate have different crystal structures) also involves short range atomic movement at the interface. Long range transport is, however, required to change the local solute concentrations, for example, from $N_\alpha$ to $N_\beta$ in a precipitation reaction (fig. 1.5). Long range atom transport requires atom movement over interparticle distances. These are much larger than the interatomic spacing. Such long range atomic movement, usually called diffusion, is characteristic of any reaction in which a change of composition occurs. The major reactions showing composition changes include: (i) precipitation of second phases from supersaturated solid solutions (§2.2); (ii) matrix phase changes in alloys such as the much studied fcc to bcc transformation in steels; (iii) many of the coarsening processes discussed in chapter 5; and (iv) homogenisation of cored solid solutions (§2.1) and the opposite reaction, spinodal decomposition (§2.2.1). As its name suggests the reaction called diffusion-induced grain boundary migration (DIGM), also involves long range atomic transport mainly along the grain boundaries. A major feature in the process of precipitation behind a moving matrix grain boundary, discontinuous precipitation, now appears to have much in common with DIGM.

The detailed processes of diffusion within crystals, bulk or volume diffusion, is well discussed in many places, for example by Shewmon (1989). Little more needs to be said here about this topic other than pointing out that the reaction is thermally activated with an activation energy that for self diffusion in fcc metals and bcc transition elements has a value usually of about $18RT_m$(J mole$^{-1}$). $T_m$ is the absolute (K) melting temperature of the metal. For bcc alkali metals the activation energy is significantly less, typically about $15RT_m$(J mole$^{-1}$), and for semiconducting elements with the diamond cubic structure, such as germanium and silicon, the energy barrier is significantly higher, $33RT_m$(J mole$^{-1}$) (Brown and Ashby 1982). Bulk diffusion of substitutional alloy additions occurs with similar activation energies to that of the matrix in many cases. However, for alloy additions with either a significantly higher or lower melting temperature than the matrix the activation energy for their diffusion in the alloy is raised or lowered. High melting point transition elements, such as iron, manganese, chromium and zirconium, diffuse much more slowly than the aluminium matrix in aluminium transition metal alloys. Elements with melting points close to aluminium such as zinc, magnesium and copper, however, diffuse in aluminium at about the same rate (to within a factor of 10). In many binary alloys of metals with somewhat different melting temperatures, for example, Zn–Cu, Ag–Au and Cu–Ni, the lower melting point component always has the higher rate of diffusion. A direct result of these differences in rates of diffusion is the Kirkendall effect, the shift of inert markers towards the slower diffusing component in a diffusion couple, a phenomenon clearly described, for example, by Shewmon (1989).

Diffusion of an interstitial solute, such as carbon or nitrogen in iron, occurs with a much lower activation energy than shown by the diffusion of the host metal. For the smallest atom, hydrogen, the activation energy is very much lower. As a result of this difference we can often have the situation called 'para-equilibrium' in which the interstitial elements can move rapidly and achieve equilibrium. In this circumstance equilibrium of the interstitial element means a constant value of its partial molar free energy, $G_i$ (the chemical potential, $\mu_i$). At the same temperature, however, the host matrix, for example, iron, and any substitutional solute atoms move at negligible rates. This low diffusivity means that large local differences in the chemical potentials of the iron and substitutional atoms can develop over short distances. For example, at 673 K the diffusion coefficient of carbon in $\alpha$ Fe is $4\times10^{-10}$ m$^2$ s$^{-1}$, and so in 1 s the mean atomic displacement of a carbon atom, $(Dt)^{0.5}$, is 20 $\mu$m. The value of $D$ for the self diffusion of iron at 673 K, is only $6\times10^{-23}$ m$^2$ s$^{-1}$, so the equivalent displacement of an iron atom is 0.08 nm, much less than the interatomic spacing of iron atoms, 0.25 nm.

Diffusion along crystal interfaces (interfacial or boundary diffusion) has been discussed, for example, by Balluffi (1982) and is known to occur with activation energies that are only 50–60% of that of bulk self diffusion. Boundary activation energies 50–60% of the bulk activation values are characteristic of both fcc and bcc transition metals. Brown and Ashby (1982) suggest the following general equations for the bulk, $D$, and boundary, $wD_B$, diffusion coefficients, as a function of absolute temperature $T$:

fcc metals: $D=5.4\times10^{-5}\exp[-(18.4T_m/T)]$ m$^2$ s$^{-1}$

$wD_B=9\times10^{-15}\exp[-(10.0T_m/T)]$ m$^3$ s$^{-1}$

bcc transition metals: $D=1.6\times10^{-4}\exp[-(17.8T_m/T)]$ m$^2$ s$^{-1}$

$wD_B=3.4\times10^{-13}\exp[-(11.7T_m/T)]$ m$^3$ s$^{-1}$

bcc alkali metals: $D=2.5\times10^{-5}\exp[-(14.7T_m/T)]$ m$^2$ s$^{-1}$

where $w$ is the effective thickness of the boundary along which the atoms are being transported, typically $w$ is expected to be 1–2 atom diameters, that is, about 0.5 nm.

The activation energy of atomic movement *along* interfaces is expected to be equivalent to that for atomic transport *across* the interfaces, the mechanism of interface migration at least for the cases where the interface migration mechanism is thermally activated. For grain boundary motion in very high purity materials (see, for example, Haessner and Hofmann (1978)) this appears to be true. In the presence of traces of solutes that are adsorbed on the interface, as discussed in §4.3.3, the process of *solute drag* slows the migration rate of the interface and increases the apparent activation energy of the process of boundary movement in the solid solution.

A simple analysis of the kinetics of interfacial migration by thermally activated atomic motion, in the absence of solute drag, is readily given by the standard analysis. For the situation shown in fig. 1.2, where the atoms in the initial state on one side of the interface have a higher energy, by $\Delta G$, than the atoms on the other side, then the rate of atom transfer from the initial to the final state is:

$$\text{Rate of atom transfer, I}\rightarrow\text{F},=A\nu\exp(-\Delta G_A/kT)\ (\text{atoms s}^{-1}) \qquad (1.14)$$

This comes from eq. (1.5) with one atom per site in the initial state and the rate that the activated atoms decay to the final state being the vibration frequency, $\nu$. The accommodation factor $A$ is the probability that an atom having left one crystal structure and having passed through the activated state can find a site to receive it on the growing side of the interface. There is, in addition, a reverse rate given by eq. (1.14a):

$$\text{Rate of atom transfer, F}\rightarrow\text{I},=A\nu\exp[-(\Delta G_A+\Delta G)/kT]\ (\text{atoms s}^{-1}) \qquad (1.14a)$$

It is assumed that the same value of the accommodation factor applies to both reactions. The net rate of atom transfer is the difference between these two rates:

$$\text{Net rate of atom transfer}=A\nu\exp[-(\Delta G_A/kT)]\,\{1-\exp[-(\Delta G/kT)]\}\ (\text{atoms s}^{-1}) \qquad (1.15)$$

For the usual situation, in which $\Delta G \ll kT$, so that $\exp[-(\Delta G/kT)]$ becomes $1-\Delta G/kT$, we obtain for the velocity of the interface migration

$$v=b\times\text{net rate of atom migration}$$

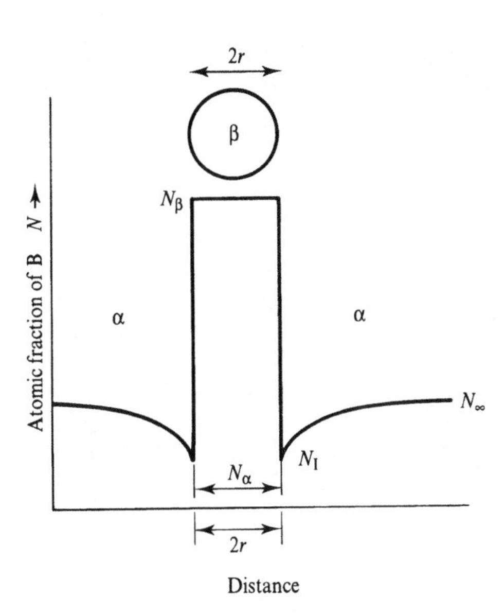


**Figure 1.6** Distribution of solid around a growing spherical precipitate of $\beta$, radius $r$, in an $\alpha$ matrix.

So:

$$v = A(b\nu/kT)\exp[-(\Delta G_A/kT)]\Delta G \text{ (m s}^{-1}) \tag{1.16}$$

This arises since each atom transferred moves the interface forward locally by a distance $b$, the atomic diameter. This result, for a constant value of the accommodation factor, $A$, gives the usual *linear* relationship between velocity and driving force per atom, $\Delta G$:

$$v = M_i \Delta G \tag{1.16a}$$

Here $M_i$ is called the interface mobility and is given by:

$$M_i = A(b\nu/kT)\exp[-(\Delta G_A/kT)] \tag{1.16b}$$

In § 5.5.3, it is shown that for diffusion controlled growth of a spherical precipitate of radius $r$, at low supersaturations, the velocity of growth, $v = \mathrm{d}r/\mathrm{d}t$, is given by eq. (1.17) where the solute content at the interface is $N_I$ and that far away from the interface is $N_\infty$, see fig. 1.6.

$$v = (D/r)(N_\infty - N_I)/(N_\beta - N_I) \tag{1.17}$$

In order to compare eq. (1.16) with eq. (1.17) we must put $\Delta G$ in eq. (1.16a), in terms of differences in atomic fraction, $\Delta N$. This can be done by use of figs. 1.5 and 1.6. With the alloy composition, $N_0$ in fig. 1.5, set at the interface concentration $N_I$ (fig. 1.6) with $N_I > N_\alpha$, the free energy per atom, $\Delta G$, driving the *growth* of the precipitate is then $\Delta G_2$ of eq. (1.11) divided by the Avogadro's number, $N_{Av}$, the number of atoms in a gram mole. That is:

$$\Delta G=-\Delta G_2/N_{Av}=-[(N_\beta-N_\alpha)/(1-N_\alpha)](R/N_A)T\ln(N_\alpha/N_I)$$

So, since Boltzmann's constant $k=R/N_{Av}$:

$$\Delta G=[(N_\beta-N_\alpha)/(1-N_\alpha)]kT\ln(N_I/N_\alpha) \quad (1.18)$$

Since $N_I/N_\alpha$ is close to 1, $\ln(N_I/N_\alpha)$ becomes $(N_I-N_\alpha)/N_\alpha$. We can then write:

$$v=M_i\Delta G=M_i'(N_I-N_\alpha)/(N_\beta-N_I) \quad 1.16c)$$

where the modified interface mobility, $M_i'$, is given by eq. (1.16d):

$$M_i'=[Ab\nu(N_\beta-N_\alpha)(N_\beta-N_I)/(1-N_\alpha)N_\alpha]\exp[-\Delta G_A/kT)] \quad (1.16d)$$

We can now compare the two expressions for the interface velocity, $v$, in eqs. (1.16c) and (1.17). This will demonstrate the parameter that determines which process provides the *rate controlling step* in any reaction involving both long range solute transport (diffusion) and short range interfacial transport. The interface velocity is the same value in both processes, so we obtain:

$$v=M_i'(N_I-N_\alpha)/(N_\beta-N_I)=(D/r)(N_\infty-N_I)/(N_\beta-N_I) \quad (1.19)$$

Rearrangement of eq. (1.19) gives:

$$M_i'r/D=(N_\infty-N_I)/(N_I-N_\alpha) \quad (1.19a)$$

When the interface composition in the $\alpha$ matrix, $N_I$, tends to $N_\alpha$, the equilibrium solubility value, we have local equilibrium at the matrix–precipitate interface. To get such 'local' equilibrium, the required dimensionless parameter, $M_i'r/D$, is $\gg 1$. For $M_i'r \gg D$, we find a high, radius-modified, mobility relative to the diffusion coefficient. Under these conditions, the composition difference driving the total growth process, $N_\infty-N_\alpha$, is almost fully applied to the *diffusion* process and this gives a 'local equilibrium' or 'diffusion controlled' process. For such a diffusion controlled process, where $N_I=N_\alpha$, we have a velocity given by eq. (1.17):

$$v=(D/r)(N_\infty-N_\alpha)/(N_\beta-N_\alpha) \quad (1.17a)$$

In such a case the velocity is the independent of any changes in mobility, provided that $M_i'\,r/D$ remains $\gg 1$.

Alternatively when the interface composition $N_I$ tends towards $N_\alpha$, the composition far from the precipitate, the dimensionless parameter $M_i'r/D$ is $\ll 1$. That is, $M_i' r \ll D$, so we have a low interface mobility relative to the diffusion coefficient and the composition difference driving the total growth process, $N_\infty - N_\alpha$, is fully applied to the *interface* reaction giving an 'interface controlled' process where the velocity is given by

$$v = M_i'(N_\infty - N_\alpha)/(N_\beta - N_\infty) \quad (1.16e)$$

With diffusion control the growth velocity is independent of any increase in mobility. Under interface control, similarly any increase in $D$ will have no detectable effect on the kinetics. In the intermediate case where $N_\alpha < N_I < N_\infty$ we have 'mixed control' and there will be a change from near interface control at very small values of the precipitate radius $r$, when $M_i' r < D$, to near diffusion control at larger values of the precipitate size when $M_i' r > D$.

Long range solute diffusion has an activation energy significantly larger than the activation energy of boundary mobility, which is expected to be that of boundary diffusion. So diffusion control is expected to occur for most precipitation reactions except at very small values of precipitate size, when $r \rightarrow 0$. This is found to be generally true, see, for example, Doherty (1983), except when the accommodation factor $A$ in the equations for boundary mobility is very much less than 1. Very small values of $A$ will greatly reduce the values of the interface mobility, $M_i$ or $M_i'$, so allowing the condition for diffusion control, $M_i' r \gg D$, to occur at larger values of $r$. To examine one case where a low value of the accommodation factor $A$ is expected to occur, some consideration must be given to the atomic structure of the interface between the two crystals.

### Interfacial structure – the need for ledges

An interface is said to be fully coherent if each atomic plane in one crystal that intersects the interface is matched by another plane on the opposite side of the interface. A simple example is the fully coherent interface of fig. 1.7 between two crystals of the same crystal structure, orientation and lattice parameter. For the fully coherent interface of fig. 1.7 then we no longer have any structural interface – merely a change of the type of atoms occupying equivalent sites. That is, we have only a *chemical* interface. Migration of this type of interface merely requires a change in composition by the exchange of B atoms for A atoms by long range atomic diffusion. Replacement of one layer of A atoms by B atoms causes the upper crystal, the B rich $\beta$ phase, to grow from $DD'$ to $CC'$. Under these circumstances no separate 'interfacial' reaction is needed. As a consequence, the migration of the interface must be controlled by the only process required – long range diffusion. The chemical change need not occur in the atomically sharp way shown in fig. 1.7. The change from mainly A to mainly B atoms may occur over

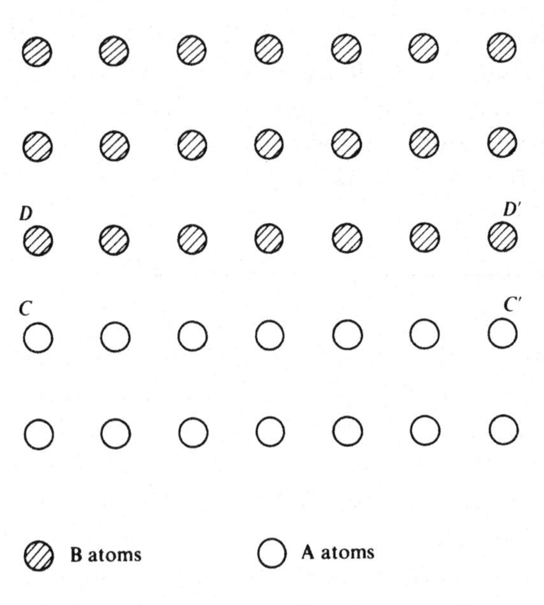


**Figure 1.7** *Coherent* interface between two phases with the same structure, orientation and lattice parameter, differing only in composition.

several planes giving a *diffuse* interface; see, for example, Cahn and Hilliard (1958, 1959). This decrease in the atomic sharpness of the interface should not change the conclusion that no interfacial reaction is required for the movement of a coherent interface *between two phases with the same atomic structure.*

This situation of full coherency between precipitate and matrix having the same structure occurs in the formation of Guinier–Preston (GP) zones seen in many precipitation reactions, for example, in the Cu–Co system investigated by Servi and Turnbull (1966). Full diffusion controlled growth kinetics was indeed found in that study. Similarly the much studied case of coarsening of coherent fcc precipitates of the $\gamma'$ phase (ordered $Ni_3Al$) in $\gamma$ fcc Ni, discussed in chapter 5, was also found to show the expected diffusion controlled kinetics.

The opposite structural extreme occurs in a situation such as a high angle grain boundary where there is no matching between the arbitrarily oriented crystals and thus no coherence between the atomic planes of the two crystals. A further example in which there should be no coherence, is the case of a $\beta$ precipitate that may have nucleated in one parent $\alpha$ crystal, and thereby acquired a particular orientation relationship, but then any potential coherency will be lost by passage of a matrix grain boundary. Such a passage of a matrix grain boundary will replace the initial parent $\alpha$ matrix by the differently orientated matrix grain growing *around* the $\beta$ precipitate, which in most cases does not change in orientation. A detailed consideration of this type of change has been given, for example, by Doherty (1982). In each of these cases, we will have a fully incoherent interface, see fig. 1.8. For such an incoherent interface the value of the activation energy, $\Delta G_A$, is expected to be that of interfacial diffusion and the value of the accommodation factor, $A$, is expected to be close to unity. This upper limit-

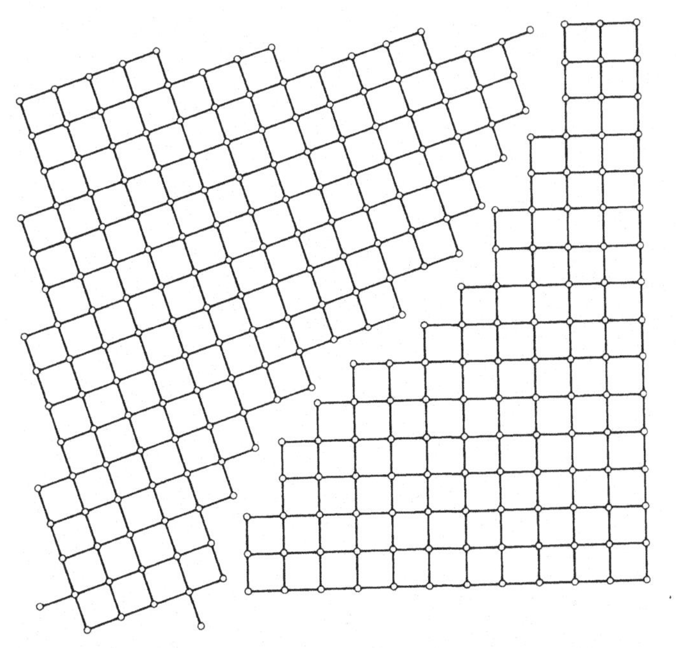

**Figure 1.8** Structure of an *incoherent* boundary in two dimensions if the two crystals conform as closely as they can without lattice distortion. (After Gleiter and Chalmers 1972.)

ing value of $A$ occurs since there are many sites at which atomic detachment or attachment can take place readily. Such a reaction involving the migration of an incoherent interface occurs in grain boundary migration, at least with high angle grain boundaries in pure materials as is discussed in chapter 4. For grain boundary migration in a pure material, the only atomic process required is atom transport across the interface and so eq. (1.16) gives the expected interface controlled kinetics. This result has indeed been demonstrated; see, for example, the review by Haessner and Hofmann (1978).

For growth of an incoherent, solute rich, $\beta$ precipitate in an $\alpha$ matrix, we can again expect the accommodation factor $A$ to approach 1 and high mobility, $M_i'$, to result. Such mobility should give diffusion control except possibly for the smallest precipitates. It should be noted, however, that the interfacial reaction, eq. (1.16), requires *all* the atoms to make *one* jump across the interface while the diffusion reaction requires only the fraction of the atoms needed to make the change in composition, $N_\beta - N_I$, to diffuse up to the interface. The extra solute atoms still have to make many successive diffusive jumps, of the order of $(r/b)^2$, in the bulk crystal. The radius of the precipitate, $r$, is in almost all cases much larger than the interatomic spacing in the matrix, $b$.

If the two crystals have the same structure and orientation but a slightly different lattice parameter then a *semicoherent* interface will be expected, fig. 1.9, with the misfit being taken up by an array of edge dislocations. In three dimensions

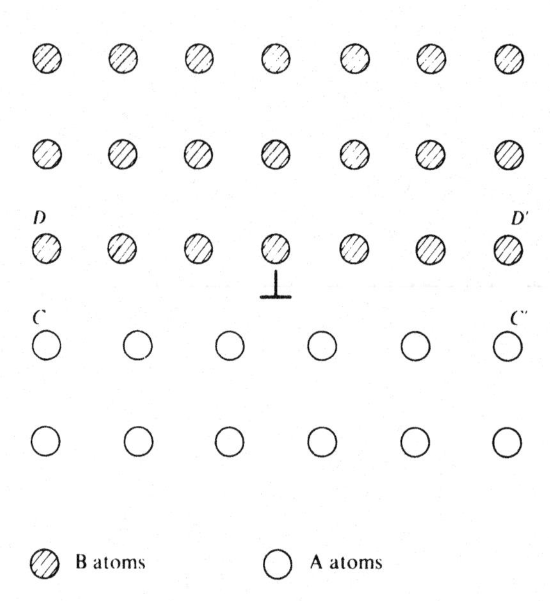


**Figure 1.9** *Semicoherent* interface: migration from $CC'$ to $DD'$ requires vacancy to allow interface dislocation to climb.

the two-dimensional interface will require a *cross-grid* of two sets of edge dislocations, with non-parallel Burgers vectors, in the interface plane. For the migration of such a semicoherent interface, a flux of point defects, usually vacancies, will be needed for the climb of the edge dislocations normal to their respective Burgers vectors. It would not be expected that this additional flux of vacancies, also requiring an activation energy of bulk diffusion, would significantly inhibit the diffusion reaction. The vacancies will come from adjacent dislocations spaced $(1/\rho)^{1/2}$ apart, where $\rho$ is the dislocation density, typically $10^{12}$ $m^{-2}$, so giving a spacing of about 1 $\mu$m. If the diffusion distances are larger than this then diffusion control should still be expected. The flux of vacancies may, however, modify the actual value of the solute diffusion coefficient. This was found for precipitation of ordered bcc zinc rich $\gamma$ from bcc $\beta$ in Cu–Zn brass (Bainbridge 1972; Fairs 1978). In this case the faster diffusion of zinc up to the precipitate compared to copper away from the precipitate created a flux of vacancies away from the precipitate that was in conflict with the vacancies needed for the climb of the dislocations at the interface.

It might be supposed that if fully coherent interfaces between phases of the same structure and fully incoherent interfaces can both migrate at rates unaffected by the interfacial structure then interface control should never be expected. This conclusion is, however, wrong. A major difficulty is now known to occur at coherent (or semicoherent) interfaces between two phases, $\alpha$ and $\beta$, which have, in at least one case, different orientations or the very common situation of *different crystal structures*. Coherent twin boundaries in fcc metals have perfect coherence at the common $\{111\}$ interfacial plane despite a misorientation of 60° about the rotation axis, $\langle 111 \rangle$ normal to the plane. Coherent twin

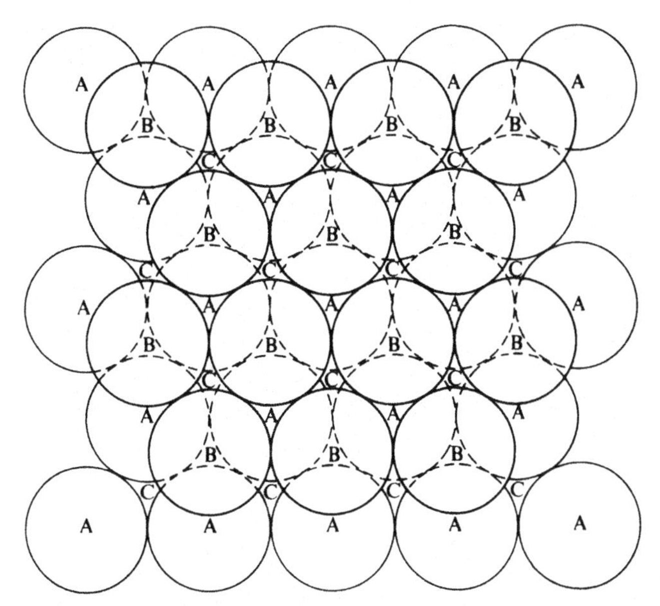


**Figure 1.10** Stacking sequence in fcc structure.

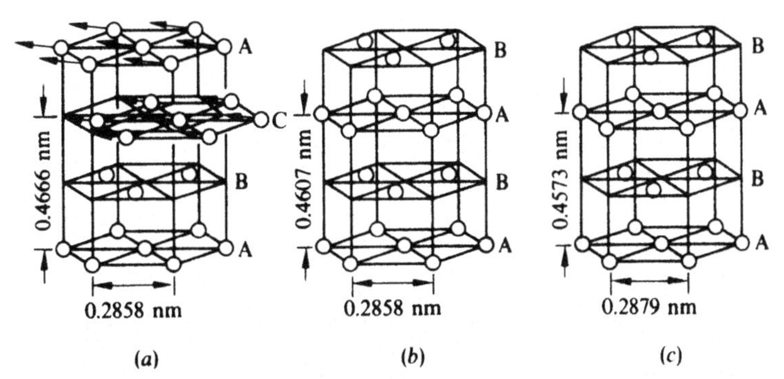


**Figure 1.11** Crystal lattice of phases in aluminium–silver. (*a*) Matrix, $\delta$ (fcc). (*b*) Transition phase, $\gamma'$ (cph). (*c*) Equilibrium precipitate, $\gamma$ (cph). (After Barrett 1952. Courtesy of McGraw-Hill.)

boundaries are indeed known to be completely immobile, against thermally activated atom transport.† Aaronson and coworkers (Aaronson 1962; Aaronson, Laird and Kinsman 1970) suggested that an exactly equivalent problem occurs at the good fit interfaces between crystals of different structures for the usual conditions of *the special orientation relationship*, produced by nucleation. The simplest example of this difficulty is the interface between fcc and cph crystals in, for example, Al–Ag which as shown in figs. 1.10 and 1.11 have identical atomic packings on the {111} fcc Al interface matching the (0001)

† *Mechanical* twinning, see, for example, Reed-Hill and Abbaschian (1992), does allow twin boundary migration by a dislocation glide process.

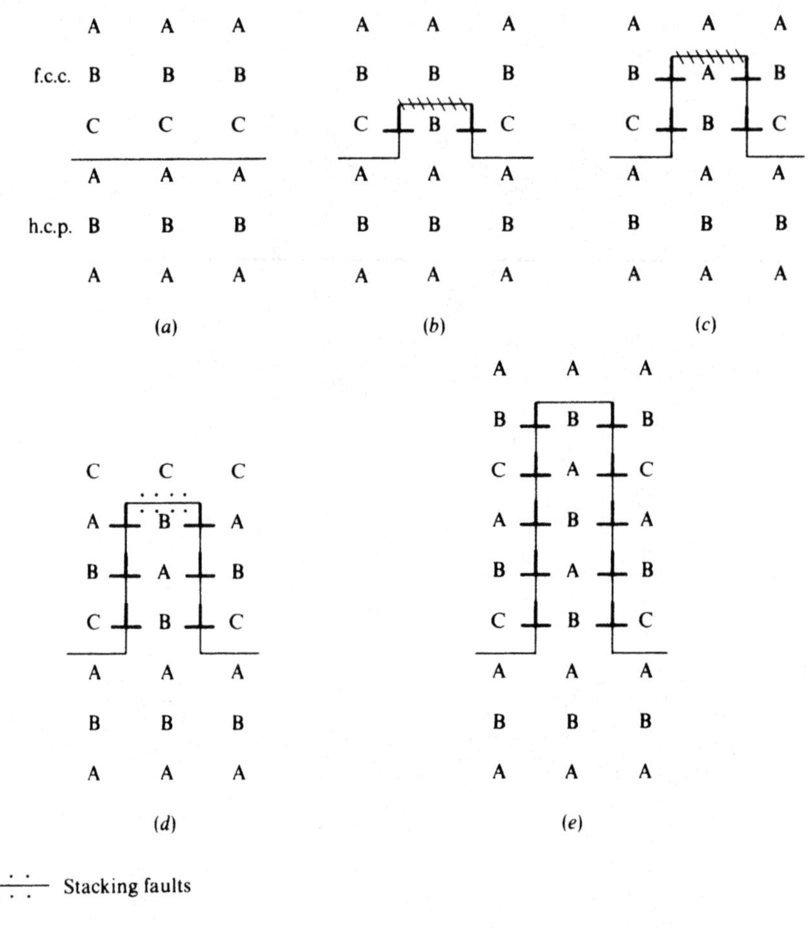


**Figure 1.12** Problems in the growth of hexagonal phase from fcc by single atom jumps. The letters indicate the stacking of atoms as illustrated in figs. 1.10 and 1.11.

cph $AlAg_2$ interface. Such an interface has much in common with a coherent {111} twin interface in fcc polycrystal. For the fcc/cph system the fully coherent interface has a stacking sequence ABCA in fcc that matches the ABABA cph sequence. A typical good fit interface is seen in fig. 1.12 where the stacking sequence is . . . ABABABCABC. . . with the highlighted **A** layer being common to both cph and fcc structures. For a coherent fcc twin the stacking sequence is . . .ABCABCACBACB. . .. These good fit interfaces have low energy and as a

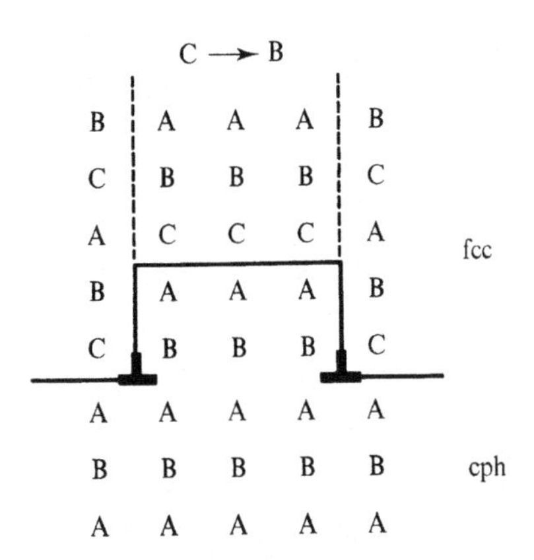


**Figure 1.13**
Transformation of two layers (C,B) of fcc to cph (B,A) by formation of a loop of Shockley partial dislocation, $a/6\langle 112\rangle = BC$, that grows by glide outwards from its points of formation.

result are almost universally produced by the process of nucleation, discussed in chapter 2.

The problems of boundary migration in these interfaces are very similar. As shown in fig. 1.12 for the cph/fcc case, if growth of the lower hexagonal structure is to occur by individual atomic jumps, as assumed in deriving eq. (1.16), then an atom in the stable C position in fcc, fig. 1.12(*a*), must displace to a B position, fig. 1.12(*b*), and this creates a highly unstable situation. There is both the type of misfit seen at the core of a Shockley partial dislocation around the atom in the new B position in the interface plane and *the very unstable* configuration of one atom on top of the other both in equivalent B positions. Displacement of a second atom B→A in fig. 1.12(*c*) does not improve matters. Only with the third displacement, A→B in fig. 1.12(*d*) is the unstable position removed. Perfect matching, however, requires five atoms to displace as seen in (*e*). It does not appear, however, that this is how the fcc→ cph transformation occurs. As has been frequently shown, for example, by Howe, Aaronson and Gronsky (1985) and by Singh and Doherty (1992), the transformation is readily accomplished by the nucleation of a loop of a Shockley partial dislocation with a Burgers vector, with C→B, that is, $a/6\langle 112\rangle$, fig. 1.13. Formation of such a dislocation loop and its glide across the interface transforms two layers, C→B and B→A, from the fcc sequence to the required cph sequence. The third layer moves A→C and restores the starting sequence. That is, the glide of an appropriate Shockley partial transforms *two layers* of fcc to cph. So this transformation can take place by the formation and glide of a partial dislocation on every second $\{111\}$ fcc plane. For the equivalent twinning transformation, a Shockley partial dislocation only transforms one layer of fcc to the twin, that is, C→B, B→A and A→C carries . . .ABCABCACBACB. . . to . . .ABCABCA**B**ACBA. . .. That is mechanical twinning requires the glide of a Shockley partial dislocation across *each* $\{111\}$ fcc plane.

This type of interface motion of good fit interfaces then occurs by the passage

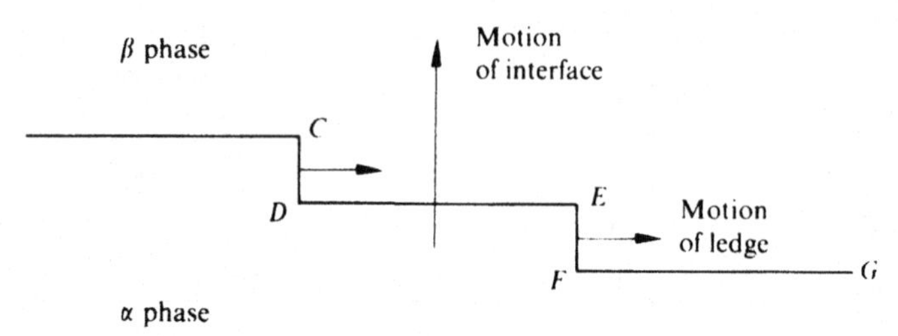


**Figure 1.14** Growth of $\alpha$ phase by motion of ledges $CD$ and $EF$, rather than by normal migration of $DE$, $FG$.

of 'steps' or 'ledges' across the interface, fig. 1.14. In these circumstances $A$ is given by the fraction of atom sites on the interfaces that are ledges and so, unless the ledge density is very high, $A \ll 1$ and the condition for diffusion control, $M'_i r \gg D$, is difficult to achieve especially for small precipitates.

For solid–solid interfaces the growth ledges are usually dislocations, as in the fcc/cph situation described above. Similar ledge growth is well established for solid–vapour and solid–liquid interfaces and is known to be required when there are high entropy changes as the crystal grows, see Jackson (1958), Jackson and Hunt (1965) and Woodruff (1973). The steps in facetted crystal–fluid interfaces do not have any dislocation character. Solidifying metals, however, by virtue of their small entropy changes on solidification, usually have atomically diffuse interfaces so giving $A \rightarrow 1$ and consequently they are found to have high mobility (Woodruff 1973). For ledged crystal growth from a fluid matrix the steps are commonly created by emerging screw dislocations in the crystal–fluid interface, but the resulting surface *ledges* are strain-free away from the core of the screw dislocation. Apart from causing a measure of interface inhibition in crystal growth and promoting the formation of 'facetted' crystal shapes, that is, crystals bounded by close-packed slow growing planes, the availability of screw dislocations in general, means there is no major difficulty for facetted crystal growth from fluids. Chapter 6 provides some examples of the microstructural and kinetics effects of the need for growth ledges in crystal–fluid interfaces.

Solid–solid interfaces between phases of different structures, especially those produced by precipitation, have long been known to occur with definite orientation relationships; see, for example, Barrett (1952). The resulting growth shapes of the precipitates were known to be highly facetted – in the characteristic 'Widmanstätten' form of thin elongated plates or rods.† The extended surface of the plate or rod was the slow growing good fit interface such as the four hexagonal 'variants' on the four {111} planes in fcc crystals. Aaronson's original suggestion of the need for growth ledges at solid–solid Widmanstätten interfaces

† For plate-like precipitates, the radius $r$ in the $M'_i\, r/D$ parameter is replaced by $a_x$, the half-thickness of the plate.

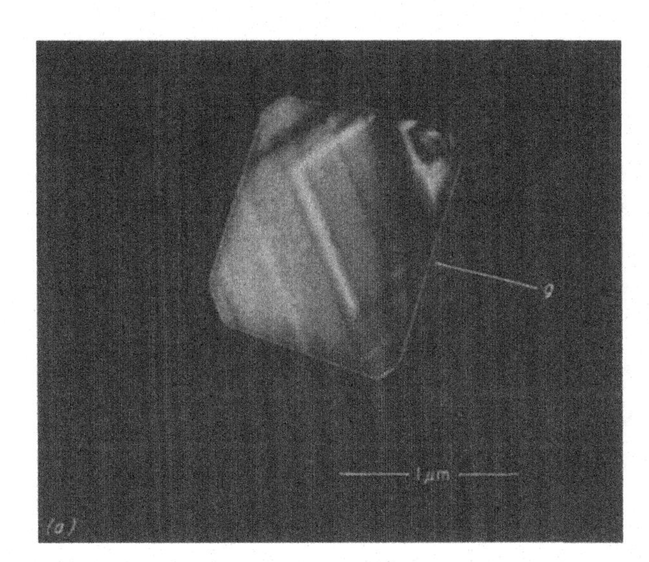


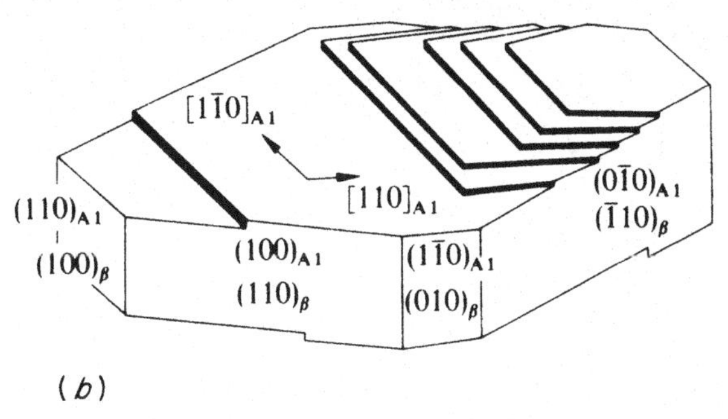


**Figure 1.15** (*a*) Growth ledges at an $Mg_2Si$ plate in Al–1.5 wt% $Mg_2Si$, solution treated and aged 2 h at 350 °C. Dark field micrograph. (*b*) Schematic diagram of (*a*) showing ledges on $Mg_2Si$ plate. (After Weatherly 1971. Courtesy of *Acta Met.*)

(Aaronson 1962) was initially greeted with scepticism. However, his ideas were confirmed by subsequent observations of the growth ledges that he predicted. A very clear early demonstration of the growth ledges was the study by Weatherly (1971). Weatherly observed growth ledges in various systems, see fig. 1.15, and he showed that these solid–solid growth ledges all had the characteristics of dislocation contrast as seen by electron microscopy. Howe *et al.* (1985, 1987) confirmed these ideas by beautiful high resolution electron micrographs of the fcc – cph interfaces discussed here, fig. 1.16. In a series of studies, one of the present authors has shown experimentally, see Ferrante and Doherty (1979) and Rajab and Doherty (1989), that in Al–Ag the initial lack of growth ledges did, as predicted by Aaronson (1962) inhibit the initial thickening of the hexagonal plates. Very elongated shapes with diameter-to-thickness *aspect* ratios larger than 100 appear to arise from this initial interface inhibition. Rajab and Doherty (1989) showed evidence to suggest that the inhibition of the thickening reaction lasted until the individual precipitates intersected each other. Subsequent thickening was then found to occur at diffusion controlled rates. This observed diffusion control indicated that precipitate intersections have been able to generate a suf-

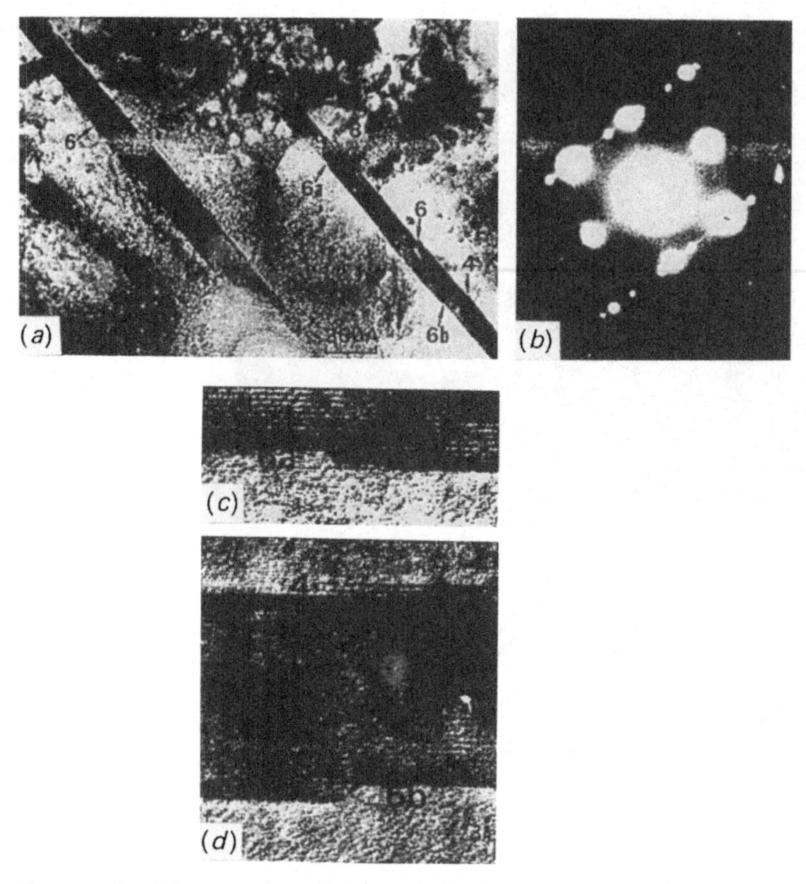


**Figure 1.16** Coherent cph precipitates of $\gamma'$ $Ag_2Al$, in fcc aluminium showing growth ledges. (*a*) Low magnification image of two precipitates with the ledge heights indicated in numbers of atomic planes; (*b*) the diffraction pattern used for the lattice imaging; (*c*) and (*d*) high magnification lattice images of some of the steps seen in (*a*). (After Howe *et al.* 1985. Courtesy of *Acta. Met.*)

ficient density of growth ledges to avoid interface mobility limiting the rate of growth. Howe *et al.* (1985, 1987) have also demonstrated ledge nucleation by precipitate–precipitate intersection. The details of ledge nucleation by precipitate intersection are still, however, an unsolved kinetic problem (Doherty and Rajab 1989). Rajab and Doherty (1989) also found inhibited thickening of the hexagonal silver precipitates during interfacial energy driven coarsening, §5.5. It now appears clear, from both growth and coarsening studies, that growth at rates slower than expected for diffusion control can and does occur for Widmanstätten precipitates due to an apparent lack of growth ledges in certain conditions. However, diffusion controlled growth of ledged precipitates is known to be possible if a sufficiently high density of ledges is present. This result has been shown both from direct kinetic studies of precipitate growth, see, for example, Ferrante and Doherty (1979) and Rajab and Doherty (1989) and also from numerical modelling of diffusion fields near growth ledges by Doherty and Cantor (1982, 1988) and Enomoto (1987) and from analytical studies by Atkinson (1981, 1982). With a low density of ledges, however, the numerical studies confirm that inhibited thickening at the earliest stages of precipitate growth is predicted as has been found experimentally. Experimental studies on the coarsening of Widmanstätten precipitates discussed in chapter 5 show clear evidence of the importance of interface limited kinetics – but this can in some cases give rather unexpected results. At a conference on the role of ledges in phase transformations, Stobbs (1991) reviewed the rather ill-defined current state of the topic in which much remains to be clarified in many aspects of this ledged growth of coherent interfaces in solids.

## 1.4 Quantitative microscopy

A central topic in any discussion of microstructural change is a consideration of the methods used for measuring various features of the microstructure, such as grain size, volume fraction of second-phase precipitates, interparticle spacing, etc. Much of this book will be concerned with how such microstructural features vary as the structure achieves a more stable form, and obviously all experimental investigations of such microstructural change will depend on accurate and reliable measurements. Fortunately this subject has been well discussed in the literature, for example, by Hilliard (1966, 1969), DeHoff and Rhines (1968) and Underwood (1970), and more recently Vander Voort (1984); these authors discuss the subject very thoroughly and give full references to all the earlier studies. Vander Voort (1984) also reviews the development of automatic image analysers which have greatly reduced the tedium associated with manual measurements and which, with on-line computer analysis capability, provide a thorough statistical analysis of the measured data.

# Chapter 2

## Structural instability due to chemical free energy

### 2.1 Instability due to non-uniform solute distribution

The simplest instability in a metallic microstructure is that produced by a non-uniform distribution of solute in an otherwise stable single phase. Such a distribution always raises the free energy of the alloy and so it will decay to a uniform distribution at a rate determined by the thermodynamics and kinetics of diffusion. The kinetics of diffusion and its relationship to the concentration and mobility of point defects is one of the best-established topics in materials science, see, for example, Shewmon (1989), and so these topics need not be repeated here. However, the thermodynamics of diffusion is less widely discussed and is described below. Some new ideas on diffusion in alloys showing different rates of atomic motion in binary alloys as indicated by the Kirkendall effect are also described.

#### 2.1.1 Thermodynamics of diffusion

Fig. 2.1 shows the free energy–composition diagram for a binary alloy. In stable regions of the system, where the second differential of the free energy with composition, $\partial^2G/\partial C^2$, is less than zero, the free energy of composition $C_3$, is increased from $G_3$ to $G'_3$ if it exists as a mixture of $C'_3$ and $C''_3$ rather than as a single uniform composition. The rise in $G$ caused by any non-uniform solute distribution in stable regions of any phase provides the driving 'force' for the diffusion that homogenises the distribution. Inside the 'spinodal' region where $\partial^2G_A/\partial C^2<0$, the free energy *falls* from $G_4$ to $G'_4$ if the composition changes from

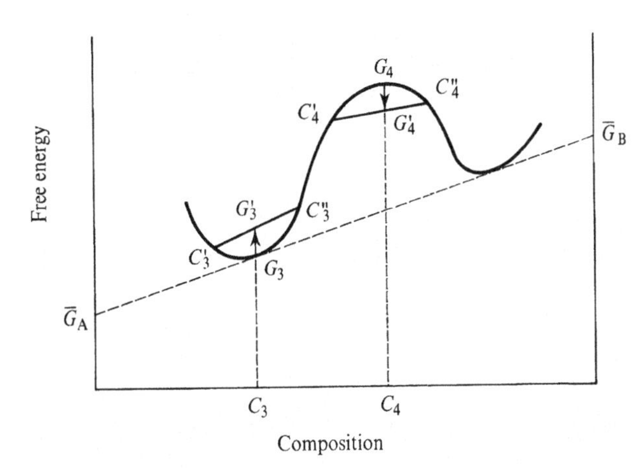


**Figure 2.1** Free energy compositional relationship for a conventional nucleation and growth process.

the constant value of $C'_4$ to a mixture of $C'_4$ and $C''_4$. Within the spinodal 'uphill' diffusion can then occur. This is discussed later in this section and more fully in §2.2.1.

In either an ideal or a dilute solution there is no change in energy and/or enthalpy ($H$) on changing the composition of the alloy. In an ideal solution there is no change of enthalpy on mixing, see, for example, Gaskell (1983), and in a dilute non-ideal solution the minority solute atoms are in contact only with solvent atoms so there is no change of atomic bonding, and thus enthalpy, as the concentration changes in this dilute solution range. However, there is a change in the 'configurational' entropy – the entropy due to atomic positional order. For these situations the usually quoted *statistical* derivation of the laws of diffusion is valid. In this derivation, illustrated in fig. 2.2, there is a variation in the concentration of $C_A$, given in atoms of A per unit volume, in the $x$ direction so that:

$$C_A = \text{function of } x$$

Therefore the concentration of A atoms on a plane, $C_{A_{(2)}}$, is given in terms of that on the preceding plane, $C_{A_{(1)}}$, by the Taylor expansion as:

$$C_{A_{(1)}} = C_{A_{(1)}} + (\partial C_A/\partial x)_1 b + (\partial^2 C_A/\partial x^2)_1 (b^2/2!) + \ldots$$

The spacing between adjacent atomic planes, 1 and 2, is $b$ and, with the concentration given as atoms per unit volume, the number of A atoms on unit area of a plane is then $C_A b$. If the energy of an atom is independent of the concentration, the jump frequency, $f_A$, of A atoms from both planes will be the same. Under these circumstances it is readily seen that, for most concentration profiles where terms higher than $\partial C_A/\partial x$ in the Taylor expansion can be neglected, we obtain the following relationship for the net flux of A atoms in the $x$ direction:

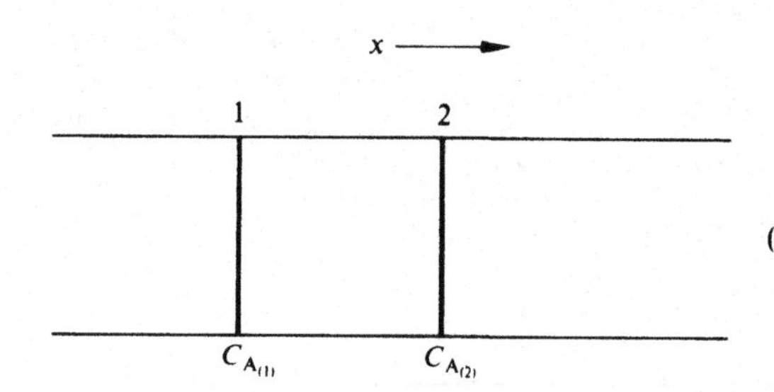


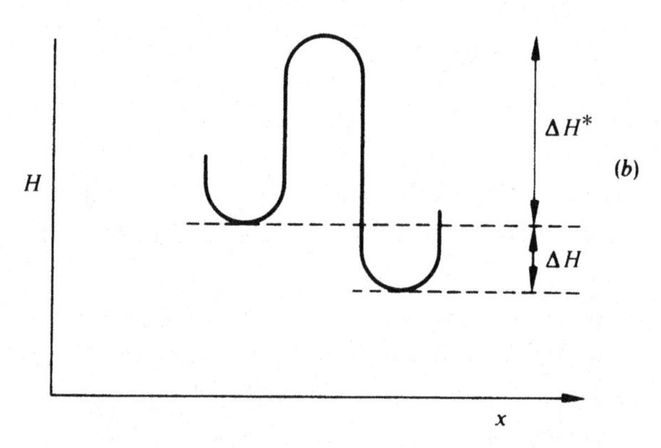


**Figure 2.2** (*a*) Variation of the concentration of A. (*b*) For non-ideal solution, the variation in enthalpy *H* between adjacent planes (much exaggerated).

$$\text{Flux } J_A = -D_A(\partial C_A/\partial x)_1 \tag{2.1}$$

The diffusion coefficient of A atoms is given by:

$$\text{Diffusion coefficient } D_A = 0.5b^2 f_A \tag{2.2}$$

The average number of atoms of A, $S_A$, in unit area of length, $\Delta x$, is:

$$S_A = \Delta x C_A \tag{2.3}$$

The rate of change of $S_A$ with time is then:

$$\partial S_A/\partial t = (\partial C_A/\partial t)\Delta x + [\partial(\Delta x)/\partial t]C_A \tag{2.3a}$$

In the usual coordinate system, that of the specimen, when the length of each region remains constant so that $\partial(\Delta x)/\partial t = 0$, the rate of change of solute becomes:

$$\partial S_A/\partial t = (\partial C_A/\partial t)\Delta x = J_A(+) - J_A(-) = -(\partial J_A/\partial x)\Delta x \tag{2.3b}$$

The flux of solute into a region of fixed length $\Delta x$ is $J_A(+)$ and that out of each region is $J_A(-)$. Substitution of eq. (2.1) then yields the usual form of the diffusion equation, given by

$$(\partial C_A/\partial t)_x = -(\partial J_A/\partial x) = [\partial(D_A \partial C_A/\partial x)/\partial x]_t = D_A(\partial^2 C_A/\partial x^2)_t \qquad (2.4)$$

The simplification of the final part of eq. (2.4) comes when the diffusion coefficient, $D_A$, does not vary with composition, $C_A$, and thus with distance, $x$.

In an alloy, which is neither ideal nor dilute, there can be an enthalpy difference, $\Delta H$, between atoms on adjacent planes with different solute contents (fig. 2.2) so that the atomic jump frequency from 1 to 2, $f_{12}$, is different from the jump frequency in the reverse direction, $f_{21}$. ($\Delta H$ can be either $<0$ or $>0$.)

$$f_{12} = \nu \exp[-(\Delta H^*/kT)]$$

$$f_{21} = \nu \exp[-(\Delta H^* + \Delta H)/kT)]$$

Here $\nu$ is the vibration frequency of the A atoms, and $\Delta H^*$ the activation energy.

The resulting variation of jump frequency will affect the value and even the sign of the diffusion coefficient. The usual way of discussing this problem is by means of the partial molar free energy, $\bar{G}_A$, also known as the chemical potential $\mu_A$.

$$G_A = (\partial G/\partial n_A)_{T,P,C_A}$$

where $n_A$ is the number of gram moles of component A. The values of $\bar{G}_A$ and $\bar{G}_B$ can be found, as shown in fig. 2.1, by extrapolating the tangent to the free energy–composition curve at $C_A$ to the pure components A and B.

Darken (1948) suggested that the drift velocity of a diffusing A atom will be proportional to the gradient of the partial molal free energy of A:

$$v_A = -\beta_A(\partial \bar{G}_A/\partial x)$$

The proportionality constant, $\beta_A$, is the inherently positive atomic mobility of the A atoms. This is the exact equivalent of Ohm's law of conductivity where the drift velocity of an electron is given by the product of the electron mobility, $\mu_e$, and the electric field – the gradient of electrical potential:

$$v_e = -\mu_e(\partial V/\partial x)$$

The flux of A atoms is then given by the product of the mean drift velocity and the concentration of A atoms:

$$J_A = v_A C_A = -C_A \beta_A(\partial \bar{G}_A/\partial x) \qquad (2.5)$$

Comparison of eq. (2.1) and eq. (2.5) shows that the diffusion coefficient is given by:

$$D_A = -C_A\beta_A(\partial\bar{G}_A/\partial C_A) = \beta_A(\partial G_A/\partial \ln C_A) = \beta_A(\partial\bar{G}_A/\partial \ln N_A) \quad (2.6)$$

Here $N_A$ is the molar or atomic fraction of A – given by $C_A/C$, where $C$ is the total number of atoms per unit volume so that $C = 1/V_m$: $V_m$ is the molar volume that, for simplicity, is usually assumed to be constant.

As discussed in any account of solution thermodynamics, for example, Gaskell (1983), the partial molar free energy of A is related to the activity of A, $a_A$, and thus the activity coefficient of A, $\gamma_A$, by the equation:

$$\bar{G}_A - G_A^\circ = RT \ln a_A = RT \ln N_A + RT \ln \gamma_A \quad (2.7)$$

$G_A^\circ$ is the constant molar free energy of A in its standard state – usually that of pure A. So:

$$\partial\bar{G}_A/\partial \ln N_A = RT(1 + \partial \ln \gamma_A/\partial \ln N_A)$$

and therefore:

$$D_A = \beta_A RT(1 + \partial \ln \gamma_A/\partial \ln N_A) \quad (2.6a)$$

The intrinsic diffusion coefficient of A, $D_A^*$, in a chemically homogeneous alloy of composition $C_A$ can be measured by the diffusion of a radioactive isotope of component A.

$$D_A^* = \beta_A^* RT(1 + \partial \ln \gamma_A^*/\partial \ln N_A^*)$$

All the terms marked with an asterisk refer to the radioisotope.

Since isotopes of an element are chemically identical their solutions are ideal, so that even in the *chemically* non-ideal alloy the term in brackets above will be unity in a homogeneous alloy in which $D_A^*$ is determined. The mobility terms, $\beta_A^*$ and $\beta_A$, will also be the same, so:

$$D_A^* = \beta_A^* RT = \beta_A RT$$

and therefore, as first shown by Darken (1948):

$$D_A = D_A^*(1 + \partial \ln \gamma_A^*/\partial \ln N_A^*) \quad (2.8)$$

In a binary alloy where $N_A + N_B = 1$ and $C_A + C_B$ is constant, the concentration gradients of the two components are related by:

$$\partial C_A/\partial x = -\partial C_B/\partial x$$

The flux of each type of atom will then occur in opposite directions but both of the fluxes contribute to the change of composition with time in the inhomogeneous alloy. In alloy systems where the thermodynamic effects are significant then the appropriate form of the diffusion equation is that given by Darken (1948), that is:

$$(\partial C_A/\partial t)_x=[\partial(\mathbf{D}\partial C_A/\partial x)/\partial x]_t \quad (2.4a)$$

The interdiffusion coefficient **D** is then given by:

$$\mathbf{D}=N_A D_B+N_B D_A \quad (2.9)$$

and therefore

$$\mathbf{D}=(N_A D_B^*+N_B D_A^*)(1+\partial \ln \gamma_A/\partial \ln N_A) \quad (2.9a)$$

This result comes from one of the standard equations of solution thermodynamics, the Gibbs–Duhem equation (see Gaskell (1983)):

$$N_A \partial \ln G_A+N_B \partial \ln G_B=0 \quad (2.10)$$

The Gibbs–Duhem equation can be manipulated, using the definition of the partial molar free energy, eq. (2.7), and that $N_A+N_B=1$, to give:

$$N_A \partial \ln \gamma_A+N_B \partial \ln \gamma_B=0 \quad (2.10a)$$

and

$$\partial \ln \gamma_A/\partial \ln N_A=\partial \ln \gamma_B/\partial \ln N_B \quad (2.10b)$$

Cahn (1968) in his review of spinodal decomposition where both $\partial^2 G/\partial C_A^2<0$ and $\partial^2 G/\partial N_A^2<0$, pointed out that the sign of $\partial^2 G/\partial N_A^2$ determines the sign of the interdiffusion coefficient **D**. This is readily seen since the free energy per mole of solution, **G**, is given by:

$$G=N_A \bar{G}_A+N_B \bar{G}_B$$

$$G=N_A(G_A^\circ+RT\ln a_A)+(1-N_A)(G_B^\circ+RT\ln a_B)$$

Differentiation of this twice with respect to $N_A$ using eq. (2.10a) gives:

$$\partial^2 G/\partial N_A^2=RT(1/N_A+1/N_B)(1+\partial \ln \gamma_A/\partial \ln N_A) \quad (2.11)$$

So, inside the spinodal where $\partial^2 G/\partial N_A{}^2<0$, the Darken thermodynamic factor $(1+\partial \ln \gamma_A/\partial \ln N_A)$ is also less than 0 and so by eq. (2.9a) is the interdiffusion coefficient **D**.

Eqs.(2.6) and (2.8) show that diffusion in condensed phases is controlled by two factors. The first is the intrinsic atomic mobility, $\beta_A$ or $D^*_A$, which for a solid is usually determined by the concentration and mobility of point defects. The second factor is the thermodynamics of the solid solution commonly given by the Darken factor $(1+\partial \ln \gamma_A/\partial \ln N_A)$. The thermodynamic effect can accelerate diffusion when there is a negative deviation from ideality such that $\gamma_A$ is less than 1 but increasing to $\gamma_A=1$ as $N_A \rightarrow 1$, so that both $\partial\gamma_A/\partial N_A>0$ and $\partial \ln\gamma_A/\partial \ln N_A>0$. The thermodynamic term, where there is a positive deviation from ideality and $\gamma_A$ is then more than 1, can equivalently retard diffusion and in the limit when $\partial^2 G/\partial N_A^2<0$ and $\partial \ln \gamma_A/\partial \ln N_A<-1$, leads to a negative diffusion and unmixing of unstable solid solutions.

In an investigation of homogenisation of microsegregated Al–Zn, in alloys whose range of compositions included 37 at% Zn (Ciach, Dukiet-Zawadzka and Ciach 1978), it was found, exactly as expected from the discussion above, that the rate of homogenisation was very much slower than would have been expected from the atomic mobility of the aluminium and zinc atoms at the homogenisation temperature. The reduction in the rate of homogenisation arises since, as the temperature falls below a critical temperature, $T_c=351.5$ °C, the alloy system undergoes phase separation, $\alpha \rightarrow \alpha' + \alpha''$. Below $T_c$, $\partial^2 G_A/\partial C^2$ and the Darken thermodynamic term $(1+\partial \ln \gamma_{Zn}/\partial \ln N_{Zn})$ are both *negative* leading to the spinodal decomposition reaction discussed in §2.2.1. At the critical unmixing temperature, the thermodynamic term, in the absence of effects due to elastic strain (§2.2.1) is 0. At the investigated homogenisation temperature of 360 °C, the Darken term was reported to be only 0.02 at the composition of $N_{Zn}=0.37$. The interdiffusion coefficient, **D** (eq. (2.10a)), was found at 360 °C to be only $6\times10^{-16}$ $m^2\,s^{-1}$, smaller by the Darken factor of 0.02 than the intrinsic atomic mobility, $N_A D^*_B + N_B D^*_A$, which was estimated to be somewhat less than $10^{-13}$ $m^2\,s^{-1}$ at the same temperature. Ciach *et al.* found that on annealing as-cast Al–60 wt% Zn the non-equilibrium zinc eutectic phase, $\eta$, dissolved at 360 °C. This is as expected since the zinc rich solid solution, at $N_{Zn}=0.6$, in equilibrium with $\eta$, has a much larger value of the thermodynamic term at 360 °C, than the value of only 0.02 found at $N_{Zn}=0.37$. Continued annealing at 360 °C for 50 h did not, however, completely remove the non-uniform zinc distribution in the resulting solid solution that included the critical composition.

The authors suggested that a well-designed homogenisation process would be one involving a preliminary low temperature initial anneal at 360 °C to remove the low melting zinc phase followed by final homogenisation at higher temperatures such as 450 °C where, due to the rise in the thermodynamic term as well as in the mobility, the value of the interdiffusion coefficient, **D**, increased by two orders of magnitude to $6\times10^{-14}$ m$^2$ s$^{-1}$. The purpose of the low temperature first step is to remove the low temperature melting zinc inclusions before annealing at a temperature above the melting point of the non-equilibrium eutectic inclusions. Doherty (1979a), on the basis of the ideas presented above, pointed out that equivalent homogenisation difficulties are to be expected in any alloy system that undergoes, at lower temperatures, spinodal decomposition.

### 2.1.2 Modified diffusion equation in the presence of a Kirkendall effect

Doherty (1993) has explored, by numerical simulation, the diffusional processes occurring in a system, Cu–Ni, that was nearly ideal but in which the two diffusion coefficients, $D_{Cu}$ and $D_{Ni}$, were significantly different; $D_{Cu}=3D_{Ni}$. This means that significant Kirkendall shifts are expected (Smigelskas and Kirkendall 1947) and are found on interdiffusing pure copper and nickel (da Silva and Mehl 1951). Numerical study revealed, as first shown by Darken (1948), that the diffusion analysis will be significantly different if is carried out in the moving lattice coordinate system rather than in the stationary specimen coordinate system. In one-dimensional diffusion in the *specimen* coordinate system, a small volume element at a distance $x$ from the end of the specimen will have, by definition, a constant length, $\Delta x$, a fixed cross-sectional area and so a constant volume. Under these circumstances, eq. (2.4a) is the correct diffusion equation to use in the specimen coordinate system and, as shown by Darken, the interdiffusion coefficient **D** of eq. (2.8) is the correct coefficient to use. If, however, the analysis or simulation is to be carried out in the *lattice* coordinate system ($x'$) then as shown below the full form of eq. (2.3a) must be used and this gives a modified diffusion equation.

The lattice coordinate system, $x'$, is one measured from a set, or sets, of inert markers placed in the lattice and as shown by the Kirkendall effect these markers move with a local velocity $v'$. Darken (1948) noted that the flux of A atoms in the *lattice* coordinate system, $x'$, is given by the usual form of the first law, given by:

$$J_A(x')=-D_A\partial C_A/\partial x' \tag{2.1a}$$

In the *specimen* coordinate system, there is an additional contribution to the flux due to the Kirkendall velocity of the lattice plane. So the flux, measured in the specimen coordinate system, $x$, at the same position as in eq. (2.1a) ($x=x'$) is that given by:

$$J_A(x')=-D_A\partial C_A/\partial x'+v'C_A \tag{2.1b}$$

Darken's clear analogy was the difference between the flux of moving objects in a river as measured by an observer in a boat on the river and the same flux as seen by a stationary observer on the river bank opposite the observer in the boat. The observer on the river bank sees both the movement due to diffusion and that due to the velocity, $v'$, of the river. Darken showed from eqs. (2.1a) and (2.1b) that the Kirkendahl velocity $v'$ is given by:

$$v'=(D_A-D_B)(dN_A/dx') \tag{2.12}$$

This velocity is the velocity of the lattice planes relative to the end of the specimen. The lattice planes at the ends of the specimen are, by eq. (2.12), stationary since at the ends of the sample the concentration gradients are zero. This condition occurs when the length of the sample is much greater than the distance over which diffusion occurs, $(\mathbf{D}t)^{0.5}$. The integration of the changing velocity, given by (eq. 2.12), over the whole diffusion run for the lattice plane *at the original interface* gives the observed Kirkendall shift of inert interface markers. Such shifts were first described by Smigelskas and Kirkendall (1947). From eq. (2.12), Darken derived the correct interdiffusion coefficient, **D**, eq. (2.9), for use in the usual experimental coordinate system – that of the specimen.

In order to simulate the Kirkendall shifts in the Cu–Ni system, Doherty (1993) used the lattice coordinate system. In the lattice system the change of solute content is dependent on both terms in eq. (2.3):

$$\partial S_A/\partial t=(\delta C_A/\partial t)\Delta x'+[\partial(\Delta x')/\partial t]C_A \tag{2.3a}$$

So

$$\partial S_A/\partial t=(\delta C_A/\partial t)\Delta x'+[\partial(\Delta x')/\partial t]C_A=-\Delta x'(\partial J_A/\partial x') \tag{2.3c}$$

In the lattice coordinate system with a finite and variable Kirkendall velocity, $v'$, the element lengths, $\Delta x'$, must change, so then the diffusion equation derived from eqs. (2.1) and (2.3c) becomes:

$$(\partial C_A/\partial t)_{x'}=[\partial(D_A\,\partial C_A/\partial x')/\partial x']_t-(\partial\ln\Delta x'/\partial t)_{x'}C_A \tag{2.4b}$$

Note that in the one-dimensional diffusional analysis here the area of the sample is assumed constant, at $A=1$, so that the length $\Delta x'$ is actually the volume, $\Delta V'$, of the region. So a more general form for the diffusion equation is given by:

$$(\partial C_A/\partial t)_{x'}=[\partial(D_A\,\partial C_A/\partial x')/\partial x']_t-(\partial\ln\Delta V'/\partial t)_{x'}C_A \tag{2.4c}$$

This form of the diffusion equation includes the additional contribution to the change of concentration, $\partial C_A/\partial t$ ($C_A$ is the number of B atoms per unit volume), arising from the change of the local volume of the region.

The local volume, $\Delta x'$ or $\Delta V'$, changes by virtue of creation or destruction of vacancies, when $D_A \neq D_B$ in substitutional diffusion. So in a system such as Ni–Cu, where $D_{Cu} > D_{Ni}$, there is a reduction of the local lengths in the copper rich regions due to the net loss of atoms or, more accurately, atom sites. The net loss of atom sites results in negative values of $d(\Delta x')/dt$ so producing an *increase* in the solute composition, eq. (2.4b). Figs. 2.3–2.5 show the results of two numerical simulations of Ni–Cu interdiffusion at 1323 K for 312 h, a situation previously studied experimentally by da Silva and Mehl (1951). In fig. 2.3 the interdiffusion had been simulated both using the standard equation, eq. (2.4a), with the Darken form of the interdiffusion coefficient, **D**, eq. (2.10), and also using the modified (lattice) diffusion equation, eq. (2.4b), but with the individual diffusion coefficients, $D_{Cu}$ or $D_{Ni}$, for profiles calculated for either element as the solute A. It is clear that both simulations give identical profiles and these profiles were found to be almost identical to the experimental profile. Fig. 2.4 gives the simulated Kirkendall marker shifts for markers placed at the initial interface – these markers move into copper with a displacement that scales with the square root of time with a constant composition as seen experimentally. The magnitude of both the shift and the local composition at the interface markers also matched the experimental values reported by da Silva and Mehl (1951). Movement of

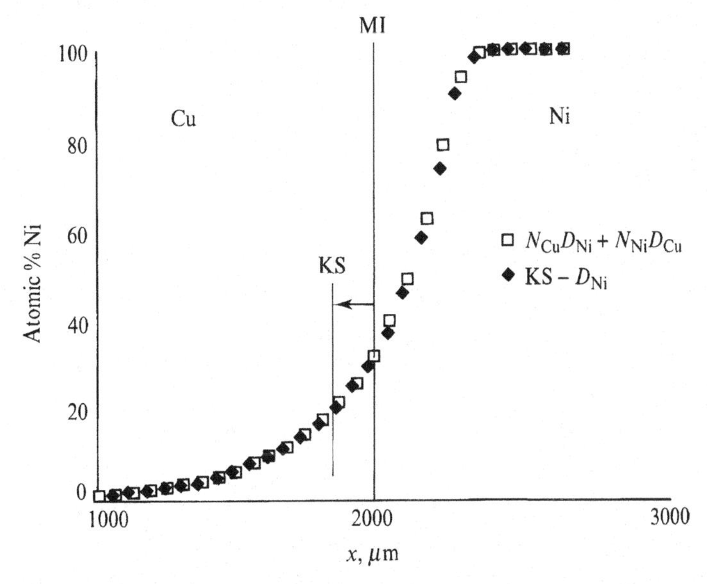


**Figure 2.3** Composition profile of nickel in the interdiffusion at 1323 K for 312 h of pure copper initially at $x \leq 1995$ μm, and pure nickel, at $x \geq 1995$ μm, simulated (i) by the individual diffusion coefficient, $D_{Ni}$ in the Kirkendall shift modified diffusion equation, eqs. (2.4b) as 'KS' $D_{Ni}$ and (ii) by use of Darken's expression for the interdiffusion coefficient, eq. (2.10) in the standard diffusion equation, eq. (2.4a), as $N_{Cu}D_{Ni} + N_{Ni}D_{Cu}$. The Matano interface, found at the initial interface at 1995 μm, is shown as the vertical line and the shift of the interface markers is indicated by the arrow marked KS. The two profiles are identical. Doherty (1993).

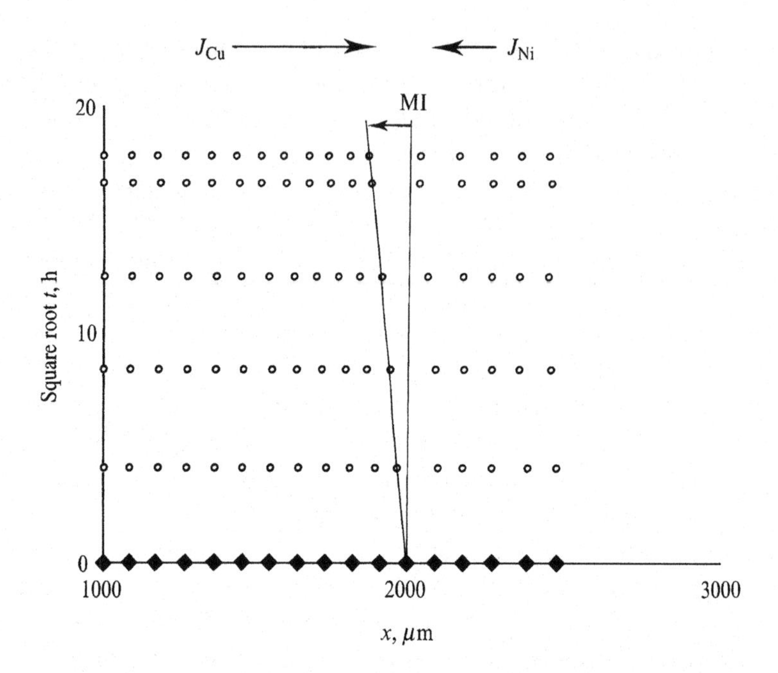


**Figure 2.4** Positions of initially equally spaced markers at different values of the square root of the diffusion time, in hours, simulated by use of the modified diffusion equation during interdiffusion of Cu–Ni at 1323 K for times up to 312 h. The initial, Matano interface (MI) is shown by the vertical line while the displacements of interface markers are indicated by the sloping line. The marker spacings in copper are reduced while those in nickel are expanded. The interface markers are correctly predicted to displace, into the faster diffusing copper, linearly with $(t)^{0.5}$. Doherty (1993).

markers placed at different distances from the initial interface was simulated and the results also plotted in fig. 2.4. Movement of markers away from the initial interface seems, however, never to have been studied experimentally or at least not reported. Fig. 2.4 clearly shows that the 'mesh' lengths, $\Delta x'$, in the copper rich regions of the interface markers all *decrease* while those on the nickel rich side all grow larger. This means that according to the modified diffusion equation, eq. (2.4a), the solute content will be increased by this effect on the copper rich regions and decreased in the nickel rich regions. Fig. 2.5 demonstrates this result. The figure shows the results of running two simulations – both using $D_{Ni}$ for the diffusion coefficient. For the simulation based on the modified lattice diffusion equation, eq. (2.4b), we obtain the same profile as was given in fig. 2.3, but calculated in two ways. The incorrect use of $D_{Ni}$ rather than the Darken $\mathbf{D}$ in the standard diffusion equation, eq. (2.4a), is seen to underestimate significantly the change of composition *on both sides of the diffusion profile*. In the copper rich region the incorrect use of $D_{Ni}$ in eq. (2.4a) leads to an *underestimate* of the content of the solute, A=Ni. The additional flux of copper atoms out of the

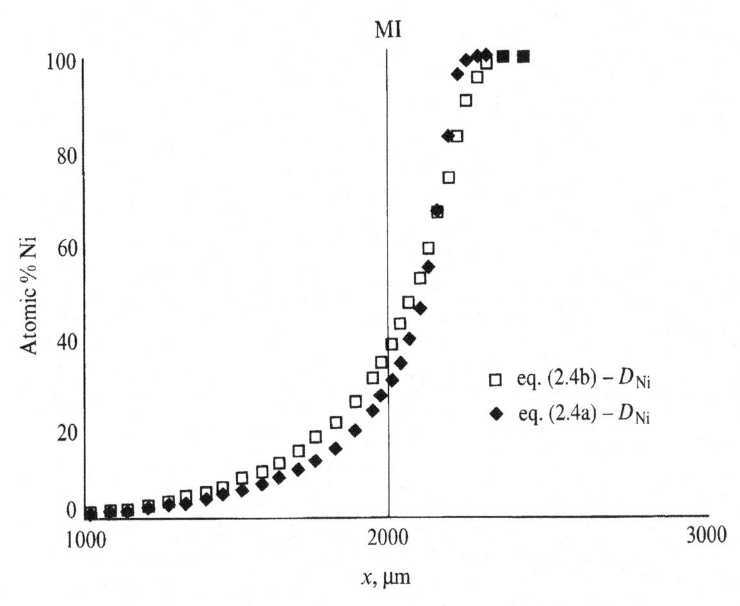


**Figure 2.5** Simulated composition profiles of Ni during interdiffusion of Cu–Ni at 1323 K for 312 h: (i) correctly predicted using $D_{Ni}$ in the KS modified diffusion equation, eq. (2.4b) (this is the same profile as in fig. 2.3) and (ii) incorrectly, by use of $D_{Ni}$ in eq. (2.4a). The incorrect equation here underpredicts the changes in composition. The full vertical line indicates, for both profiles, the initial position of the interface which is also the Matano interface (MI). Doherty (1993).

copper region, neglected by the use of $D_{Ni}$ in eq. (2.4a), leads to the observed *increase* in solute content on the low nickel (copper rich) side of the diffusion couple. An analogy is the increased concentration of the salt content of a salt solution by evaporation of the solvent water. The equivalent extra flux of copper atoms into the nickel rich side of the interface leads to the observed decrease in solute (nickel) contents in the nickel rich regions. The fall of solute (nickel) content comes from the further dilution of the alloy by the additional flux of the solvent – copper. These pairs of simulations thus serve to illustrate the *physical* reason behind the increase of effective diffusion coefficient from $D_{Ni}$ to $\mathbf{D}$ required mathematically by Darken's analysis leading to $\mathbf{D}=N_{Ni}D_{Cu}+N_{Cu}D_{Ni}$ (eq. (2.10)) (noting that $D_{Ni}=0.3D_{Cu}$).

### 2.1.3 The origin of non-uniform solid solutions – coring or interdendritic segregation

The most usual cause of an inhomogeneous solid solution is the coring that results from solidification. Any alloy that freezes over a range of temperature forms a solid with a range of composition. The name of *coring* for this phenom-

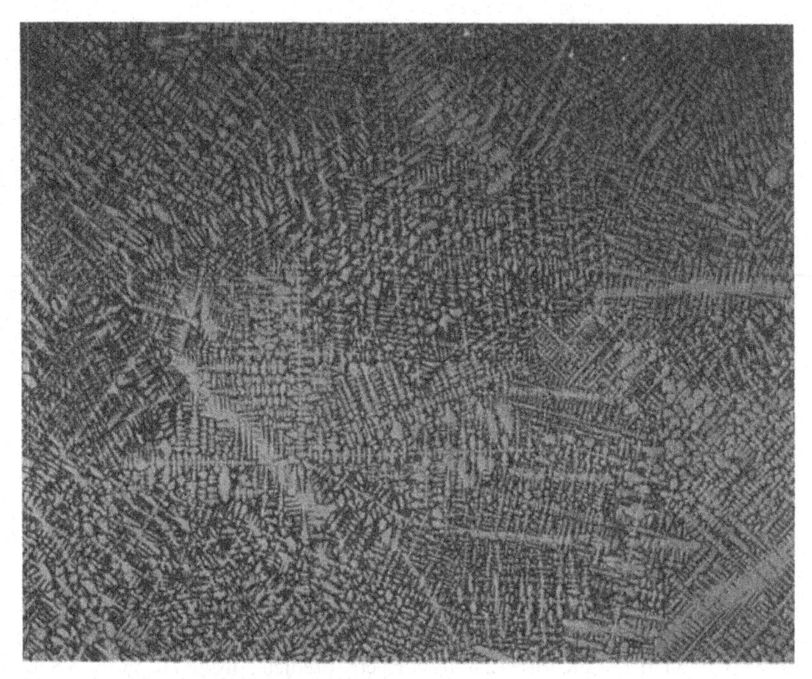

**Figure 2.6** Section from equiaxed region of an ingot.

enon arises since the core of the solid, the first region to solidify, has a different composition from the periphery which freezes last. The alternative description of this as interdendritic segregation comes from the dendritic or tree-like morphology of the solid (fig. 2.6). The dendritic mode of solidification occurs for all metal alloys frozen in the absence of an imposed temperature gradient, for example, in most ingots and castings (Chalmers 1964). This dendritic pattern has the important result that the scale or wavelength of the segregation is that of the dendrite arm spacing. The dendrite itself determines the grain size. Cole (1971) has reviewed the subject of dendritic segregation and Doherty (1979b) the process of dendritic growth itself.

The scale of the segregation, the dendrite arm spacing, and the diffusion coefficient are the most important factors in determining the kinetics of homogenisation of the cored solid solution, since this spacing is the distance over which atomic migration must occur. Despite its importance there are conflicting accounts of how this spacing is related to the composition of the alloy but the major effect of the cooling rate on the spacing is now well established. Kattamis, Coughlin and Flemings (1967) have shown that the time, $t_s$, that an alloy spends in the freezing range, from the liquidus temperature to the solidus, is related to the dendrite arm spacing $d$ by the empirical equation

$$d = d_0 t_s^n$$

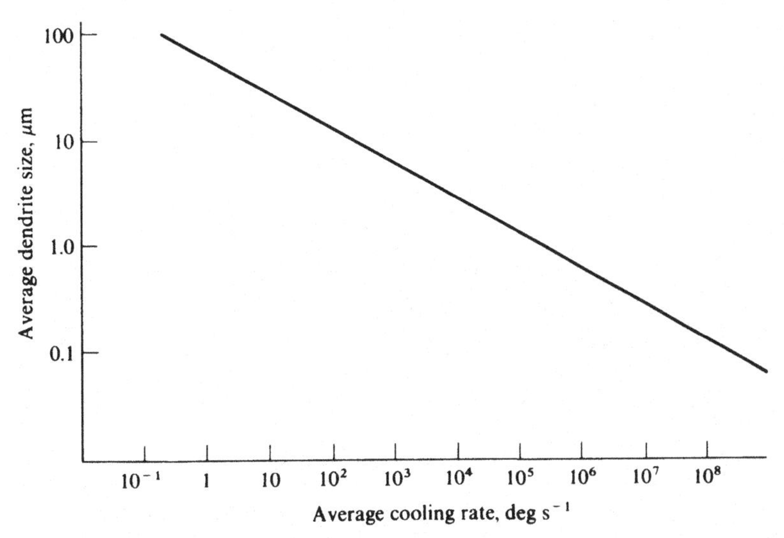


**Figure 2.7** An approximate curve of dendrite size versus cooling rate of an aluminium alloy.

The exponent, $n$, in this equation lies between about 0.3 and 0.5. This relationship has been investigated over many orders of magnitude, from times of the order of hours characteristic of the centre of large ingots (e.g., Doherty and Melford (1966)) to a few microseconds in 'splat cooled' droplets (Matyja, Giessen and Grant 1968). Fig. 2.7 shows the results for an aluminium alloy where the spacing decreases from over 100 $\mu$m to less than 0.1 $\mu$m.

The form of the equation suggests a connection with the surface energy driven coarsening processes described in chapter 5, and this is supported by the direct observations made by Kahlweit (1968) on the arm coarsening in dendritic crystals of ammonium chloride in water.

The amplitude of dendritic segregation in cast alloys has been well studied; for example, by Kohn and Philibert (1960) and by Feest and Doherty (1973). It is usually found that for a binary alloy the composition will range from a minimum of somewhat more than the solidus $C$, to a maximum that for a eutectic alloy (fig. 2.8) will reach the eutectic composition $C_e$.

## 2.1.4 Homogenisation of segregation in cast alloys

As discussed by Shewmon (1966) and Purdy and Kirkaldy (1971), any segregation profile can be represented by a Fourier series of cosine functions of which the longest wavelength will be that of the dendrite arm spacing $d$. This Fourier component will be

$$C(x)=C_0+C_a \cos(2\pi x/d) \tag{2.13}$$

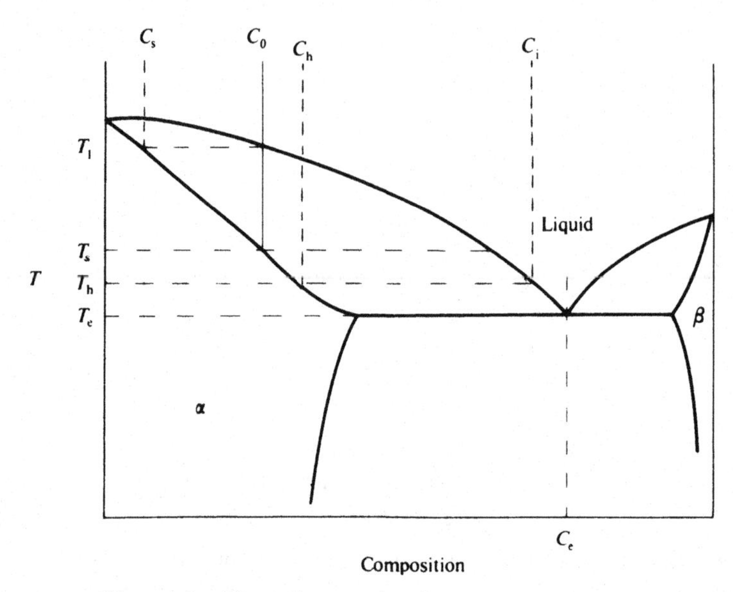


**Figure 2.8** Phase diagram showing temperatures and compositions relevant to the discussion of the removal of dendritic segregation.

where $C_0$ is the average composition, $C_a$ the amplitude and $d$ the fundamental wavelength, the arm spacing. Other components will have shorter wavelengths. The usual type of segregation pattern is illustrated in fig. 2.9 where it is obvious that the higher harmonics are important. Each harmonic can decay independently at a rate given, for the fundamental, by the equation

$$C(x)=C_0+C_a \cos(2\pi x/d) \exp(-t/\tau)$$

where

$$\tau=4d^2/\pi^2 D \tag{2.14}$$

The higher harmonics all have smaller values of the relaxation time than the fundamental and so will decay faster than the fundamental cosine function. This means that the pattern of segregation will decay from the form shown in fig. 2.9 to a simple cosine function of wavelength $d$, and so the major factor determining the kinetics of homogenisation will be the value of the fundamental relaxation time. Homogenisation for times of the order of $3\tau$ $(12d^2/\pi^2 D \approx d^2/D)$ will eliminate segregation almost completely.

This analysis has been found to be successful both qualitatively and quantitatively, for example, by Ward (1965), but other studies by Weinberg and Buhr (1969) found that exact agreement was not achieved. Despite the obvious importance of the subject, there are remarkably few investigations of the effect of

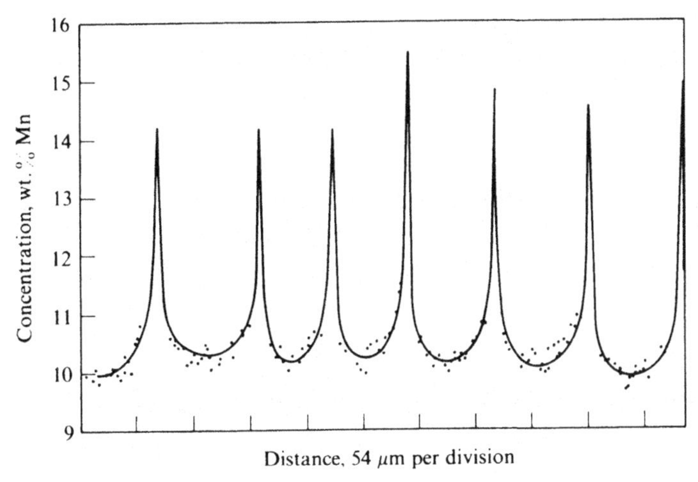


**Figure 2.9** Microprobe scan across a series of side branches. (After Hone and Purdy 1969.)

homogenisation on mechanical properties. In one such study, however, Ahearn and Quigley (1966) found that, as might have been expected, the room temperature ductility in some cast steel alloys improved steadily with homogenisation as measured by the dimensionless homogenisation parameter $Dt/d^2$, where $t$ was the homogenisation time.

These ideas combined with the previous discussion of the factors controlling the dendrite arm spacing lend support to the usual practice in alloy production. This consists of chill casting the liquid alloy in as thin a section as is practical to achieve a fine dendrite arm spacing ($d$) followed by homogenisation at as high a temperature as possible to give a maximum value for $D$. This gives the shortest times for complete homogenisation. There are clear advantages therefore in the practice of continuous casting thin sections of alloys rather than the alternative practice of casting large ingots followed by deformation. (This is considered further in §2.1.6.)

Purdy and Kirkaldy (1971) have drawn attention to the advantages to be gained by homogenisation at temperatures above the melting point of the segregated alloy, such as at $T_h$ in fig. 2.8 where $T_h$ lies between the eutectic temperature $T_e$ and the equilibrium solidus temperature $T_s$. There are two advantages for the use of this temperature as distinct from annealing below the eutectic temperature. Firstly the value of $D$ will be larger, and secondly the second solid phase will have melted. Atom transfer is usually very rapid across solid–liquid interfaces in metals, but this may not always be the case for solid–solid interfaces as is discussed later. This practice has been successfully used to eliminate non-equilibrium eutectic carbides in an alloy steel by Proud and Wynne (1968). The presence of these carbides greatly reduced the hot ductility of these steels which were to be deformed above $T_e$.

The thermodynamic factor in homogenisation would not usually be significant in the dilute alloys that make up the bulk of commercial usage. However, this is likely to be a problem in alloys such as aluminium–zinc where just below the melting point in the middle of the diagram there is an unmixing into two phases of the same structure. Below such a temperature $d^2F/dN^2$ will be negative, and so diffusion even above the critical temperature will be very sluggish, since the Darken factor $(1+\partial\ln \gamma/\partial\ln N)$ will, though not yet negative, be very close to zero. All potential spinodally decomposing alloys will be in this situation.

## 2.1.5 Homogenisation above the non-equilibrium melting temperature

In a range of solidification studies, Vogel, Cantor and Doherty (1979), Doherty, Feest and Lee (1984), Annavarapu, Liu and Doherty (1992) and Annavarapu and Doherty (1993, 1995) investigated the microstructural changes occurring on annealing polycrystalline alloys just above the solidus temperature. Many of the grain boundaries were rapidly wetted by thin layers of liquid. The condition for this was the usual result for high angle grain boundaries (Gündoz and Hunt 1985) that:

$$\sigma_{gb} \geq 2\sigma_{sl}$$

Here $\sigma_{gb}$ is the energy of a high angle grain boundary (with misorientation greater than about 20°) and $\sigma_{sl}$ is the energy of the solid–liquid interface. Certain lower energy grain boundaries, notably those with low angles of misorientation and some special 'coincident site' boundaries, see, for example, Shewmon (1966), were found not to be wetted by liquid. Longer periods of holding in the two-phase, solid plus liquid phase, field at low volume fractions of liquid caused the liquid to concentrate at three grain triple points and four grain corners as a result of the change of solubility at curved grain boundaries – the Gibbs–Thomson effect, §5.5.2. The major microstructural change was the migration of the wetted and partially wetted grain boundaries driven by grain growth at rates limited by solute transport across the small liquid regions.

Of more interest to studies of homogenisation was that the usual process of dendrite arm coarsening, eq. (2.13), did not occur significantly on reheating just into the solid plus liquid region. The reason for this was clear from the studies of dendritically segregated Al–Cu alloys by Annavarapu and coworkers (Annavarapu *et al.* 1992; Annavarapu and Doherty 1993, 1995). The isolated dendrite arms within a single grain in the polycrystalline as-cast structure sintered together since with little or no misorientation the interface between adjacent arms could not be rewetted on remelting. Fig. 2.10 shows that after

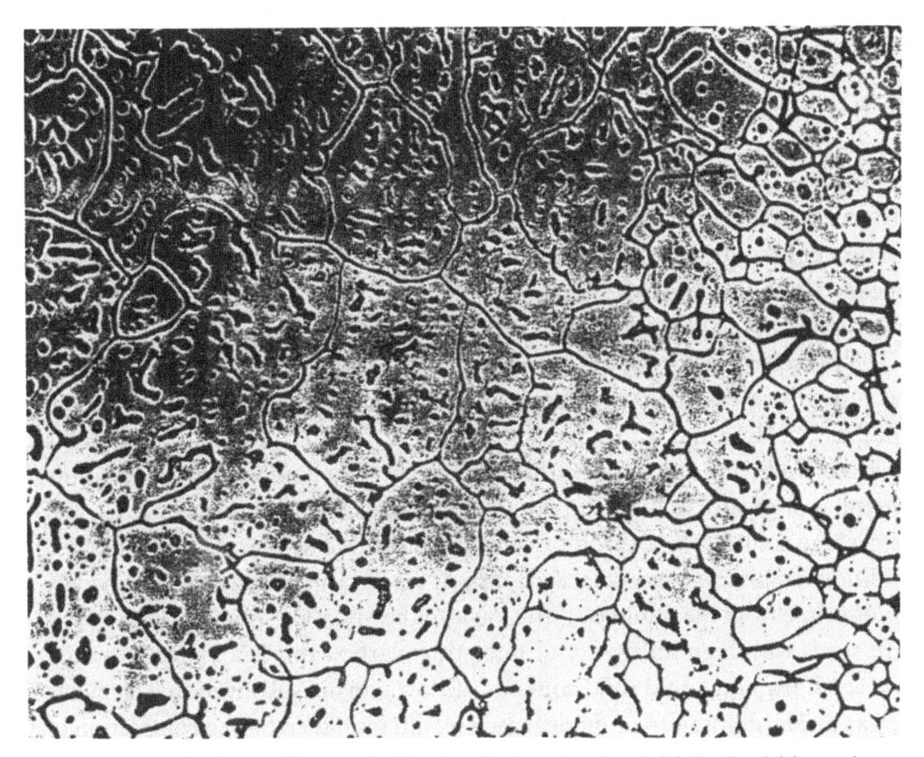

**Figure 2.10** Chill cast Al–6.7 wt% Cu showing fine, initially dendritic, grains after reheating to 612 °C (above the eutectic temperature) holding for 60 seconds and quenching, the dendrite arms have coalesced together trapping Cu rich liquid droplets within the grains and wetting all the grain boundaries with a thin liquid film. The coarsening of the liquid droplets can only occur by slow solid state diffusion. The droplets can be swept-up by the grain boundaries as these migrate. Annavarapu and Doherty (see Annavarapu *et al.*, 1992, 1993).

exposure to temperatures below the eutectic temperature the continuous liquid regions between dendrite arms is broken up into isolated eutectic regions that on remelting remain isolated from each other. This process of structural inversion from solid in a matrix of liquid to liquid in a solid matrix dramatically slows the coarsening kinetics. The kinetics change from a rapid process of dendrite arm coarsening using liquid state diffusion $D_l$, where $D_l$ is typically about $10^{-9}$ m$^2$ s$^{-1}$, to a much slower coarsening process for the now isolated liquid droplets based on the much slower solid state diffusion process through the continuous solid phase. This process using solid state diffusion, with $D_s$ typically $10^{-13}$ m$^2$ s$^{-1}$ or less, is very much slower and occurs at rates expected for solid state 'Ostwald ripening', §5.5. With negligible arm coarsening on reheating to the partially molten state, the mean diffusion distances used in eq. (2.15) do not rise so that acceleration of homogenisation by a rise of $D_s$ is not offset by rises of segregation spacing, *d*, by arm coarsening. Grain coarsening, however, does lead to an increase of the spacing between the liquid films at grain boundaries and grain corners. This type of growth of the spacing between solute rich liquid droplets will be difficult to remove by post-solidification annealing. It would thus appear that the optimum homogenisation will be expected at temperatures above the eutectic melting point, $T_e$, in fig. 2.8 but below the equilibrium solidus temperature, $T_s$. Under these circumstances the liquid phase will have only a transient existence from the time when the as-cast sample is heated above the eutectic temperature until solid state diffusion removes the last trace of liquid. The liquid disappears when the highest remaining solute content is less than the solidus composition at the homogenisation temperature, $C_h$, in fig. 2.8.

The processes occurring during homogenisation at temperatures between $T_e$ and $T_s$ are those that also occur during 'diffusional solidification'. In diffusional solidification low solute solid particles are infiltrated with high solute liquid. If the process is occurring at a temperature $T_h$ in fig. 2.8 the liquid must have a solute content at or greater than the liquidus composition at that temperature, $C'_l$, and the average solute composition of the alloy must be less than $C_h$. Under these conditions solidification occurs at a constant temperature at rates determined by loss of solute from the liquid by diffusion into the low alloy solid particles.

In steels with high diffusivity of interstitial carbon this is a viable process – however, if the liquid and solid contents of slow moving substitutional elements such as nickel, chromium or manganese are different, then, at the completion of solidification, non-uniform distribution of these elements will remain in the microstructure. This type of diffusional solidification will have been experienced by anyone who has received amalgam dental fillings. In such alloys high mercury content liquid reacts with metallic elements with a significant solubility for mercury at 37 °C. Diffusional solidification then converts a pasty solid–liquid mixture into a fully solid 'casting' in the recipient's dental cavity.

### 2.1.6 Deformation of segregated alloys and its effect on homogenisation

In large ingots, of more than a tonne in mass, the slow cooling rate gives dendrite arm spacings so large that homogenisation is almost impossible. (If the arm spacing is 1 mm and the diffusion coefficient is of the order of $10^{-12}$ m$^2$ s$^{-1}$ the time for homogenisation, $d^2/D$, will be $10^7$ s or several months.) Such ingots are usually heavily deformed to reduce their section and since the deformation will change the segregation wavelengths, this process is likely to influence strongly the homogenisation possibilities.

An extrusion process that reduces the diameter of an ingot to $1/R$ will change a cube into a tetragonal shape whose dimensions will be $R^2$, $1/R$, $1/R$. It is difficult to describe the results of such a change on the complex geometry of a solidification dendrite but some insight may be obtained from a simple model illustrated in fig. 2.11. In this model there is a 'simple cubic' array of regions of high solute content ($C_{max}$) with the regions of low solute ($C_{min}$) in the centre of the cube forming a 'CsCl' structure. The wavelength of the distribution $d$ will lie along '⟨111⟩'. On this model all 'planes' parallel to {110} should have the average composition $C_0$ (since they have an equal density of $C_{max}$ and $C_{min}$ regions), while 'planes' parallel to {100} will have a *periodic* variation in concentration from a maximum of less than $C_{max}$ to a minimum of more than $C_{min}$. Other planes will show some periodicity but the amplitude will be less than that of '{100}'.

The effect of deformation can now be seen by extrusion in two alternative directions '⟨110⟩' and '⟨100⟩'. After an extrusion in the '⟨110⟩' direction, fig. 2.11($c$), all distances in the (011) plane will be reduced by $1/R$ and so the time for homogenisation of this plane will be reduced by $1/R^2$. After this treatment the *whole structure* will be at a uniform composition. Unfortunately, although extrusion in the '⟨100⟩' direction, fig. 2.11($d$), will enable planes parallel to {100} to

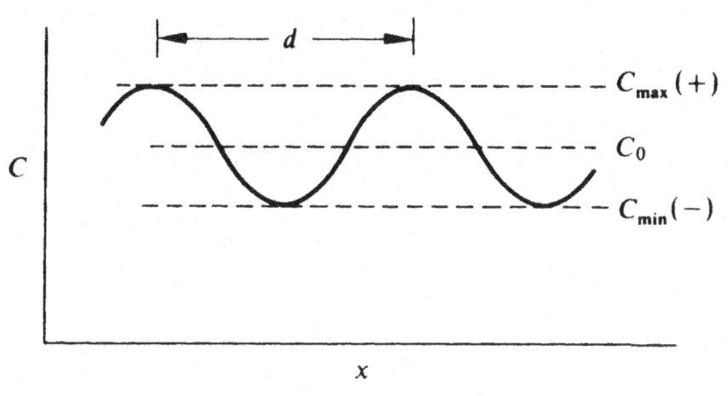


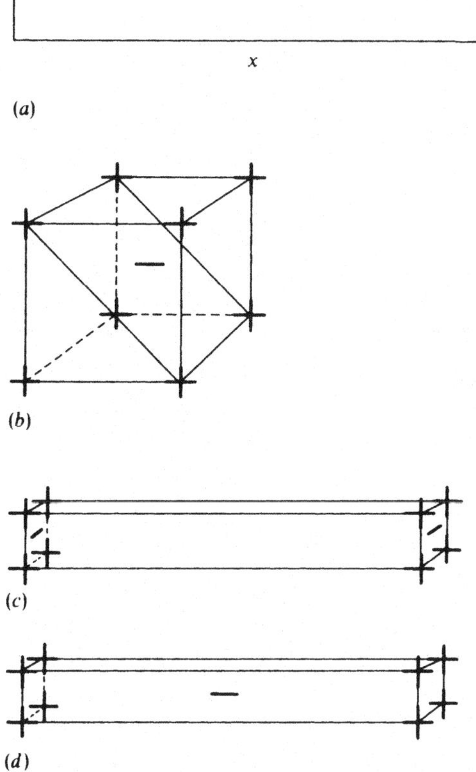


**Figure 2.11**
($a$) Schematic segregation over a wavelength $d$.
($b$) 'CsCl' model of segregation showing $C_{max}$(+) and $C_{min}$(−) position. ($c$) The model after extrusion in a '⟨110⟩' direction ($d$) The model after extrusion in a '⟨100⟩' direction.

achieve a uniform composition after short annealing times, the periodicity in composition along '⟨100⟩' will have its wavelength increased by $R^2$, and so the time for complete homogenisation will *increase* by $R^4$. The average situation will lie between these extremes so that in deformation the initial stages of homogenisation will be accelerated but the last stages of homogenisation will be greatly delayed.

There are few experimental studies of this subject that allow these ideas to be tested. Kohn and Doumerc (1955) reported that the segregation of phosphorus in steel was removed more rapidly after deformation. However, Weinberg and Buhr (1969) reported that deformation had no significant effect on the homogenisation of nickel and chromium in steel. This discrepancy may arise from the experiments looking at different stages of the process but this is merely speculation.

The clearest evidence appears to be the study of Heckel and Balasubramanian (1971) who studied theoretically and experimentally the homogenisation of sintered alloys made from different pure metal powders. These provide well-defined starting materials as well as being technologically important. The spherical, three-dimensional diffusion problem in unextruded powder compacts becomes, after extrusion, a cylindrical problem and after rolling a one-dimensional problem. These changes in geometry reduce the number of diffusion paths which slows homogenisation and the reduction of diffusion distances accelerates the process. The cited authors did not consider the diffusion distances that, as pointed out above, will be increased by deformation. The two effects they considered gave an approximate solution to the homogenisation problem in terms of the dimensionless parameter

$$\alpha = \mathbf{D}tn^2 Y/l_0^2$$

In this equation, $l_0$ is the radius of the spherical powder in the undeformed compact, or the radius of the deformed cylindrical powder after extrusion, or the thickness of the deformed strip of powder after rolling; $n$ is the dimensionality of the problem (three for spheres, two for cylinders and one for strips); $t$ is the diffusion time; and $Y$ is a geometrical constant determined by the volume ratio of the two powders. $Y$ is unity for an equal mixture of the two powders but falls steadily as the volume ratio of minority component falls, since each particle of the minority component is surrounded by an increasing thickness of the majority component. Some of the Heckel and Balasubramanian results are shown in fig. 2.12, where their dimensionless parameter is plotted against rolling reduction $R$. The solid lines are their theoretical predictions for their homogenisation parameter $F$ which approaches unity at complete homogenisation, and the dotted lines show the experimental results. The sloping lines represent rolling deformations at constant annealing time (cold deformation) and the vertical

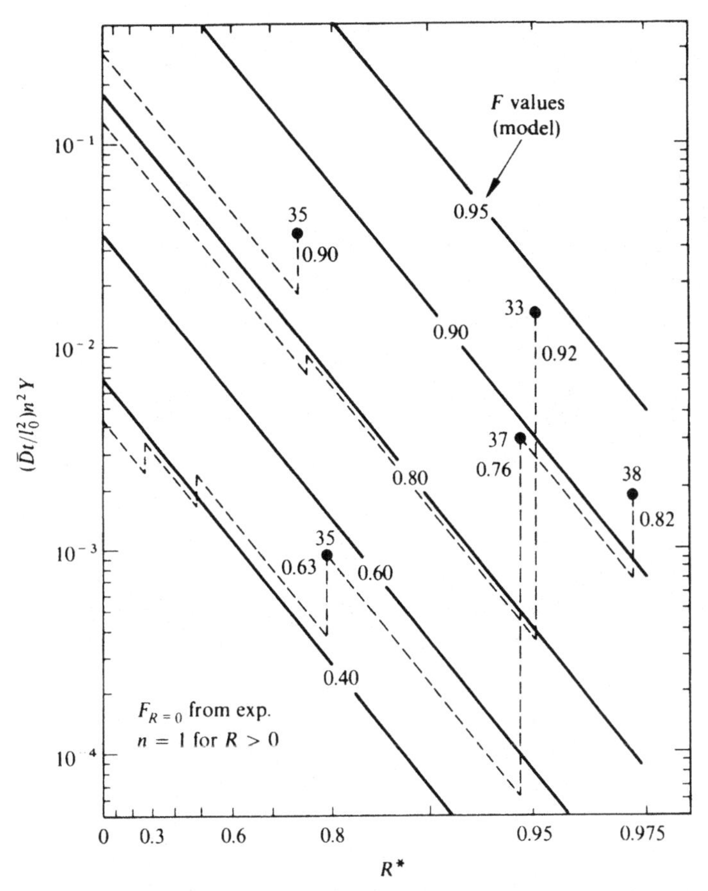


**Figure 2.12** Summary of homogenisation–deformation sequence data and comparison with the predictions of the model. Data points are marked by specimen numbers and experimentally determined $F$ values. (20 at% Cu powder/80 at% Ni powder.)

lines homogenisation anneals. The fractional numbers on the dotted lines are experimental determinations of the homogenisation $F$ at different times. It is assumed that $n=1$ for all deformations, however small, and at low deformations this leads to their model *underestimating* the amount of homogenisation, in that the experimental values of $F$ are greater than the predicted values. However, it is clear that after heavy deformations (points 33, 37 and 38) the measured homogenisation is *less* than that predicted. This is precisely the result expected on the basis of the ideas presented at the start of this section. This is because after heavy deformation ($R>0.95$, 95% reduction) and long annealing times (e.g., $Dt/l_0^2>0.1$), diffusion over those distances reduced by the deformation will be almost complete, but the inhomogeneities over those distances increased by the deformation will be largely unaffected. This aspect of deformation was not considered by Heckel and Balasubramanian but appears to be demonstrated by their study.

## 2.1.7 Banding in rolled steels

Large steel ingots which have significant casting segregation frequently show alternate layers of ferrite and pearlite arranged in bands: this is illustrated in fig. 2.13. This phenomenon of 'banding' arises after heavy deformation, usually by rolling, that spreads out in bands the regions in the cast structure that were either rich or poor in alloying additions. Since the diffusion coefficient of interstitial carbon is very high, the banding is unlikely to result from the casting segregation of carbon itself but comes from those more slowly diffusing alloy additions which effect the austenite to ferrite transition temperature. As described by Bastien (1957) and by Kirkaldy, von Destinon-Forstmann and Brigham (1962), regions which have low concentrations of austenite stabilising elements such as manganese will, on slow cooling, transform first to ferrite and reject their carbon content to the regions richer in manganese that still have the austenite structure. Further cooling concentrates the carbon content in these manganese rich regions until they transform to the eutectoid pearlite structure.

If the difference in the austenite to ferrite transformation temperature in different regions is $\Delta T$ and the cooling rate is $\mathrm{d}T/\mathrm{d}t$, then the time for carbon diffusion is $\Delta T/(\mathrm{d}T/\mathrm{d}t)$, so if the interband spacing is $d$ and the diffusion coefficient of carbon is $D_c$, then the condition for the development of ferrite–pearlite bands is

$$\mathrm{d}T/\mathrm{d}t < D_c \Delta T/d^2$$

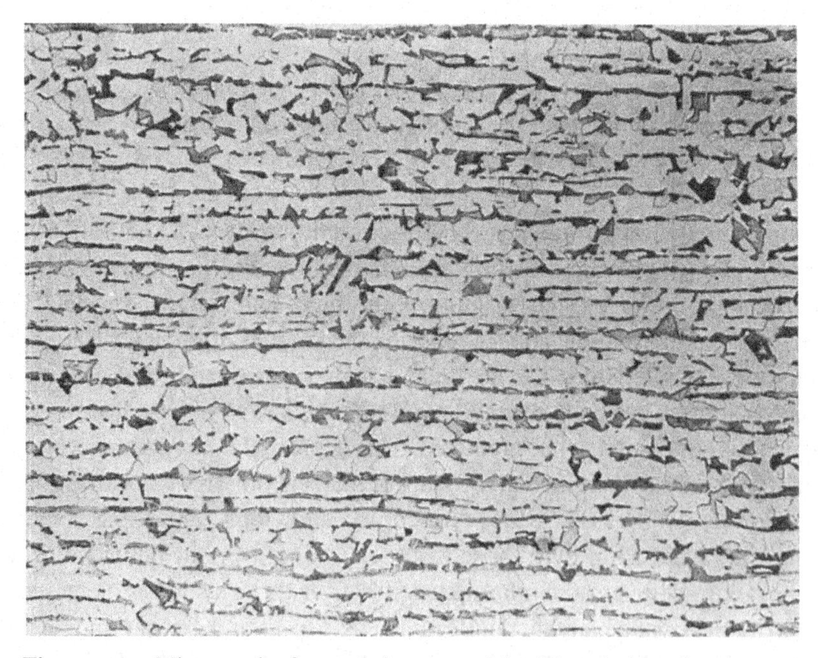

**Figure 2.13** Micrograph of annealed commercial mild steel plate, showing highly developed pearlite banding.

If the segregated material is cooled faster than this, the banding will not develop, but the potentiality for banding remains and can develop on subsequent heat-treatment.

Grange (1971) reported that a short high temperature anneal (10 min at 1350 °C) of segregated 10W alloy steels removed the tendency for banding on slow cooling. This treatment is unlikely to remove the interdendritic segregation that causes banding and this was confirmed by the equally mysterious return of banding during hot deformation. However, this annealing treatment allowed Grange to demonstrate the effect of banding on reducing the transverse ductility and fracture toughness of the alloy. This was most marked for a 'clean' steel; for a 'dirty' steel containing a large number of oxide and other inclusions which would also be drawn out in the rolling direction, the damaging effect of banding was less as the properties were already degraded by the inclusions.

The general conclusion appears to be that the problem of the removal of segregation by diffusion is largely understood at least semiquantitatively, and, though this is much less well established, segregation always reduces the mechanical properties, especially the ductility. In view of the ever-present character of this problem, however, the effect of segregation on all the mechanical properties of alloys seems to have been astonishingly neglected.

## 2.2 Nucleation of precipitates from supersaturated solid solution

Christian (1975) has reviewed the subject of nucleation in solids comprehensively, and the topic is dealt with at an introductory level in the standard undergraduate texts on phase transformations in condensed systems. We will survey here the nature of the microstructures developed during precipitation from a generalised mechanistic point of view.

### 2.2.1 Spinodal decomposition

Gibbs (1876) showed that two types of phase change exist in solid solutions, and it is important to distinguish between them. Firstly there is the normal precipitation reaction which involves a thermally activated nucleation step, and secondly there is spinodal decomposition in which the material is inherently unstable to small fluctuations in composition and hence decomposes spontaneously. The difference is illustrated in figs. 2.14 and 2.1 which give the free energy composition curves for a homogeneous phase which is metastable with respect to phases A and B between compositions $C_A$ and $C_B$. In this type of system no structural change is involved in the transformation, but compositional changes do occur.

In fig. 2.14(*a*), an alloy of initial composition $C_1$ and free energy $G_1$ decom-

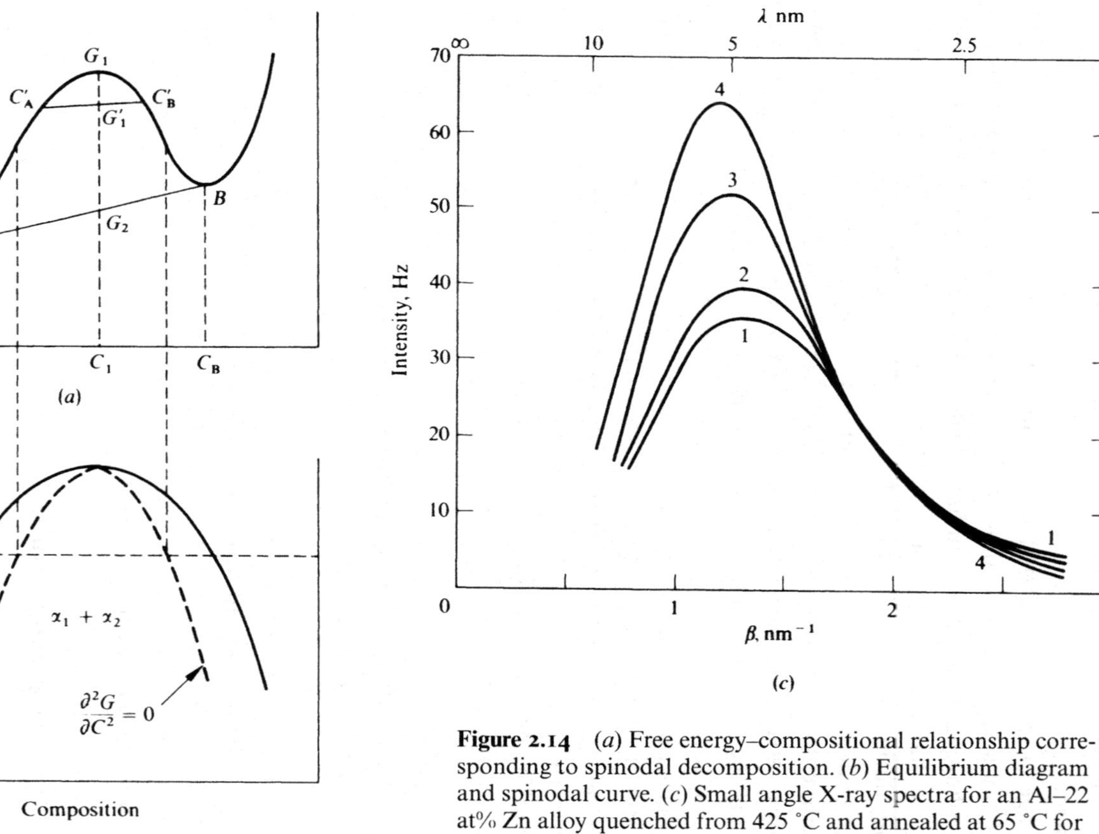


**Figure 2.14** (*a*) Free energy–compositional relationship corresponding to spinodal decomposition. (*b*) Equilibrium diagram and spinodal curve. (*c*) Small angle X-ray spectra for an Al–22 at% Zn alloy quenched from 425 °C and annealed at 65 °C for (1) 0.5 min, (2) 5.0 min, (3) 34.5 min, (4) 65.0 min. (After Rundman and Hilliard 1967.)

poses into a mixture of two phases, composition $C_A$ and $C_B$ and average free energy $G_2$, resulting in a decrease of free energy. At an early stage in the transformation the compositions of the two separating phases may be $C'_A$ and $C'_B$ with an average free energy $G'_1$ which is always lower than $G'_1$. The decomposition thus takes place with a continual decrease in free energy and there is no thermodynamic barrier to decomposition of the solid solution. This type of phase change is known as spinodal decomposition and has the characteristic that the transformation occurs simultaneously throughout the matrix, although there is no precise stage at which the new phases appear.

In fig. 2.1 (p. 29), on the other hand, where an alloy of initial composition $C_3$ and free energy $G_3$ is unstable with respect to a mixture of two phases of composition $C_A$ and $C_B$ and average free energy $G_4$ the initial stages of the decomposition to $C'_A$ and $C'_B$ result in an increased free energy $G'_3$, and it is not until a large composition difference is obtained that further decomposition can take place with a net drop in free energy. Thus there is a nucleation barrier to overcome before the transformation can proceed and it is this type of phase change that gives rise to the conventional nucleation and growth behaviour typical of most age-hardening systems.

The transition between the two types of phase change occurs at the points on the free energy–composition curve in fig. 2.14(*a*), that is, where $d^2G/dC^2=0$. The loci of these points at different temperatures give rise to the 'spinodal curve' on a phase diagram (fig. 2.14(*b*)), whereas the solvus line will be given approximately by the loci of the minima of the curve, that is, where $dG/dC=0$. Theories of the spinodal decomposition process have been developed by Hillert (1961), Cahn and Hilliard (1958, 1959), Cahn (1961, 1962, 1964, 1968) and Hilliard (1970), and a review of the phenomenon has been given by Soffa and Laughlin (1983).

Theories of the transformation take into account the positive free energy contributions arising from the bond energies and elastic strains across the diffuse interface between the phases. In Cahn and Hilliard's treatment, the free energy of the material is given by the expression

$$G=\int_v\left[f'(C)+K(\nabla C)^2+\frac{\eta^2E}{(1-\nu)}(C-C_0)^2\right]dV \qquad (2.15)$$

where $f'(C)$ is the Helmholtz free energy per unit volume of homogeneous material of composition $C$ and $K(\nabla C)^2$ is a measure of the increase in energy due to the non-uniform environment of atoms in a concentration gradient. $\eta^2E(C-C_0)^2/(1-\nu)$ is a measure of the strain energy associated with the interface where $E$ is Young's modulus, $\nu$ is Poisson's ratio, $C_0$ is the average composition of the solid solution and $\eta$ is the linear expansion of the lattice per unit composition change.

Assuming that $C$ varies sinusoidally with distance $x$, that is, assuming

$C - C_0 = B\cos\beta x$, the wavelength of sinusoidal fluctuations will be given by $2\pi/\beta$. It is found that the alloy is unstable to fluctuations above a critical wavelength which depends on the supersaturation and is given by:

$$\frac{2\pi}{\beta_c} = \left\{ -\frac{8\pi^2 K}{\left[\frac{d^2 f'}{dC^2} + \frac{2\eta^2 E}{(1-\nu)}\right]} \right\}^{1/2} = \lambda_c \qquad (2.16)$$

If, therefore, $d^2f'/dC^2$ is sufficiently negative, the solution is unstable with respect to all (very small) composition fluctuations spread over a wavelength greater than that given by eq. (2.16). The kinetic process may be regarded as one of up-hill diffusion (although, of course, the diffusive flux is still down the chemical potential gradient); the wavelength $\lambda_c\sqrt{2}$ is supposed quickly to outgrow all others and dominate the structure – shorter wavelengths having too sharp a concentration gradient and longer wavelengths having too large a diffusion distance. Additionally, the elastic anisotropy of most materials means that the alloy is first unstable to fluctuations in particular directions (usually $\langle 100 \rangle$ in cubic materials) where $E$ is a minimum.

Unambiguous observations of spinodal decomposition are rare, but an experimental test may result from the build-up of the amplitude of the composition fluctuations with a preferred wavelength. This results in the production of 'side bands' in small angle, X-ray scattering experiments, first studied by Daniel and Lipson (1943, 1944) in copper–nickel–iron alloys. Fig. 2.14(*c*) shows how the small angle, X-ray scattering varies with time in an Al–Zn alloy: there is seen to be a cross-over point in the spectra. The small angle region corresponding to small $\beta$ is growing in intensity, while the large angle region is shrinking: this is to be expected if there is a critical wavenumber ($2\pi/\beta$) as proposed in the theory.

An experimental study of the kinetics of phase separation by spinodal decomposition in Fe – 45 at% Cr has been carried out by Cerezo *et al.* (1992). Atom probe microanalysis demonstrated that both amplitude and wavelength of the phases increased during the early stages of separation, which conflicts with the Cahn and Hilliard (1958) model, in which the wavelength of the microstructure should remain approximately constant. Cerezo *et al.* (1992) employed a Monte Carlo simulation of phase separation to model the process, which convincingly predicted the rate of growth of both wavelength and amplitude of the composition fluctuations.

The copper–nickel–iron system appears to satisfy all the criteria for spinodal decomposition, and the transformed structure consists of regularly spaced quasi-spherical particles at small volume fractions and a lattice of interconnected rods lying along $\langle 100 \rangle$ directions at large volume fractions. It is also striking that there is no preferential precipitation on the grain boundaries, which is most unusual in precipitating systems. Such preferential nucleation arises (in the absence of spinodal decomposition) because of the reduced activation energy for nucleation at such defects, and such an effect would not be expected to exist in spinodal alloys owing to the absence of a nucleation barrier to precipitation.

Ham, Kirkaldy and Plewes (1967) examined a series of copper–nickel–iron alloys, being particularly interested in the fatigue strength, since the thermodynamic conditions of spinodal decomposition imply that the alloy should have unusually good mechanical stability under conditions of fatigue. Alloys conventionally strengthened by precipitates which are sheared during deformation are often mechanically unstable under conditions of fatigue at normal temperatures, and it has been shown that the softening responsible for the instability may occur either by the resolution of precipitates or by the nucleation of a later precipitate in the sequence during fatigue. Such alloys thus have fatigue properties which are poor relative to their tensile properties. The process of re-solution is thought to depend upon the existence of a critical size for the precipitate concerned; if the precipitate is effectively cut by the local cumulative shear due to fatigue to a size smaller than critical at the test temperature then it may redissolve. It has furthermore been argued that a high concentration of lattice vacancies is produced by fatigue so that local diffusion rates are generally high in substitutional alloys.

Ham *et al.* (1967) showed that a spinodal microstructure in copper–nickel–iron has unusually good mechanical stability under conditions of fatigue, since sheared precipitates would withstand very severe distortion before they developed an effective wavelength which is unstable at room temperature. Whereas alloys strengthened by precipitates have a fatigue ratio (fatigue strength for a given life/ultimate tensile stress) of only about 0.3 for a life of $10^6$ cycles, the fatigue ratio of the copper–nickel–iron alloys was very high, being 0.5 or over for this fatigue life, which is higher than in any other copper alloy. The effect clearly arises from the inability of the precipitates either to redissolve or to overage in the usual way during fatigue (§5.6.1).

## 2.2.2 Homogeneous nucleation

In the second type of phase change, illustrated in fig. 2.1 (p. 29), precipitation of the second phase occurs by nucleation and growth processes. By homogeneous nucleation, we refer to nucleation that occurs without benefit of pre-existing heterogeneities. It is customary to apply the thermodynamic treatment of Volmer and Weber (1925) and Becker and Döring (1935) to nucleation in solids: here the change in free energy $\Delta G$ associated with the homogeneous nucleation of an embryo of the new phase is equated with the sum of the change in volume free energy $\Delta G_v$ and the interfacial energy $\sigma$ required to form the new surface. With solid systems a strain energy term $\epsilon$ must be introduced giving:

$$\Delta G = \Delta G_v + \sigma + \epsilon \qquad (2.17)$$

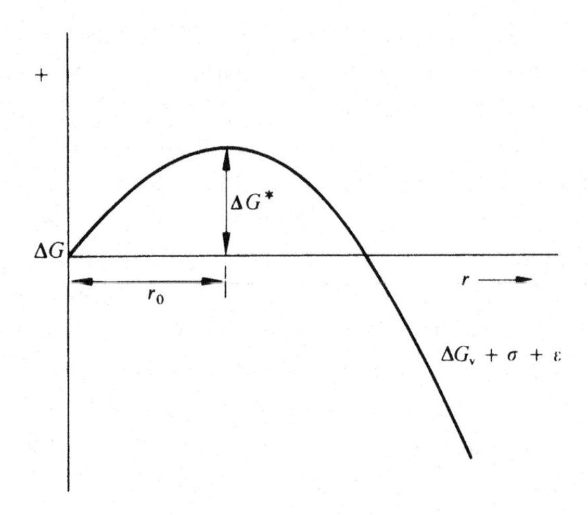


**Figure 2.15** Schematic representation of the free energy of a nucleus plotted against its radius.

For a spherical nucleus of radius $r$

$$\Delta G = 4\pi r^2 + \tfrac{4}{3}\pi r^3(\Delta G_v + \epsilon) \tag{2.18}$$

and the relationship between $\Delta G$ and $r$ will have the form shown in fig. 2.15. The condition for continued growth of an embryo is that the radius should exceed $r_0$, where $d(\Delta G)/dr = 0$, that is,

$$r_0 = 2\sigma / -(\Delta G_v + \epsilon)^2 \tag{2.19}$$

and the critical free energy, or activation energy for nucleation, is given by

$$\Delta G^* = (16\pi/3)\, \sigma^3/(\Delta G_v + \epsilon)^2 \tag{2.20}$$

$\Delta G_v$ is strongly temperature dependent (while $\sigma$ and $\epsilon$ are only slightly affected by temperature) so the temperature dependence of $r_0$ and $\Delta G$ may be assessed in terms of the temperature dependence of $\Delta G_v$. At $T_E$ (the equilibrium transformation temperature), $\Delta G_v$ is zero, and becomes increasingly more negative as the temperature is lowered below $T_E$, so we can write

$$r_0 \propto \frac{1}{(T_E - T)} \qquad \Delta G^* \propto \frac{1}{(T_E - T)^2} \tag{2.21}$$

A steady state nucleation rate $N_v$ may be defined as the number of stable nuclei produced in unit time in unit volume of untransformed solid, and the theories give the result that $N_v$ is proportional to $\exp(-\Delta G^*/kT)$. The rate at which individual nuclei grow will also be dependent on the frequency with which atoms adjacent to the nucleus can join it and this will be proportional to $\exp(-\Delta G_D/kT)$, where $\Delta G_D$ is the free energy of activation for diffusion. One may therefore write a simplified representation of the rate of nucleation of a precipitate as:

$$N_v = K \exp\left(\frac{-\{[A\sigma^3/(\Delta G_v + \epsilon)^2] + \Delta G\}}{kT}\right) \qquad (2.22)$$

where $A$ is a geometrical constant and $K$ another constant.

From a consideration of eq. (2.22) it will be seen that $N_v$ increases rapidly from zero at $T_E$, but eventually decreases again. The decrease is due to the factor $\exp(-\Delta G_D/kT)$, since $\Delta G_D$ is nearly independent of temperature. When combined with a growth rate temperature dependence which is based mainly on the activation energy for diffusion, the resulting rate of transformation will show a temperature dependence of the *C* curve type.

Servi and Turnbull (1966) have studied the precipitation of the fcc cobalt phase from fcc Cu–Co alloys (one of the few equilibrium systems which decompose by homogeneous nucleation) and found quantitative agreement between theory and experiment. Although there were several orders of magnitude difference between measured and predicted nucleation rates (a significant but not gross discrepancy), the theory predicts the temperature (at fixed composition, or vice versa) for a given nucleation rate very accurately.

### 2.2.3 Heterogeneous nucleation

Considering now the technologically more important subject of heterogeneous nucleation, the basic idea is quite simple. Heterogeneities such as dislocations or grain boundaries represent regions of higher than average free energy, so when a nucleus forms such that it replaces part of a boundary area or dislocation line the associated free energy is available to drive the reaction. Likely nucleation catalysts include grain boundaries, grain edges (three-grain intersections), grain corners (four-grain intersections), dislocations, stacking faults and the surfaces of impurity particles. The normal segregation of solute atoms to grain boundaries and dislocations also facilitates the formation of the appropriate group of atoms which constitutes a nucleus, and nucleation is further promoted by the enhanced diffusion in the vicinity of each type of defect. The rate of heterogeneous nucleation, $N_s$, of a precipitate will be given by an expression of similar form to that of eq. (2.22); Cahn (1956, 1957) has derived equations for the rates of heterogeneous nucleation at various singularities in the solid. He has compared the barriers for nucleation on grain boundaries, corners and edges on the basis of incoherent nuclei and isotropic surface tension, and fig. 2.16 illustrates the conditions giving the highest volume nucleation rate for each kind of defect. The ordinate represents the opposing factors: $D/a$, a ratio of grain diameter to grain boundary thickness, and the homogeneous nucleation barrier, $16\pi\sigma^3/3\Delta G_v^2$. The abscissa represents the catalytic effectiveness of the grain boundary, with

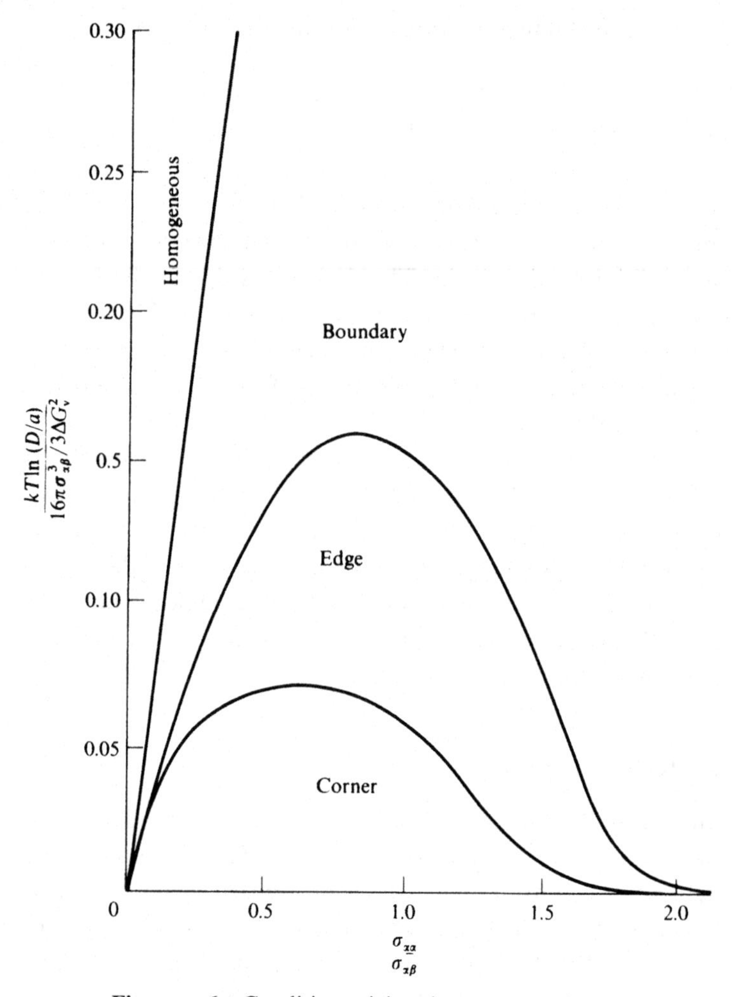


**Figure 2.16** Conditions giving the greatest volume nucleation rates for various kinds of nucleation. (After Cahn 1956.)

$\sigma_{\alpha\alpha}/\sigma_{\alpha\beta}=0$ denoting zero catalytic effect (a contact angle of $\pi$) and $\sigma_{\alpha\alpha}/\sigma_{\alpha\beta}=2$ giving perfect catalysis (nucleation as soon as $\Delta G_v>0$). For substrates of low potency and high $\Delta G_v$, homogeneous nucleation dominates, whereas corners dominate only at the lowest $\Delta G_v$. With regard to nucleation on dislocations, Cahn assumed elastic and surface isotropy and an incoherent nucleus of the form shown in fig. 2.17. Cahn expresses his results in terms of a dimensionless parameter

$$\alpha=\Delta G_v \mu b^2/2\pi^2\sigma^2 \tag{2.23}$$

where $\mu$ is the shear modulus and $b$ the Burgers vector. In fig. 2.18 $\Delta G^*/\Delta G_h^*$ is plotted versus $\alpha$: $\Delta G_h^*$ is the activation energy for homogeneous nucleation (eq.

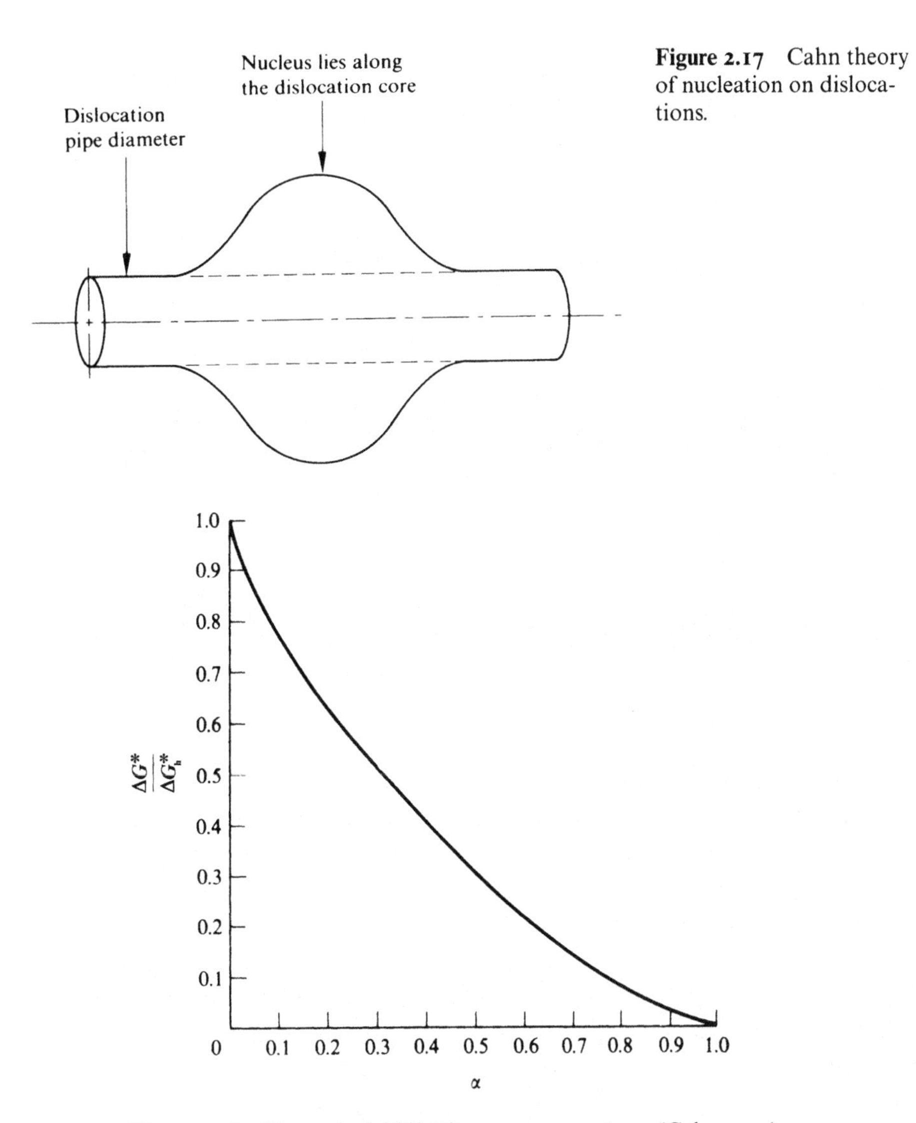


**Figure 2.17** Cahn theory of nucleation on dislocations.

**Figure 2.18** The ratio $\Delta G^*/\Delta G_h^*$ versus parameter $\alpha$ (Cahn 1957).

(2.20)) and it is seen that the ratio $\Delta G^*/\Delta G_h^*$ decreases as $\alpha$ increases; that is, the effectiveness of a dislocation in catalysing nucleation increases with the Burgers vector since the self energy of the dislocation increases with the Burgers vector.

Nicholson (1970) has surveyed the experimental evidence on the nucleation of precipitates at imperfections in solids. He showed that although the present theory of heterogeneous nucleation is qualitatively satisfactory, it cannot be applied in a quantitative form to real systems because of the lack of interfacial energy data relating to a defect and a nucleus, which would enable the relevant nucleation parameters to be calculated.

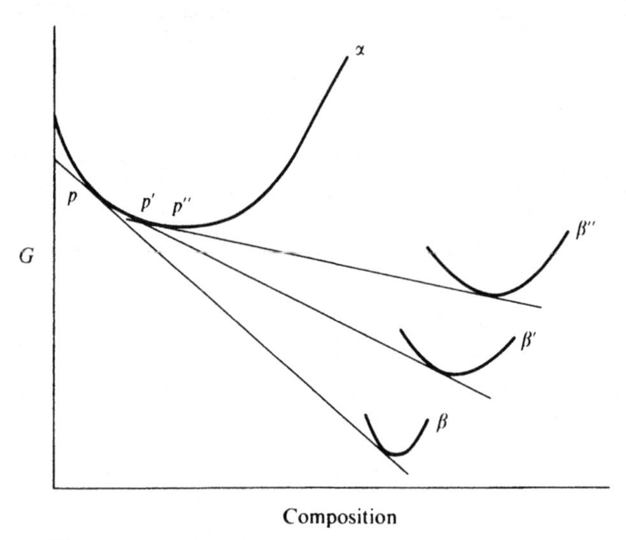


**Figure 2.19** Free energy composition diagrams for precipitation sequences in which transition precipitates intervene.

## 2.2.4 The formation of transition phases

When supersaturated solid solutions decompose, one or more metastable or transition phases (not indicated on the phase diagram) may appear prior to or in addition to the equilibrium precipitate. By definition, the free energy per atom of a transition phase is less negative than that of the equilibrium phase at a given temperature and alloy composition; fig. 2.19 represents at a given temperature the free energy relations between the parent solid solution ($\alpha$), the equilibrium precipitate ($\beta$) and metastable precipitates ($\beta'$ and $\beta''$), all precipitates having higher solute contents than the parent phase. From the common tangents it is evident that the solute concentration $p''$ of the solid solution in (metastable) equilibrium with $\beta''$ is higher than that ($p'$) for $\beta'$, which in turn is higher than that ($p$) for the equilibrium precipitate $\beta$. Thus, only alloys of solute content greater than $p''$ can form the $\beta''$ transition phase; alloys of solute content between $p'$ and $p''$ must decompose directly to $\beta'$, and alloys in the composition range $p$ to $p'$ must decompose directly to the stable phase, $\beta$. This illustrates a general rule that as supersaturation decreases, the number of intermediate reaction stages decreases.

The changes in fig. 2.19 with respect to temperature may be represented on the appropriate phase diagram by constructing upon it extra solubility lines to denote the temperatures and compositions at which the $\alpha$ phase becomes supersaturated with respect to the precipitation of the $\beta'$ and $\beta''$ phases. An example of this in the Al–Cu system is shown in fig. 2.20, where it is seen that the metastable equilibrium solvus curves fall in a lower temperature range the lower the

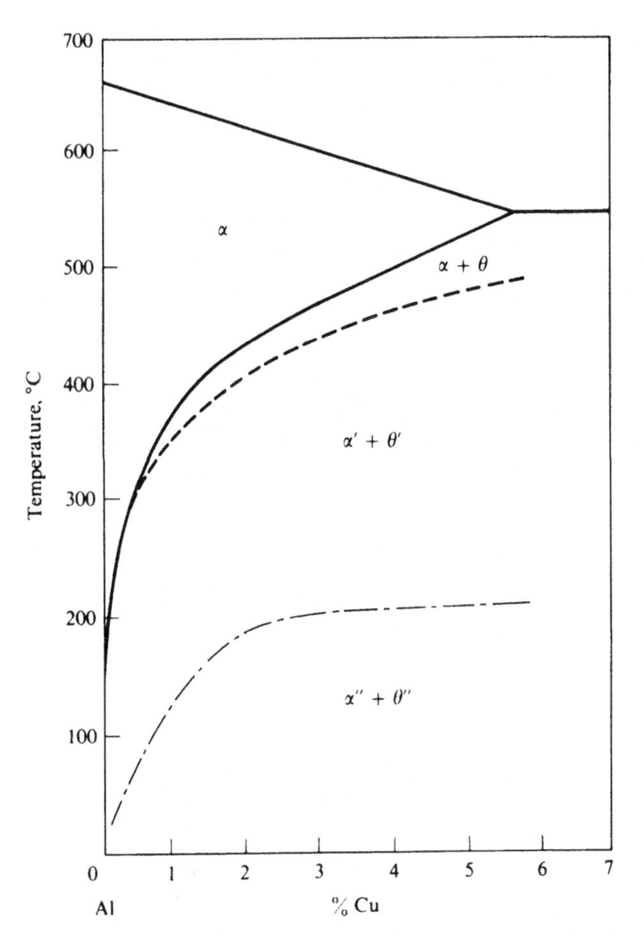


**Figure 2.20** Metastable equilibrium solvus curves for aluminium rich Al–Cu alloys. (After Hornbogen 1967.)

stability of the transition phases to which they correspond. The prediction of the number of transition phases, their crystal structure and composition has not yet been achieved for any alloy system, even on an empirical basis. A study of the structures of the metastable transition phases in solid state precipitation (like those produced by splat cooling, §3.1.1) provide an almost untouched field of alloy theory.

Since the basic heat-treatment of precipitation-hardening alloys is solution treatment followed by quenching, this is effective in retaining at the low temperature the concentration of vacancies which existed in thermal equilibrium at the solution heat-treatment temperature. It is commonly observed that the early stages of precipitation in alloys at low ageing temperatures occurs at a rate seven or eight orders of magnitude greater than that expected from the extrapolation of high temperature diffusion data, due to the presence of these 'quenched-in' vacancies. If, on the other hand, the quenching is followed by a 'reversion treatment' (for example, by heating for a few minutes to an intermediate temperature) a remarkable decrease in the rate of precipitation at low temperature is observed, which results from the rapid annihilation of vacancies at sinks such as grain boundaries and dislocations at the intermediate temperature.

Aaronson, Aaron and Kinsman (1971) have, in reviewing the origins of microstructures resulting from precipitation, discussed the nucleation kinetics of transition phases. We have already noted that the rate of heterogeneous nucleation ($N_s$) of a precipitate may be represented by an equation similar to (2.20). Since the volume free energy change accompanying nucleation ($\Delta G_v$) may be less negative for a transition phase, the most obvious means through which such a phase can achieve a value of $N_s$ competitive with that of the equilibrium phase is to have a smaller value of $\sigma$, the net interfacial free energy of the critical nucleus. Aaronson *et al.* observe that this leads directly to the most important general characteristic of transition phases: their crystal structure and habit plane allow them to achieve exceptionally good lattice matching with the matrix, thereby markedly reducing $\sigma$. The data of Hornbogen (1967) shown in fig. 2.21 for the aluminium rich Al–Cu system permits a comparison of the lattice matching between precipitate and matrix in the case of the $\theta$, $\theta'$ and $\theta''$ phases. Fig. 2.21

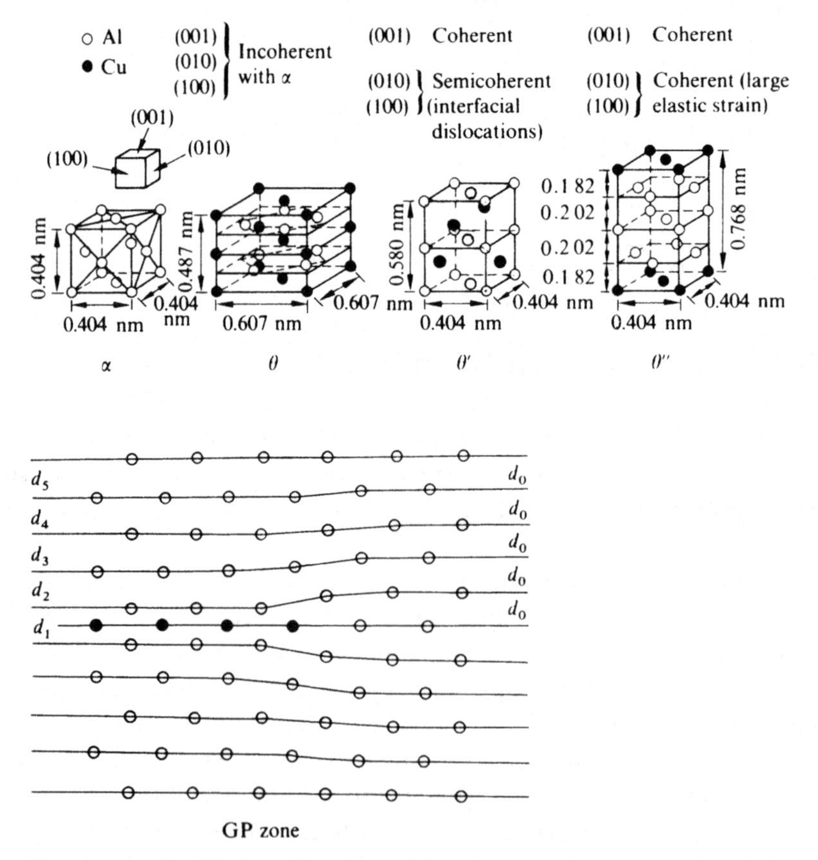


**Figure 2.21** Equilibrium ($\theta$) and transition phase crystal structures in aluminium rich Al–Cu alloys. (After Hornbogen 1967.)

suggests that the better a transition phase matches the matrix lattice, the less it may resemble the equilibrium precipitate.

In many systems of technological importance a transition phase completely coherent with the matrix may form (see p. 18). A simple example of this is the Guinier–Preston (GP) zones in the Al–Cu system (fig. 2.21) which consist of a plate-like cluster of copper atoms on the {100} aluminium matrix, each plate being about 10 nm in diameter. In this situation, the value of $\sigma$ is so low that homogeneous nucleation of the phase may take place. Lorimer and Nicholson (1969) have shown that such zones can subsequently act as nucleating sites for other metastable precipitates, so that by appropriate heat-treatment a uniform distribution of these latter precipitates may be produced in the microstructure.

A second general aspect of the comparative nucleation kinetics of transition phases observed by Aaronson *et al.* (1971) is that semicoherent transition phases nucleate primarily at dislocations (both isolated and in arrays) and at sites not resolved by transmission electron microscopy (for example, at vacancy clusters), whereas equilibrium precipitates tend to nucleate at large angle grain boundaries. In the latter case the grain boundary contribution to $\sigma$ tends to counterbalance the lesser degree of lattice matching between precipitate and matrix relative to that of the transition phases. At a dislocation, the reduction in $\sigma$ is small and thus the lesser $\sigma$ of the transition precipitate makes its nucleation kinetics more rapid.

The general process that occurs, therefore, when a supersaturated solid solution of the type represented by fig. 2.20 is aged at a temperature just below the $\theta''$ phase solvus, is that the rate of formation of $\theta''$ is greater than for the other possible products. $\theta''$ precipitates thus nucleate preferentially, thereby achieving the maximum rate of decrease of free energy in the system. The $\theta'$ phase has the next highest rate of nucleation, and nuclei of this phase will form on prolonged ageing. By analogy with fig. 2.19 it can be seen that the regions of the matrix around these new nuclei will be of a composition less than the solubility of the $\theta''$ phase (for example, $p'$ instead of $p''$), and hence the $\theta''$ phase will dissolve. The multi-stage precipitation process will thus continue until the most stable state ($\theta$ precipitate) is produced: the formation of a more stable product resulting in the re-solution of less stable phases formed in the earlier stages.

## 2.2.5 Trace element effects

It is well known that minor or trace additions of elements may modify the nucleation of precipitates and thus exert large effects upon the structure and properties of age hardening alloys. Several mechanisms have been recognised by which trace additions may modify precipitate nucleation, and these have been reviewed by Polmear (1966).

(i) Trace elements may interact with vacancies, leading to a reduction in the rate of lattice diffusion of substitutional elements. Thus the marked effects of small amounts of cadmium, indium and tin in reducing the rate of GP zone formation in Al–Cu alloys may be attributed to this type of interaction.

(ii) Trace elements may modify the interfacial energy between a precipitate and the matrix, thereby changing the nucleation rate of that phase (eq. (2.22)) through the change in the value of $\sigma$.

(iii) Trace elements may change the free energy relationships in an alloy system, so that precipitation of a different phase is favoured. Thus Vietz and Polmear (1966) have shown that the addition of approximately 0.1 at% Ag to Al–Cu–Mg alloys induces homogeneous nucleation at a ternary $T$ phase instead of the normal GP zones and the $S'$ phase ($Al_2CuMg$) which is nucleated on dislocations.

(iv) Trace elements may segregate to grain boundaries and inhibit discontinuous precipitation.

Although trace element effects are being widely studied, only limited progress has been made in establishing rules governing their behaviour. In some cases the trace element effect appears unique to a particular alloy system whereas in others the same trace element may stimulate precipitate nucleation in several systems.

## 2.3 Growth of precipitates from supersaturated solid solution

In a great number of precipitation processes in metallic systems the nucleation occurs very rapidly at the beginning of the transformation at sites of catalysts, so that subsequent nucleation may be neglected. A very small amount of precipitation will change the mean concentration of solute in the matrix by an amount sufficient to decrease $N_s$ by one or two orders of magnitude. Almost all results on continuous precipitation can best be interpreted by assuming that all nuclei are present from the beginning of the transformation.

A full analysis of growth rates taking into account all the variables has never been attempted because of the complexity of the problem. The usual practice is to assume that the rate of growth is determined by only one or two factors, thus in the growth of precipitates from supersaturated solid solutions the possible limiting factors usually considered are the rate at which atoms are brought to or removed from the interface by *diffusion*, and the rate at which they *cross the interface*. The interface reaction is likely to be the slower step during the early stage of growth, since the diffusion distance tends to zero in this situation, while at large particle sizes diffusion is likely to be rate controlling since the continuous removal of solute from solution reduces the concentration gradient, that is, the

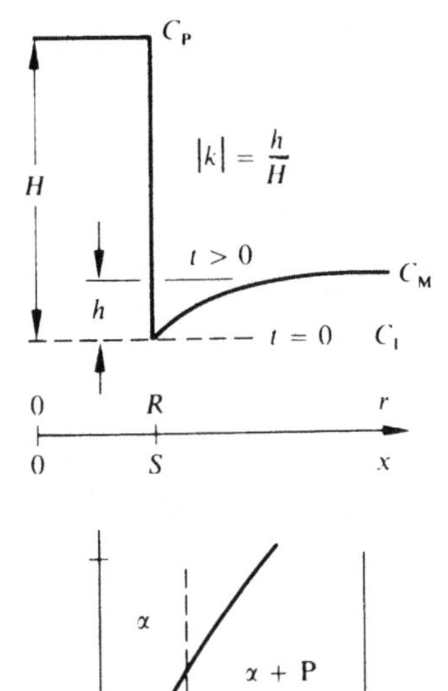


**Figure 2.22** Schematic concentration field for the growth of precipitate (P) at temperature $T_G$, and corresponding position of the phase diagram showing the single-phase region ($\alpha$) and the two-phase region ($\alpha$+P).

driving force for diffusion, whereas the interface area (and thus the flux across it) is increasing. Both of these situations are discussed in the standard textbooks on phase transformations: in the present discussion we shall confine ourselves to a consideration of diffusion controlled growth, since this is the process which has been analysed in the greater detail.

### 2.3.1 Diffusion controlled growth

Exact analytical solutions for the kinetics of a given diffusion controlled phase transformation are often difficult if not impossible to obtain, and several approximate analyses have been proposed in order to simplify the mathematics. Aaron, Fainstein and Kotler (1970) have critically compared and evaluated the approximations made for diffusion limited growth (and dissolution) of spherical and planar precipitates.

In any transformations involving long range transport, the diffusion equation

$$D\nabla^2 C = \partial C/\partial t \tag{2.24}$$

must be satisfied. Considering an isolated precipitate in an infinite matrix, $D$ (assumed independent of composition) is the volume diffusion coefficient in the matrix and $C = C(r,t)$ is the concentration field in the matrix surrounding the precipitate, subject to the conditions illustrated in fig. 2.22 that

$$C(r=R,t)=C_I \qquad 0<t\geq\infty$$

and

$$C(r,\, t=0)=C_M \qquad r\geq R$$

or

$$C(r=\infty,t)=C_M \qquad 0\leq t\leq\infty$$

where $r=R$ at the precipitate–matrix interface, $C_I$ is the concentration in the matrix at the precipitate–matrix interface, and $C_M$ is the matrix composition at a point remote from the precipitate. The flux balance must also be satisfied:

$$(C_P-C_I)\mathrm{d}R/\mathrm{d}t=D(\partial C/\partial r)_{r=R} \tag{2.25}$$

where $C_P$, the composition of the precipitate, is assumed constant and independent of $r$ and $t$; $R=R_0$ at $t=0$.

The three approximations considered by Aaron *et al.* (1970) were: (*a*) The invariant field approximation, in which $\partial C/\partial t$ (eq. (2.24)) is set equal to zero, $R$ is considered constant leaving the time-independent diffusion problem in the matrix to be solved. (*b*) The invariant size approximation, again requiring a stationary interface. The diffusion field around the precipitate is assumed to be the same as that which would have existed if the precipitate–matrix interface had been fixed at $R$ initially, and ignores the effect on the diffusion field of interface motion. (*c*) The linearised gradients approximation which simplifies the diffusion field by assuming the concentration gradient immediately surrounding the precipitate is linear with respect to $r$.

For the growth of *spherical precipitates*, the radius $R$ after time $t$ is given by an equation of the form

$$R=\lambda_j(Dt)^{1/2} \tag{2.26}$$

and fig. 2.23 illustrates how the growth kinetics predicted depend upon the approximation assumed. Here $\lambda_j$ is plotted as a function of $k$, which describes the supersaturation:

$$k=2(C_I-C_M)/C_P-C_I \tag{2.27}$$

A similar calculation made for the growth of planar precipitates again predicts a parabolic relationship between half-thickness ($S$) and time:

$$S=\lambda_j(Dt)^{1/2} \tag{2.28}$$

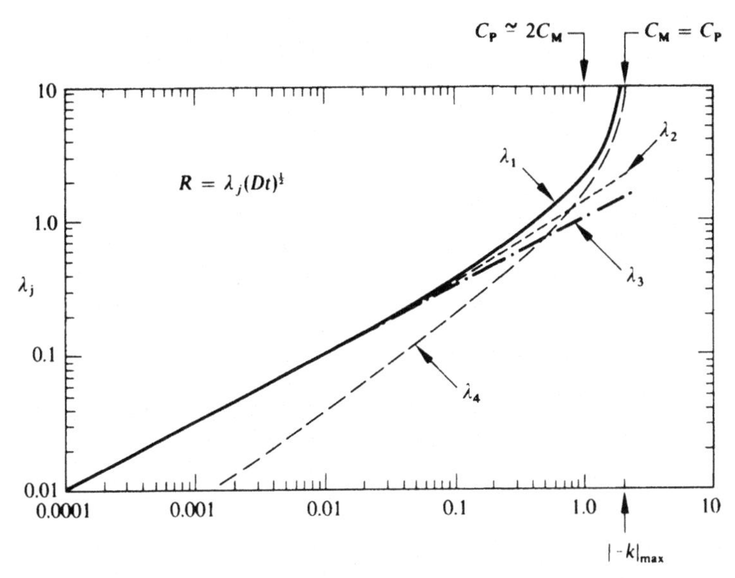


**Figure 2.23** Growth kinetics for spherical precipitates as a function of $k$.
$\lambda_1$ Exact solution:

$$\lambda_1 = 2\lambda \quad \lambda^2 e^{\lambda^2}[e^{-\lambda^2} - \lambda\pi^{1/2}\text{erfc}\,\lambda] = -k/4$$

$$\lim_{k\to 0}\lambda_1 = \lambda_2 \quad \lim_{k\to 0}\lambda_1 = \lambda_3 \quad k<0;\ \lambda_1>\lambda_2>\lambda_3$$

$$\lim_{k\to 0}\lambda_1 = \lambda_4;\quad k\neq -2;\ \lambda_1>\lambda_4$$

$\lambda_2$ Invariant size approximation:

$$\lambda_2 = \frac{-k}{2\pi^{1/2}} + \left(\frac{k^2}{4\pi} - k\right)^{1/2} \quad \lim_{k\to 0}\lambda_2 = \lambda_3$$

$\lambda_3$ Invariant field approximation: $\lambda_3 = (-k)^{1/2}$
$\lambda_4$ Linearised gradients approximation:

$$f(k) = \left(\frac{-4k}{1+4k}\right)^{1/3}$$

$$\gamma = \frac{1-f(k)}{f(k)}$$

$$\lambda_4 = (-k/\gamma)^{1/2}$$

Fig. 2.24 shows the growth kinetics as a function of $k$, $\lambda_1$ being the exact solution, $\lambda_2$ the invariant size approximation and $\lambda_4$ the linear gradient approximation.

Aaron *et al.* (1970) observe that since the stationary interface and the linearised gradient approximations are convenient and tractable and are better than the invariant field approximation, the latter should not be used. Since for many alloy systems of interest $k<0.3$, the invariant size approximation should probably be used exclusively for the analysis of precipitate growth.

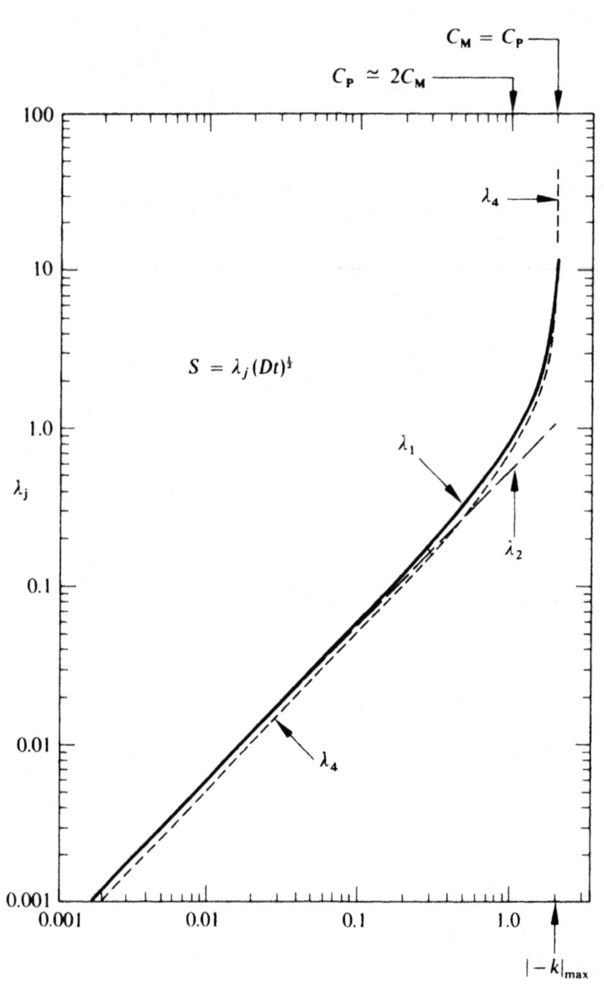


**Figure 2.24** Growth kinetics for planar precipitates as a function of $k$.
$\lambda_1$ Exact solution:

$$\lambda_1 = 2\lambda \quad \pi^{1/2}\lambda e^{\lambda^2} \operatorname{erfc} \lambda = -k/2$$

$$\lim_{k \to 0} \lambda_1 = \lambda_2 \quad k<0;\ \lambda_1 > \lambda_2 \quad |k|>1.98;\ \lambda_1 < \lambda_4$$

$\lambda_2$ Invariant size approximation: $\lambda_2 = -k\pi^{1/2}$
$\lambda_4$ Linearised gradients approximation:

$$\lambda_4 = \frac{k}{2(1+\frac{1}{2}k)^{1/2}} \quad \lim_{k \to 0} \lambda_4 = -k/2$$

$\lambda_3$ (invariant field approximation) is not defined, as the far field condition cannot be fulfilled.

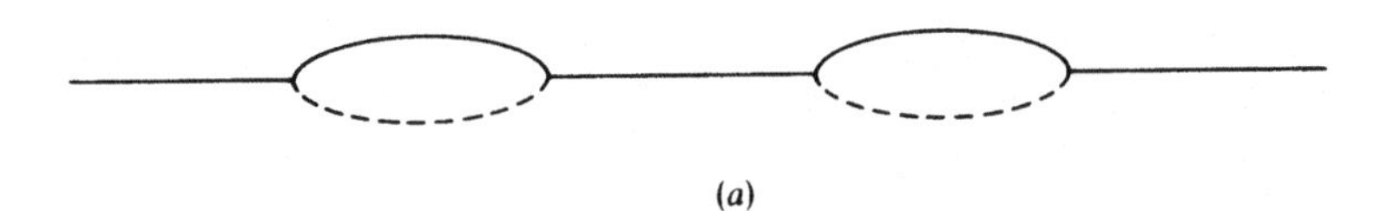

(a)

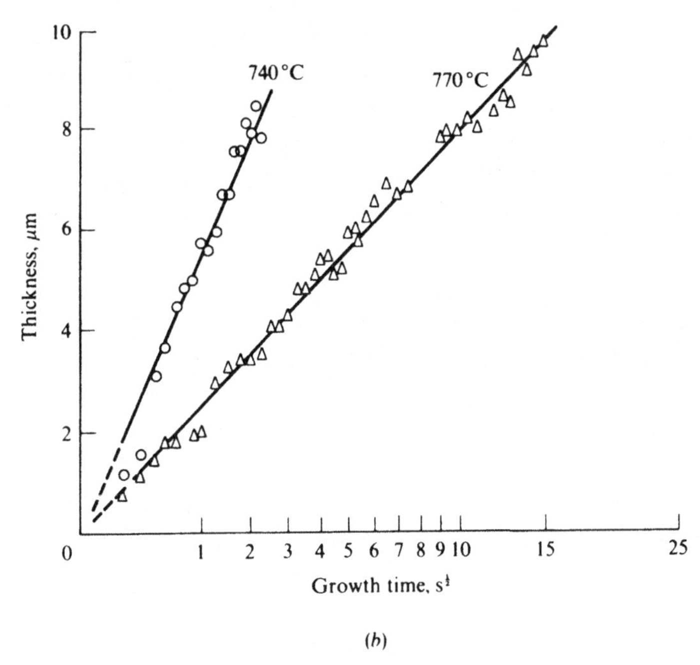


(b)

**Figure 2.25** (*a*) Grain boundary allotriomorphs: these are crystals which nucleate at grain boundaries of the matrix and grow preferentially along these boundaries. Solid line indicates disordered boundaries, dashed line indicates disordered plus dislocation boundaries. (*b*) Representative plots of pro-eutectoid ferrite allotriomorph thickness versus $(\text{time})^{1/2}$ in Fe – 0.11% C. (Kinsman and Aaronson 1967.)

## 2.3.2 Growth kinetics of grain boundary phases

Grain boundary allotriomorphs (fig. 2.25 (*a*)) provide the best source of disordered interphase boundaries upon which to conduct measurements to test the above theories. Most of the data available in this category have been obtained on the pro-eutectoid ferrite reaction, and typical data are illustrated in fig. 2.25(*b*). Thickening of the particles is seen to be parabolic in accord with eq. (2.28), and fig. 2.26 shows the temperature dependence of the parabolic rate constant ($\alpha$) measured experimentally and also that calculated from

$$(x_\beta - x_\beta^{\beta\alpha} / x_\alpha^{\alpha\beta} - x_\beta^{\beta\alpha})(D_v/\pi)^{1/2} = (\alpha/2)\text{erfc}(\alpha/2D_v^{1/2})\exp(\alpha^2/4D_v) \qquad (2.29)$$

where $x_\beta$ is the mole fraction of solute in the matrix prior to transformation; $x_\beta^{\beta\alpha}$ is the mole fraction of solute in the matrix ($\beta$) at the $\beta(\alpha+\beta)$ phase boundary; $x_\alpha^{\alpha\beta}$ is the mole fraction of solute in $\alpha$ at the $\alpha(\alpha+\beta)$ phase boundary; and $D$ is the volume interdiffusion coefficient.

This expression was derived by Dubé (1948) and Zener (1949), assuming that the interphase boundary is planar and disordered, and that $D_v$ is independent of carbon content. Hannerz (1968) has computed a *TTT* diagram describing the kinetics of the precipitation of a grain boundary phase AB, a simple nitride or carbide, which will be controlled by the diffusion of A (the substitutional solute) since B (the interstitially dissolved alloying element) can be assumed to diffuse fast enough to maintain para-equilibrium. Calculations for the precipitation of NbC and AlN in ordinary carbon steels have been carried out, and it has been shown, for example, that the rate of growth in thickness of plates of NbC in austenite grain boundaries is very slow – for example, $10^5$ s are required at a temperature of 1300 °C to grow a plate of thickness 0.2 $\mu$m. It was concluded, however, that such conditions exist when heavy castings of niobium-bearing steels are cooled through the austenite region, and the resultant precipitation gives rise to embrittlement, causing 'rock candy' fracture. The theoretically derived formula, requiring only a knowledge of the diffusion coefficient and of the solubility product for precipitation, can thus in favourable circumstances be used to calculate the risk of development of undesirable properties associated with precipitation.

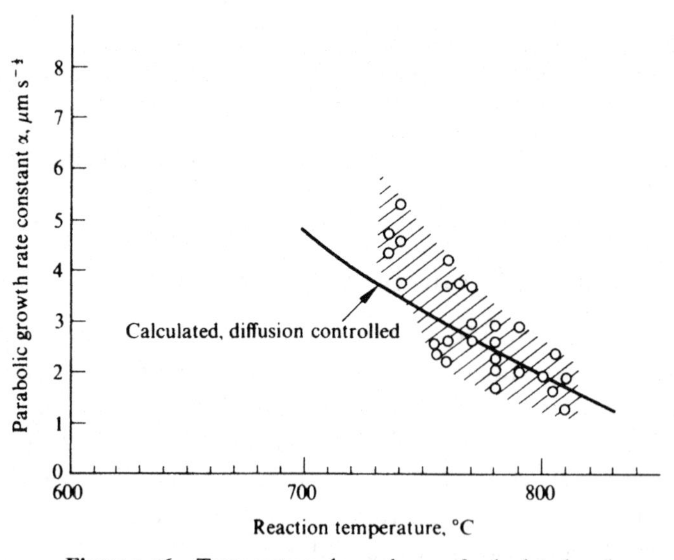


**Figure 2.26** Temperature dependence of calculated and measured values for the growth of pro-eutectoid ferrite allotriomorphs: Fe – 0.11% C. (After Kinsman and Aaronson 1967.)

## 2.3.3 Growth kinetics of Widmanstätten plates

Fig. 2.27 shows a typical plot of half-thickness versus time for plates of ferrite precipitated from austenite at 710 °C, and a curve calculated from eq. (2.28) is included so that the experimentally determined kinetics can be compared with those expected of a disordered boundary at the same temperature and alloy composition. It is obvious that the kinetics of thickening of Widmanstätten plates are very different from those of disordered boundaries.

This effect arises when a precipitate particle grows wholly within the interior of a parent grain. Provided the crystal structures of the precipitate and the matrix are reasonably compatible, one would anticipate that a rational orientation relationship would exist between the two *lattices* involved in order to minimise the interfacial free energy on forming a critical nucleus. The orientation of the interphase *boundary* must, on the other hand, vary widely in order to enclose completely the precipitate crystal, and certain of these boundaries may possess structures of high stability so that the *habit* of the crystal which forms may reflect this stability.

If the surface tension $\gamma$ of an interface between two crystals of specified orientations is plotted as the radial coordinate of a graph where the other variable is the orientation of the crystal interface certain orientations may be found to possess a low energy or cusp orientation (see chapter 5). There will be, of course, a different three-dimensional $\gamma$-plot for each possible relative orientation between the crystal *lattices*.

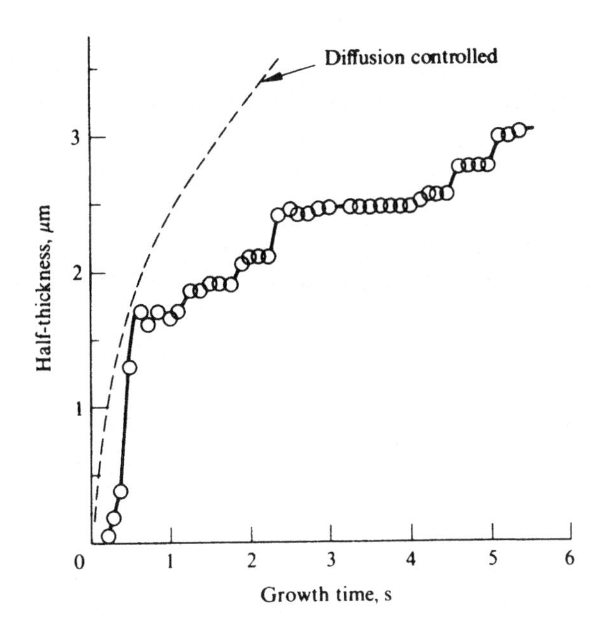


**Figure 2.27** Half-thickness versus reaction time of Widmanstätten ferrite plates: Fe – 0.22% C at 710 °C. (After Kinsman, Eichen and Aaronson 1970.)

These cusps may be quite deep for certain crystal interfaces. Hu and Smith (1956) showed, for example, that interfaces between fcc $\alpha$ brass and bcc $\beta$ brass, where the interface plane contains both the close packed $\langle 110 \rangle_\alpha$ and $\langle 111 \rangle_\beta$ directions, have a surface tension of one-third of the general surface tension between these crystalline phases. Both Shewmon (1965) and Townsend and Kirkaldy (1968) developed precipitates of ferrite in austenite in a dilute Fe–C alloy which they allowed to come to solute equilibrium. The ferrite precipitates were then undercooled by rapidly cooling a small amount. Precipitates that had had only a short time to come to equilibrium (12 min) underwent growth by developing surface perturbations, but those precipitates that had had longer equilibrating times (36 h) were very resistant to perturbation. The explanation offered for this difference is that after long annealing times the precipitate–matrix interfaces had rotated to achieve low energy (cusp) orientations. This explanation would account for the common observation of the precipitation of needles or plates (for example, Purdy (1971)) with the long axis of the needle or the plane of the plate having interfaces with good crystallographic fit.

Aaronson *et al.* (1970) have offered the alternative suggestion of low interface *mobility* to account for the stability of these needle- or plate-like precipitates – on the basis that their interfaces, by virtue of their good crystallographic fit, would be expected to be immobile. As yet there is no general agreement as to which of the two suggestions is likely to be correct.

### 2.3.4 Formation of misfit dislocations at interphase boundaries

As discussed in §2.2.4, nucleation of precipitates commonly occurs by the formation of coherent transition phases, whose low interfacial free energy, $\sigma$, permits a high rate of nucleation. Any misfit in the atomic spacings of the two structures in the interfacial plane is accommodated by elastic 'coherency strains'. As a coherent precipitate grows, the magnitude of its elastic energy is proportional to the volume of the precipitate; ultimately it may be energetically more favourable for the interfacial misfit to be accommodated by dislocations lying in the interface, hence reducing the elastic energy of the system and replacing it with a 'surface energy', that is, the energy of the interface dislocations. A precipitate which contains dislocations in its interface is termed 'semicoherent' or 'partially coherent' and the process of the introduction of the interface dislocations is known as the *loss of coherency*.

Since the energy of the particle with a dislocated interface is proportional to the area of interface of the precipitate, whereas that with coherency strains depends upon its volume, it follows that there will be a critical size above which the semicoherent state will have the lower energy. If a coherent precipitate grows beyond this size it will have a tendency to lose coherency and thus be metastable.

The theories and observations of the generation at growing precipitates of misfit dislocations have been reviewed in detail by Aaronson, Laird and Kinsman (1970); it seems likely that the loss of coherency of a growing particle could occur by one of three possible mechanisms.

### Prismatic loop punching

This mechanism consists of the formation of a pair of prismatic loops, one interstitial and the other of the vacancy type, at the interphase boundary. If the crystal has a larger specific volume than its matrix, the vacancy loops will be left at the boundary and the interstitial loop will be expelled into the matrix, and vice versa, as illustrated diagrammatically in fig. 2.28.

This suggestion was first put forward by Brooks (1952), and using Eshelby's (1957) theory of a misfitting inclusion, Weatherly (1968) showed that the maximum stress at the surface of an elastically isotropic ellipsoidal precipitate is a function only of the misfit (and of the elastic moduli of the precipitate and the matrix) and is not affected by the precipitate size. The theory successfully predicts the absence of such punching, but not always its presence, although one successful case is that of copper containing particles of MgO produced by the internal oxidation of Cu–Mg solid solutions. Fig. 2.29 shows prismatic loops (or degenerate dislocation helices with ⟨110⟩ glide axes) in association with MgO particles which have lost coherency by this process.

### Adsorption of matrix dislocations at the interphase boundary

The driving force for this mechanism is again the excess elastic strain energy at interphase boundaries resulting from coherency, and Gleiter (1967a,b,c) has made a theoretical analysis of the problem. He showed that when a dislocation encounters the strain field of a precipitate, it undergoes a series of manoeuvres designed to introduce into the interface a dislocation loop of such a sign that it cancels the strain field of the precipitate. The first observations of coherency loss

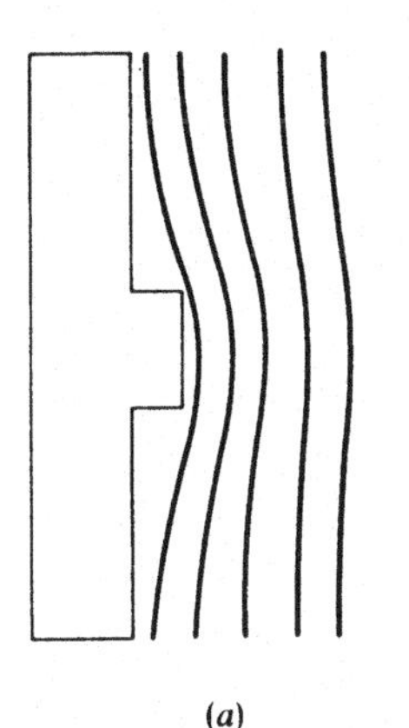

(a)

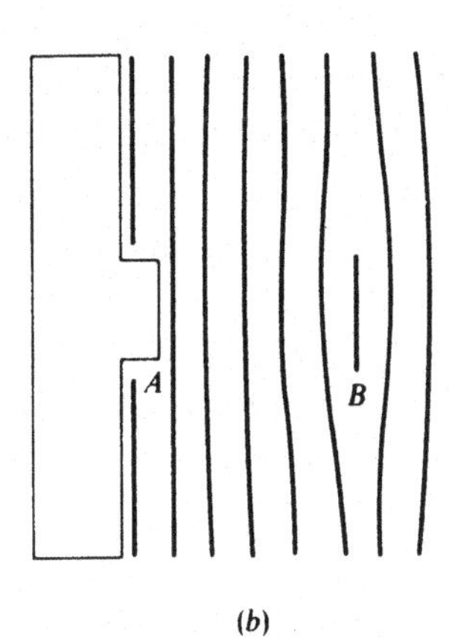


(b)

**Figure 2.28** 'Punching' of prismatic interstitial dislocation loop into matrix by particle of larger specific volume than the matrix.

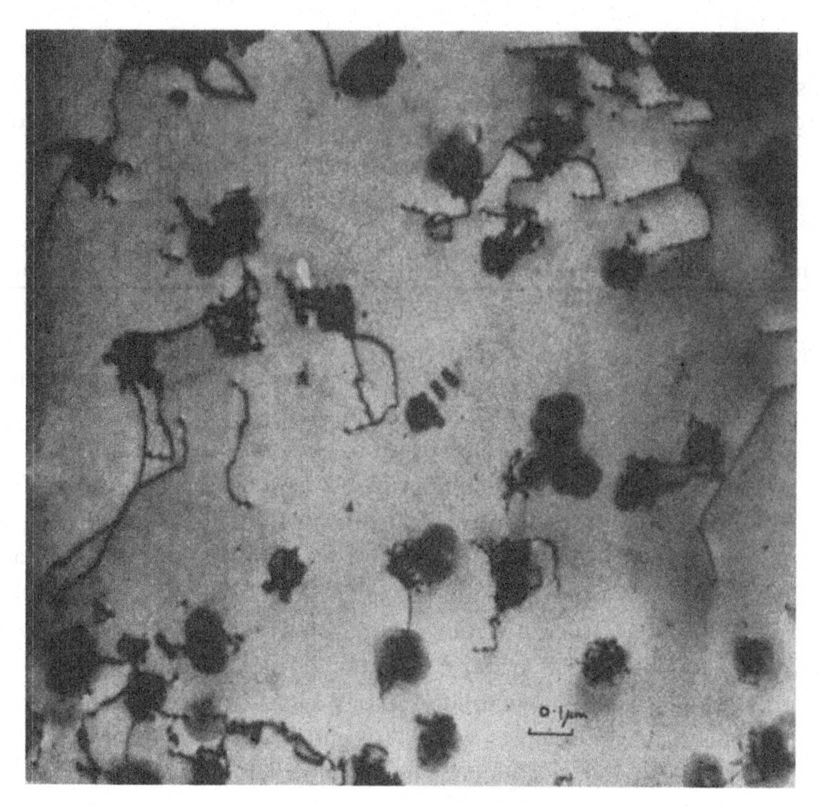


**Figure 2.29** Prismatic loops punched by growing MgO particles in copper produced by internal oxidation at 800 °C. (After Lewis and Martin 1964.)

resulting from deformation were made by Bonar (1962) and Phillips and Livingstone (1962) in the system Cu–Co, and micrographs of this system, fig. 2.30, illustrate clearly the change in contrast at cobalt particles when coherency is lost by this process.

Climb of dislocations from a source in the matrix is an alternative mechanism of interaction to that of dislocation glide. Weatherly and Nicholson (1968) have observed coherency loss in the lath shaped *S*-precipitates in the Al–Cu–Mg system. They observe that initially a dislocation with a Burgers vector component along the lath axis climbs to a particle and lies along one interface of the particle or loops around it. The dislocation then breaks up to form prismatic loops, as shown in fig. 2.31. One significant implication of these observations is that if alloys containing such precipitates were simultaneously aged and plastically deformed under creep conditions, rapid climb of dislocations to particle–matrix interfaces would be expected until coherency is completely lost. Weatherly and Nicholson (1968) observed in fact that all laths had lost coherency after 20 minutes creep at 533 K, whereas 25 days ageing without stress at the same temperature still preserved the coherency of many of the laths; they

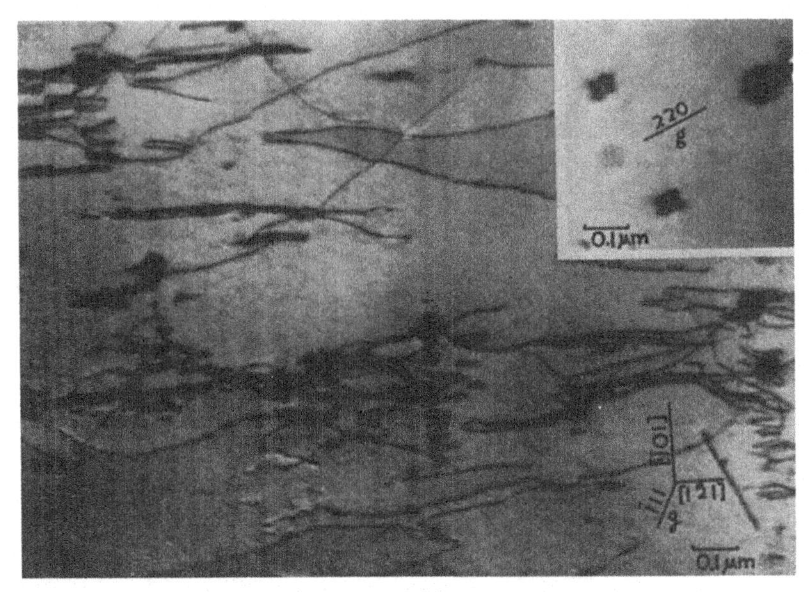


**Figure 2.30** (Inset) A Cu – 0.9 wt% Co crystal has been aged to produce a uniformly distributed precipitate of coherent cobalt particles of radius 25±1 nm; the (111) section of the crystal after a shear strain of 0.21 shows that the particles no longer exhibit coherency strains. (Courtesy F.J. Humphreys.)

found an acceleration of rate of loss of coherency of at least three orders of magnitude under creep conditions.

### Nucleation of dislocation loops within precipitates

One mechanism applicable to precipitates with an fcc structure has been proposed by Baker, Brandon and Nutting (1959); it involves a double shear of a normal stacking fault loop. Alternatively, a small cluster of interstitials (if the precipitate is under tension) or vacancies (if under compression) is expected to collapse, and the dislocation loop to grow in the particle stress field. Weatherly and Nicholson observed that the flat faces of the $\theta'$ phase in Al–Cu alloys lose coherency by the nucleation and growth of small dislocation loops inside the precipitate, which then climb into the precipitate–matrix interface.

It should be pointed out in conclusion that the coherency loss resulting from the deformation (or other disruption) of metastable precipitates can have consequences for the stability of the microstructure. The driving force for the competitive growth of an assembly of precipitates is the interfacial energy (see chapter 5), and if this is increased by the addition of a dislocation network, the growth rate may be enhanced.

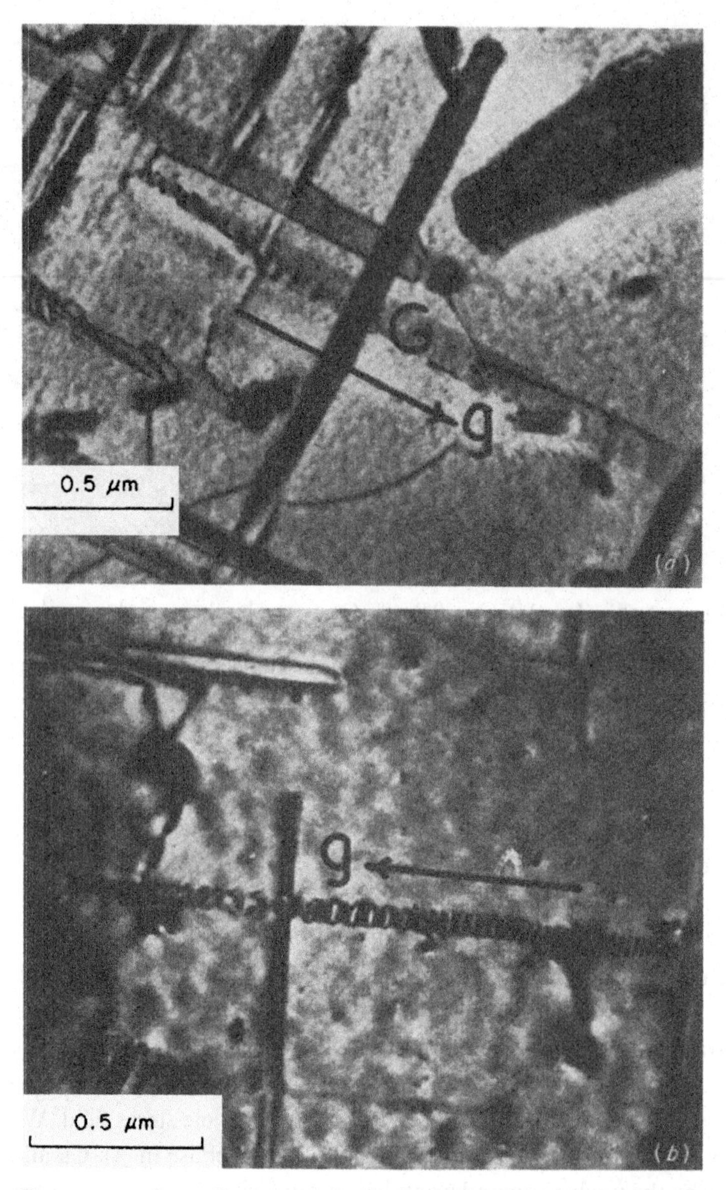


**Figure 2.31** Loss of coherency by the formation from matrix dislocations of prismatic dislocation loops surrounding laths of *S* phase in the Al–Cu–Mg system. (*a*) Specimen aged one day at 325 °C: dislocations are climbing round the precipitates – a dipole is formed at *G*. (*b*) Specimen aged 60 h at 325 °C: a well-developed array of dislocation loops on an *S*-lath. (Courtesy of G.C. Weatherly and R.B. Nicholson.)

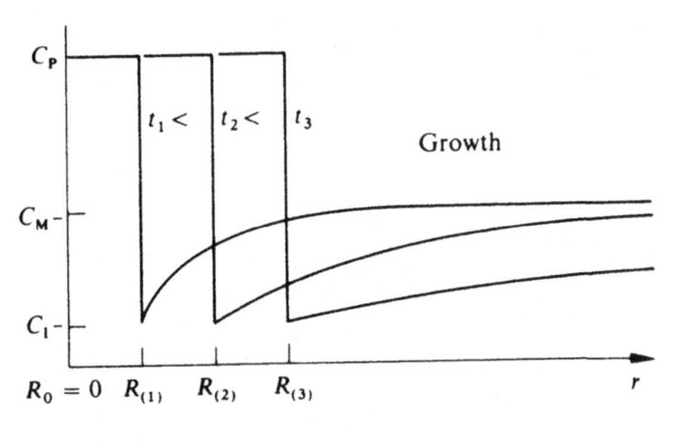


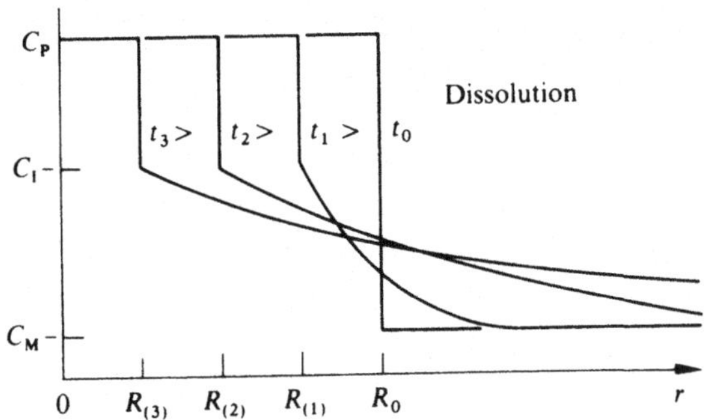


**Figure 2.32** Schematic comparison between precipitate growth and dissolution showing the differences in the time evolution of the concentration fields around the precipitate resulting from the difference in initial radius ($R_0$). (After Aaron and Kotler 1971.)

## 2.4 Second-phase dissolution

When an alloy containing dispersed particles of a second phase is held above the solvus temperature of the precipitate, dissolution of the precipitate crystals takes place under the chemical driving force arising from their instability at the higher temperature. Although this process of mass transport from an unstable second phase (of limited extent) to a stable matrix phase has not received the attention accorded to the related processes of precipitate growth, the process is of considerable practical importance. Where the presence of a second phase results in improved properties one wants to prolong precipitate life, whereas when the precipitate is detrimental (for example, during an homogenisation anneal) one wants to decrease precipitate life.

Aaron and Kotler (1971) have made a detailed theoretical study of the dissolution of precipitates, and the discussion that follows is based essentially upon their analysis. It must be initially pointed out that it will not, in general, be valid to treat dissolution as simply the reverse of growth: fig. 2.32 makes a schematic comparison between precipitate growth and dissolution. It shows the differences in the time evolution of the concentration fields around the precipitate resulting from the difference in initial radius ($R_0$). Essentially, whereas during growth the matrix is depleted of solute immediately ahead of the advancing interface, during dissolution the solute concentration in the matrix at a position $r$ far from the precipitate (so that $r \gg R(t)$) increases with time, although close to the precipitate ($r \gtrsim R(t)$) the solute concentration decreases with time. The complexities in the concentration profiles are the reasons why it has not been possible to find a closed form analytic solution for the problem of the volume diffusion controlled dissolution of a single, spherical precipitate in an infinite matrix. Aaron and Kotler (1971) are thus concerned with certain special cases where analytic solutions are obtainable.

## 2.4.1 Diffusion limited dissolution

As in the case of precipitate growth, there are two distinct possibilities for the rate controlling mechanism governing precipitate dissolution–diffusion in the matrix or reaction at the matrix–particle interface. By employing the technique of electron-probe microanalysis, Hall and Haworth (1970) have measured the diffusion fields around large Widmanstätten plates of $\theta$ phase in Al–Cu alloys during partial dissolution of the particles. The diffusion profile was able to be described in terms of the classical diffusion coefficient for the system at the concentration and temperature involved, thus showing that the process was diffusion controlled.

The concentration field for the dissolution of an isolated precipitate in an infinite matrix is shown in fig. 2.33 together with the corresponding portion of the phase diagram. The field equation (eq. (2.24)) applies to dissolution as well as to growth, subject to the same boundary conditions stated on p. 66; the independent flux balance stated in (2.23) must also necessarily be satisfied, $R_0 > 0$ for dissolution.

Of the three approximations considered by Aaron *et al.* (1970), the stationary interface analysis again appears to be the most accurate approximation, since it

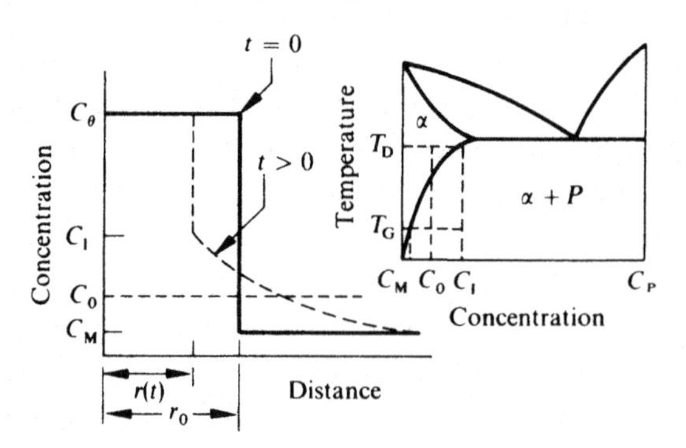


**Figure 2.33** Solute concentration profile around a dissolving precipitate (which was initially in equilibrium with its depleted matrix) with its corresponding binary phase diagram. (After Aaron and Kotler 1971.)

alone satisfies all of the boundary conditions without further assumptions. The dissolution velocity of a sphere may be obtained from eq. (2.25) as

$$\frac{dR}{dt} = -k\left[\frac{D}{R} + \left(\frac{D}{\pi t}\right)^{1/2}\right] \tag{2.30}$$

where $k$ (eq. (2.27)) describes the supersaturation with the values shown in fig. 2.33.

This stationary interface approximation for dissolution has been integrated by Whelan (1969); eq. (2.30) differs from that for a planar precipitate by the $D/R$ term. The result of this difference is that the dissolution velocities of spherical and planar precipitates are similar only at early times (fig. 2.34). The basic difference between the two particle geometries is that the area available for transport increases with radial distance from the centre of a sphere, but not from a planar interface. Thus as the sphere becomes small, the dissolution rate increases (fig. 2.34) as the interfacial area of the sphere decreases markedly, decreasing the size of the solute source with respect to the area of the diffusion zone surrounding it. Aaron and Kotler (1971) have also considered the effects of interface curvature, since the Gibbs–Thompson equation (as modified by Hillert (1957)) shows that the composition in the matrix at a curved precipitate–matrix interface, $C_I(R)$, varies with precipitate radius, $R$, as

$$C_I(R) = C_I(\infty)\exp\left(\sigma_{\alpha p} V_p / kTRC_p\right) \tag{2.31}$$

where $\sigma_{\alpha p}$ is the specific interfacial energy of the matrix–precipitate boundary, $V_p$ is the molar volume of the precipitate, $C_p$ is the mole fraction of solute in the precipitate, $k$ is Boltzmann's constant, and $T$ is the absolute temperature. They conclude, however, that in most alloy systems curvature effects upon dissolution

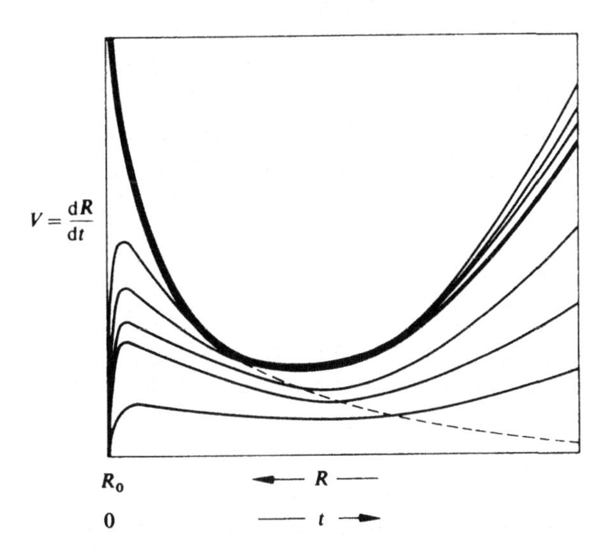


**Figure 2.34** Dissolution velocity as a function of precipitate size (or dissolution time) for a spherical precipitate (heavy solid line) and a planar precipitate (dashed line). In the spherical case, an interface reaction drastically reduces the velocity at early times whereas curvature effects slightly increase the velocity at long times (solid lines). (After Aaron and Kotler 1971.)

kinetics will be negligible, unless the concentration difference $(C_I - C_M)$ is extremely small.

### 2.4.2 Interface reaction kinetics

In some situations it is possible to have an interface process which proceeds more slowly than volume diffusion controlled dissolution. Since an interface reaction will reduce the flux of atoms which crosses the interface from the precipitate into the matrix, the actual interface concentration, $C_I$, during dissolution will be less than the 'equilibrium' interface concentration, $C_I(R)$; a positive $\Delta C_K$ for dissolution may be defined:

$$\Delta C_K = C_I(R) - C_I \qquad (2.32)$$

The relationship between the interface growth velocity ($V$) and $\Delta C_K$ will depend upon the the mechanism of atomic detachment at the interface during dissolution. In the case of uniform atomic detachment, we write

$$V = -K_0 \, \Delta C_K \qquad (2.33)$$

Abbott and Haworth (1973) have examined the dissolution of large Widmanstätten particles of $\gamma$ phase in Al–Ag alloys at temperatures between 393 and 433 °C. They showed by electron-probe microanalysis that the concentration of silver in the matrix at the matrix–precipitate interface was less than the equilibrium value, indicating that an interface reaction controls the rate of dissolution of the precipitates. Aaron and Kotler (1971) include the effects of both curvature and interface kinetics by coupling these conditions to the diffusion field approximation to obtain by computation a numerical solution. Fig. 2.35 shows the results obtained when $P=(k/4\pi)^{1/2} \approx 0.09$, a value typical of many alloys of interest, and putting $\beta = 2\sigma_{\alpha p} V_p / RTC_p = 10^{-6}$ mm.

The results show that, in contradistinction to the effects of curvature (§2.4.1), interface reaction kinetics may produce quite large alterations in dissolution kinetics and increases to total time needed for dissolution. The interface reaction will have no sensible effect for $K_0 > 10^{-2}$ mm s$^{-1}$, but for $K_0 < 10^{-6}$ mm s$^{-1}$ the particle is, effectively, insoluble due to the extremely sluggish kinetics.

Several attempts have been made at accounting for the contribution of vacancy flow or plastic deformation to particle dissolution which arises as a result of local volume changes. Nemoto (1974) has conducted *in situ* studies of the dissolution of cementite in steel; at high dissolution rates, the dissolution of cementite particles was accompanied by the generation of dislocation helices at the particle interface that acted to relieve the build-up of strain around the particle during dissolution. At slower dissolution rates, however, the strain field

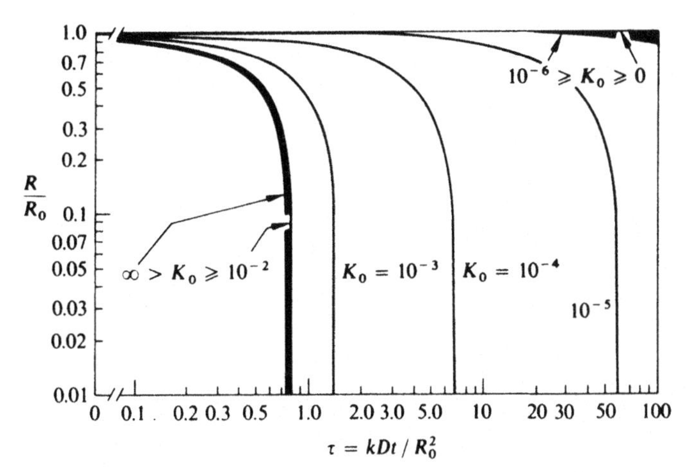


**Figure 2.35** Dissolution kinetics for $P=(k/4\pi)^{1/2}\approx 0.09$ and $\beta=2\sigma_{\alpha p}V_p/RTC_p=10^{-6}$ mm showing how a slow interface reaction (of the uniform detachment type) can dramatically increase the dissolution time. The kinetic coefficient $K_0$ is a measure of the difficulty of 'detaching' an atom from the precipitate – detachment becoming progressively more difficult as $K_0$ decreases towards zero. (After Aaron and Kotler 1971.)

build-up appeared to be relieved by a vacancy flux moving away from the dissolving precipitate.

Hewitt and Butler (1986) have studied the kinetics of dissolution of $\theta'$ in Al–3% Cu by an *in situ* hot stage technique, and modified Nemoto's analysis to interpret their data. The key observation was that the stress field caused by the local volume change gave rise to the generation and movement of dislocations. By assuming that the flow of vacancies away from a dissolving particle is induced by a stress field of shear character, Nemoto concluded that the effect should be closely related to Nabarro–Herring creep. Hewitt and Butler used this approach to write the shrinkage rate of a plate shaped $\theta'$ particle when controlled completely by a vacancy flow as:

$$dr/dt=(\sigma v D/\lambda kT)/(1-V_{mAl}/V_{m\theta'})$$

where $D$ is the self-diffusivity of aluminium, $v$ is the atomic volume of Al, $\lambda$ is the effective diffusion distance and may be estimated as the radius of the particle, $\sigma$ is the tensile stress in the matrix close to the particle (evaluated from a knowledge of the values of the elastic modulus and the critical misfit strain at the interface), $V_{m\theta'}$ is the molar volume of $\theta'$ and $V_{mAl}$ that of aluminium. This relationship is in agreement with the authors' observation of a linear relationship between plate radius and dissolution time.

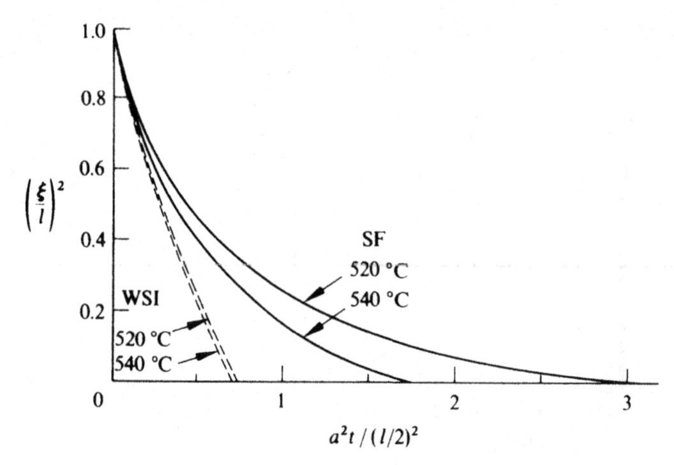


**Figure 2.36** Comparison of the spherical–infinite model of Whelan (1969) with the spherical–finite model of Baty *et al.* (1970). The plot of reduced radius versus reduced time shows the increased dissolution time required in the finite matrix geometry. (After Baty, Tanzilli and Heckel 1970.)

## 2.4.3 Precipitate arrays

There are many practical situations in which the distribution of precipitates is such that the diffusion fields of individual precipitates do not sensibly overlap during most of the dissolution process. In these cases the preceding analyses are directly applicable, but Aaron and Kotler (1971) have extended their study to the situation in which the particles are not sufficiently separated so as to be non-interacting. It is assumed that the precipitates have equal radii (or thickness) and are uniformly spaced in the matrix, so that the matrix may be regarded as divided into identical, symmetrical cells, each cell having a precipitate particle at its centre.

The major effect of the finite matrix (as compared with the infinite matrix) is that when the diffusion fields overlap, the dissolution kinetics are severely slowed down. This is shown clearly in fig. 2.36, where comparison is made between the spherical-infinite model of Whelan (1969) and the spherical-finite model.

## 2.4.4 Application of the theory

Aaron and Kotler (1971), in the final section of their paper, used their theoretical analyses to interpret the results of an experimental investigation of the technologically important problem of the dissolution of 'reluctantly soluble' carbides in the Fe–C–V system. They demonstrated that the lack of reliable phase diagram information can result in misinterpretations of experimental observations, but showed that vanadium carbide and iron carbide were readily dissolved while complex alloy sulphides, silicates and oxides persisted, with little if any dissolution.

Alloying a steel with vanadium may be accompanied by a reduction in hardenability. In view of the ready solubility of vanadium carbide itself, the effect is not thus explicable in terms of nucleation of austenite decomposition products on vanadium carbide. It is suggested that the sulphides, silicates and oxides perform this nucleation function; alternatively, refinement of the austenite grain size which these inclusions produce (see §5.8.3) may be responsible. It appears likely in fact that both of these factors contribute to the observed effect.

# Chapter 3

## Highly metastable alloys

Almost all metallurgical materials are metastable in one way or another. Manipulating the metastability in alloy microstructures has proved to be essential in order to obtain the wide range of properties needed for different kinds of manufactured component. The conventional metallurgical processing methods of casting, deformation and heat treatment are used to control microstructural features such as chemical homogeneity, grain size, extent of precipitation and dislocation substructure. These are associated with relatively slight deviations from equilibrium, and are discussed in the other chapters of this book. In recent years a variety of processing methods have been developed to manufacture alloys with highly metastable microstructures, that is, with greater deviations from equilibrium. These highly metastable alloys are the subject of the present chapter.

The different methods of manufacturing highly metastable alloys all depend upon manoeuvring the material into a condition far from equilibrium, and simultaneously removing its thermal energy to freeze it into a metastable state. The microstructures that can then develop depend upon both thermodynamic and kinetic factors. Thermodynamic conditions define a set of possible alloy microstructures with lower free energy than the starting state. Kinetic behaviour determines which of these microstructures actually develops during manufacture. The main kinds of metastable material that can be manufactured are microcrystalline and nanocrystalline alloys with ultra-fine grain sizes, segregation-free highly supersaturated solid solutions, new metastable crystalline alloy compounds, amorphous alloys with non-crystalline disordered atomic structures, and quasi-crystalline alloys with ordered but non-periodic atomic structures.

This chapter outlines the manufacturing methods, thermodynamic and kinetic constraints and observed microstructures and stability of these different metastable crystalline, amorphous and quasi-crystalline alloys.

## 3.1 Manufacturing methods

### 3.1.1 Rapid solidification

Rapid solidification methods have been reviewed by Jones (1982), Liebermann (1983), Cahn (1991) and Suryanarayana (1991). The main microstructural features of rapidly solidified alloys were first discovered by Duwez and coworkers in 1960 (Duwez, Willens and Klement 1960; Klement, Willens and Duwez 1960). They used a rapid solidification technique called splat quenching, in which a liquid droplet is propelled at high velocity towards the surface of a cold substrate. The droplet spreads out thinly over the substrate surface, heat is extracted efficiently, and the liquid cools and solidifies rapidly. Splat quenched Ag–Cu alloys were found to show complete fcc solid solubility at all alloy compositions, compared with maximum equilibrium solid solubilities of 5 at% Ag in fcc Cu and 14 at% Cu in fcc Ag. Splat quenched Ag–Ge alloys were found to contain a metastable hcp $Ag_3Ge$ compound which does not exist at equilibrium, and a splat quenched Au–25 at% Si alloy did not crystallise at all, that is, it solidified as a metallic glass with an amorphous structure.

Splat quenching is only suitable for rapidly solidifying a small quantity of material, and rapidly solidified alloys were at first regarded as a laboratory curiosity. A variety of other methods have since been developed to manufacture substantial amounts of rapidly solidified material. In processes such as melt spinning, melt extraction and planar flow casting, a liquid stream is spread continuously across the surface of a rotating wheel, to manufacture wire, strip and sheet products (Strange and Pim 1908; Pond 1958; Pond and Maddin 1969; Mobley, Clauer and Wilcox 1972; Takayama and Oi 1969; Narasimhan 1979). In atomisation processes, liquid droplets are sprayed continuously into a cooling fluid, to manufacture powder for subsequent consolidation into bulk components (Klar and Shaefer 1972). In spray forming, the droplet spray is deposited onto a substrate for direct consolidation in a single stage process (Singer 1970). Fig. 3.1 shows a schematic diagram of the melt spinning process which is used to manufacture soft ferromagnetic amorphous alloy sheet and strip for transformers, and fig. 3.2 shows a schematic diagram of the gas atomising process which is used to manufacture high speed steel powders for consolidation into tool bits and dies.

In all these rapid solidification processes, the liquid is manipulated to be thin in at least one dimension and in good thermal contact with a cooling solid substrate or fluid jet. The liquid cools rapidly so that the nucleation of solidification

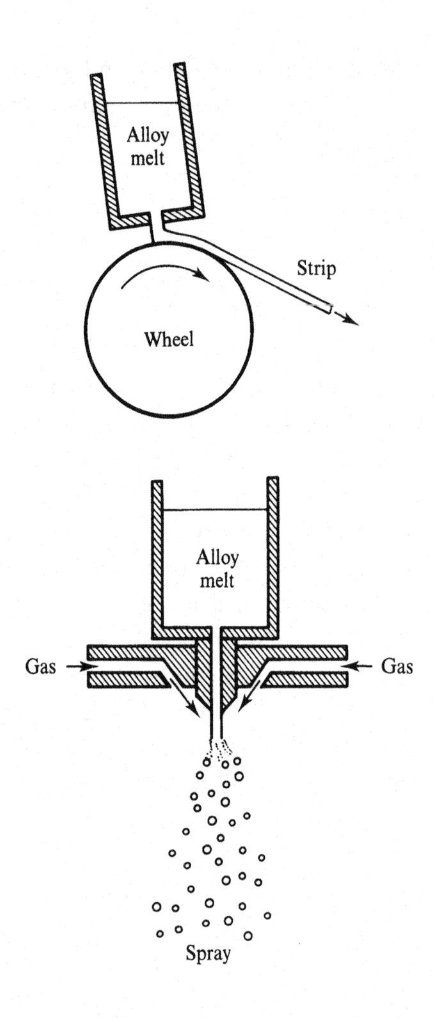


**Figure 3.1** The melt spinning process for manufacturing strip.

**Figure 3.2** The gas atomisation process for manufacturing powder.

is delayed to a high undercooling below the equilibrium melting point. Solidification then takes place at high speed, far from equilibrium, leading to metastable microstructures such as those found by Duwez and coworkers. The thin dimension of the liquid is typically $<100\ \mu m$ with a corresponding cooling rate of $>10^4$ K $s^{-1}$, and cooling and solidification are complete within a few milliseconds. Fig. 3.3 shows pyrometric temperature measurements during melt spinning of a Ni–5 wt% Al alloy with a melt superheat of 200 °C, melt ejection pressure of 42 kPa and wheel surface velocities of 12 and 24 m $s^{-1}$. The average cooling rate is $\sim 3\times 10^5$ K $s^{-1}$ and the solidification arrest shows a solidification time of ~0.5 ms. The melt spun strip thickness is 70 $\mu m$, corresponding to a solidification velocity of ~14 cm $s^{-1}$. Rapid solidification can also be achieved without the need for quenching to avoid nucleation. Fluxing and droplet emulsification and levitation processes remove heterogeneous nuclei, so that the liquid can be cooled relatively slowly to a high undercooling before initiating solidification (Bardenheuer and Bleckmann 1939; Ojha, Ramachandrarao and

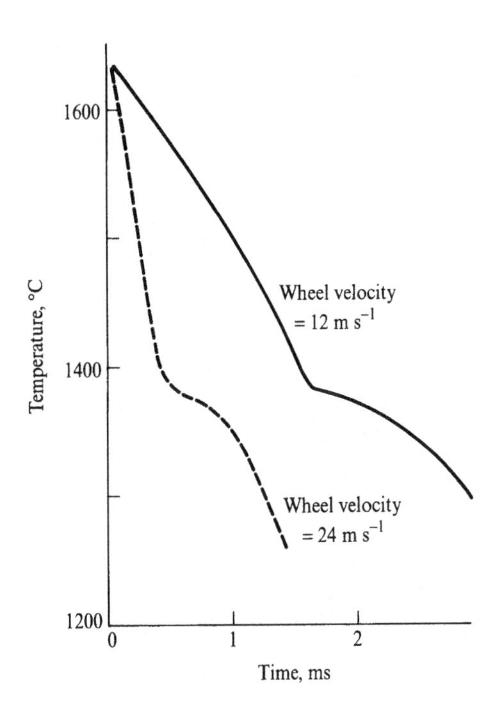


**Figure 3.3** Temperature measurements during melt spinning of Ni – 5 wt% Al. (After Gillen and Cantor 1985.)

Anantharaman 1983; Vonnegut 1948; Perepezko 1980; Herlach *et al.* 1993). Alternatively, an intense laser or electron beam can be pulsed or scanned over a solid surface, melting and then resolidifying a thin layer epitaxially and at high speed (Breinan, Kear and Banas 1976; Lux and Hiller 1977).

## 3.1.2 Cooling rate and undercooling

Heat flow during rapid solidification has been reviewed by Jones (1982) and Cantor (1986). Consider a thin alloy layer of thickness $X$ at an initial temperature $T_I$, brought into contact with a cold substrate (or gas jet) at a temperature $T_S$. The resulting temperature profiles in the alloy $T(x,t)$ are shown schematically in fig. 3.4. Heat flow within the alloy obeys the one dimensional thermal diffusion equation:

$$\frac{\partial T}{\partial t}=\alpha\frac{\partial^2 T}{\partial x^2} \tag{3.1}$$

where $\alpha=k/\rho C$ is the thermal diffusivity, and $k$, $\rho$ and $C$ are the thermal conductivity, density and specific heat respectively. Thermal contact with the substrate is described by Newton's law of cooling, with the heat flux $q$ through the alloy–substrate interface given by:

$$q=h(T_X-T_S)=-k\left(\frac{\partial T}{\partial x}\right)_X \tag{3.2}$$

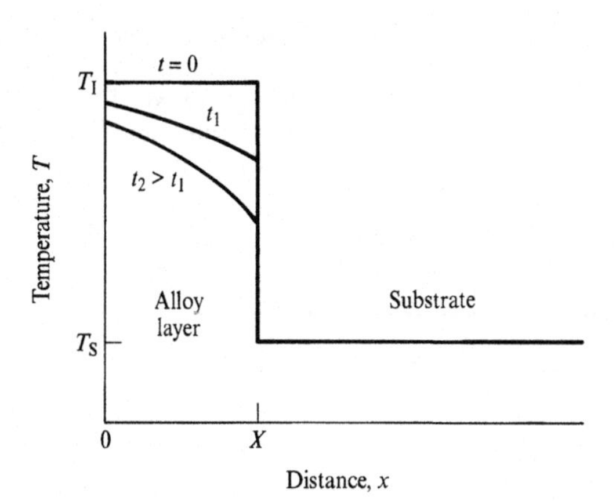


**Figure 3.4** Temperature profiles in an alloy layer in contact with a cold substrate.

where $h$ is the heat transfer coefficient, $T_X$ and $(\partial T/\partial x)_X$ are the temperature and thermal gradient adjacent to the interface, and the substrate is assumed to be an efficient heat sink so its temperature stays constant at $T_S$. The heat removed into the substrate leads to a combination of cooling and solidification, with associated heats $q_1$ and $q_2$ given by:

$$q_1 = -\rho C \int_0^X \left(\frac{\partial T}{\partial t}\right)_x \mathrm{d}x = -\rho C X \frac{\mathrm{d}T}{\mathrm{d}t} \tag{3.3}$$

$$q_2 = \rho L X \frac{\mathrm{d}f}{\mathrm{d}t} = \rho L v = \rho L \mu \Delta T \tag{3.4}$$

where $-\mathrm{d}T/\mathrm{d}t$ is the average cooling rate, $L$ is the latent heat of fusion, $f$ is the fraction solidified, $v$ is the solid–liquid interface velocity, and solidification kinetics are linear, that is, $v$ is proportional to undercooling $\Delta T$ with an interface mobility $\mu$. The overall heat balance $q = q_1 + q_2$ gives the boundary condition for eq. (3.1):

$$h(T_X - T_S) = -\rho C X \frac{\mathrm{d}T}{\mathrm{d}t} + \rho L X \frac{\mathrm{d}f}{\mathrm{d}t} \tag{3.5}$$

Unfortunately, eqs. (3.1) and (3.5) cannot be solved analytically without further assumptions.

The alloy cooling behaviour can be obtained analytically from eq. (3.1) by using an effective specific heat to simplify the heat balance in eq. (3.5):

$$h(T_X - T_S) = -\rho C' X \frac{\mathrm{d}T}{\mathrm{d}t} \tag{3.6}$$

where $C' = C - (L/\Delta T_0)$ and $\Delta T_0$ is the effective freezing range. This corresponds to neglecting undercooling before the onset of solidification, and assuming that solidification is linear over the freezing range. Carslaw and Jaeger (1959) have given the solution to eqs. (3.1) and (3.6):

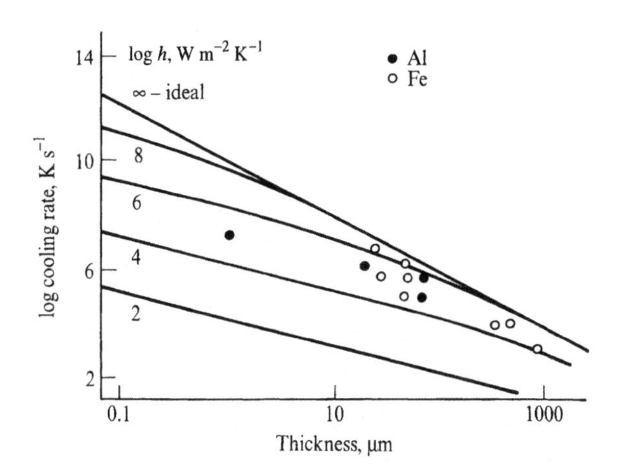


**Figure 3.5** Cooling rate $-dT/dt$ versus thickness $X$ and heat transfer coefficient $h$ for Al in contact with a Cu substrate together with measurements for pure Al and Fe (After Jones 1982 and Cantor 1986.)

$$\frac{T-T_S}{T_l-T_S}=\sum_1^n \frac{2Bi\cos(A_n x/X)\sec(A_n)\exp(-A_n^2\alpha t/X^2)}{Bi(Bi+1)+A_n^2} \tag{3.7}$$

where $A_n$ are the roots of $A\tan A = Bi$ and the Biot number $Bi = hX/k$ measures the relative efficiency of heat transport within the alloy layer and across the alloy–substrate interface. Newtonian cooling conditions correspond to the extreme case of $Bi \ll 1$, with negligible thermal gradients in the alloy, and heat transfer into the substrate providing the only resistance to heat flow. Eq. (3.7) then becomes:

$$\frac{T-T_S}{T_l-T_S}=\exp(-Bi\alpha t/X^2)=\exp(-ht/\rho C'X) \tag{3.8}$$

Ideal cooling conditions correspond to the other extreme case of $Bi \gg 1$, with perfect alloy–substrate thermal contact, and conduction in the alloy providing the only resistance to heat flow. Eq. (3.7) then becomes:

$$\frac{T-T_S}{T_l-T_S}=\sum_1^n \frac{2\cos(B_n x/X)\exp(-B_n^2\alpha t/X^2)}{B_n^2} \tag{3.9}$$

where $B_n=(n+0.5)\pi$. Fig. 3.5 shows average calculated cooling rates for the solidification of Al in contact with a Cu chill, as a function of thickness $X$ and heat transfer coefficient $h$ together with corresponding measurements for rapidly solidified pure Al and Fe. Cooling rates increase with increasing heat transfer coefficient $h$ and decreasing thickness $X$, as expected from eq. (3.6), but become limited by thermal conduction as $Bi$ tends to 1. Measured cooling rates of $10^4$–$10^6$ K s$^{-1}$ during rapid solidification correspond to heat transfer coefficients of $10^4$–$10^5$ W m$^{-2}$ K$^{-1}$ and Biot numbers of 0.1–1, that is, heat flow conditions are intermediate between Newtonian and ideal.

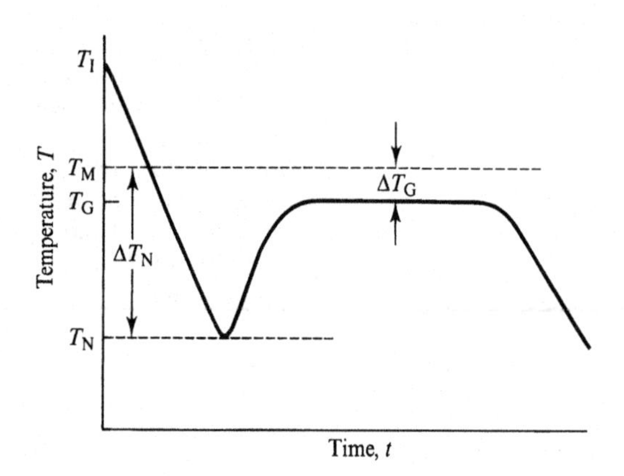


**Figure 3.6** Cooling curve showing undercooling and recalescence under full Newtonian conditions. (After Cantor 1986).

Undercooling effects can be obtained analytically from eq. (3.5) by assuming full Newtonian conditions with no thermal gradients in the alloy. Fig. 3.6 shows a corresponding schematic cooling curve. The liquid undercools to $T_N$ before solidification is nucleated, followed by recalescence to a steady state growth plateau at $T_G$. On the steady state growth plateau, $\mathrm{d}T/\mathrm{d}t=0$ and the first term on the right hand side of eq. (3.5) can be ignored:

$$h(T_G-T_S)=\rho Lv=\rho L\mu\Delta T_G \tag{3.10}$$

In other words, extraction of heat into the substrate drives the solid–liquid interface across the alloy layer with a velocity $v$. With $h\simeq10^5$ W m$^{-2}$ K$^{-1}$, $(T_G-T_S)\simeq1000$ K and $\rho L\simeq10^9$ J m$^{-3}$, the steady state solid–liquid interface velocity $v$ is typically 10 cm s$^{-1}$, corresponding to a steady state growth undercooling of $\Delta T_G\approx1$ K. During recalescence, $\mathrm{d}T/\mathrm{d}t$ is large so that solidification is adiabatic and the term on the left hand side of eq. (3.5) can be ignored:

$$C\frac{\mathrm{d}T}{\mathrm{d}t}=L\frac{\mathrm{d}f}{\mathrm{d}t} \tag{3.11}$$

Integrating from $f=0$ at $T=T_N$:

$$T-T_N=\frac{L}{C}f \tag{3.12}$$

that is, the solid fraction increases linearly with increasing temperature. The value of $L/C$ is the temperature rise caused by latent heat evolution when a material solidifies fully under adiabatic conditions. $L/C$ is called the hypercooling limit, and is typically 200–500 K for different metals, as shown in table 3.1. Eq. (3.12) shows that steady state growth at low undercooling only takes place when the initial nucleation undercooling $\Delta T_N$ is less than $L/C$. Otherwise, solidification is completed adiabatically and at high undercooling during recalescence, before reaching the steady state growth plateau.

### 3.1.3 Condensation, chemical reaction and mechanical disruption

Rapid solidification processes are not the only methods for manufacturing highly metastable alloys. Condensing atoms from a vapour or solution phase and quenching them onto a cold substrate to form a thin or thick solid film can also be very effective. Buckel and Hilsch (1952, 1954, 1956) discovered metastable microstructures similar to those found later by Duwez and coworkers in thin films manufactured by thermal evaporation followed by condensation onto a cold substrate. Thermally evaporated Sn–Cu and Sn–Bi thin films condensed onto substrates cooled to 4 K consisted of fine-scale grains of a single phase bct solid solution, with maximum solid solubilities of ~20 at% Cu and ~45 at% Bi, compared with very low values at equilibrium. Thermally evaporated Sn–Cu thin films were amorphous at compositions >20 at% Cu. Fig. 3.7 shows a schematic diagram of the thermal evaporation process which is used widely in the manufacture of thin film semiconductor, interconnect and packaging components in integrated circuits. Similar metastable thin film alloy microstructures are also achieved when the vapour is obtained by sputter evaporation in a gas discharge, again followed by condensation onto a cold substrate (Cantor and Cahn 1976a; Dahlgren 1978). Amorphous Ni–P and other transition metal alloys are readily manufactured by electrodeposition or electroless deposition, that is, by condensation from a supersaturated aqueous solution with or without electrolytic stimulation (Brenner, Couch and Williams 1950; Brenner 1963).

The microstructure of an alloy manufactured by condensation methods such as evaporation, sputtering or electrodeposition depends on the relative rates of deposition and surface diffusion. Depositing atoms lose their excess thermal energy and equilibrate with the substrate within a few atomic vibrations, that is, almost instantaneously. They then diffuse across the surface until surrounded by adjacent depositing atoms. The time available for surface diffusion is approxi-

| Metal | $L/C$ (K) |
|---|---|
| Al | 382 |
| Cu | 433 |
| Fe | 344 |
| Mg | 302 |
| Ni | 491 |
| Zn | 250 |

**Table 3.1** *Hypercooling limit $L/C$ for pure metals. (After Cohen, Kear and Mehrabian 1980.)*

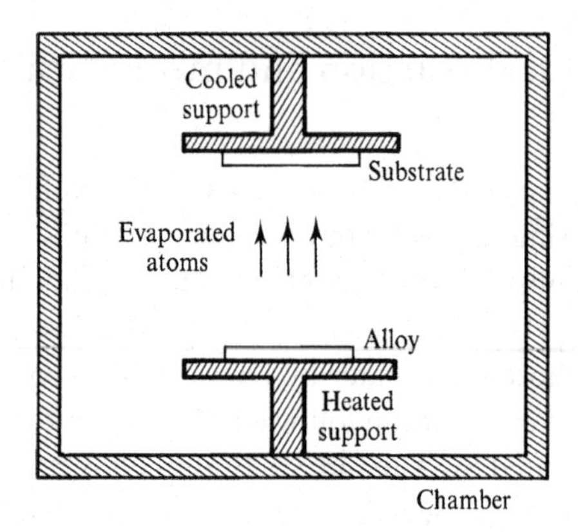


**Figure 3.7** The thermal evaporation process for manufacturing thin films.

mately equal to the time taken to deposit a monolayer of atoms, $t \approx a/R$, where $a$ is the interatomic spacing and $R$ is the deposition rate. The average surface diffusion distance $x$ is given by the Einstein relation:

$$x^2 = 2Dt \approx \frac{2\nu a^3}{R} \exp\left(\frac{Q}{kT}\right) \tag{3.13}$$

where $D$ is the surface diffusion coefficient, $\nu$ is the atomic vibration frequency, $Q$ is the activation energy for surface diffusion, $k$ is Boltzmann's constant and $T$ is the substrate temperature. Below a critical substrate temperature $T_C$, the average surface diffusion distance is less than the interatomic spacing, i.e., is effectively zero:

$$T_C \simeq \frac{Q}{k \ln(2\nu a/R)} \tag{3.14}$$

Below $T_C$, there is insufficient time for atomic ordering or clustering, and the deposited alloy structure is a disordered solid solution. Below some lower critical temperature, there is insufficient time for a deposited atom even to find a local energy minimum, and the deposited alloy structure is amorphous. Eq. (3.14) shows that $T_C$ is controlled by the activation energy for surface diffusion $Q$, and therefore scales with the alloy melting point (Brown and Ashby 1980). With $\nu \approx 10^{12}$ Hz, $a \approx 0.3$ nm and $R \approx 0.1$ nm s$^{-1}$, typical values of $Q$ range from 20–150 kJ mol$^{-1}$, with corresponding critical substrate temperatures $T_C$ varying from $-200$ °C to 350 °C. Thus, supersaturated solid solutions and amorphous alloys require substrate cooling well below room temperature for low melting point metals such as Sn (Buckel and Hilsch 1952, 1954, 1956), but are obtained with substrates heated well above room temperature in refractory metals such as Ta and W (Hunt, Cantor and Kijek 1985).

In 1983, Johnson and coworkers (Schwarz and Johnson 1983; Yeh, Samwer and Johnson 1983) discovered that amorphous alloys can be manufactured without quenching or rapid solidification, simply by interdiffusion of the alloy components. Diffusion of gaseous hydrogen into partially ordered fcc $Zr_3Rh$ at 150–225 °C was found to break down the crystalline structure, resulting in a $Zr_{36}Rh_{12}H_{52}$ amorphous alloy (amorphous alloy compositions are usually quoted as $A_xB_y$ with $x$, $y$ in at%). Similarly, solid state interdiffusion within a multi-layer composite of alternating 10–50 nm thick La and Au films produced La–Au amorphous alloys with compositions in the range ~25–55 at% Au, depending on the initial ratio of La–Au film thicknesses. Solid state amorphisation reactions of this type have been reviewed by Johnson (1986). They are examples of the more general phenomenon of a chemical reaction proceeding via a metastable intermediate phase. The metastable alloy structure has lower free energy than the separate components, but higher free energy than at equilibrium. It can be crystalline or amorphous, and forms because of kinetic difficulties in nucleating the equilibrium structure. A simple reaction in a general A–B diffusion couple can be represented schematically by:

$$A+B \rightarrow \alpha'(A_xB_y) \rightarrow \alpha(A_xB_y) \tag{3.15}$$

where $\alpha'$ is the metastable crystalline or amorphous intermediate $A_xB_y$ structure, and $\alpha$ is the equilibrium crystalline $A_xB_y$ structure. More complex reactions can proceed with several intermediate stages, each involving several metastable phases.

Metastable alloys can also be manufactured by mechanically disrupting the atomic structure at the surface of a material by irradiation and implantation with high energy electrons (Holland, Mansur and Potter 1981), heavy ions (Grant *et al.* 1978; Picraux and Choyke 1982) or fission fragments (Bloch 1962). Interdiffusion and chemical reaction methods of forming metastable alloys are considerably enhanced either by simultaneous irradiation (Tsaur, Lau and Mayer 1980), or by simultaneous mechanical deformation using techniques such as mechanical alloying of elemental powders in a high energy ball mill (Benjamin 1976; Yermakov, Yurchikov and Barinov 1981; Koch 1991), or co-rolling of elemental multi-layer composites (Bevk 1983; Atzmon, Veerhoven, Gibson and Johnson 1984). Fig. 3.8 shows a schematic diagram of the ball milling process for mechanical alloying, which is used to manufacture superalloy powders for consolidation into aeroengine components. Direct application of high pressure can also be used to manufacture metastable alloys (Battezzati 1990).

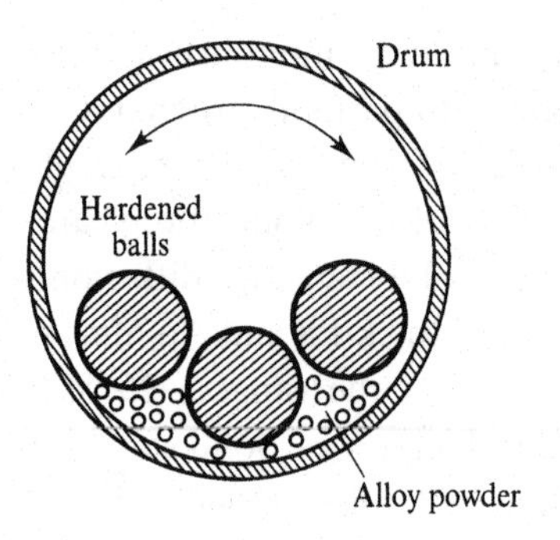


**Figure 3.8** The mechanical alloying process for manufacturing powder.

## 3.2 Metastable phase diagrams

Conventional phase diagrams show the range of stability of the equilibrium phases in an alloy as a function of the temperature, pressure and alloy composition. Phase diagrams can also be extended to show the range of stability of the metastable phases in an alloy. This corresponds to including constraints which prevent equilibrium being achieved.

### 3.2.1 $T_0$ lines

Fig. 3.9(*a*) shows the phase diagram for a binary A–B alloy system with complete miscibility in both the solid and liquid state. Fig. 3.9(*b*) shows the corresponding free energy–composition curves at a temperature $T$ within the freezing range. Common tangents between the solid and liquid free energy–composition curves define the freezing range between $c_\alpha$ and $c_L$. If the alloy is constrained to prevent diffusion and segregation, the intersection of the liquid and solid free energy–composition curves defines a metastable equilibrium between liquid and solid of the same composition $c_0$. For any given alloy composition, the temperature at which the liquid and solid have the same free energy is called the $T_0$ temperature for that alloy. The metastable phase diagram consists of a line of $T_0$ temperatures, as shown by the dashed line in fig. 3.9(*a*). Segregation-free solidification can take place below the $T_0$ temperature, often with a kinetic advantage, since solid–liquid interface motion is not then limited by solute diffusion. Fig. 3.10 shows an example of a segregation-free solidification microstructure of columnar fcc grains in a melt spun Fe–35 wt% Ni alloy.

The $T_0$ temperature is also important in alloys of limited solid miscibility. Figs. 3.11(*a*)–(*c*) show three examples of a simple eutectic phase diagram. In fig. 3.11(*a*), the terminal solid solutions $\alpha$ and $\beta$ have the same crystal structure. The $T_0$ line extends continuously across the phase diagram, and segregation-free solidification is possible for all alloy compositions. In fig. 3.11(*b*), the terminal solid solutions have different crystal structures, $\alpha$ and $\beta$. The $\alpha$ and $\beta$ $T_0$ lines

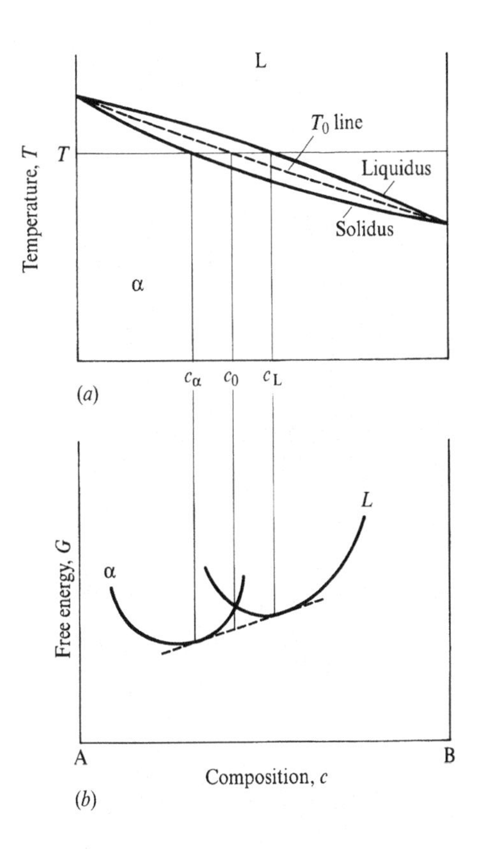


**Figure 3.9** (*a*) Phase diagram and (*b*) free energy–composition curves showing $T_0$ lines for segregation-free solidification

extend continuously across the phase diagram, and segregation-free solidification of both $\alpha$ and $\beta$ is possible for all alloy compositions. In fig. 3.11(*c*), the terminal solid solutions again have different crystal structures, but the $\alpha$ and $\beta$ $T_0$ lines both fall rapidly to absolute zero, and segregation-free solidification is not possible within a composition range in the centre of the phase diagram.

## 3.2.2 Submerged phases and transformations

Fig. 3.12(*a*) shows another example of a simple eutectic phase diagram. In this case, there is a big difference in A and B melting points and limited A–B terminal solid solubilities, leading to a wide freezing range and a flat liquidus. Fig. 3.12(*b*) shows the corresponding free energy–composition curves at a temperature $T$. The flat liquidus in fig. 3.12(*a*) corresponds to an incipient immiscibility in the liquid. Near the liquidus, the liquid free energy–composition curve is broad and flat, so that the liquidus composition changes rapidly with temperature. The common tangent between the solid and liquid free energy–composition curves defines the equilibrium solid and liquid compositions $c_\alpha$ and $c_L$. Below the liquidus, the liquid free energy–composition curve develops a miscibility gap. If the alloy is constrained to prevent solidification, a metastable liquid miscibility

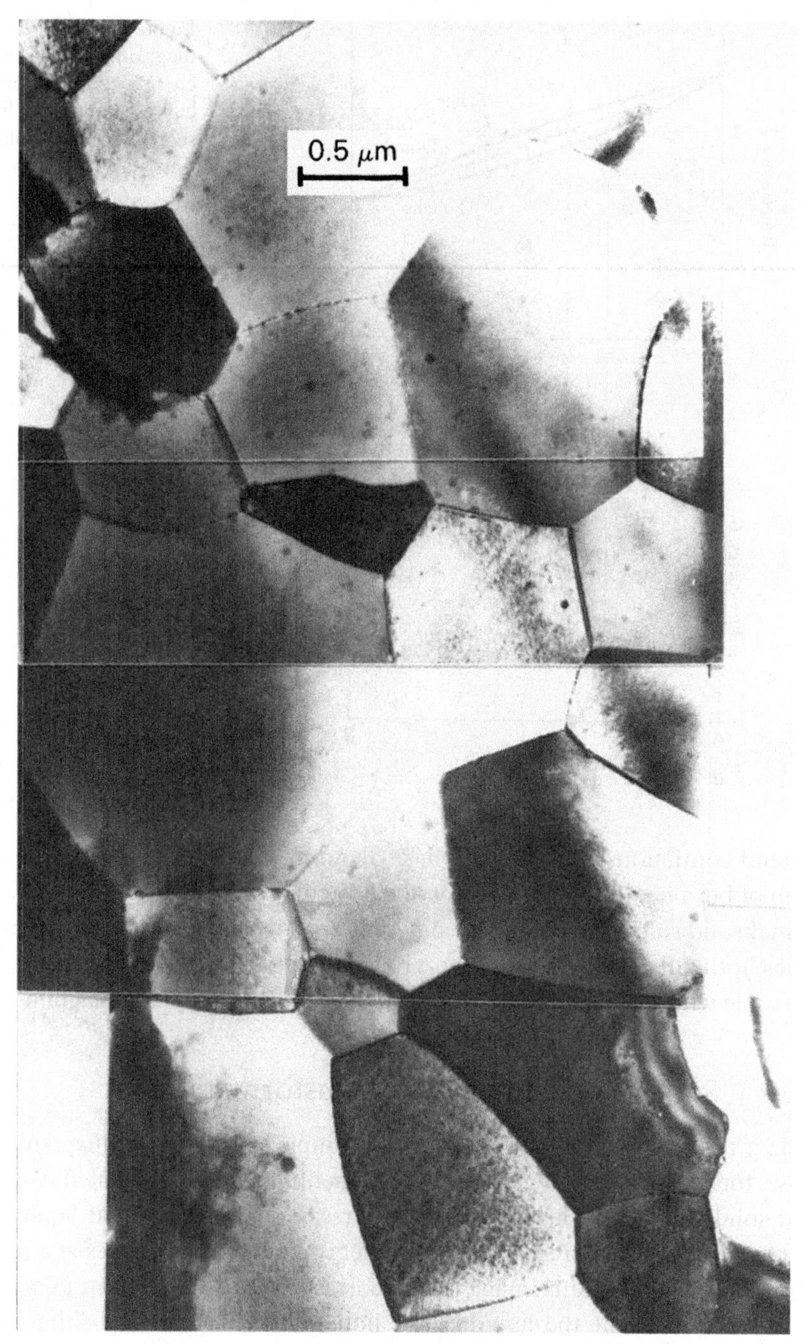


**Figure 3.10** Segregation-free solidification microstructure in melt spun Fe – 35 wt% Ni. (After Hayzelden, Rayment and Canter 1983.)

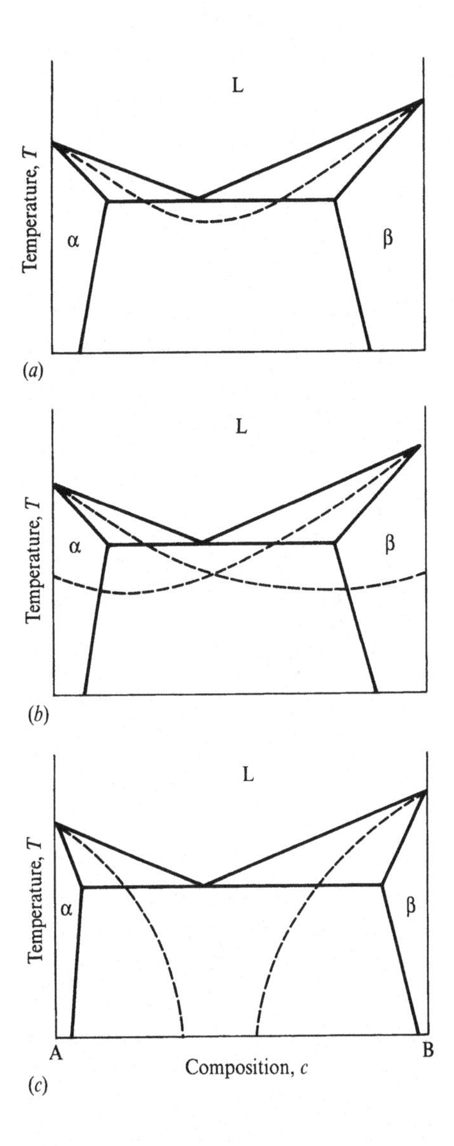


**Figure 3.11** Phase diagrams showing $T_0$ lines in eutectic alloys.

gap appears in the undercooled liquid, with phase separated liquid compositions $c_{L1}$ and $c_{L2}$ as given by the common tangent in fig. 3.12(*b*). The metastable phase diagram consists of a submerged liquid miscibility gap, as shown by the dashed lines in fig. 3.12(*a*). Fig. 3.13 shows a typical liquid phase separated microstructure from the submerged liquid miscibility gap in a melt spun Al–10 wt% Sn alloy.

Fig. 3.14(*a*) shows the phase diagram for a binary A–B alloy system with two eutectics between the terminal $\alpha$ and $\beta$ solid solutions and an equiatomic AB intermetallic compound $\theta$. Fig. 3.14(*b*) shows the corresponding free energy–composition curves at a temperature $T$ well below the melting point. Common tangents between the $\alpha$, $\theta$ and $\beta$ free energy–composition curves

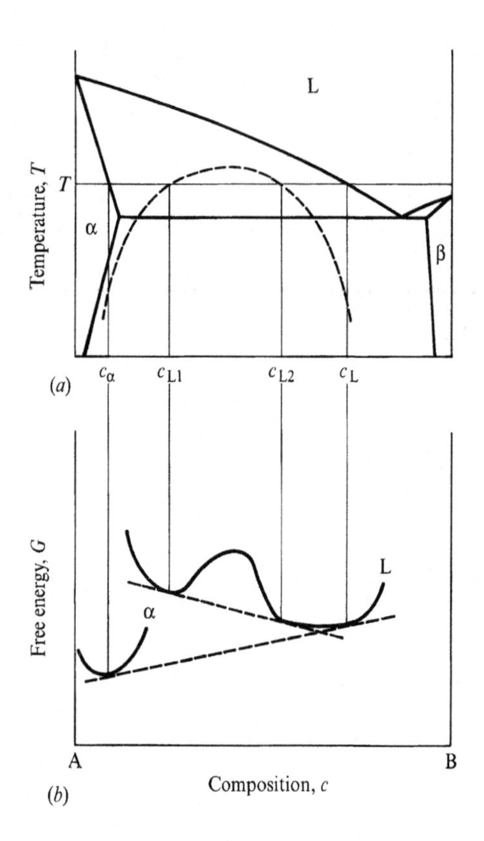


**Figure 3.12** (*a*) Phase diagram and (*b*) free energy–composition curves showing submerged liquid miscibility gap.

define the $\alpha$ and $\beta$ solubility limits $c_\alpha$ and $c_\beta$, as well as the $\theta$ homogeneity range between $c_1$ and $c_2$. If the alloy is constrained to prevent formation of the AB compound $\theta$, metastable equilibrium is then represented by a common tangent between the $\alpha$ and $\beta$ free energy–composition curves, giving extended solubility limits $c'_\alpha$ and $c'_\beta$ as shown in fig. 3.14(*b*). The metastable phase diagram consists of a submerged eutectic between $\alpha$ and $\beta$, as shown by the dashed lines in fig. 3.14(*a*). This kind of suppression of an intermetallic compound is quite common, since complex ordered intermetallic structures are often difficult to nucleate. Fig. 3.15 shows an example of a thin film manufactured by sputtering sequential layers of Al and Zr, followed by heat treatment for 300 min at 370 °C. An $Al_3Zr$ reaction layer nucleates and grows at the Al–Zr interface but other equilibrium intermetallic compounds such as $Al_2Zr$, AlZr and $AlZr_3$ cannot nucleate and are suppressed.

Fig. 3.16(*a*) shows the phase diagram of another alloy system similar to that shown in fig. 3.14(*a*), with two eutectics between the terminal $\alpha$ and $\beta$ solid solutions and an equiatomic AB intermetallic compound $\theta$. Fig. 3.16(*b*) shows the corresponding free energy–composition curves at a temperature $T$ well below the melting point. Common tangents between the $\alpha$, $\theta$ and $\beta$ free energy–composition curves again define the $\alpha$ and $\beta$ solubility limits $c_\alpha$ and $c_\beta$, and the $\theta$ homo-

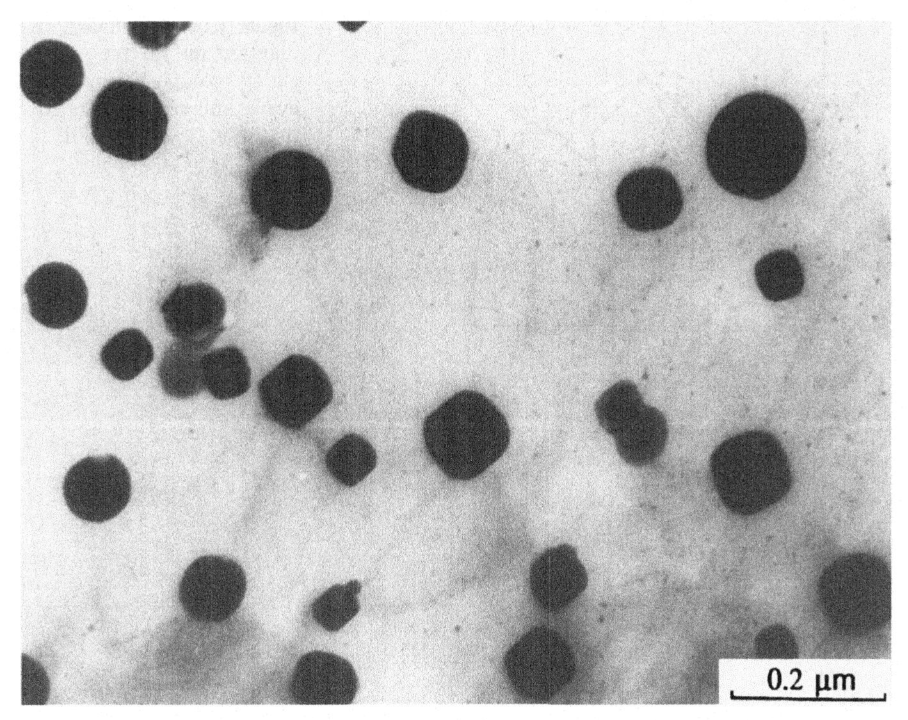


**Figure 3.13** Liquid phase separated microstructure in melt spun Al–10 wt% Sn. (After Kim and Cantor 1991.)

geneity range between $c_1$ and $c_2$. In this case, the AB compound can adopt an alternative metastable crystal structure $\theta'$. If the alloy is constrained to prevent formation of $\theta$, metastable equilibrium is then represented by common tangents between the $\alpha$, $\theta'$ and $\beta$ free energy–composition curves, giving extended solubility limits $c'_\alpha$ and $c'_\beta$ and the $\theta'$ homogeneity range between $c'_1$ and $c'_2$ as shown in fig. 3.16(*b*). The metastable phase diagram consists of two submerged eutectics between $\alpha$, metastable $\theta'$ and $\beta$, as shown by the dashed lines in fig. 3.16(*a*). This kind of submerged intermetallic compound is again quite common, often with a disordered variant of an ordered equilibrium structure. Fig. 3.17 shows an example of a melt spun Al–Ti alloy, containing particles of $L1_2$ ordered metastable $TiAl_3$, replacing the more complex $DO_{22}$ ordered equilibrium $TiAl_3$.

### 3.2.3 Glass transition temperature

An amorphous structure is obtained when an alloy is constrained to prevent crystallisation. This is often achieved by quenching from the liquid state, and the alloy is then called a glass. Plunging $T_0$ lines, deep eutectics and compound suppression such as shown in figs. 3.11(*c*) and 3.14(*a*) are all conducive to forming an amorphous alloy or glass, since they disfavour competing crystallisation reactions. When an alloy is cooled from the liquid state, its density gradu-

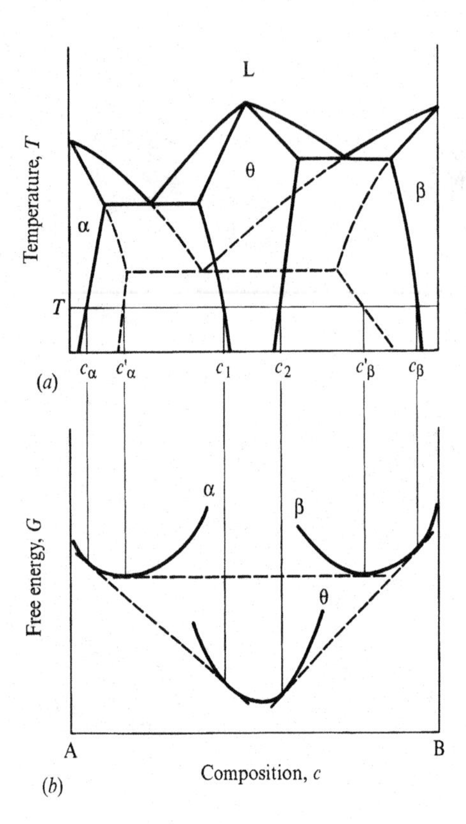


**Figure 3.14** (*a*) Phase diagram and (*b*) free energy–composition curves showing suppressed intermetallic compound.

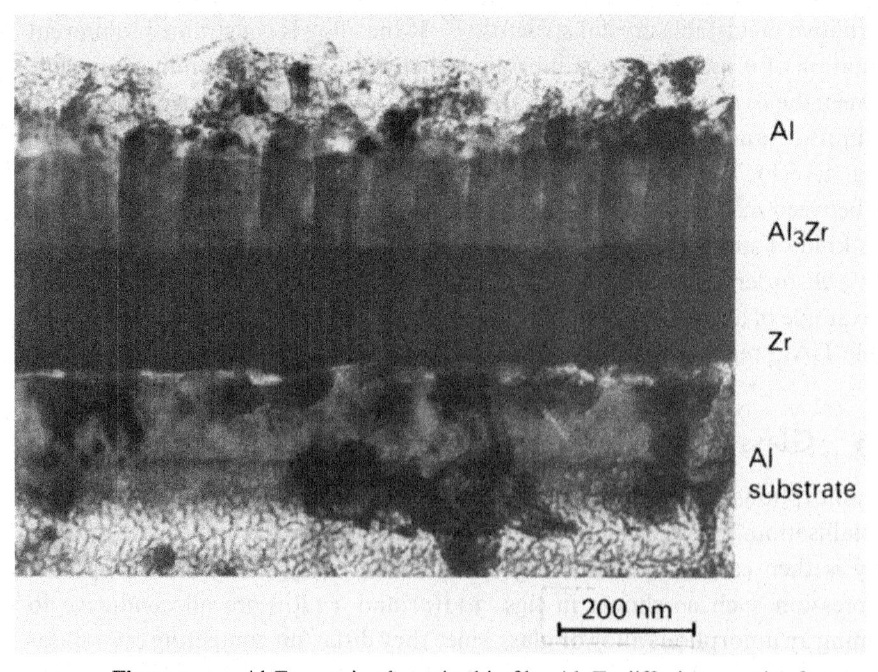


**Figure 3.15** $Al_3Zr$ reaction layer in thin film Al–Zr diffusion couple after annealing for 300 min at 370 °C. (After Mingard and Cantor 1993.)

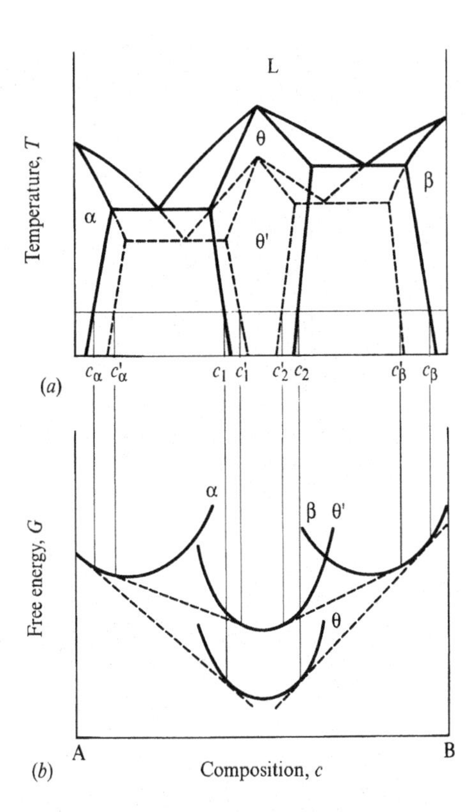


**Figure 3.16** (*a*) Phase diagram and (*b*) free energy–composition curves showing metastable intermetallic compound.

ally falls and its viscosity gradually rises, making macroscopic flow progressively more difficult. The glass transition temperature $T_G$ is the point at which the alloy finally solidifies with a frozen-in amorphous atomic structure. The glass transition temperature is usually defined operationally as the temperature at which the liquid viscosity reaches $10^{11}$ poise. However, its significance is shown more clearly by considering the behaviour of thermodynamic properties such as density and specific heat rather than kinetic properties such as viscosity.

Figs. 3.18(*a*) and (*b*) show schematic variations of the molar free energy $G$ with temperature $T$ and pressure $P$ for a glass forming alloy in the vicinity of the melting point $T_M$. The crystalline solid has lower free energy than the liquid, i.e. $G_S < G_L$ at low temperatures and high pressures. In fig. 3.18(*a*), $G_S$ and $G_L$ are equal at the melting point, and the driving force for solidification $\Delta G_V = G_L - G_S$ increases progressively with increasing undercooling $\Delta T$:

$$\Delta G_V = \frac{L\Delta T}{T_M} + \int_{T_M}^{T} \Delta C \mathrm{d}T - T\int_{T_M}^{T} \frac{\Delta C}{T}\mathrm{d}T \simeq \frac{L\Delta T}{T_M} \tag{3.16}$$

where $\Delta C = C_L - C_S$, and $C_L$ and $C_S$ are the liquid and solid specific heats respectively. The molar volume $V$, density $\rho$ and entropy $S$ are first derivatives of the free energy with pressure and temperature, and therefore give the slopes of the free energy curves in figs. 3.18(*a*) and (*b*):

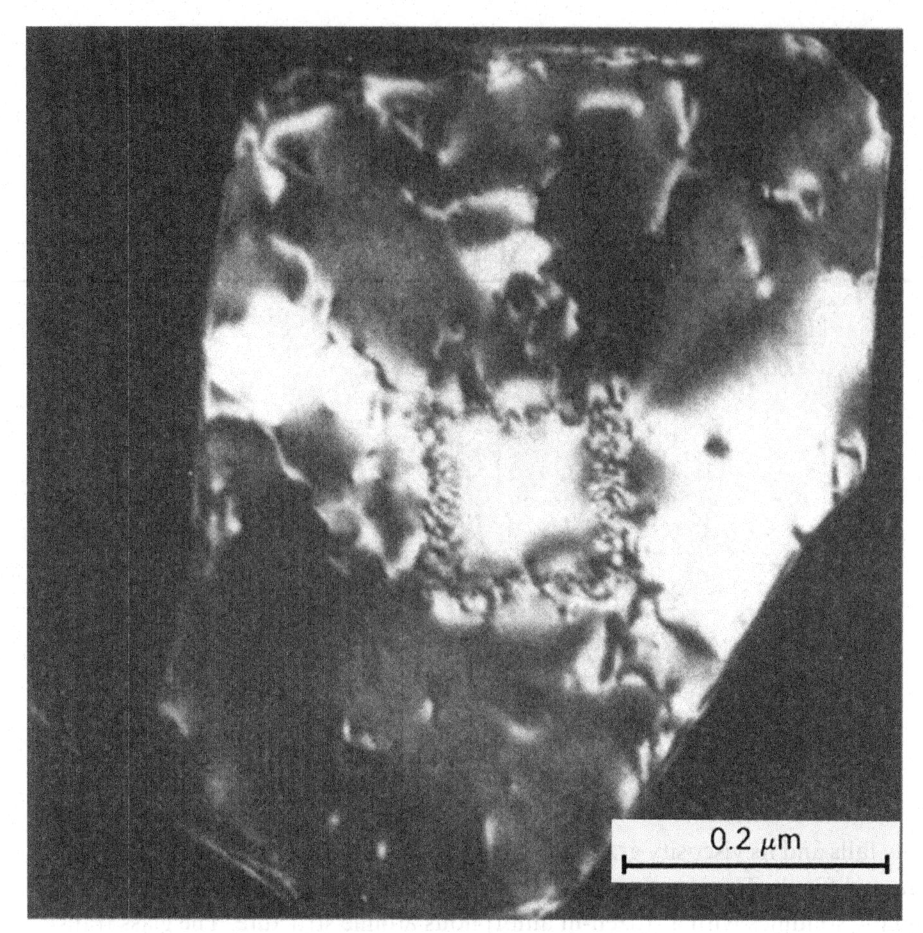


**Figure 3.17** Metastable $L1_2$-$TiAl_3$ particle at the centre of an fcc Al grain in melt spun Al–5 wt% Ti–1 wt% B. (After Kim, Cantor, Griffith and Jolly, 1992.)

$$V=\frac{1}{\rho}=\left(\frac{\partial G}{\partial P}\right)_T \tag{3.17}$$

$$S=-\left(\frac{\partial G}{\partial T}\right)_P \tag{3.18}$$

Similarly, the bulk modulus $K$, specific heat $C$ and expansion coefficient $\alpha_V$ are second derivatives of the free energy:

$$K=\left(\frac{\partial V}{\partial P}\right)_T=\left(\frac{\partial^2 G}{\partial P^2}\right)_T \tag{3.19}$$

$$C=T\left(\frac{\partial S}{\partial T}\right)_P=-T\left(\frac{\partial^2 G}{\partial T^2}\right)_P \tag{3.20}$$

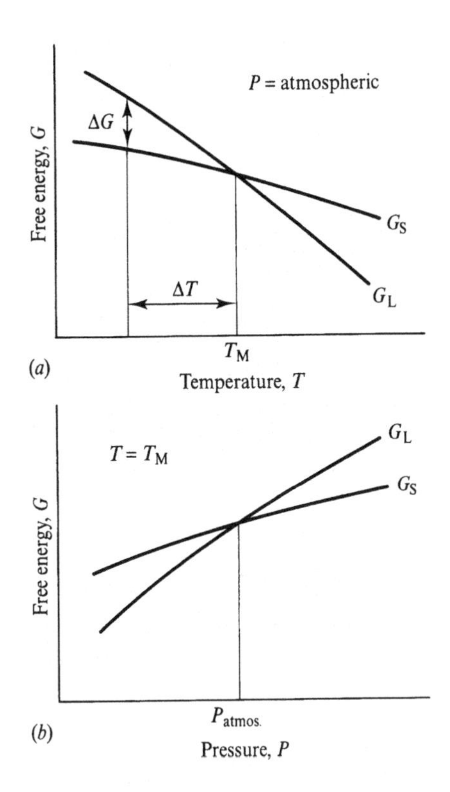


**Figure 3.18** Free energy $G$ versus ($a$) temperature $T$ and ($b$) presure $P$.

$$\alpha_V = \frac{1}{V}\left(\frac{\partial V}{\partial T}\right)_P = \frac{1}{V}\left(\frac{\partial^2 G}{\partial P \partial T}\right) \tag{3.21}$$

Figs. 3.19($a$)–($d$) show schematic temperature variations of the molar volume $V$, entropy $S$, expansion coefficient $\alpha_V$ and specific heat $C$ for a glass forming alloy.

Crystalline alloys usually expand by a few per cent when they melt, that is, $\Delta V = V_L - V_S = 2$–$4\%$ at the melting point $T_M$, as shown in fig. 3.19($a$). The expansion coefficient of a crystal arises from anharmonicity in the interatomic potentials. The expansion coefficient of the corresponding liquid is greater, as shown in fig. 3.19($c$), because of the changes in liquid structure which also take place. The liquid density therefore rises more rapidly than the solid, that is, $\Delta V$ decreases with increasing undercooling below the melting point. At the glass transition temperature $T_G$, structural changes become impossible and the amorphous liquid structure is frozen in to form a glass. The expansion coefficient of the glass is again determined by anharmonicity in the interatomic potentials, and is therefore similar to the crystalline solid. The residual density difference between the crystal and the glass is typically $\sim 1\%$ as shown in fig. 3.19($a$). The disordered structure of a liquid has an entropy considerably greater than the corresponding crystalline solid, with $\Delta S = S_L - S_S$ typically 8–10 J mol$^{-1}$ K$^{-1}$ at the melting point $T_M$, as shown in fig. 3.19($b$). The specific heat of the liquid is also

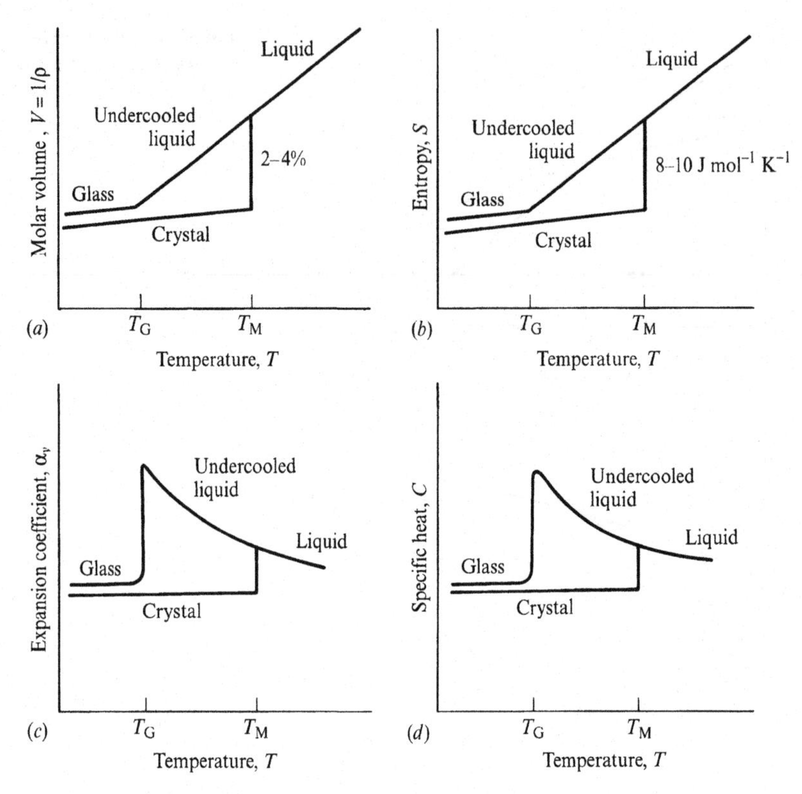


**Figure 3.19** (*a*) Molar volume $V$ and density $\rho = 1/V$, (*b*) entropy $S$, (*c*) expansion coefficient $\alpha_v$ and (*d*) specific heat $C$ versus temperature $T$.

greater than the solid, as shown in fig. 3.19(*d*), with translational rather than vibrational modes of storing thermal energy. The entropy of the liquid therefore falls more rapidly than that of the solid, that is, $\Delta S$ decreases with increasing undercooling below the melting point. At the glass transition temperature $T_G$, the amorphous liquid structure is frozen in to form a glass. The specific heat of the glass is again determined by vibrational modes, and is therefore similar to the crystalline solid. The residual entropy difference between crystal and glass is very small, as shown in fig. 3.19(*b*).

Freezing an amorphous structure at the glass transition temperature is a kinetic rather than a thermodynamic transition. Decreasing the cooling rate gives more time for atomic motion, allowing the liquid to maintain its equilibrium structure to lower temperatures during cooling. The glass is then frozen in at a lower glass transition temperature, with a denser, lower entropy structure, as shown in fig. 3.20. The differences in volume and entropy between the liquid and the crystal $\Delta V$ and $\Delta S$ decrease progressively as the temperature falls, and approach zero at a temperature called the ideal glass transition temperature $T_G^0$. Kauzmann (1948) was the first to point out that an amorphous structure with higher density and

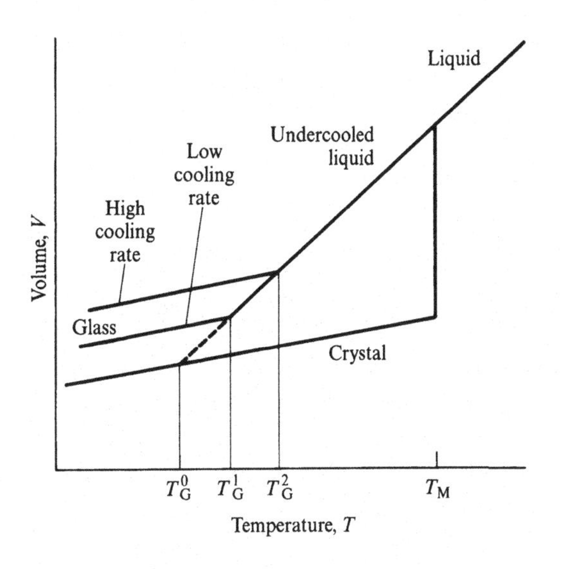


**Figure 3.20** Variation of glass transition temperature $T_G$ with cooling rate.

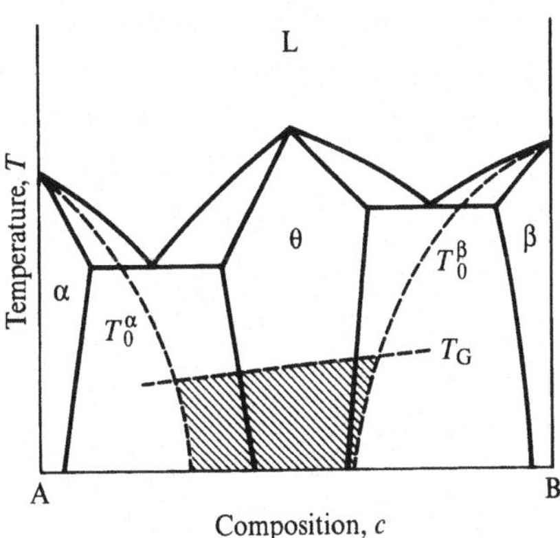


**Figure 3.21** Phase diagram of a glass forming alloy.

lower entropy than the corresponding crystal seems impossible, so $T_G^0$ is a natural lower limit to the range of possible glass transition temperatures, corresponding to an infinitely slow cooling rate. In practice, measured glass transition temperatures vary by ~10–20 K over the range of accessible cooling rates, and are always $>T_G^0$ because of the need to cool fast enough to avoid crystallisation.

Fig. 3.21 shows the phase diagram for a glass forming binary A–B alloy system with two eutectics between the terminal $\alpha$ and $\beta$ solid solutions and an equiatomic AB intermetallic compound $\theta$, again similar to the alloy system shown in fig. 3.14(*a*). If the alloy is constrained to prevent segregation and also to prevent formation of the AB compound $\theta$, the metastable phase diagram consists of plunging $\alpha$ and $\beta$ $T_0$ lines, as shown by the dashed lines in fig. 3.21. The glass transition temperature $T_G$ usually varies only weakly with alloy composition, as also shown by the dashed lines in fig. 3.21. Glass formation is unlikely below the $\alpha$ and $\beta$ $T_0$ lines, where segregation-free solidification is thermodynamically possible and kinetically favourable. The possible region of glass formation is therefore below $T_G$ and between the $\alpha$ and $\beta$ $T_0$ lines, as shown shaded in fig. 3.21 (Saunders and Miodownik 1986, 1988).

## 3.3 Metastable crystalline phases

### 3.3.1 Microcrystalline and nanocrystalline alloys

The grain size and structure in rapidly solidified crystalline alloys have been discussed by Cantor, Kim, Bewlay and Gillen (1991) and Greer (1991). In general, grain size is determined by a balance between the nucleation and growth rates during manufacture. Rapidly solidified alloys manufactured by processes such as splat quenching and melt spinning often exhibit a microstructure of through-thickness columnar grains, as shown in fig. 3.10. The liquid undercools until solidification is nucleated on the cold substrate surface. The nucleated grains grow laterally to cover the substrate surface, and then a stable set of columnar grains continues to solidify through the thickness of the alloy layer. The average grain diameter $d=vt_S$, where $v$ is the solid–liquid interface velocity at the nucleation temperature and $t_S$ is the local solidification time, that is, the time taken to cover the substrate surface (notice that on average a grain grows for $\sim\frac{1}{2}t_S$). The number of grains per unit area on the substrate surface is $\frac{1}{2}Jt_S$ where $J$ is the nucleation rate (notice that on average only $\sim\frac{1}{2}$ of the substrate surface area is available for nucleation). The average cross-sectional area of a columnar grain can therefore be expressed independently in terms of $d$, $v$ and $J$:

$$\pi\left(\frac{d}{2}\right)^2=\pi\left(\frac{vt_S}{2}\right)^2=\frac{2}{Jt_S} \tag{3.22}$$

Eliminating $t_S$, the grain size $d$ is given by:

$$d^3=\frac{8v}{\pi J} \tag{3.23}$$

The growth rate $v$ and nucleation rate $J$ both depend upon the nucleation temperature:

$$v=a\nu\exp\left(-\frac{Q}{kT}\right)\left[1-\exp\left(-\frac{\Delta G_V}{kT}\right)\right]\approx\frac{D}{a}(1-\exp(-A\Delta T)) \tag{3.24}$$

$$J=n\nu\exp\left(-\frac{Q}{kT}\right)\exp\left[-\frac{16\pi\sigma^3 f(\theta)}{3\Delta G_V^2 kT}\right]\approx\frac{nD}{a^2}\exp\left(-\frac{B}{\Delta T^2}\right) \tag{3.25}$$

where $a$ is the interatomic spacing, $\nu$ is the atomic vibration frequency, $Q$ is the activation energy for an atom to transfer across the solid–liquid interface,

$\Delta G_V = L\Delta T/T_M$ is the driving force for solidification as given in fig. 3.18(*a*) and eq. (3.16), $D$ is the diffusion coefficient in the liquid, $n$ is the nucleation site density, $\sigma$ is the solid–liquid surface energy, $f(\theta)=\frac{1}{4}(2-3\cos\theta+\cos^3\theta)$, $\theta$ is the contact angle for a solid nucleus on the substrate surface, and $A$ and $B$ are approximately constant. Combining eqs. (3.23)–(3.25) gives the grain size as a function of nucleation undercooling:

$$d^3 = \frac{8a[1-\exp(-A\Delta T)]}{\pi n \exp(-B/\Delta T^2)} \tag{3.26}$$

Fig. 3.22 shows the variations of growth rate $v$, nucleation rate $J$ and grain size $d \propto (v/J)^{1/3}$ with nucleation undercooling $\Delta T$. Close to the melting point $T_M$, $\exp(-A\Delta T) \approx 1 - A\Delta T$ so that $v \approx DA\Delta T/a$ and the solid–liquid interface velocity increases linearly with increasing driving force and undercooling. Close to the glass transition temperature $T_G$, however, $\exp(-A\Delta T) \approx 0$ so that $v \approx D/a$, and the solid–liquid interface velocity reaches a maximum, and then falls exponentially as the thermal energy becomes insufficient to sustain atomic attachment at the interface. Unless the substrate is an exceptionally good catalyst, nucleation cannot take place close to $T_M$, where the driving force is too low to overcome the solid–liquid surface energy barrier, that is, $\exp(-B/\Delta T^2) \approx 0$ and $J \approx 0$. Below a critical nucleation onset temperature $T_N$, however, the nucleation rate rises rapidly with increasing driving force and undercooling. At lower temperatures close to $T_G$, $\exp(-B/\Delta T^2) \approx 1$ so that $J \approx nD/a^2$ and the nucleation rate also reaches a maximum and then falls exponentially as the thermal energy again becomes insufficient to sustain atomic attachment to the growing nucleus. Small undercoolings are effective at driving solid–liquid interface motion but not at

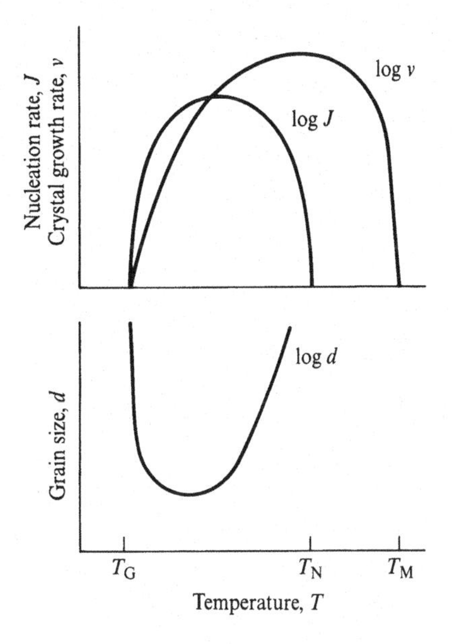


**Figure 3.22** Crystal growth rate $v$, nucleation rate $J$ and grain size $d$ versus undercooling.

stimulating nucleation, so the growth rate maximum is at a higher temperature than the nucleation rate maximum.

Grains cannot nucleate above $T_N$ and cannot grow below $T_G$. From eqs. (3.23) and (3.26), the grain size $d$ depends upon $v/J$ and therefore decreases rapidly, reaches a minimum and then increases rapidly again with increasing undercooling between $T_N$ and $T_G$ as shown in fig. 3.22. Fine scale microcrystalline and nanocrystalline grain structures can be obtained therefore either by rapid quenching to a high undercooling during solidification of a liquid alloy, or by rapid heating to a high temperature during crystallisation of an amorphous alloy. Metastable alloys manufactured by condensation methods such as evaporation and sputtering also often exhibit a nanocrystalline structure of through-thickness columnar grains. Solidification is again nucleated on the cold substrate surface, followed by lateral growth to cover the substrate surface, and then columnar solidification through the thickness of the alloy film. The grain size $d$ is again given by eq. (3.23), with $J$ and $v$ representing surface nucleation and growth rates at the substrate temperature. Fig. 3.23 shows an example of ~5 nm sized fcc Ag particles embedded in an amorphous Si matrix in a sputtered Si–51 wt% Ag thin film.

### 3.3.2 Segregation and supersaturated solid solutions

The internal grain structure in a rapidly solidified alloy is determined by the solidification growth mechanism, which in turn depends sensitively on the solute partitioning which takes place at the moving solid–liquid interface. Growth mechanisms and kinetics have been reviewed by Boettinger and Perepezko (1993), Kurz and Gilgien (1994) and Jones (1991, 1994), and the resulting rapidly solidified microstructures have been reviewed by Suryanarayana, Froes and Rowe (1991) and Lavernia, Ayers and Srivatsan (1992). Different solidification conditions can lead to complete suppression of partitioning, a dendritic solid–liquid interface with interdendritic microsegregation, or a planar solid–liquid interface with macrosegregation between the initial and final solidified regions.

Fig. 3.24(*a*) shows the A rich end of the phase diagram of a binary A–B alloy system, which solidifies as a single crystalline phase α. Consider an alloy of composition $c_0$ undergoing steady state planar or cellular/dendritic solidification. The alloy has liquidus and solidus temperatures $T_L$ and $T_S$, freezing range $\Delta T_0 = T_L - T_S$, liquidus slope $m$ and equilibrium partition coefficient $k = c_S/c_L$, which in fig. 3.24(*a*) is independent of temperature. Fig. 3.24(*b*) shows a composition profile through the solidifying interface, with solid and liquid interface compositions $c_S^*$ and $c_L^*$ respectively, an interface temperature $T$ and a positive thermal gradient $G$, that is, latent heat generated by the solidifying interface is extracted through the freshly formed solid. Fig. 3.24(*c*) shows the relationship

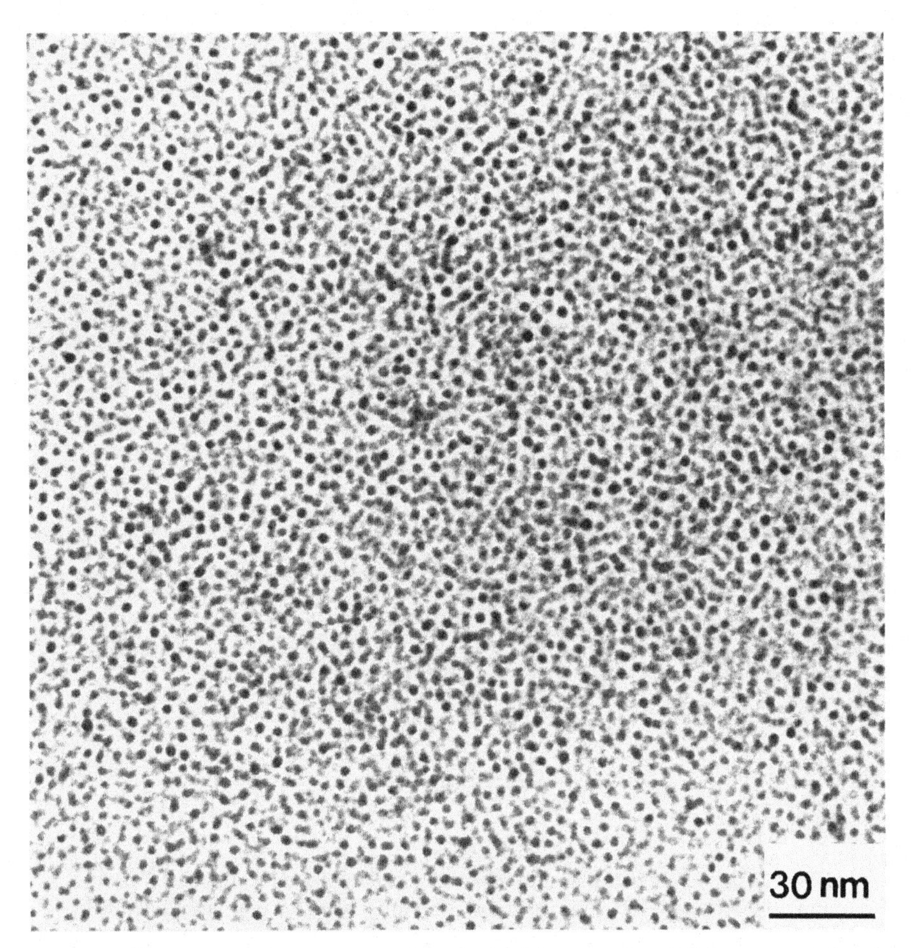


**Figure 3.23** Ag nanocrystals embedded in amorphous Si in a sputtered Si – 51 wt% Ag thin film. (After Lee, Chang, Dobson and Cantor 1994.)

between the solid–liquid interface velocity $v$ and interface temperature $T$. The driving force for solidification $\Delta G_V$ is divided into three components: capillarity energy to maintain interface curvature $\Delta G_C$, diffusional energy to drive atomic motion to and from the interface $\Delta G_D$, and kinetic energy to drive atom attachment processes at the interface $\Delta G_K$:

$$\Delta G_V = \Delta G_C + \Delta G_D + \Delta G_K \tag{3.27}$$

The undercooling at the interface $\Delta T$ is correspondingly divided into three components $\Delta T_C$, $\Delta T_D$ and $\Delta T_K$ (Boettinger and Coriell 1986, Kurz, Giovanola and Trivedi 1986):

$$\Delta T = \Delta T_C + \Delta T_D + \Delta T_K = \frac{2\sigma T_L}{rL} + m_v(c_L^* - c_0) + \frac{v}{\mu} \tag{3.28}$$

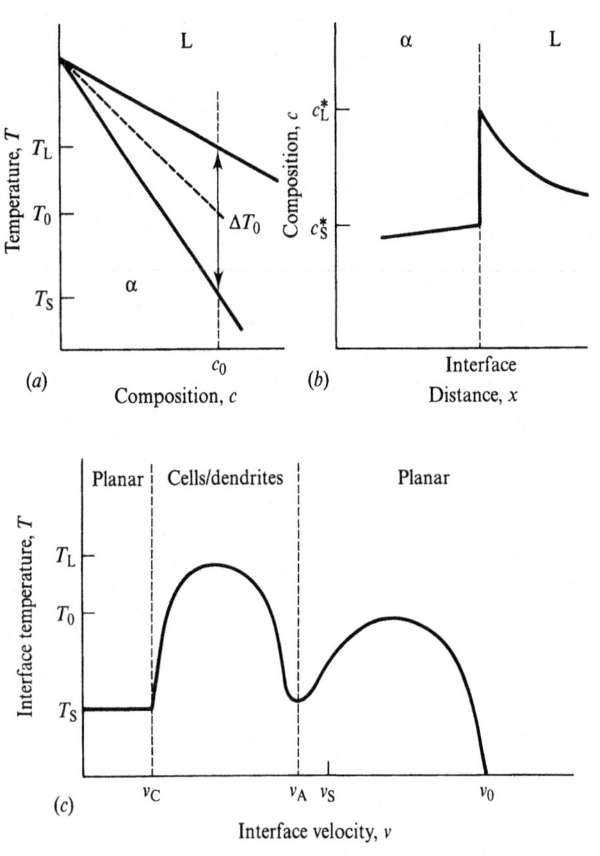


**Figure 3.24** (*a*) Phase diagram, (*b*) solid–liquid interface composition profile and (*c*) interface velocity versus interface temperature for planar and dendritic single phase solidification.

where $\sigma$ is the solid–liquid surface energy, $r$ is the radius of curvature at the cell or dendrite tip, $L$ is the latent heat of fusion, $m_v = m\{1-k_v[1-\ln(k_v/k)]\}/(1-k)$ is the kinetic liquidus slope, $k_v=(k+va/D)/(1+va/D)$ is the kinetic partition coefficient, $a$ is the interatomic spacing, $D$ is the diffusion coefficient in the liquid, and $\mu$ is the interface mobility.

As the interface velocity increases in eq. (3.28) and fig. 3.24(*c*), the solidification mechanism undergoes a series of transitions, becoming dominated in turn by thermal conditions, solute diffusion, interface curvature and finally interface attachment kinetics. Planar solidification is stable at low interface velocities, below the constitutional undercooling limit $v<v_C$. Curvature and kinetic undercoolings can be ignored in eq. (3.28) so that the interface liquid composition is $c_L^* = c_0/k$, the interface temperature is the alloy solidus temperature $T=T_S$, and the undercooling equals the freezing range $\Delta T = \Delta T_0 = m(c_L^* - c_0) = mc_0(1-k)/k$. Solute partitioning at the solidifying interface is balanced by diffusion into the bulk of the liquid, pushing a solute rich layer ahead of the interface. This layer solidifies last, leaving a normal macrosegregation profile in the final solidified alloy. The solid–liquid

interface remains planar when the thermal gradient $G$ is steep enough to prevent local solidification into the solute rich layer. The size of the undercooled region ahead of the interface $\Delta T_0/G$ must be smaller than the size of the solute rich layer $D/v$, giving the constitutional undercooling limit $v_C$ (Trivedi 1994):

$$v_C = \frac{GD}{\Delta T_0} \tag{3.29}$$

At higher interface velocities, $v > v_C$, a planar solid–liquid interface becomes unstable, leading to cellular and dendritic solidification. Dendrites grow into the solute rich layer, and the dendrite tip temperature rises rapidly towards the alloy liquidus, $T = T_L$ and $\Delta T = 0$. Solute partitioning at the solidifying interface takes place between the dendrites rather than ahead of them, leaving an intercellular or interdendritic microsegregation profile in the final solidified alloy. With increasing interface velocity, the dendrite tips sharpen and the curvature undercooling increases in eq. (3.28), so the dendrite tip temperature reaches a maximum and then falls back towards $T = T_S$ and $\Delta T = \Delta T_0$, as shown in fig. 3.24(*c*). The solid–liquid interface again becomes planar when there is too much surface energy associated with the curvature at the dendrite tips. The size of the solute rich layer $D/v$ must be smaller than the radius of curvature at the dendrite tip $k\sigma T_L/L\Delta T_0$, that is, the interface velocity must be above the absolute stability velocity $v_A$ (Trivedi 1994):

$$v_A = \frac{L\Delta T_0 D}{k\sigma T_L} \tag{3.30}$$

A solute rich layer is again pushed ahead of the interface, leaving a normal macrosegregation profile in the final solidified alloy. With increasing interface velocity, however, the interface begins to move as fast as the rate at which solute atoms can diffuse ahead of the interface. The size of the solute rich layer $D/v$ approaches the atomic size $a$, that is, the interface velocity approaches the solute trapping velocity $v_S$ (Trivedi 1994):

$$v_S = \frac{D}{a} \tag{3.31}$$

Partitioning and segregation are then fully suppressed, the solute rich layer cannot build up, and solute atoms are trapped in place by the solidifying interface, with equal solid and liquid interface compositions, $c^*_S = c^*_L = c_0$ at the $T_0$ temperature. At higher interface velocities $v > v_S$, the kinetic undercooling continues to increase in eq. (3.28), and the interface temperature again reaches a maximum and then gradually falls to zero as the interface velocity approaches its final limiting value of the sound velocity $v_0$, as shown in fig. 3.24(*c*).

Solidification takes place without microsegregation when the interface velocity is either below $v_C$ in near-equilibrium single crystal growth, or above $v_A$ or $v_S$ in non-equilibrium rapid solidification. The absolute stability velocity $v_A$ decreases with decreasing freezing range $\Delta T_0$, so $v_A < v_S$ when the freezing range is narrow, that is, in dilute alloys and in alloys with a large partition coefficient $k \rightarrow 1$. On a cooling curve such as in fig. 3.6, segregation-free recalescence is then followed by cellular or dendritic segregation, when the interface velocity falls below $v_A$ at relatively low undercooling on the steady state growth plateau. Fig. 3.25 shows an example of the transition from segregation-free to cellular solidification at the end of recalescence in a melt spun Ni–11 wt% Al alloy. In concentrated alloys with a small partition coefficient, the freezing range is wide and $v_A > v_S$. Dendritic segregation now sets in when the interface velocity falls below $v_S$ at greater undercooling during recalescence and before steady state growth.

The scale of intercellular and interdendritic segregation profiles as measured by the cell or dendrite arm spacing $\lambda_D$ decreases with increasing cooling rate $dT/dt$ and solid–liquid interface velocity $v$. Dendrite arms develop by Ostwald ripening while solidification takes place, so $\lambda_D$ shows cube law coarsening kinetics with the local solidification time $t_S \approx \Delta T_0/(dT/dt)$, so that, (Jones 1982):

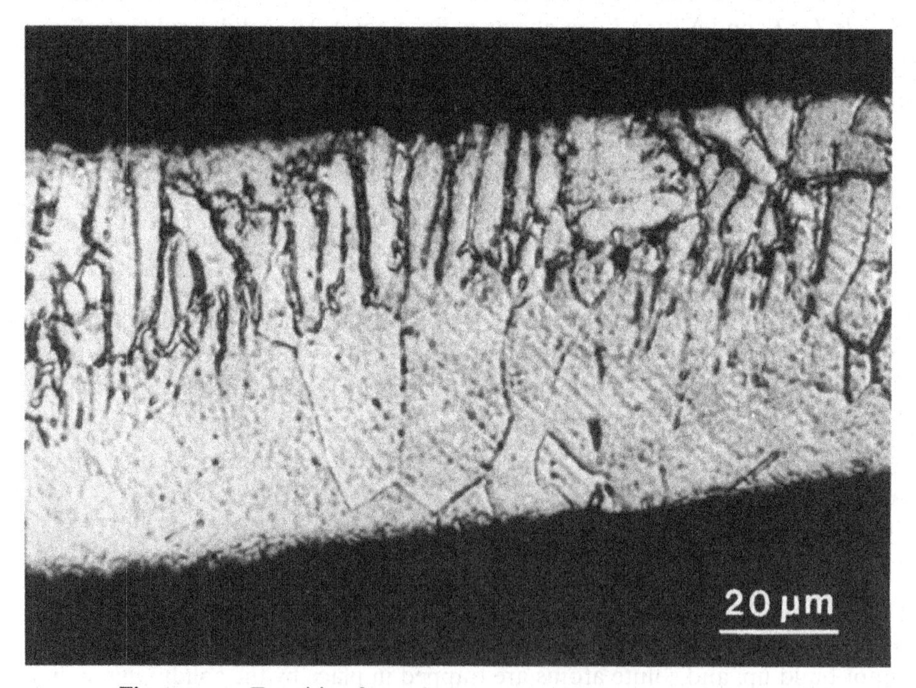


**Figure 3.25** Transition from planar segregation-free to segregated cellular solidification at the end of recalescence in a melt spun Ni – 5 wt% Al alloy. (After Gillen and Cantor 1985.)

$$\lambda_D^3 \frac{dT}{dt} \simeq \text{constant} \tag{3.32}$$

Fig. 3.26 shows typical results for the variation of dendrite arm spacing with cooling rate in Al–Cu and Al–Si alloys, with a slope of $-\frac{1}{3}$ for log $\lambda_D$ versus log($dT/dt$). In concentrated alloys, single phase solidification of $\alpha$ is often terminated by an equilibrium eutectic reaction, $L \rightarrow \alpha + \beta$, with a corresponding maximum equilibrium solubility, as shown on phase diagrams such as in figs. 3.11(*a*)–(*c*). If primary $\alpha$ solid is not an effective catalyst for nucleating $\beta$, segregation-free solidification conditions lead to a supersaturated $\alpha$ solid solution, following the extended solidus and $T_0$ line, as shown in figs. 3.14(*a*) and 3.11(*a*)–(*c*) respectively. Table 3.2 gives typical values of supersaturated solubility extensions in rapidly solidified Al and Ti based alloys.

Coupled two-phase eutectic growth takes place when $\alpha$ is able to act as a catalyst for nucleating $\beta$, usually in highly concentrated alloys close to the eutectic

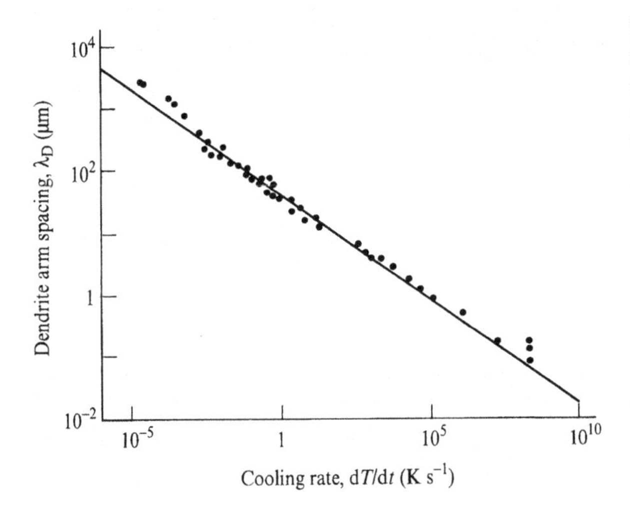


**Figure 3.26** Dendrite arm spacing $\lambda_D$ versus cooling rate $dT/dt$ in Al – 4–5 wt% Cu and Al – 7–11 wt% Si alloys. (After Jones 1982.)

**Table 3.2** *Solubility extensions in rapidly solidified Al and Ti alloys. (After Lavernia et al. 1992 and Suryanarayana et al. 1991.)*

| Alloy | Equilibrium solubility (at%) | Extended solubility (at%) |
|---|---|---|
| Al–Cu | 2.5 | 17 |
| Al–Fe | 0.02 | 4 |
| Al–Mg | 19 | 37 |
| Al–Si | 1.6 | 10 |
| Ti–B | 0.5 | 3 |
| Ti–C | 0.6 | 10 |
| Ti–Fe | 22 | 35 |

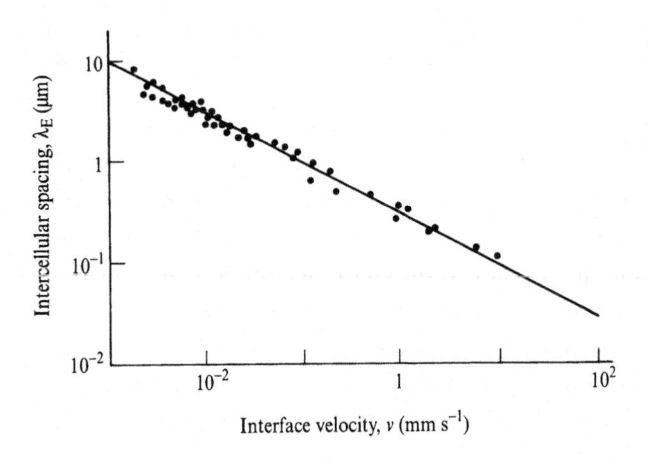


**Figure 3.27** Eutectic interlamellar spacing $\lambda_E$ versus solid–liquid interface velocity $v$ in Al – 33 wt% Cu. (After Jones 1982.)

composition, and in interdendritic segregated regions also at high concentration. The scale of the eutectic microstructure as measured by the interlamellar or interfibre spacing $\lambda_E$ again decreases with increasing cooling rate $dT/dt$ and solid–liquid interface velocity $v$. Coupled eutectic growth kinetics represent a balance between small $\lambda_E$ which speeds up lateral diffusion supplying the growing $\alpha$ and $\beta$ phases at the solidifying interface, and large $\lambda_E$ which reduces solid–liquid interface curvature and $\alpha$–$\beta$ surface energy. This leads to (Jackson and Hunt 1966; Kurz and Trivedi 1991):

$$\lambda_E^2 v \simeq \text{constant} \tag{3.33}$$

Fig. 3.27 shows typical results for the variation of eutectic interlamellar spacing with solid–liquid interface velocity for the fcc $\alpha$ Al(Cu)/tetragonal $\theta$ $Al_2Cu$ eutectic in Al–33 wt% Cu, with a slope of $-\frac{1}{2}$ for log $\lambda_E$ versus log $v$. Banded structures are often observed, consisting of either alternating $\alpha$ and $\beta$, or alternating segregated and segregation-free $\alpha$. Fig. 3.28 shows an example of alternating fcc $\alpha$ Al(Cu) and tetragonal $\theta$ $Al_2Cu$ phases in a banded eutectic nodule in melt spun Al–38 wt% Cu. Overall rapid solidification behaviour can be represented on a microstructure selection map of solid–liquid interface velocity versus alloy concentration, showing regions of formation of segregation-free, cellular/dendritic, lamellar eutectic, banded and other microstructures. An example is shown for the Ag–Cu alloy system in Fig. 3.29.

## 3.3.3 Metastable crystalline and quasicrystalline compounds

A wide range of metastable crystalline compounds can be manufactured, and table 3.3 gives some examples. Metastable crystalline compounds often have simple atomic structures, but the overall pattern of metastable compound forma-

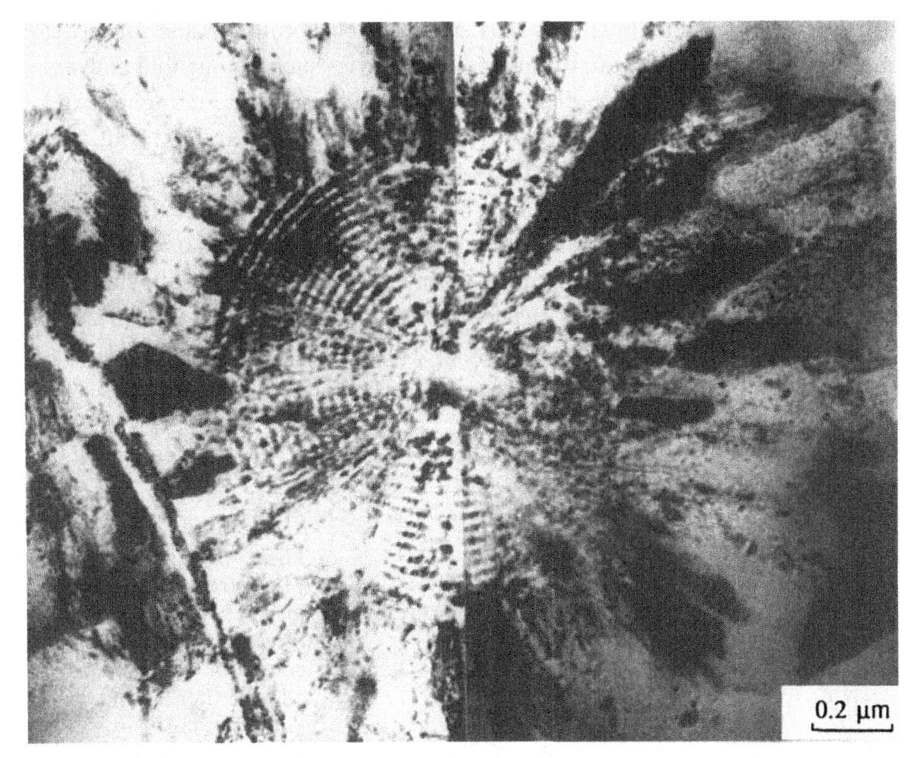


**Figure 3.28** Alternating fcc $\alpha$ Al(Cu), and tetragonal $\theta$ $Al_2Cu$ phases in a banded eutectic nodule in melt spun Al – 38 wt% Cu. (After Kim, Cantor and Kim 1990.)

tion is quite varied and has not been fully classified. Metastable phases such as hcp $Ag_3Ge$ are missing Hume–Rothery compounds, with an electron/atom ratio of 1.5, equivalent to the corresponding equilibrium hcp $\zeta$ compounds found in similar alloy systems such as $Cu_3Ge$ and $Ag_3Sn$ (Duwez *et al.* 1960; Klement *et al.* 1960). Simple fcc and bcc compounds are formed in alloys such as Cd–Sn and Ti–Fe respectively (Giessen 1969; Dong, Chattopadhyay and Kuo 1987). Complex ordered equilibrium compounds are often replaced with simpler structures, either with the same or different stoichiometry, as in $L1_2$ $Al_3Ti$ and orthorhombic $Al_6Fe$ respectively (Kim *et al* 1992; Jones 1982). Other metastable compounds such as bct $(Fe,Mo)_3B_2$ are found as intermediate precipitates during heat-treatment of supersaturated solid solutions or during crystallisation of amorphous alloys (Kim, Clay, Small and Cantor 1991).

In 1984, Schechtman, Blech, Gratias and Cahn discovered that melt spun Al–Mn alloys contain an $Al_6Mn$ metastable compound, which exhibits five-fold symmetric diffraction patterns. This was the first observation of a quasicrystalline compound, that is, a material with non-crystallographic symmetry such as a five-fold rotation axis. The atomic structure of a crystalline material shows rotational symmetry combined with translational periodicity. The crystal can be built up by stacking identical unit cells, each of which contains the full point group symmetry of the crystal. Rotational symmetries such as four-fold and six-fold axes can be repeated periodically, and form the basis of the tetragonal and hexagonal crystal structures respectively. However, five-fold and ten-fold axes are excluded since they are inconsistent with translational periodicity. A variety of different quasicrystalline compounds have been found, and typical examples are given in table 3.4. Most quasicrystalline compounds are based on Al and Ti, with stoichiometric additions of transition metal and rare earth alloying elements. At first, only metastable quasicrystalline compounds were discovered, such as $Al_6Mn$, $Ti_2Fe$ and $Mg_{32}(Al,Zn,Cu)_{49}$, but a range of stable quasicrystalline compounds have now also been discovered, such as $Al_{65}Cu_{20}Cr_{15}$ and $Al_6CuLi_3$.

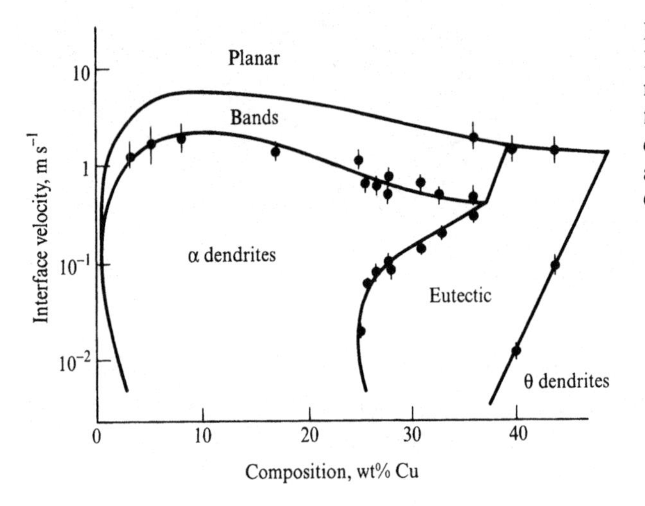


**Figure 3.29** Microstructure selection map of solid–liquid interface velocity $v$ versus alloy concentration $c$ for Ag–Cu alloys. (After Kurz and Gilgien 1994.)

**Table 3.3** *Typical metastable crystalline compounds. (After Jones 1982, Lavernia et al 1992 and Suryanarayana et al. 1991.)*

| Alloy | Structure |
|---|---|
| Ag–Ge, Au–Sb | hcp |
| Au–Sn | complex cubic |
| Ti–Fe, $Al_3Fe$ | bcc |
| $Al_6Fe$ | orthorhombic |
| Cd–Sn, Zn–Ga | fcc |
| Al–Ga, In–Bi, $(Fe,Mo)_3B_2$ | bct |
| $Al_3Zr$, $Al_3Ti$ | $L1_2$ |

Icosahedral compounds are three-dimensional quasicrystals, with full icosahedral point group symmetry m$\overline{35}$. Fig. 3.30 shows a typical icosahedral $Al_6Mn$ particle embedded in an fcc Al matrix in melt spun Al–10 wt% Mn; fig. 3.31 shows an icosahedral stereogram with 6 five-fold, 10 three-fold and 15 two-fold rotation axes; and fig. 3.32 shows a five-fold diffraction pattern from melt spun $Al_6CuLi_3$. Quasicrystalline diffraction patterns such as in fig. 3.32 show non-crystallographic rotational symmetries with non-periodic diffraction spots. Adjacent spot spacings scale by $\tau$ or $\tau^3$ where $\tau=\frac{1}{2}(1+\sqrt{5})\approx1.618$. The icosahedral quasilattice constant is defined as $a=\frac{1}{2}\tau^3 d$, where $d$ is the spacing corresponding to the strongest diffraction spot (Elser 1985). Analysis of diffraction patterns such as in fig. 3.32 indicates that icosahedral compounds can be classified into two types, i-$Al_6Mn$ and i-$Al_6CuLi_3$ with quasilattice constants of $a=0.45$ and 0.52 nm respectively. Decagonal compounds are two-dimensional quasicrystals, with point group symmetry 10/m or 10/mmm. The ten-fold symmetric quasicrystalline planes are stacked periodically along a single ten-fold rotation axis. Fig. 3.33 shows a typical decagonal $Al_4Fe$ particle embedded in an fcc Al matrix in melt spun Al–19 wt% Fe; fig. 3.34 shows a decagonal stereogram with 1 ten-fold and 10-each pseudo five-fold, three-fold and two-fold rotation axes; and fig. 3.35 shows a ten-fold diffraction pattern from melt spun $Al_4Fe$. The decagonal quasilattice constant is defined as the periodic repeat distance along

**Table 3.4** *Quasicrystalline compounds. (After Kelton 1993 and Steurer 1995.)*

| Type | Quasilattice constant (nm) | Metastable compounds | Stable compounds |
|---|---|---|---|
| icosahedral | 0.45 | $Al_6Mn$; $Al_6Fe$; $Al_6Cr$; $Ti_2Fe$; $Ti_2Mn$; $Ti_2Co$ | $Al_{65}Cu_{20}Cr_{15}$; $Al_{65}Cu_{20}Fe_{15}$ |
| icosahedral | 0.52 | $Al_6AuLi_3$; $Al_6CuMg_4$; $Mg_{32}(Al,Zn,Cu)_{49}$ | $Al_6CuLi_3$ |
| decagonal | 0.4 | $Al_4Ni$ | $Al_{65}Cu_{20}Co_{15}$; $Al_{70}Ni_{15}Co_{15}$ |
| decagonal | 0.8 | | $Al_{10}Co_4$ |
| decagonal | 1.2 | $Al_4Mn$; $Al_{65}Cu_{20}Mn_{15}$ | |
| decagonal | 1.6 | $Al_4Fe$; $Al_5Ir$ | $Al_{80}Fe_{10}Pd_{10}$ |
| octagonal | 0.63 | $Mn_4Si$; $Cr_5Ni_3Si_2$ | |
| dodecagonal | 0.45 | $Cr_{70}Ni_{30}$; $Ni_2V_3$ | $Ta_XTe$ |

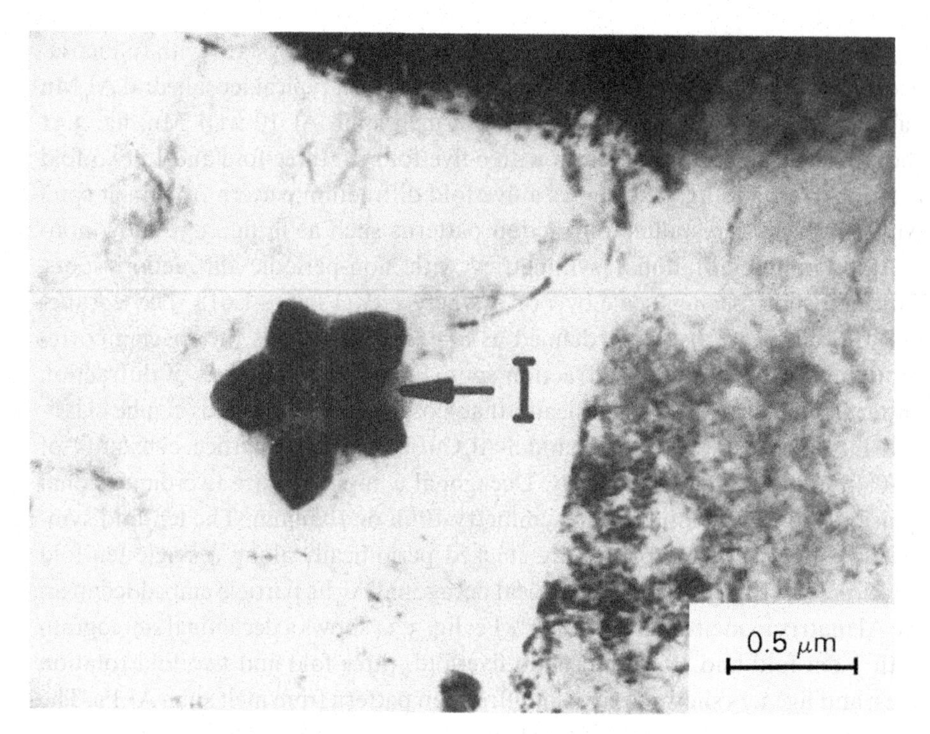


**Figure 3.30** Icosahedral $Al_6Mn$ particle embedded in an fcc Al matrix in melt spun Al – 10 wt% Mn. (After Kim and Cantor 1990.)

the ten-fold axis. Analysis of diffraction patterns such as in fig. 3.35 indicates that all decagonal compounds have similar atomic structures within the two-dimensional quasicrystalline planes, but that different stacking sequences give different quasilattice constants of 0.4, 0.8, 1.2 and 1.6 nm for $Al_4Ni$, $Al_{10}Co_4$, $Al_4Mn$ and $Al_4Fe$ respectively. Octagonal and dodecagonal compounds are also two-dimensional quasicrystals, with eight-fold or twelve-fold symmetric quasicrystalline planes stacked periodically along a single eight-fold or twelve-fold rotation axis respectively.

There has been considerable dispute about the atomic structure of quasicrystalline compounds. Diffraction patterns such as in figs. 3.32 and 3.35 can arise from multiple twinning or from periodic packing of icosahedral molecules (Pauling 1985, 1987; Field and Fraser 1985; Vecchio and Williams 1988; Anantharaman 1988). Fig. 3.36 shows an example of a ten-fold twinned monoclinic $Al_{13}Fe_4$ dendritic particle in melt spun Al–19 wt% Fe. However, high resolution electron microscope images of quasicrystalline compounds show no evidence of twin boundaries or of periodic lattice fringes (Schechtman and Blech 1985; Bursill and Lin 1985; Knowles, Greer, Saxton and Stobbs 1985). Quasicrystalline structures are now usually described as three-dimensional Penrose tilings (Penrose 1974a,b; Mackay 1976, 1981) or as projections from a higher dimension (de Bruijn 1981a,b; Elser 1985, 1986). Fig. 3.37 shows acute and obtuse

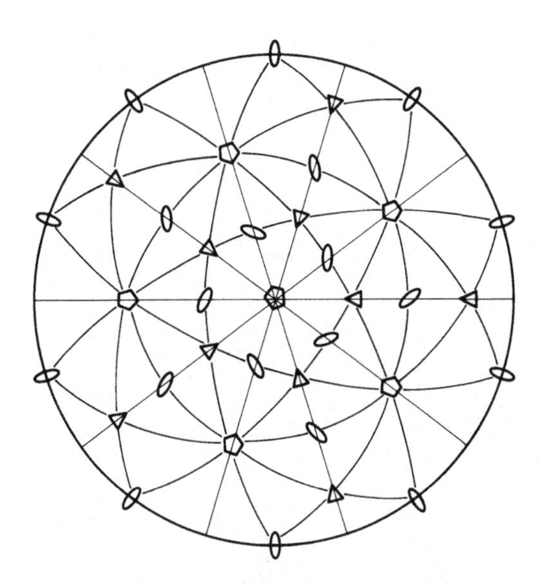

**Figure 3.31** $m\overline{35}$ icosahedral stereogram.

**Figure 3.32** Five-fold diffraction pattern from melt spun $Al_6CuLi_3$. (After Kim, Hutchison and Cantor 1990.)

parallelograms which can be used as tiles to fill a two-dimensional plane. Periodicity is avoided by placing the tiles according to matching rules, that is, by ensuring consistent edge arrows, leading to the quasiperiodic two-dimensional Penrose tiling shown in fig. 3.38. A two-dimensional Penrose tiling has a ten-fold symmetric diffraction pattern, and is a suitable quasilattice for describing the atomic structure of the non-periodic planes in decagonal compounds. Similar three-dimensional Penrose tilings are suitable quasilattices for describing the atomic structure in icosahedral compounds. Fig. 3.39(*a*) shows a two-dimensional square lattice, with the lattice points in a narrow shaded strip between two

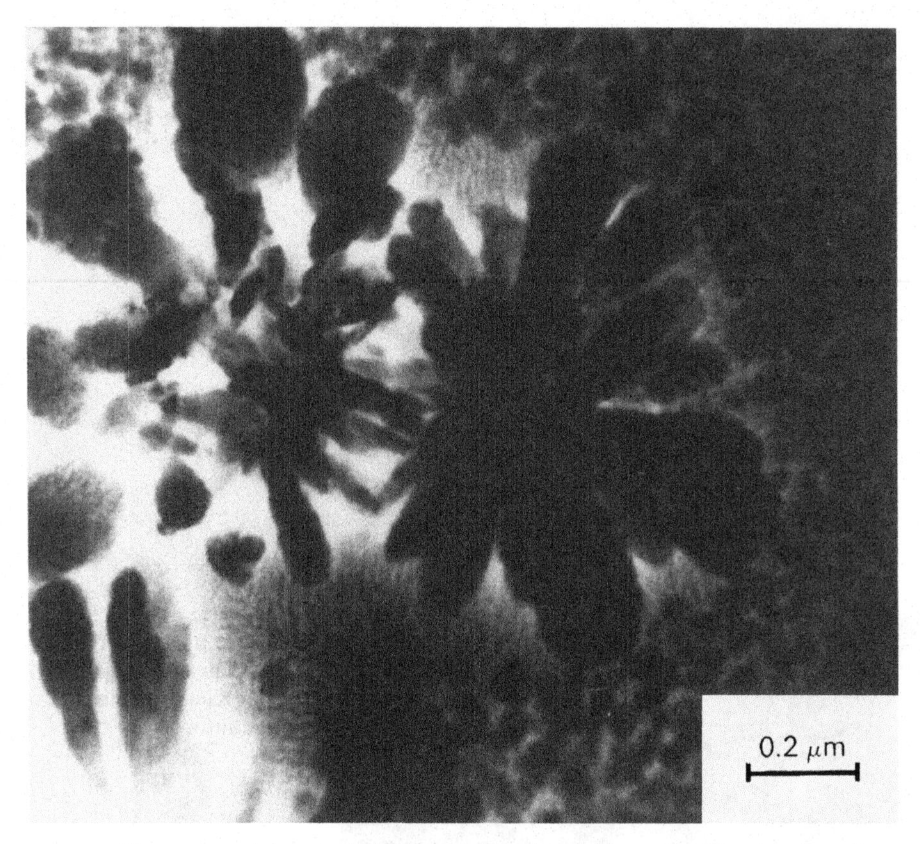


**Figure 3.33** Decagonal $Al_4Fe$ particle embedded in an fcc Al matrix in melt spun Al – 19 wt% Fe (After Kim and Cantor 1994.)

dashed lines projected onto a solid line inclined at the irrational slope of $1/\tau$. The effect of the projection is to tile the solid line with a non-periodic sequence of short and long line segments, in the ratio $1/\tau$ short/long. Fig. 3.39(*b*) shows the same two-dimensional square lattice, with the strip of lattice points now projected onto a solid line inclined at the rational slope of 1/2. This tiles the solid line with a periodic sequence of 1/2 short/long segments. The periodic tiling in fig. 3.39(*b*) is called a 1/2 approximant to the quasi-periodic tiling in fig. 3.39(*a*). Similar projections from five-dimensional and six-dimensional cubic lattices can be used to generate two-dimensional and three-dimensional quasicrystalline Penrose tilings and related approximant periodic lattices. In fact, projection methods provide a general and powerful method for generating crystalline, quasicrystalline and incommensurate lattices within a single elegant framework, as discussed in detail elsewhere (de Bruijn 1981a,b; Elser 1985, 1986; Steinhardt and Ostlund 1987; Bak 1986). Full description of the atomic structure of quasicrystalline compounds requires specification of the molecular motifs which decorate the quasilattices generated by Penrose tiling or projection. These molecular motifs have not yet been fully identified. It also remains unclear how liquids can solidify so rapidly, with complex and highly ordered quasicrystalline structures.

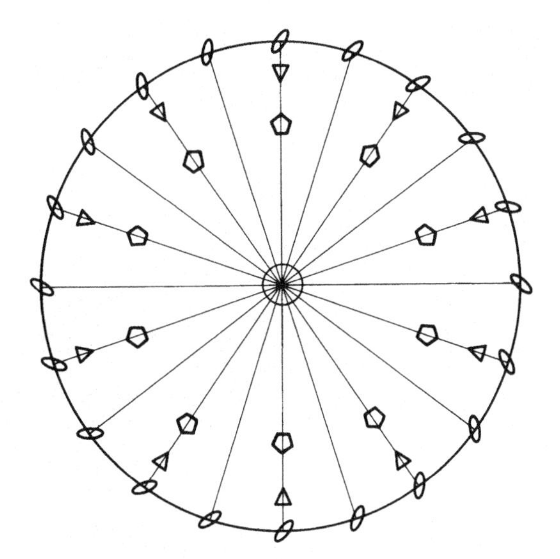

**Figure 3.34** 10/mm decagonal stereogram.

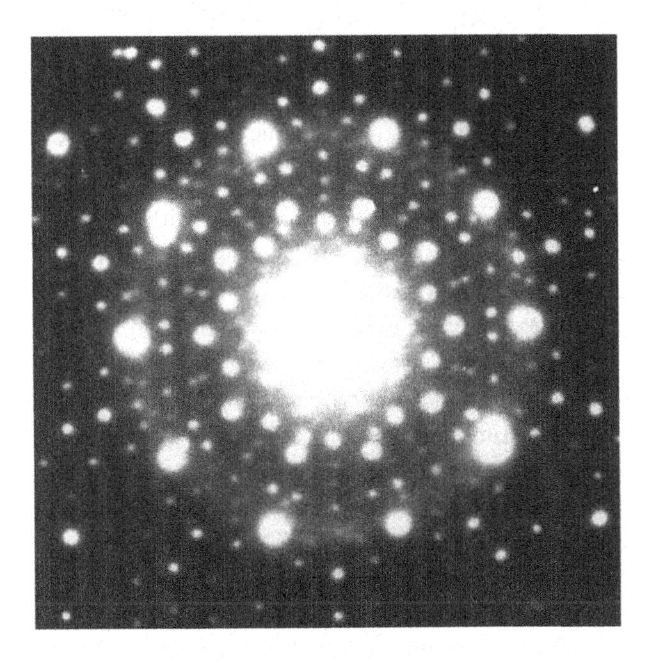

**Figure 3.35** Ten-fold diffraction pattern from melt spun $Al_4Fe$. (After Kim and Cantor 1994.)

## 3.4 Amorphous alloys

### 3.4.1 Formation of amorphous alloys

Hundreds of different amorphous alloys have been manufactured, and some typical examples are given in table 3.5. Despite considerable effort, there is no fully agreed method of classifying the wide range of different amorphous alloy systems. The first amorphous alloys manufactured by rapid solidification and

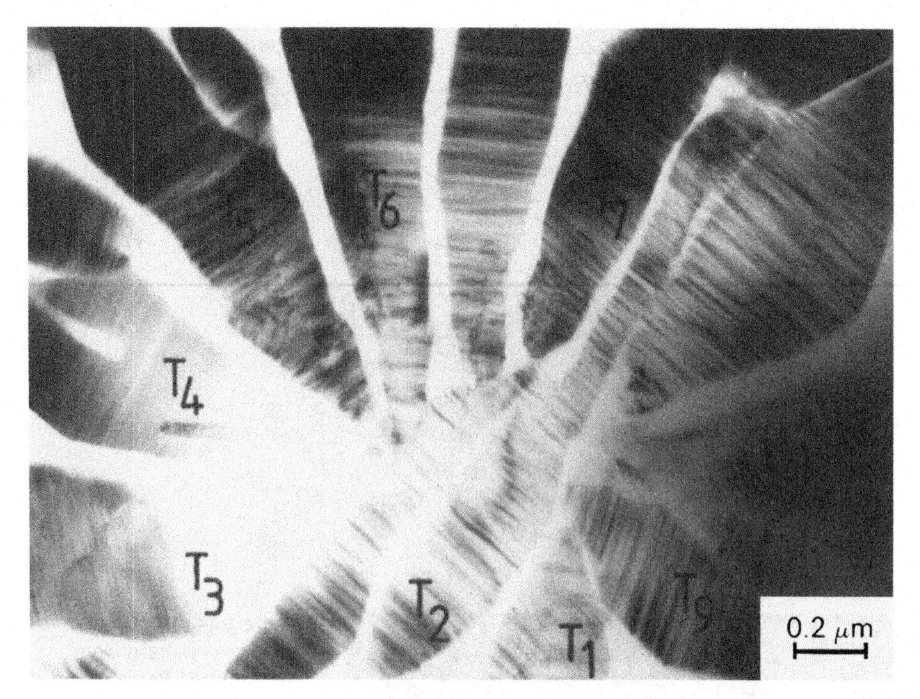


**Figure 3.36** Ten-fold twinned monoclinic $Al_{13}Fe_4$ dendritic particle in melt spun Al – 19 wt% Fe. (After Kim and Cantor 1994.)

electrodeposition were mixtures of one or more transition metal with ~15–25 at% of one or more metalloid element. Metal–metalloid glasses include Au–Si and Ni–P as discovered by Duwez *et al.* (1960) and Brenner *et al* (1950) respectively, model binary alloys such as Pd–Si and Fe–B (Hasegawa and Ray 1978), and the soft ferromagnetic Fe–B–Si alloys (Hagiwara, Inoue and Masumoto 1981). The first amorphous alloys discovered with all metal constituents were mixtures of one or more early transition metal and one or more late transition metal, with the amorphous structure often formed over a wider composition range than metal–metalloid glasses (Ruhl, Giessen, Cohen and Grant 1967). Metal–metal glasses include model binary alloys such as Cu–Zr (Ray, Giessen and Grant 1968) lanthanide and actinide systems such as La–Au (Schwarz and Johnson 1983) and the Gd–Co magnetic alloys used for bubble domain thin film memory devices (Fukamichi, Kikuchi, Masumoto and Matsura 1979). Other important amorphous alloys are based on A group rather than transition metals, including the model binary alloy Mg–Zn (Calka *et al.* 1977) and the Al based alloys such as Al–Y–Ni (Inoue and Masumoto 1991).

A wide variety of different criteria have been proposed to explain which alloys can be manufactured as glasses. These include a large difference in constituent atom sizes (Mader, Nomick and Widmer 1967; Polk 1970), a large negative deviation from ideality in the liquid alloy (Zielinski, Ostalek, Kijek and Matyja 1978; Ramachandrarao 1980), the formation of topologically close-packed equi-

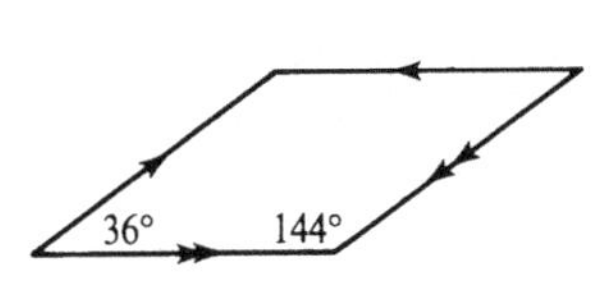

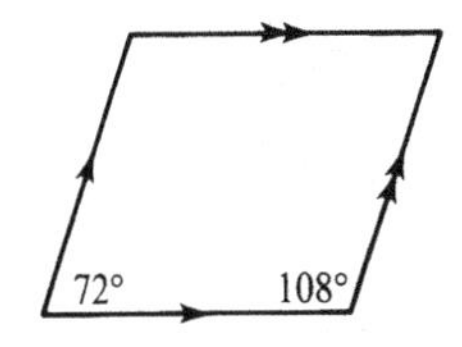


**Figure 3.37** Acute and obtuse Penrose tiles.

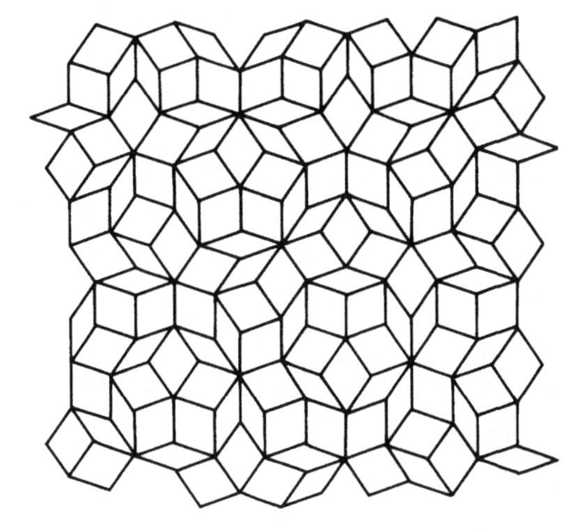

**Figure 3.38** Quasiperiodic two-dimensional Penrose tiling.

librium crystalline compounds (Hafner 1980), a deep eutectic on the equilibrium phase diagram (Jones 1973), a high viscosity in the liquid alloy (Uhlman 1972; Davies 1983) and a correlation between the electronic wave vector at the Fermi energy and the interatomic separation in the liquid (Nagel and Tauc 1977; Hafner 1983). None of these criteria works well across the full range of amorphous alloys, and none has been very successful at predicting new amorphous alloys. Nevertheless, amorphous alloys often show the following: (1) an energetic preference for mixing in the liquid, that is, a negative deviation from ideality, with a negative heat of mixing and a reduced atomic volume; (2) significant atomic size differences which prevent mixing without excessive strain in a simple crystalline solid solution, and which lead to the formation of complex ordered crystalline compounds such as Frank–Kaspar and Laves phases; and therefore (3) limited terminal solid solubilities with plunging $T_0$ lines, and deep eutectics at non-stoichiometric compositions between the crystalline compounds, as shown in fig. 3.21.

Consider the crystallisation process in an undercooled liquid. The number of crystals which nucleate at times between $\tau$ and $\tau + d\tau$ is $J d\tau$ where $J$ is the nucleation rate. At a later time $t$, each of these crystals grows to a size $(4/3)\pi v^3(t-\tau)^3$, where $v$ is the solid–liquid interface velocity, and crystal growth is assumed to be spherical. The crystal fraction $y'$ is obtained by integrating over crystals nucleated at all times $\tau$ between 0 and $t$:

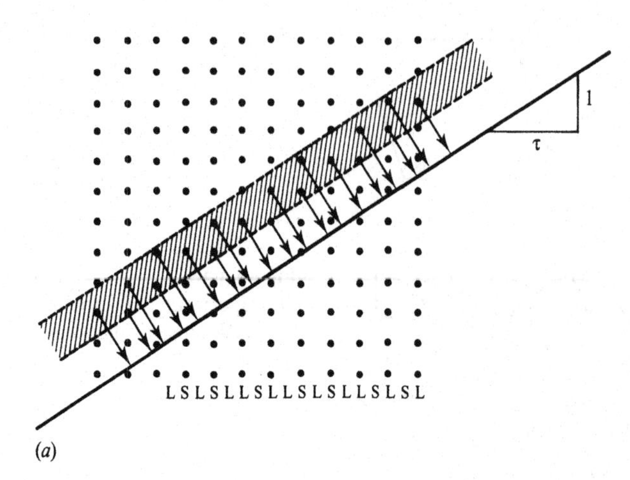


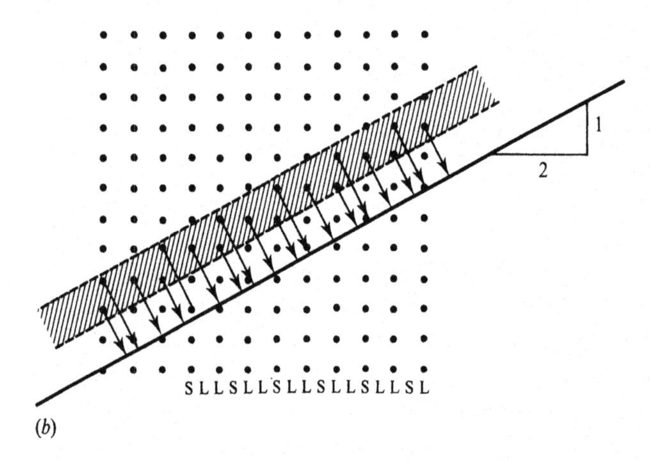


**Figure 3.39** Two-dimensional square lattice, with a strip of lattice points projected onto a line inclined at (*a*) $1/\tau$ and (*b*) 1/2.

**Table 3.5** *Glass forming alloy systems. (After Cahn 1991.)*

| Type | Typical alloy systems |
|---|---|
| metal–metalloid | Au–Si; Ni–P; Pd–Si; Fe–B; Pd–Cu–Si; Fe–Ni–B; Fe–B–Si; Pt–Ni–P; Au–Ge–Si |
| metal–metal | Ni–Nb; Ni–Ti; Ni–Ta; Cu–Zr; Ir–Ta; La–Au; Gd–Co; Gd–Fe; U–V; U–Cr; U–Co |
| A group metal based | Mg–Zn; Ca–Mg; Ca–Al; Be–Zr–Ti; Al–Y–Ni; Al–La–Fe |

$$y'=\int_0^t \frac{4}{3}\pi v^3(t-\tau)^3 J d\tau=\frac{1}{3}\pi v^3 J t^4 \tag{3.34}$$

Eq. (3.34) ignores the reduction in liquid volume as crystallisation proceeds. Only the remaining liquid fraction is available for transformation at any time, so the real crystal fraction $y$ increases more slowly than $y'$ by a factor $1-y$, that is $dy=(1-y)dy'$. Integrating:

$$y=1-\exp(-y')=1-\exp(-\frac{1}{3}\pi v^3 J t^4) \tag{3.35}$$

which is a form of the Johnson–Mehl–Avrami equation, discussed in more detail in §3.4.4. The solidification velocity $v$ and nucleation rate $J$ both depend upon the temperature, as given in eqs. (3.24) and (3.25):

$$v\simeq\frac{D}{a}[1-\exp(-A\Delta T)] \tag{3.36}$$

$$J\simeq\frac{nD}{a^2}\exp\left(-\frac{B}{\Delta T^2}\right) \tag{3.37}$$

where $a$ is the interatomic spacing, $D$ is the diffusion coefficient in the liquid, $n$ is the nucleation site density, $A$ and $B$ are approximately constant and $\Delta T$ is the undercooling.

Combining eqs. (3.55)–(3.37):

$$y=1-\exp\left\{-\frac{\pi n D^4 t^4}{3a^5}[1-\exp(-A\Delta T)]^3\exp\left(-\frac{B}{\Delta T^2}\right)\right\} \tag{3.38}$$

Fig. 3.22 shows the variations of growth rate $v$ and nucleation rate $J$ with undercooling $\Delta T$. Fig. 3.40 shows a corresponding *TTT* diagram for the onset of crystallisation, obtained from eq. (3.38) by plotting the time $t_0$ to form an undetectable fraction of crystals, typically $y\approx 0.1\%$, versus undercooling $\Delta T$. The *TTT* diagram shows characteristic *C* curve behaviour. Close to the melting point $T_M$, there is little driving force for crystallisation, so the crystal nucleation and growth rates $v$ and $J$ are small, and the crystallisation onset time $t_0$ is large. As the temperature falls, the crystallisation onset time reaches a minimum value $t_0^*$ at a temperature $T^*$, and then increases again as the thermal energy becomes insufficient for atomic motion. Close to the glass transition temperature $T_G$, atomic motion is completely stifled and the amorphous structure is frozen in, so that $v$ and $J$ are again small, and the crystallisation onset time $t_0$ becomes large. Eq. (3.38) can be used to calculate $t_0^*$ and $T^*$, and therefore the critical cooling rate $R^*$ to avoid crystallisation and cool below the glass transition temperature $T_G$:

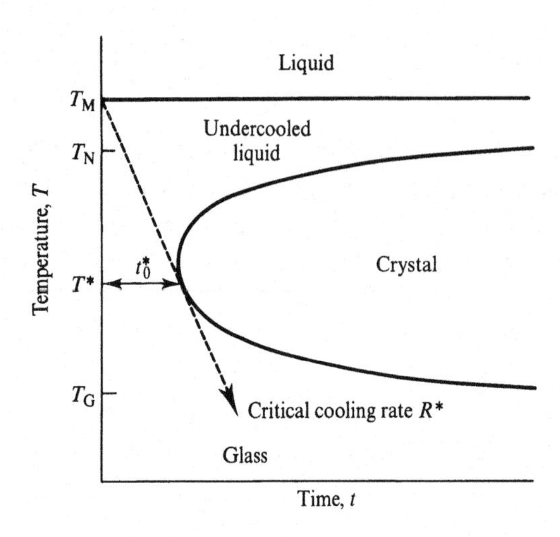


**Figure 3.40** *TTT* diagram for the onset of crystallisation.

$$R^* = \frac{T_M - T^*}{t_0^*} \tag{3.39}$$

as shown in fig. 3.40 (Uhlman 1972). Glass formation is promoted by a low value of critical cooling rate $R^*$, corresponding to a low melting point $T_M$ and a high glass transition temperature $T_G$, as shown in fig. 3.40. Typical values of critical cooling rate and reduced glass transition temperature $\theta_G = T_G/T_M$ are given in table 3.6. Glass formation, that is, a critical cooling rate $R^* < 10^6$ K s$^{-1}$ is associated with a reduced glass transition temperature $\theta_G = T_G/T_M > 0.5$ (Turnbull 1969; Davies 1983).

## 3.4.2 Amorphous alloy structure

The structure of amorphous alloys has been reviewed by Suzuki (1983), Wagner (1983) and Egami (1993). The distribution of atoms in an amorphous alloy is non-periodic and lacking in symmetry. Amorphous alloy structures cannot therefore be described by the powerful deterministic methods of mathematical crystallography. Statistical methods must be employed instead. The most important statistical characteristic of an amorphous structure is the radial distribution function $g(r)$, which is defined as the probability of finding two atoms separated by a distance $r$. Consider a general irregular array of atoms. As shown in fig. 3.41, the radial density function $\rho(r)$ is the average number of atoms per unit volume at a distance $r$ from another atom, and the function $N(r)$ is the average total number of atoms within a distance $r$ from another atom. Averaging here means measuring $\rho$ and $N$ versus $r$ for a given central atom, repeating the measurement for all atoms, and then taking the average. The average number of atoms d$N$ in the shell between $r$ and $r$+d$r$ is given by the radial distribution function multiplied by the thickness of the shell $g$d$r$, and is also given by the radial density function multiplied by the volume of the shell $4\pi r^2 \rho$d$r$:

$$g=4\pi r^2\rho=\frac{dN}{dr} \tag{3.40}$$

$$N=\int_0^r g\,dr=\int_0^r 4\pi r^2\rho\,dr \tag{3.41}$$

Notice that the radial distribution function $g$ is not a true probability, since its integrated value is the total number of atoms $N$ not 1. In a fully random, gas-like structure, the density is the same everywhere, that is, $\rho=\rho_0$, where $\rho_0$ is the average density, and the radial distribution function is then parabolic in $r$, that is, $g_0=4\pi r^2\rho_0$. Otherwise, $\rho$ and $g$ fluctuate about their random values, with peaks and troughs corresponding to the atomic coordination shells and interstitial spaces respectively, as shown in fig. 3.42. From eq. (3.41), the number of atoms in each coordination shell is obtained by integrating the radial distribution function over the corresponding peak, as shown in fig. 3.42. The near neighbour coordination number $z$, for instance is given by:

$$z=\int_0^{r^*} g\,dr \tag{3.42}$$

where $r^*$ is the first minimum in the radial distribution function, as shown in fig. 3.42. Amorphous alloy structures contain more than one kind of atom, and the partial radial distribution functions $g_{ij}$ are defined in a similar way to $g$ as the probability of finding two atoms of types $i$ and $j$ separated by a distance $r$. For a binary A–B amorphous alloy, there are three partial radial distribution functions $g_{AA}$, $g_{AB}$ and $g_{BB}$, describing the AA, AB and BB coordination respectively:

$$g=\sum_{i,j=A,B} n_{ij}g_{ij} \tag{3.43}$$

where $n_{ij}$ is the fractional number of $ij$ pairs ($n_{AA}=c_A^2$, $n_{AB}=2c_Ac_B$ and $n_{BB}=c_B^2$ for atomic concentrations $c_A$ and $c_B=1-c_A$).

The radial distribution functions in an amorphous alloy are important since they can be measured directly by X-ray and neutron scattering experiments. Let the atom positions in an amorphous alloy be defined by the vectors $\mathbf{r}_n$ relative to an arbitrary origin. Consider the scattering from the $n$th atom through an angle

**Table 3.6** *Critical cooling rates to manufacture a glass $R^*$ and reduced glass transition temperatures $\theta_G=T_G/T_M$. (After Davies 1983.)*

| Alloy | $R^*$ (K/s$^{-1}$) | $\theta_G$ |
|---|---|---|
| $Au_{78}Ge_{14}Si_8$ | $10^6$ | 0.47 |
| $Fe_{79}Si_{10}B_{11}$ | $2\times10^5$ | 0.58 |
| $Ni_{62}Nb_{38}$ | $10^3$ | 0.65 |
| $Pd_{40}Ni_{40}P_{20}$ | $10^2$ | 0.65 |

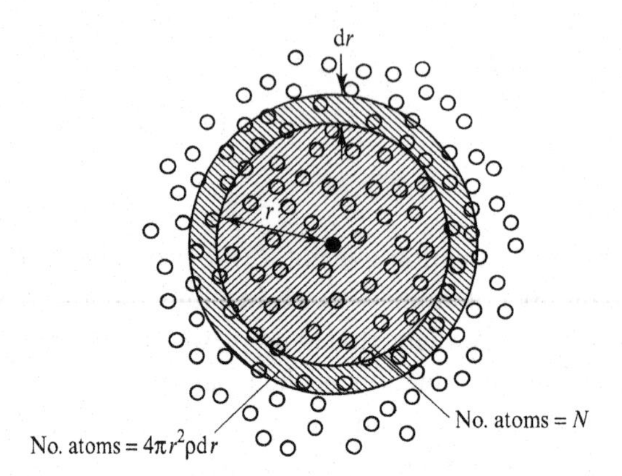


**Figure 3.41** Relation between the radial distribution and radial density functions $g$ and $\rho$.

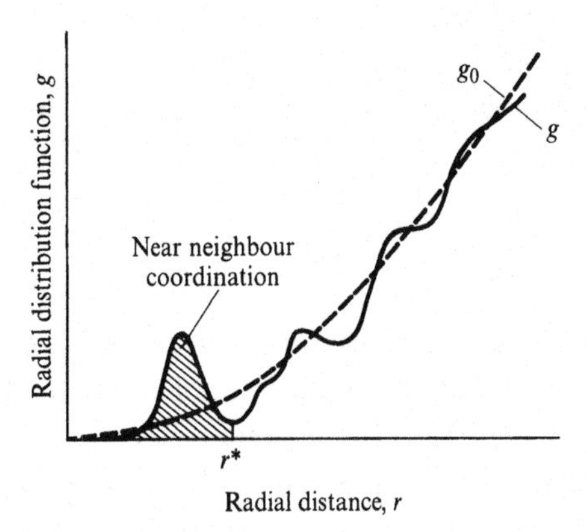


**Figure 3.42** Radial distribution function $g$ in an amorphous alloy.

$2\theta$, with incident and scattered unit wave vectors $\mathbf{k}_0$ and $\mathbf{k}$ respectively, as shown in fig. 3.43. The path difference relative to the origin is $\mathbf{r}_n\cdot(\mathbf{k}-\mathbf{k}_0)$, the phase difference is $(2\pi/\lambda)\mathbf{r}_n\cdot(\mathbf{k}-\mathbf{k}_0)=\mathbf{r}_n\cdot\mathbf{q}$, and the scattered wave is $\varphi_n=f_n\exp(i\mathbf{r}_n\cdot\mathbf{q})$, where $\lambda$ is the wavelength, $\mathbf{q}=(2\pi/\lambda)(\mathbf{k}-\mathbf{k}_0)$ is the scattering vector with modulus $q=(4\pi/\lambda)\sin\theta$, and $f_n$ is the atomic scattering factor. The total scattered intensity $I$ is obtained by summing $\varphi_n$ over all atoms and squaring:

$$I=\sum_{n,m}\phi_n\cdot\phi_m^*=\sum_{n,m}f_nf_m\exp[i(\mathbf{r}_n-\mathbf{r}_m)\cdot\mathbf{q}]$$

$$=\sum_n f_n^2+\sum_{n,m\,n\neq m}f_nf_m\exp(ir_{nm}q) \tag{3.44}$$

where $\phi^*$ is the complex conjugate of $\phi$, and $r_{nm}=|r_n-r_m|$ is the separation between the $n$th and $m$th atoms, There are $N^2$ terms in eq (3.44). The first sum consists of $N$ terms with $n=m$, and corresponds to uncorrelated scattering from the $N$ atoms independently. The second sum consists of $N(N-1)$ terms with $n\neq m$, and corresponds to scattering interference between the $N(N-1)$ atomic pairs. A sum over all $N(N-1)$ atomic pairs in volume $V$ is eqivalent to an integral over the radial distribution function $g$, spherically averaged over all $N$ atoms:

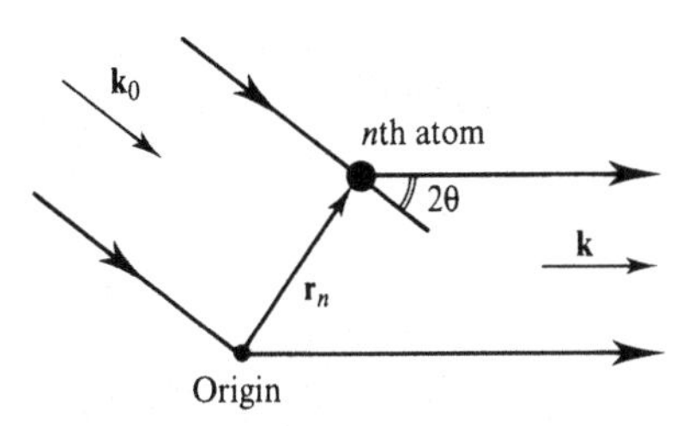


**Figure 3.43** X-ray or neutron scattering in an amorphous alloy.

$$\sum_{n,m\ n\neq m} = \frac{N}{V}\int_V g\,dV \equiv \rho_0 \int_{r=0}^{\infty}\int_{\theta=0}^{\pi}\int_{\psi=0}^{2\pi} g(r)\,dr\sin\theta\,d\theta\,d\psi \tag{3.45}$$

using spherical coordinates $(r,\theta,\psi)$. Combining eqs. (3.43)–(3.45) for $N$ atoms of a binary A–B amorphous alloy:

$$I = N\sum_{i=A,B} c_i f_i^2 + N\rho_0 \sum_{i,j=A,B} n_{ij} f_i f_j s_{ij} \tag{3.46}$$

(Notice that the sums in eq. (3.44) are over all pairs of atoms $n,m$, whereas the sums in eq. (3.46) are over all pairs of atom types $i,j$=A,B, so there are only five terms in eq. (3.46) compared with $N^2$ terms in eq (3.44).) The three terms in the second sum of eq. (3.46) contain structural information about the atomic arrangement, and represent scattering from the AA, AB and BB atom pairs, as given by the partial interference functions $s_{ij}$:

$$s_{ij} = 4\pi\int_0^{\infty} g_{ij}\left[\frac{sin(rq)}{rq}\right]\mathrm{d}r \tag{3.47}$$

Eq. (3.47) shows that the partial interference functions $s_{ij}$ and the partial radial distribution functions $g_{ij}$ are related by Fourier transformation.

Three independent scattering experiments are needed to determine the three partial structure functions $s_{ij}$ and then by inverse Fourier transformation the three corresponding partial radial distribution functions $g_{ij}$. This can be achieved by using different alloy compositions, different isotopes or different kinds of radiation. Fig. 3.44 shows partial radial distribution functions $g_{ij}$ measured by neutron scattering from Pd–Si amorphous alloys containing 15, 20 and 22 at% Si. The coordination numbers obtained by integrating over the near neighbour peaks in $g_{ij}$ show that there are no Si–Si near neighbours. This kind of chemical short range order is not unexpected, with a similar absence of Si–Si near neighbours in the equilibrium crystalline compound $Pd_3Si$ (Gaskell 1978). Metal–metalloid glasses generally show strong short range order, but metal–metal glasses are often disordered, as shown in table 3.7.

When they were first discovered, amorphous alloys were thought to have a structure corresponding to a polycrystalline aggregate with the grain size reduced to ~1 nm, but this does not agree with the interference and radial distribution functions measured by X-ray and neutron scattering. Bernal (1960) manufactured a dense random packing of ball bearings, which was later found to have a radial distribution function similar to the X-ray and neutron scattering results from amorphous alloys (Cargill 1969, 1970). Very good agreement with X-ray and neutron scattering has since been obtained by building complex computer models, which incorporate different alloy components and short range order by packing molecular fragments from a suitable crystalline compound, and which then relax the structure under the action of suitable interatomic potentials (Finney 1983; Hafner 1985). Chemical ordering, molecular structure beyond the first coordination shell and relaxation of the computer models are also essential to fit the characteristically high densities of amorphous alloys

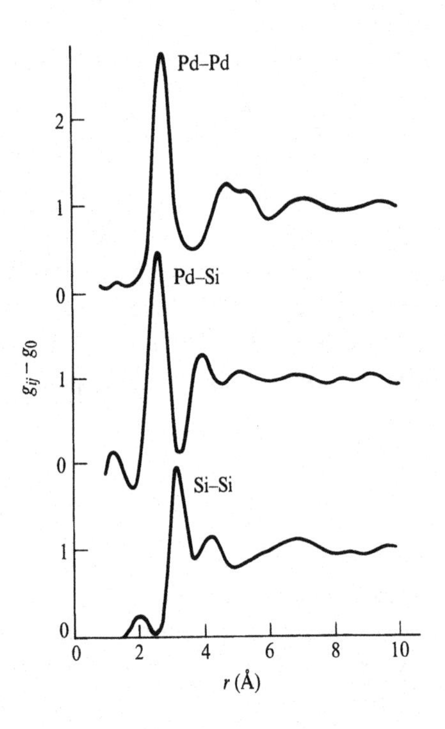


**Figure 3.44** Partial radial distribution functions $g_{ij}$ for amorphous Pd–Si. (After Suzuki 1983.)

**Table 3.7** *Amorphous alloy coordination. (After Spaepen and Cargill 1985.)*

| Alloy A–B | Measured A neighbours of B (%) | A neighbours of B if there were no ordering (%) |
|---|---|---|
| $Fe_{80}B_{20}$ | 100 | 80 |
| $Ni_{81}B_{19}$ | 100 | 81 |
| $Ni_{64}B_{36}$ | 88.8 | 64 |
| $Cu_{57}Zr_{43}$ | 56.2 | 57 |
| $Zr_{65}Ni_{35}$ | 70.1 | 65 |
| $Zr_{57}Be_{43}$ | 70 | 57 |

(Gaskell 1982). Additional evidence of localised chemical short range order and a fragmentary molecular structure is provided by high resolution electron microscopy, as shown in fig. 3.45 for a melt spun Al–Y–Ni glass. Realistic computer models of amorphous structures are, however, difficult to build, rely on prior knowledge of the amorphous alloy to guide the model building and must be rebuilt for each amorphous alloy.

The Voronoi construction is used to describe the geometry of an amorphous structure (Finney and Wallace 1981; Ahmadzadeh and Cantor 1981), as shown in two dimensions in fig. 3.46. Each atom occupies its own Voronoi cell, defined as the space closer to that atom than any other, and shown as dashed lines in fig. 3.46. The collection of all Voronoi cells forms a set of disjoint space-filling polyhedra (polygons in two dimensions). The Voronoi cells correspond to the Wigner–Seitz cells in a crystal, except that the Voronoi cells are all different. The volumes of the Voronoi cells (areas in two dimensions) give the distribution of volumes occupied

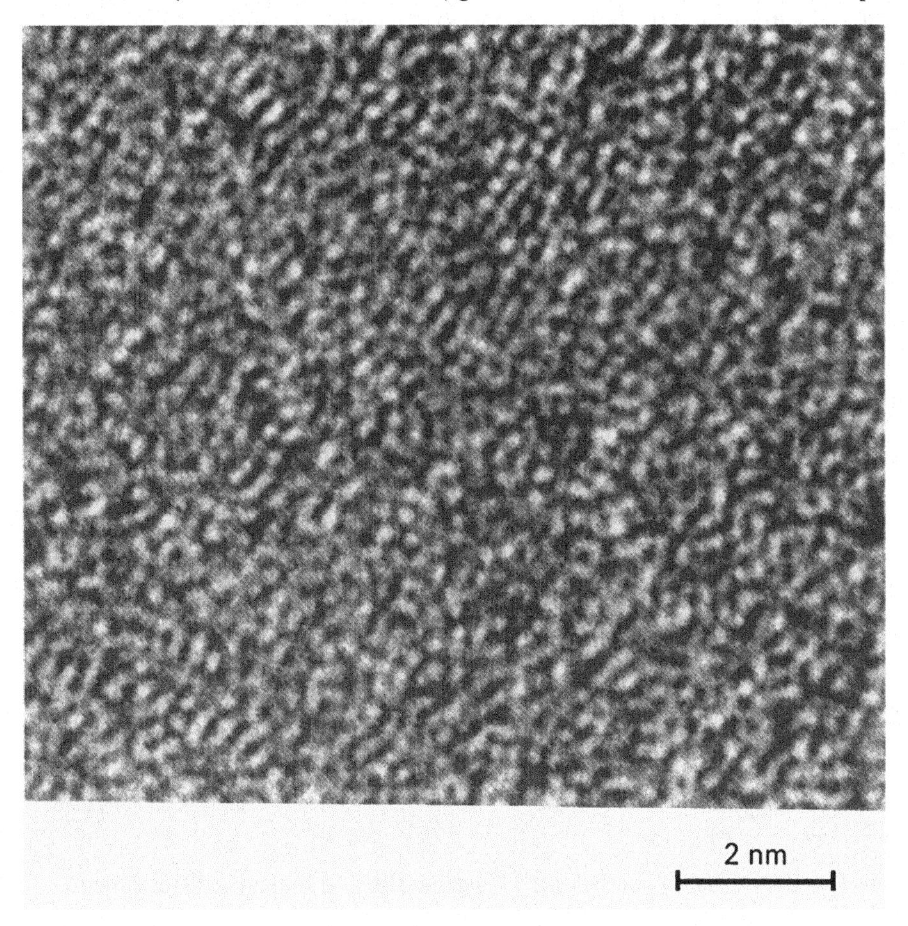


**Figure 3.45** High resolution electron micrograph showing short range order in amorphous Al–Y–Ni. (Micrograph courtesy of I.T.H. Chang.)

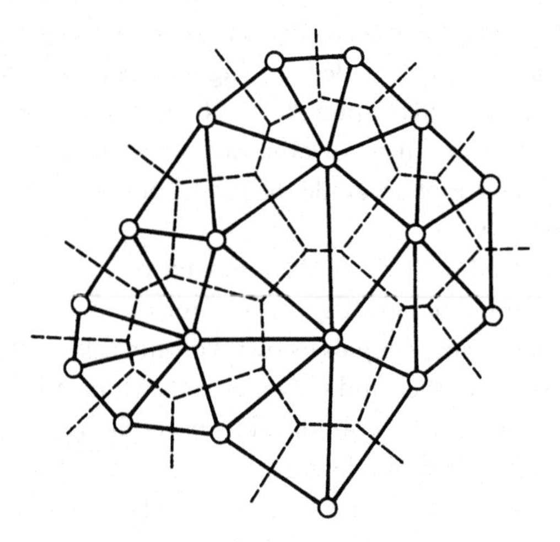

**Figure 3.46** Two-dimensional amorphous atomic structure, showing Voronoi cells (dashed lines) and Bernal holes (solid lines).

by different atoms, and the polyhedral faces (polygonal edges in two-dimensions) give the distribution of near neighbour coordinations. The average near neighbour coordination in a Bernal dense random packing is found to be ~14, but this varies widely with multiple components and with relaxation under different interatomic potentials. The near neighbour bonds in the amorphous structure define the Bernal holes, as shown in two dimensions by the solid lines in fig. 3.46. The collection of all Bernal holes forms a second set of disjoint space-filling polyhedra (polygons in two dimensions). The Bernal holes correspond to the interstitial polyhedra in a crystal, except again that the Bernal holes are all different.

### 3.4.3 Atomic mobility and relaxation

An amorphous alloy has a disordered, frozen-in liquid structure, but this does not mean that its constituent atoms are completely immobile. As in a crystalline alloy, atomic motion takes place at high temperature under the action of applied chemical and mechanical forces. This leads to a variety of complex relaxation processes in the metastable amorphous structure, followed ultimately by crystallisation at sufficiently high temperature.

Chemically driven atomic motion is described by Fick's laws of diffusion. The mass flux per unit area $\mathrm{d}m/\mathrm{d}t$ is proportional to the applied concentration gradient $\mathrm{d}c/\mathrm{d}x$:

$$\frac{\mathrm{d}m}{\mathrm{d}t} = -D\frac{\mathrm{d}c}{\mathrm{d}x} \tag{3.48}$$

where $D$ is the diffusion coefficient. This generalises to the mass diffusion equation:

$$\frac{\partial c}{\partial t} = D\frac{\partial^2 c}{\partial x^2} \tag{3.49}$$

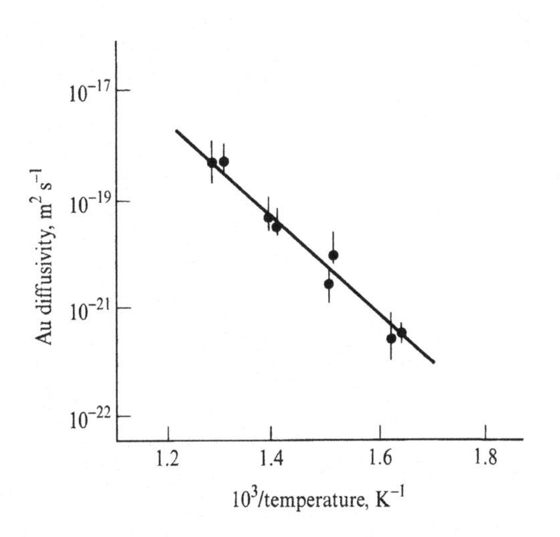


**Figure 3.47** Au diffusion coefficient in melt spun amorphous $Ni_{64}Zr_{36}$. (After Kijek, Akhtar, Cantor and Cahn 1982.)

for a general one-dimensional concentration gradient $c(x,t)$ with $D$ independent of concentration. Diffusion coefficients are obtained by fitting measured concentration profiles to a suitable solution of eq. (3.49). Exceptionally high spatial resolution is needed for an amorphous alloy, since annealing times and temperatures are severely restricted by the onset of crystallisation. Diffusion measurements in amorphous alloys have been reviewed by Cantor and Cahn (1983), Cantor (1985), Greer (1993), Bøttiger, Greer and Karpe (1994) and Frank, Horner, Schwarwaechter and Kronmuller (1994). Fig. 3.47 shows typical results obtained by Rutherford backscattering spectrometry for the diffusion of Au in melt spun amorphous $Ni_{64}Zr_{36}$, plotted on an Arrhenius graph of ln $D$ versus $T^{-1}$. At any given temperature, diffusion in an amorphous alloy is considerably faster than vacancy diffusion in a corresponding crystalline material, but considerably slower than crystalline interstitial or grain boundary diffusion. Not surprisingly, small atoms such as C diffuse considerably faster than large atoms such as Au. Diffusive jumps in a crystalline material are all equivalent, as shown schematically in fig. 3.48($a$), and the diffusion coefficient obeys the Arrhenius law:

$$D = D_0 \exp\left(-\frac{Q_D}{kT}\right) \tag{3.50}$$

where $D_0$ is the frequency factor and $Q_D$ is the activation energy barrier for each jump. Amorphous alloy diffusion coefficients also show excellent Arrhenius behaviour, even though a diffusing atom encounters a range of local atomic environments with varying jump distance and activation barrier, as shown schematically in fig. 3.48($b$). This implies that the overall diffusion process is dominated by a specific jump geometry corresponding to a specific defect in the amorphous structure. Small atoms diffuse interstitially through the Bernal holes shown in fig. 3.46 (Ahmadzadeh and Cantor 1981; Bøttiger *et al.* 1994). Large atoms diffuse through relatively loose-packed regions of the amorphous structure, but it is not clear whether this is best described as a dispersed vacancy mechanism or a

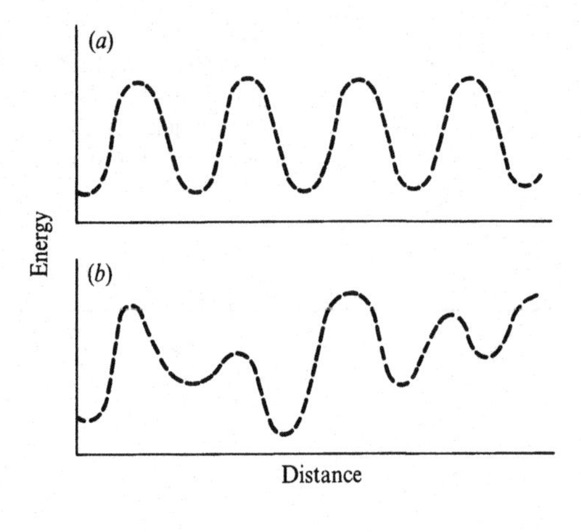


**Figure 3.48** Schematic variation of the energy of a diffusing atom in (*a*) crystalline and (*b*) amorphous alloys. (After Cantor 1985.)

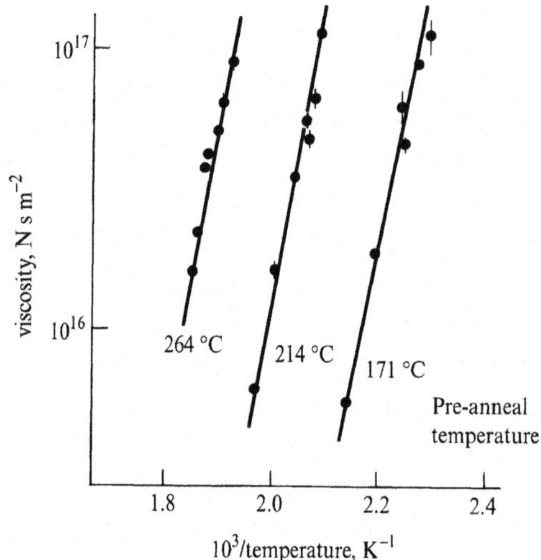


**Figure 3.49** Viscosity of melt spun amorphous $Pd_{82}Si_{18}$ after pre-annealing at 171, 214 and 264 °C. (After Spaepen and Taub 1983.)

cooperative atomic rearrangement (Cantor 1985). Unfortunately, this has not been resolved by pressure measurements of the diffusional activation volume, which give conflicting results ranging from ~0 to close to the atomic volume (Faupel, Hüppe and Rätzke 1990; Rätzke, Hüppe and Faupel 1992, Höfler, Averbach, Rummel and Mehrer 1992, Duine, Wonnell and Sietsma 1994).

Mechanically driven atomic motion is described by the Newtonian laws of creep and fluid flow. The shear strain rate $d\gamma/dt$ is proportional to the applied shear stress $\tau$:

$$\frac{d\gamma}{dt}=\frac{\tau}{\eta} \tag{3.51}$$

where $\eta$ is the viscosity. Amorphous alloy viscosities are measured directly from the ratio of stress and strain rate in a creep test. Fig. 3.49 shows typical results for melt spun amorphous $Pd_{82}Si_{18}$ plotted on an Arrhenius graph of ln $\eta$ versus $T^{-1}$. Amorphous alloy viscosities also obey the Arrhenius law:

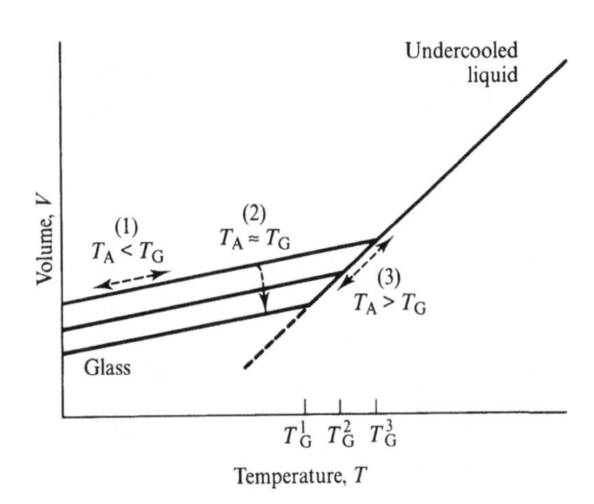


**Figure 3.50** Reversible and irreversible structural relaxation on a plot of volume $V$ versus temperature $T$.

$$\eta = \eta_0 \exp\left(\frac{Q_\eta}{kT}\right) \tag{3.52}$$

where $\eta_0$ is the frequency factor and $Q_\eta$ is the activation energy for local atomic shear. This again implies that the overall creep process is dominated by a specific atomic shear geometry, corresponding to a specific defect in the amorphous structure. Creep and diffusion mechanisms are the same in an amorphous alloy above its glass transition temperature $T_G$. From eqs. (3.50) and (3.52), $Q_\eta = Q_D$ and the amorphous alloy then obeys the Stokes–Einstein relation (Spaepen and Taub 1983):

$$T > T_G: \eta D = \text{constant} \tag{3.53}$$

This breaks down below $T_G$, and instead (Duine, Sietsma and van den Beukel 1993; Wagner and Spaepen 1994):

$$T < T_G: \eta D^2 = \text{constant} \tag{3.54}$$

From eqs. (3.50) and (3.52), $Q_\eta \propto 2Q_D$, which implies that diffusive jumps take place at a single defect, but local atomic shear needs the interaction of two defects (van den Beukel 1988).

The atomic structure of an amorphous alloy is metastable. It evolves towards states of lower energy and higher density during heat-treatment, and this behaviour is known as structural relaxation. Structural relaxation has been reviewed by Chen (1983), Egami (1984) and Greer (1993). The properties of an amorphous alloy all change substantially during structural relaxation, with typical increases of up to 5–7% in Young's modulus, 2–3% in electrical resistivity, 20–30 K in Curie temperature and two orders of magnitude in viscosity. Overall, structural relaxation is quite complex, and comprises a variety of competing and interacting phe-

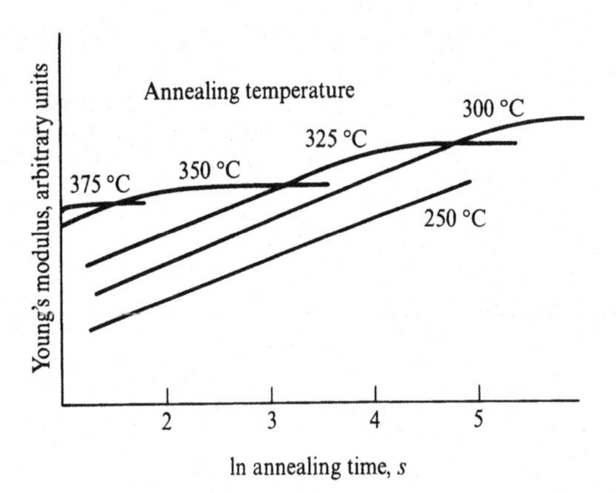


**Figure 3.51** Structural relaxation in the Young's modulus for a range of annealing times and temperatures in melt spun amorphous $Fe_{40}Ni_{40}B_{20}$. (After Scott *et al.* 1982a.)

nomena. Fig. 3.50 shows the temperature variation of molar volume for an amorphous alloy manufactured at different cooling rates. As discussed in §3.2.3, decreasing the cooling rate $dT/dt$ gives more time for atomic motion, allowing the liquid to maintain its equilibrium structure to lower temperatures during cooling. The amorphous alloy is then frozen in at a lower glass transition temperature $T_G$, with a denser, lower energy structure, as shown in fig. 3.50. Reversible structural relaxation takes place at annealing temperatures $T_A < T_G$, as shown by reaction (1) in fig. 3.50. This is usually caused by changes in chemical short range order (CSRO), that is, the chemical distribution of atomic near neighbours, superimposed on the normal thermal expansion from anharmonic atomic vibrations (Luborsky and Walter 1977; Egami 1978; Takahara and Matsuda 1994). Irreversible structural relaxation takes place at annealing temperatures $T_A \approx T_G$, as shown by reaction (2) in fig. 3.50. The amorphous alloy structure densifies gradually as it evolves towards the equilibrium liquid structure corresponding to $T_A$. This is usually dominated by changes in topological short range order (TSRO), that is, changes in atom packing independent of chemical bonding (Egami 1978; Wagner and Spaepen 1994). Reversible structural relaxation is also sometimes seen at higher annealing temperatures $T_A > T_G$, as shown by reaction (3) in fig. 3.50. The alloy follows the equilibrium liquid structure, with changes in both CSRO and TSRO (Greer and Leake 1978; Tsao and Spaepen 1985). Fig. 3.51 shows typical structural relaxation measurements of the variation of Young's modulus with annealing time and temperature in melt spun amorphous $Fe_{40}Ni_{40}B_{20}$. The modulus increases with ln time, before stabilising at a value characteristic of the annealing temperature $T_A$. Structural relaxation often exhibits ln time kinetics, caused by a broad distribution of activation energies associated with wide variations in local atomic relaxation mechanism (van den Beukel and Radelaar 1983; Gibbs and Evetts 1983; Evetts 1985).

Figs. 3.50 and 3.51 seem to show that the different atomic structures of a given amorphous alloy can be characterised by a single parameter, the glass transition temperature $T_G$. Thus, irreversible structural relaxation seems to correspond to

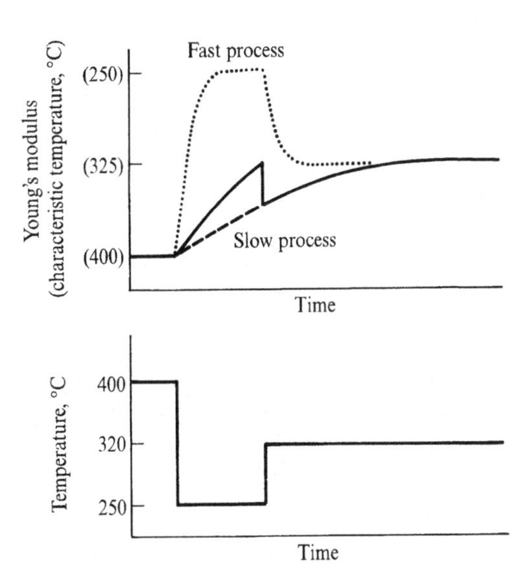


**Figure 3.52** Cross-over experiment for the Young's modulus in melt spun amorphous $Fe_{40}Ni_{40}B_{20}$. (After Scott *et al.* 1982a.)

a gradual reduction in $T_G$. Cross-over experiments show that this is incorrect, and that the range of possible amorphous atomic structures is more complex (Scott *et al.* 1982a). Fig. 3.52 shows the results of a cross-over experiment for the Young's modulus in melt spun amorphous $Fe_{40}Ni_{40}B_{20}$. The alloy is annealed for long enough to stabilise at 400 °C, and then quenched to 250 °C. The modulus begins to increase from its stable value at 400 °C towards its stable value at 250 °C. This is not allowed to go to completion, and is interrupted before the alloy stabilises at 250 °C. The alloy is upquenched to an intermediate temperature of 320 °C, just as the modulus reaches the stable value characteristic of 320 °C (that is, as it crosses over the 320 °C value on its way to the 250 °C value). However, the modulus does not remain constant, even though it has the value characteristic of the current annealing temperature. This shows that the transition in alloy structure from stability at 400 °C to stability at 250 °C does not take place by a gradual progression through the intermediate stable structures. Instead, the modulus falls sharply, and then recovers slowly, to restabilise after some time at the 320 °C value. The key to explaining this behaviour is that structural relaxation takes place with a broad range of activation energies, that is, with a broad range of speeds for the different local atomic relaxation processes. Fast processes allow some local regions to relax completely as soon as the alloy is quenched to 250 °C, as indicated by the dotted line in fig. 3.52. Slow processes mean that other regions have hardly begun to relax at the end of the 250 °C anneal, as indicated by the dashed line in fig. 3.52. The alloy crosses over the modulus characteristic of 320 °C with a balance of different degrees of local relaxation, not with a structure stabilised throughout at 320 °C. When the alloy is upquenched to 320 °C, the modulus falls sharply as relaxation is rapidly reversed by fast processes, but then rises again as slow processes are completed.

### 3.4.4 Crystallisation

The crystallisation of amorphous alloys has been reviewed by Scott (1983), Köster and Herold (1982), Köster and Schünemann (1993) and Greer (1994). When an amorphous alloy is heated to a sufficiently high temperature, thermal motion becomes sufficient for the nucleation and growth of one or more crystalline phases. Crystallisation usually takes place close to the glass transition temperature $T_G$. In some alloys the glass transition temperature $T_G$ is detected 20–50 K below the crystallisation temperature $T_X$, but in other alloys crystallisation sets in immediately on reaching the glass transition temperature, which is not therefore detected independently. Crystallisation takes place in a liquid highly undercooled below the melting point, since the reduced glass transition temperature $\theta_G = T_G/T_M \approx 0.5$–$0.6$, as shown in table 3.6. At high undercoolings, a variety of equilibrium and metastable crystal structures and morphologies can have lower energy than the amorphous structure. They compete as possible crystallisation products, leading to a wide range of different crystallisation reactions, mechanisms and final microstructures in different amorphous alloy systems.

Figs. 3.53(*a*)–(*c*) show typical free energy–composition curves for an amorphous alloy close to its crystallisation temperature, to illustrate the range of possible crystallisation mechanisms (Köster and Schünemann 1993). Polymorphic crystallisation is when an amorphous alloy decomposes into a single equilibrium or metastable crystalline phase $\alpha$ of the same composition, as shown by the reaction in fig. 3.53(*a*):

$$\text{polymorphic: } a \rightarrow \alpha(A_xB_y) \qquad (3.55)$$

where $A_xB_y$ is the $\alpha$ crystal stoichiometry. With no need for long range diffusion, polymorphic crystal growth rates are often fast and isotropic, leading to large single crystals of $\alpha$. Fig. 3.54 shows an example of polymorphic $Al_3Ni$ crystals growing in an amorphous $Al_{75}Ni_{25}$ matrix. If the crystallisation product $\alpha$ is metastable, it can undergo further decomposition, usually at higher temperature after crystallisation is complete. Eutectic crystallisation is when the amorphous alloy decomposes into a mixture of two equilibrium or metastable crystalline phases $\alpha$ and $\beta$, as shown by the reaction in fig. 3.53(*b*):

$$\text{eutectic: } a \rightarrow \alpha(A_pB_q) + \beta(A_rB_s) \qquad (3.56)$$

Individually, $\alpha$ and $\beta$ have different compositions from the amorphous alloy, $A_pB_q \neq A_rB_s \neq A_xB_y$, but the proportions of $\alpha$ and $\beta$ give an overall eutectic composition of $A_xB_y$. Coupled growth of fine scale $\alpha$ and $\beta$ needs lateral diffusion at the crystal–amorphous interface, but no long range diffusion into the

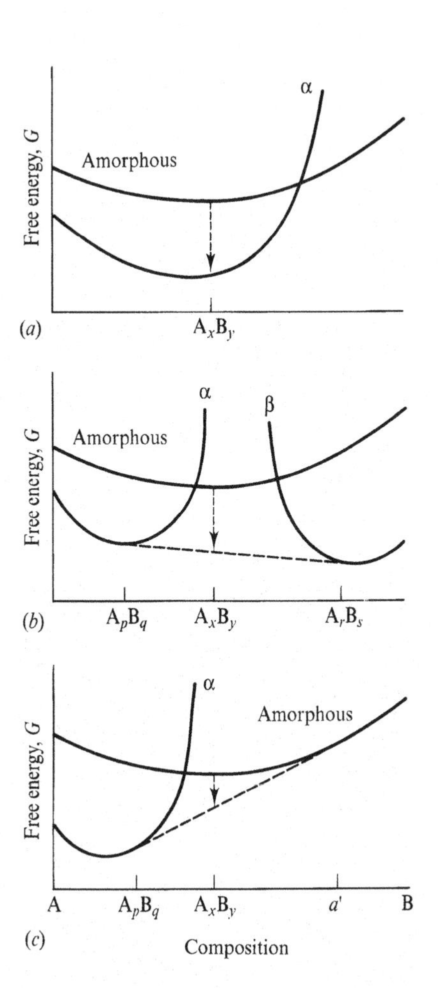


**Figure 3.53** Free energy–composition curves showing (*a*) polymorphic, (*b*) eutectic and (*c*) primary crystallisation mechanisms.

amorphous matrix. Eutectic growth rates are again fast and isotropic, leading to large duplex crystals of $\alpha$ and $\beta$. Fig. 3.55 shows an example of a duplex bcc $\alpha$ Fe(Si)/bct $Fe_3B$ eutectic crystal growing in an amorphous $Fe_{78}Si_9B_{13}$ matrix. Metastable crystallisation products $\alpha$ and $\beta$ can again decompose further, usually at higher temperature after crystallisation is complete. Primary crystallisation is when an amorphous alloy decomposes into a single equilibrium or metastable crystalline phase $\alpha$ of different composition, as shown by the reaction in fig. 3.53(*c*):

$$\text{primary: } a \rightarrow a' + \alpha(A_pB_q) \tag{3.57}$$

where $A_pB_q$ is the $\alpha$ stoichiometry. Since $A_pB_q \neq A_xB_y$, the remaining amorphous matrix $a'$ also has a different composition, and the reaction cannot go to completion until $a'$ undergoes terminal polymorphic or eutectic crystallisation. Primary crystal growth needs long range diffusion, so crystal growth rates are often slow, leading to fine scale dendritic crystals of $\alpha$. Fig. 3.56 shows an example of a bcc $\alpha$ Fe(Si) dendrite growing in an amorphous $Fe_{78}Si_9B_{13}$ matrix. A metastable crystallisation product $\alpha$ can again decompose further, usually at higher temperature after crystallisation is complete.

The number of crystals which nucleate at times between $\tau$ and $\tau+d\tau$ is $Jd\tau$ where $J$ is the nucleation rate. Crystal growth usually follows a power law with annealing time. At a later time $t$, the volume of each crystal is therefore $G(t-\tau)^m$, with $G$ and $m$ depending on the crystal morphology and the crystal–amorphous interface velocity $v$. The crystal fraction $y'$ is obtained by integrating over crystals nucleated at all times $\tau$ between 0 and $t$:

$$y' = \int_0^t G(t-\tau)^m J d\tau = Kt^n \tag{3.58}$$

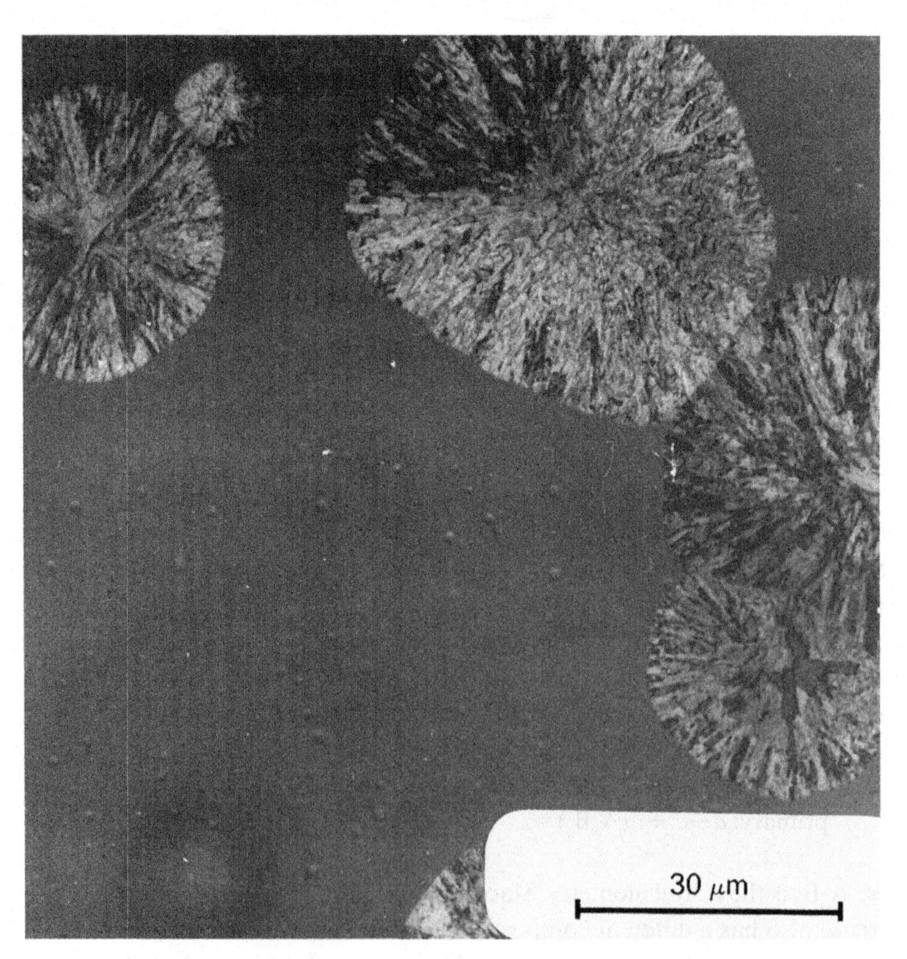


**Figure 3.54** Polymorphic $Al_3Ni$ crystals in sputter deposited amorphous $Al_{75}Ni_{25}$ after annealing for 6 min at 300 °C. (After Cantor and Cahn 1976b.)

where $K=GJ/(m+1)$ and $n=m+1$. Eq. (3.58) ignores the reduction in amorphous volume as crystallisation proceeds. Only the remaining amorphous fraction is available for crystallisation at any time, so the real crystal fraction $y$ increases more slowly than $y'$ by a factor $1-y$, that is, $dy=(1-y)dy'$. Integrating gives the Johnson–Mehl–Avrami equation:

$$y=1-\exp(-y')=1-\exp(-Kt^n) \tag{3.59}$$

with a kinetic constant $K$ which depends on the annealing temperature $T$, and an Avrami exponent $n$ which depends on the crystallisation mechanism. The annealing time $t$ in eq. (3.59) is often replaced by $t-t_I$, where $t_I$ is the incubation time for initial equilibration at the annealing temperature.

Polymorphic crystallisation often takes place with a constant homogeneous or heterogeneous nucleation rate, followed by three-dimensional linear growth

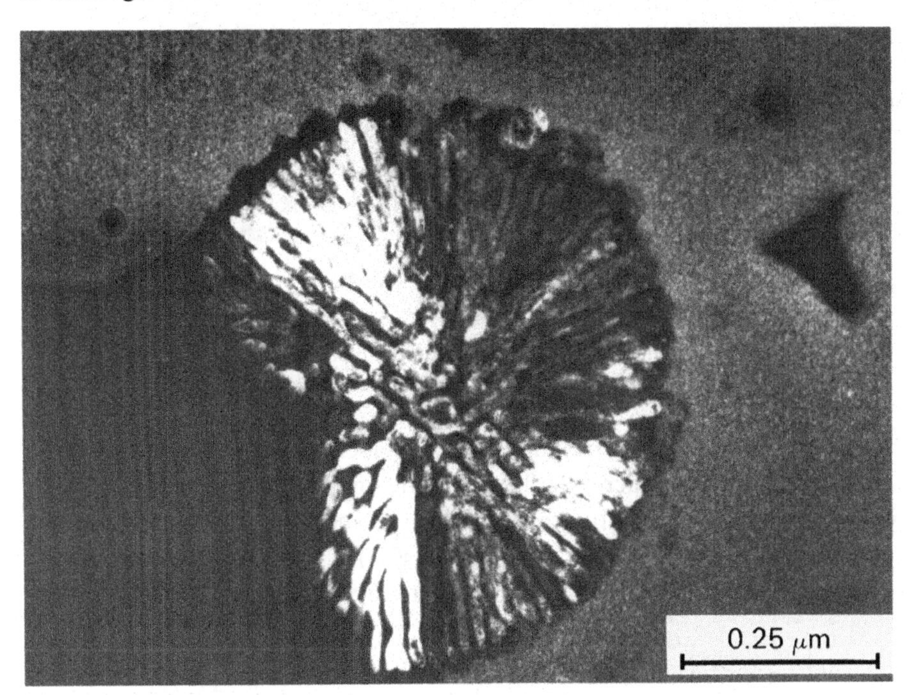


**Figure 3.55** Duplex Fe/$Fe_3B$ eutectic crystal in melt spun amorphous $Fe_{78}Si_9B_{13}$ after annealing for 13 min at 510 °C. (After Bhatti and Cantor 1994.)

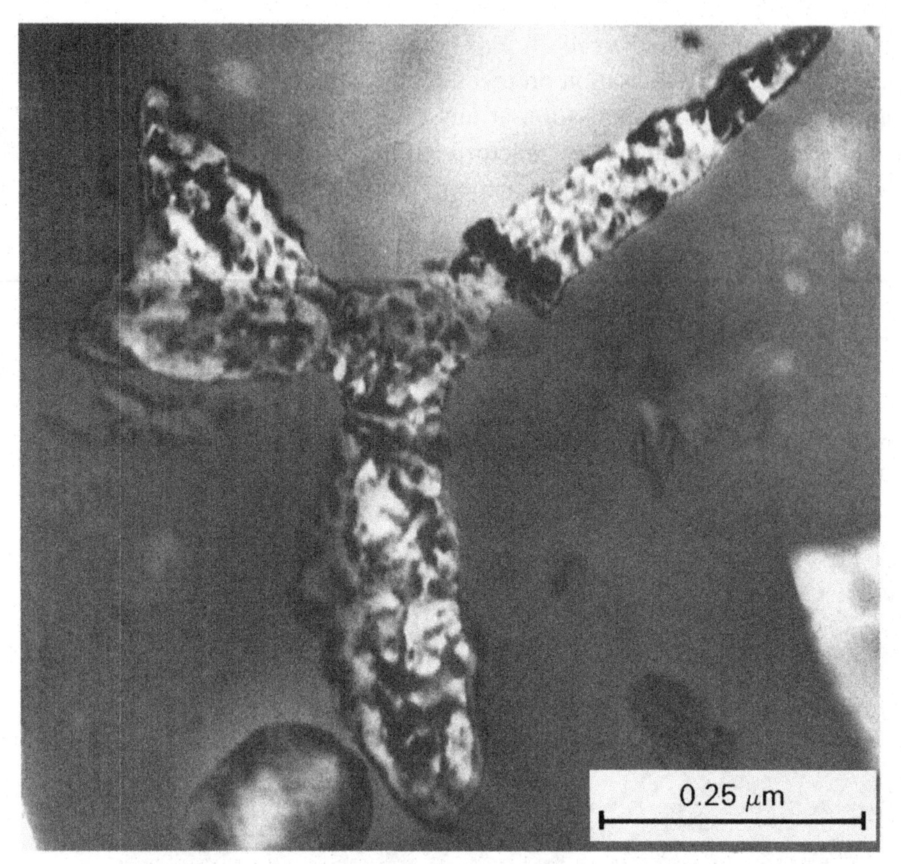


**Figure 3.56** Primary $\alpha$ Fe dendritic crystal in melt spun amorphous $Fe_{78}Si_9B_{13}$ after annealing for 13 min at 510 °C. (After Bhatti and Cantor 1994.)

of spherical crystals with a constant interface velocity. Under these conditions, the number of crystals increases linearly with time, and the volume of each crystal increases linearly with growth in each of the three dimensions, so the overall Avrami exponent is $n=4$. Eqs. (3.35)–(3.38) in §3.4.1 give the full derivation of the corresponding Johnson–Mehl–Avrami equation:

$$y=1-\exp\left(-\frac{1}{3}\pi v^3 Jt^4\right)=1-\exp\left\{-\frac{\pi n D^4 t^4}{3a^5}[1-\exp(-A\Delta T)]^3\exp\left(-\frac{B}{\Delta T^2}\right)\right\} \quad (3.60)$$

In other polymorphic reactions, the heterogeneous nucleation sites are saturated rapidly, so that nucleation effectively takes place instantaneously on a fixed number of nuclei. The corresponding Avrami exponent is then $n=3$. Similar Avrami exponents of $n=4$ and 3 are also found for eutectic crystallisation, with constant and site-saturated nucleation respectively, followed by three-dimensional linear growth of spherical eutectic nodules. Primary crystallisation requires long range diffusion, which inhibits crystal growth and gives parabolic rather than linear kinetics, that is, the crystal size increases with the square root

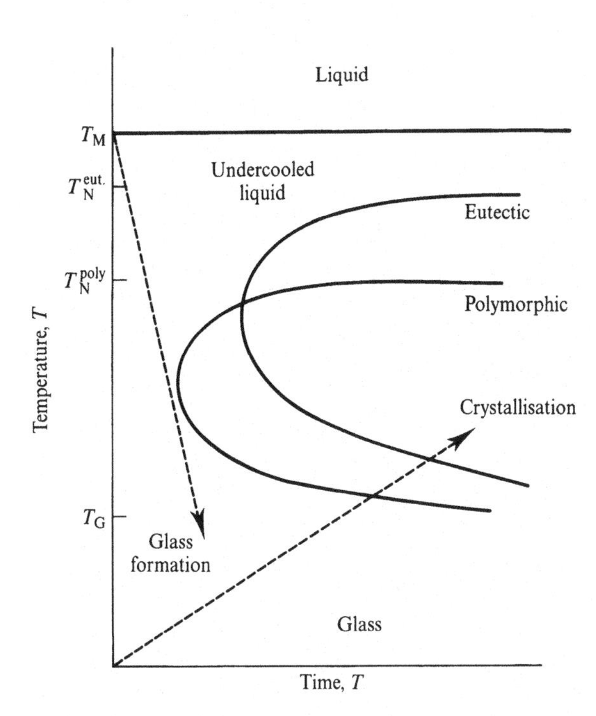


**Figure 3.57** *TTT* diagram for the onset of competing metastable polymorphic and equilibrium eutectic crystallisation reactions.

of annealing time in each dimension. The corresponding Avrami exponent is then $n=5/2$ or $3/2$ for constant and site-saturated nucleation respectively.

Fig. 3.57 shows a *TTT* diagram for the onset of crystallisation, obtained from the Johnson–Mehl–Avrami equation (3.60) by plotting the time $t_0$ to form an undetectable crystal fraction versus undercooling $\Delta T$. There are two crystallisation curves, corresponding to competing metastable polymorphic and equilibrium eutectic crystallisation reactions. Both crystallisation reactions show characteristic *C* curve behaviour. Close to the melting point $T_M$, there is little driving force for crystallisation, so the crystal nucleation and growth rates $v$ and $J$ are small, and the crystallisation onset time $t_0$ is large. As the temperature falls, $t_0$ reaches a minimum value and then increases again as the thermal energy becomes insufficient for atomic motion. Close to the glass transition temperature $T_G$, atomic motion is completely stifled, so that $v$ and $J$ become small, $t_0$ becomes large, and the amorphous structure is frozen in. Metastable polymorphic crystals are only nucleated at high undercooling since the driving force is relatively low, but they grow rapidly with no need for long range diffusion. Conversely, equilibrium eutectic crystals are nucleated at low undercooling since the driving force is relatively high, but they grow more slowly with lateral diffusion needed at the duplex crystal–amorphous interface. *TTT* diagrams such as in fig. 3.57 show the relationship between manufacturing an amorphous alloy by avoiding crystallisation during cooling, and crystallisation during subsequent heat treatment.

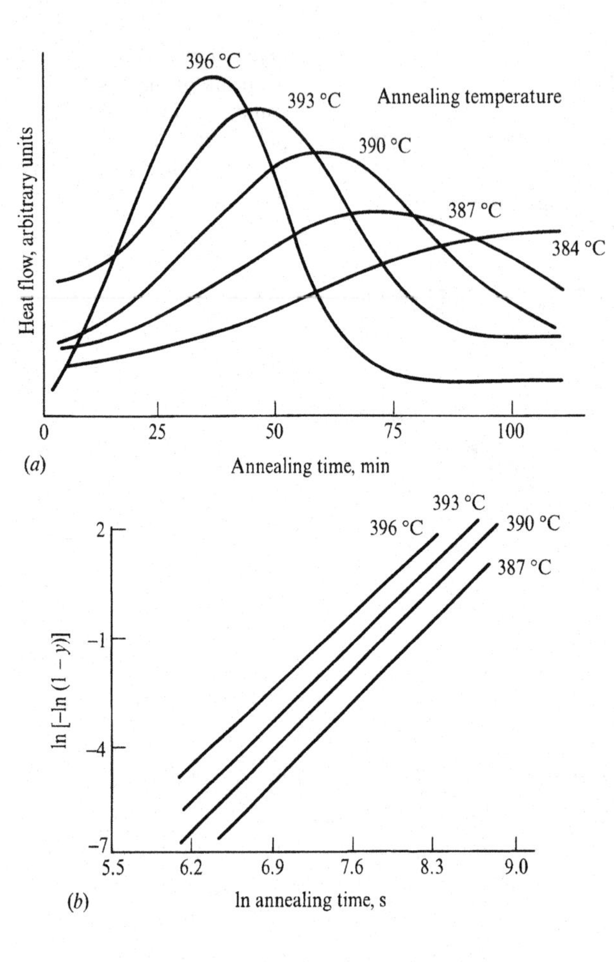


**Figure 3.58** (*a*) DSC traces and (*b*) corresponding Avrami plots for the crystallisation of melt spun amorphous $Fe_{40}Ni_{40}B_{20}$ during isothermal annealing at temperatures in the range 384–396 °C (After Gao and Cantor 1989.)

**Table 3.8** *Crystallisation mechanisms and Avrami exponents. (After Gao and Cantor 1989; Watanabe and Scott 1980; Scott et al. 1982b; Drehman and Greer 1984; Cochrane, Schumacher and Greer 1991; Thompson, Greer and Spaepen 1983.)*

| Alloy | Nucleation | Growth | Avrami exponent |
|---|---|---|---|
| $Fe_{40}Ni_{40}B_{20}$ | het | eutectic | 3 |
| $Fe_{40}Ni_{40}P_{14}B_6$ | hom | eutectic | 4 |
| $Pd_{40}Ni_{40}P_{20}$ | het | eutectic | 3 |
| $Al_{85}Fe_{10}Ce_5$ | het | primary | 1.5 |
| $Au_{68}Cu_9Si_9Ge_{14}$ | hom | polymorphic | 4 |
| $Ni_{33}Zr_{67}$ | hom | polymorphic | 4 |

Taking logs twice, eq. (3.59) gives:

$$\ln[-\ln(1-y)]=\ln K+n\ln t \quad (3.61)$$

Avrami plots of $\ln[-\ln(1-y)]$ versus $\ln t$ can therefore be used to measure values of the kinetic constant $K$ and Avrami exponent $n$. Fig. 3.58(*a*) shows a series of differential scanning calorimeter (DSC) traces from an amorphous $Fe_{40}Ni_{40}B_{20}$ alloy during isothermal annealing at temperatures in the range 384–396 °C. The exothermic peak at each temperature corresponds to evolution of the latent heat of crystallisation, and integration gives the crystal fraction $y$ versus annealing time $t$. The DSC peak broadens and crystallisation slows down with decreasing temperature, as expected from *TTT* diagrams such as fig. 3.57. Fig. 3.58(*b*) shows corresponding Avrami plots of $\ln[-\ln(1-y)]$ versus $\ln t$ for amorphous $Fe_{40}Ni_{40}B_{20}$. Straight lines show that crystallisation obeys Johnson–Mehl–Avrami kinetics with an Avrami exponent of $n=3$, in agreement with the observed crystallisation mechanism of site saturated heterogeneous nucleation, followed by three-dimensional spherical growth of eutectic fcc $\gamma$ (Fe,Ni)/orthorhombic $(Fe,Ni)_3B$ nodules (Gao and Cantor 1989).

Table 3.8 gives nucleation and growth mechanisms and corresponding Avrami exponents for some amorphous alloys which obey Johnson–Mehl–Avrami kinetics. However, crystallisation is not usually so straightforward. For instance, primary bcc Fe dendrites and eutectic bcc $\alpha$ Fe(Si)/bct $Fe_3B$ nodules grow side by side during the crystallisation of amorphous $Fe_{78}Si_9B_{13}$ at 510 °C, as shown in figs. 3.55 and 3.56, even though DSC traces show deceptively simple Johnson–Mehl–Avrami kinetics with $n=4$ (Hughes, Bhatti, Gao and Cantor 1988). As another example, polishing the surface of amorphous $Fe_{40}Ni_{40}B_{20}$ activates surface heterogeneous nuclei, so that eutectic fcc $\gamma$ (Fe,Ni)/orthorhombic $(Fe,Ni)_3B$ crystals grow directionally inwards from the surface, giving linear rather than Johnson–Mehl–Avrami kinetics. Crystallisation kinetics are also often complicated by subsequent decomposition of metastable crystallisation products. For instance, melt spun amorphous $Fe_{70}Cr_{18}Mo_2B_{10}$ undergoes a complex sequence of reactions (Kim, Clay, Small and Cantor 1991):

$$a \xrightarrow{500\,^{\circ}\mathrm{C}} \alpha+M_3B \xrightarrow{700\,^{\circ}\mathrm{C}} \alpha+M_2B+M_{23}B_6 \xrightarrow{900\,^{\circ}\mathrm{C}} \gamma+M_2B+M_{23}B_6 \xrightarrow{950\,^{\circ}\mathrm{C}} \gamma+M_2B+M_3B_2 \quad (3.62)$$

where $\alpha$ and $\gamma$ are bcc ferrite and fcc austenite respectively, and $M_2B$, $M_3B$, $M_3B_2$ and $M_{23}B_6$ are different metal boride compounds. The kinetic constant $K$ is often assumed to obey the Arrhenius law:

$$K = K_0 \exp\left(-\frac{Q}{kT}\right) \tag{3.63}$$

where $K_0$ is the frequency factor and $Q$ is the activation energy for crystallisation. However, the derivation of the Johnson–Mehl–Avrami equation (3.59), and the particular case of polymorphic crystallisation in eq. (3.60) both show that this is not usually justified.

# Chapter 4

## Instability due to strain energy

When a metal is subjected to plastic deformation by cold work, most of the work done in the deformation process is converted into heat, but the remainder of this energy is stored in the metal. Cold-worked metals are unstable and, under favourable circumstances, undergo the processes of recovery and recrystallisation, and during these processes stored energy is released. Before discussing these processes of relaxation, we will consider the stored energy of cold work – how it is measured, how it is stored, and the variables which affect its storage.

### 4.1 The stored energy of cold work

#### 4.1.1 Changes in thermodynamic functions resulting from deformation

Detailed reviews of this aspect are given by Titchener and Bever (1958) and by Bever, Holt and Titchener (1973). The free energy change associated with cold work can only be evaluated via the first law of thermodynamics:

$$\Delta U = Q + W \tag{4.1}$$

where $\Delta U$ is the change in internal energy, $Q$ is the heat associated with the deformation (positive if absorbed), and $W$ is the work (positive if done on the body).

Some methods of measuring the stored energy of cold work determine the change in heat content or enthalpy, $\Delta H$, due to cold work. At constant pressure:

$$\Delta H = \Delta U + P\Delta V \tag{4.2}$$

where $P$ is the hydrostatic pressure and $\Delta V$ the volume change associated with the process. In crystalline solids $P\Delta V$ is negligibly small at atmospheric pressure and we can write

$$\Delta H \approx \Delta U \tag{4.3}$$

As mentioned in chapter 1, the change in Gibbs free energy, $G$, is related to other thermodynamic changes by the equation:

$$\Delta G = \Delta H - T\Delta S \tag{4.4}$$

$$\approx \Delta U - T\Delta S \approx \Delta F \tag{4.5}$$

where $F$ is the Helmholtz free energy, $T$ the absolute temperature and $S$ the entropy.

It is clear, therefore, that a knowledge of entropy changes is required in order to evaluate the free energy changes associated with cold work. The total entropy change for an atomic process is given by:

$$\Delta S = \Delta S_{\text{configurational}} + \Delta S_{\text{vibrational}} \tag{4.6}$$

Cottrell (1953) has estimated the configurational entropy and vibrational entropy due to a dislocation, using Boltzmann's equation, and concludes that $-TS$ due to the configurational entropy of one dislocation is only about $-2\times10^{-8}$ eV per interplanar distance along the dislocation. The vibrational entropy at room temperature is roughly 0.1 eV per atom plane pierced by the dislocation, and both of these contributions are small compared with the strain energy per atom plane length of dislocation.

Hence, at ordinary and low temperatures the entropy change due to a dislocation can be neglected and the free energy change, $\Delta G$, may be taken as essentially equal to the strain energy or internal energy change due to the dislocation. The argument can be extended to groups of dislocations, for example, to arrays forming low angle boundaries. Although there is an appreciable difference between the free energy and internal energy changes associated with excess vacancies, these energy effects are small compared with those of dislocations. No serious error, therefore, is introduced by equating free and internal energies of cold work.

It follows, therefore, that a measurement of the change in internal energy $\Delta U$ due to the introduction of these defects will be essentially equal to $\Delta F$ or the stored energy of cold work. The experimental parameter will be the change in enthalpy, $\Delta H$, and, as we have seen, $\Delta H \approx \Delta U \approx \Delta F$.

From eq. (4.1) we could also obtain $\Delta F$ ($\approx\Delta U$) because of the above approximation, if both $Q$ and $W$ are determinable. The calculation of $W$ appears superficially straightforward since, for a finite process,

$$W=\int f\mathrm{d}x \tag{4.7}$$

where $f$ is the deforming force and $x$ the distance. If a plot of true stress versus strain is made, the area under the curve gives the work of deformation per unit volume. For most industrial deformation processes, however, very complex stress systems are involved, and a considerable fraction of the energy expended is frequently lost through various inefficiencies and therefore does not contribute to the final strain. The actual work will then be greater than the 'ideal' work given by eq. (4.7), and this approach is only likely to be successful in very simple situations of deformation.

## 4.1.2 Measurement of the stored energy of cold work

The methods for measuring the stored energy of cold work may be classified as 'one-step' methods and 'two-step' methods. The former are those in which all measurements are made during the deformation, while in the latter the deformation is carried out first and the stored energy is measured at a later time. Reviews of the techniques used for these experiments are given by Titchener and Bever (1958) and by Bever *et al.* (1973), to which reference should be made for detail. Knowledge of these methods is indispensable for evaluating the results reported in the literature.

### Single-step methods

Farren and Taylor (1925) used the principle of determining the difference between the mechanical energy put in and the energy evolved as heat during the deformation in one of the earliest stored energy investigations. The mechanical work is obtained directly from the integration of the stress–strain diagram, and the heat change calculated from the change in temperature. A plot of the stored energy versus the shear strain obtained by Williams (1964) using a refinement of Farren and Taylor's method is shown in fig. 4.1, using specimens of single-crystal copper. It is notable that only a small fraction of the work done on the crystals is stored in this way.

### Two-step methods

There are two categories of two-step methods: solution calorimetry, in which the heat of solution on dissolving deformed and annealed specimens is compared, and annealing methods, which can be isothermal (where the $\Delta U$ of a deformed sample is studied as a function of time) or anisothermal (where the temperature is increased at a controlled rate and $\Delta U$ is studied as a function of temperature).

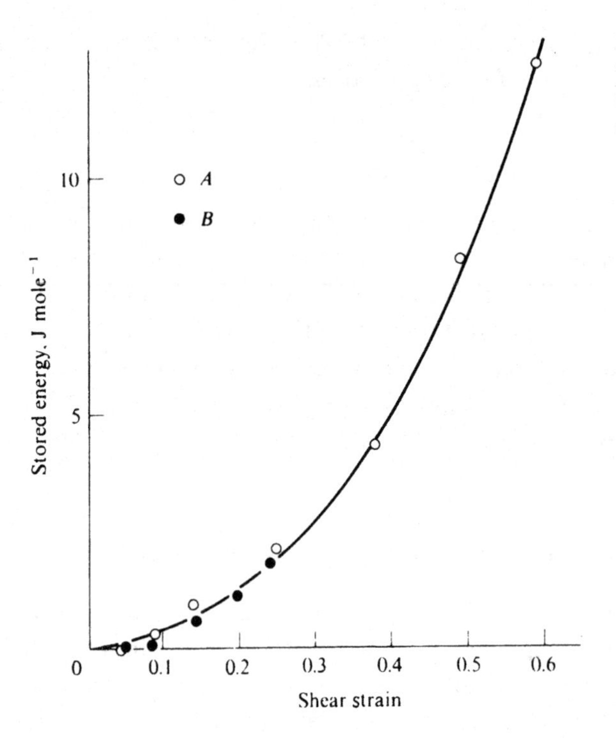


**Figure 4.1** Stored energy versus the shear strain for single crystals of copper. (After Williams 1964.)

**(i) Solution methods** The main difficulty with solution calorimetry arises from the size of the heat effects associated with aqueous solvents, which are so large that they cannot be measured with the accuracy necessary for finding the stored energy. Bever and his coworkers have, however, successfully avoided this difficulty by using liquid tin as solvent. For further details, the reader should consult the review of Titchener and Bever (1958).

**(ii) Isothermal annealing methods** An isothermal jacket calorimeter can be used to measure the energy released by a cold-worked sample during annealing at constant temperature. Gordon (1955) designed a microcalorimeter capable of holding temperature variations to within $10^{-4}$ °C, and fig. 4.2 shows his results obtained using specimens of 99.999% purity copper at temperatures between 100 and 200 °C after 30% elongation in tension, at room temperature. These experiments showed two stages in the release of energy – the first where the rate of release is most rapid at the beginning of the process but decreases quickly with time; the second release leading to a very pronounced peak which accounts for 90% of the stored energy (and is associated with recrystallisation).

**(iii) Anisothermal annealing methods** Developments in techniques are reviewed by Bever *et al.* (1973), and one method, due originally to Clarebrough, Hargreaves, Mitchell and West (1952), involves differential calorimetry, whereby two specimens (one deformed and the other annealed) are slowly heated

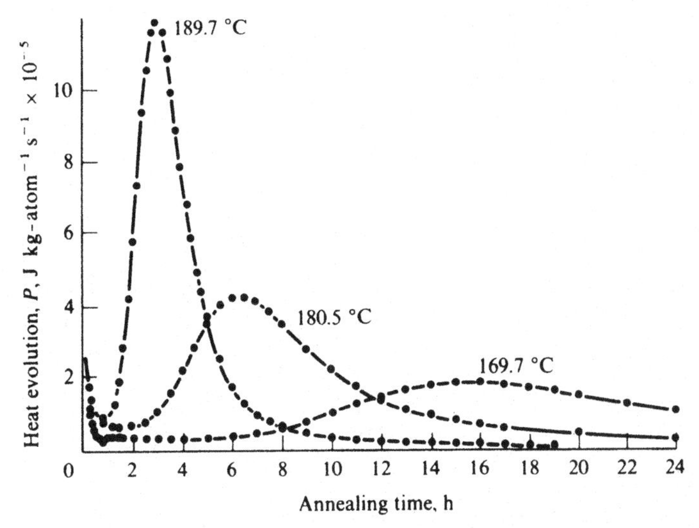


**Figure 4.2** Isothermal release of stored energy from 99.999% copper deformed 30% elongation in tension, at room temperature. (After Gordon 1955.)

at a constant rate. The two specimens are heated in separate identical furnaces and the power necessary to maintain the same rate of rise of temperature in each specimen is measured. The power needed for the deformed specimen will be less because it will gradually release its stored energy of cold work, so the power difference, $\Delta P$, is proportional to the released energy. Some typical results are shown in fig. 4.3 for the case of copper (of commercial purity) deformed 33% elongation in tension at room temperature. As well as the power difference, changes in hardness (*VHN*) resistivity (*R*) and density (*D*) are shown.

In this material it is seen that release of energy starts at about 70 °C and occurs steadily until a sharp peak is reached just below 400 °C (which has been shown to correspond with the onset of recrystallisation in these specimens). It can also be seen from fig. 4.3 that the hardness, density and electrical resistivity also all alter markedly in the recrystallisation range and undergo smaller but significant changes at lower temperatures. It should be emphasised that in all the methods used to measure stored energy it is possible to introduce serious errors if deformed specimens are allowed to remain at room temperature for an appreciable time prior to the measurement, due to the partial relaxation of the specimen at room temperature.

## 4.1.3 Mechanisms for the storage of energy

The interpretation of the mechanisms of storage of energy arising from cold work is based on two concepts, namely the presence of lattice strains and of imperfections.

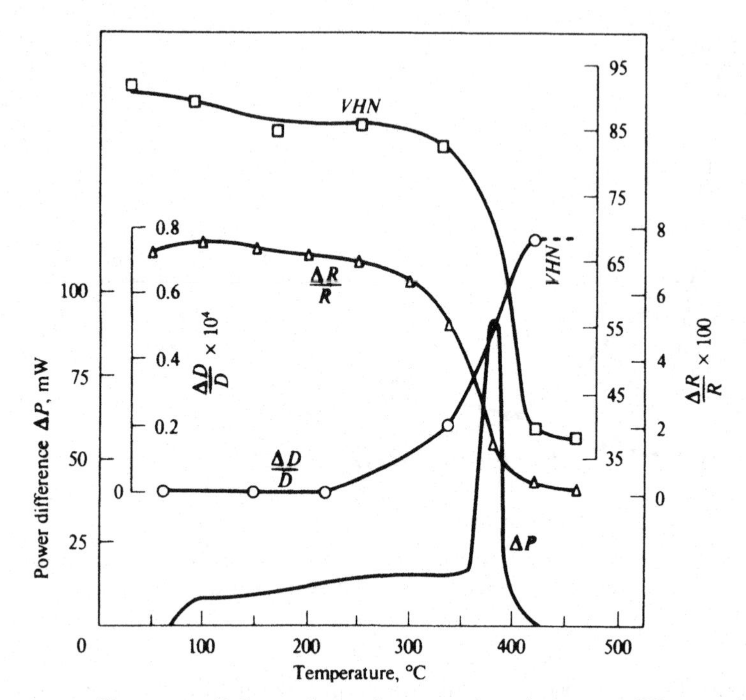


**Figure 4.3** Release of stored energy, plotted as power difference $\Delta P$, from commercial copper deformed 33% elongation in tension at room temperature. Changes in hardness (VHN), resistivity ($R$) and density ($D$) are also shown. (After Clarebrough, Hargreaves and Loretto 1963.)

## Elastic strain energy

A direct comparison of the elastic strain energy and the total stored energy is possible from the investigations of Averbach, Bever, Comerford and Leach (1956), Michell (1956) and Michell and Haig (1957) for filed 75% Au – 25% Ag alloy and ground nickel respectively. In the alloy, the strain energy (calculated from X-ray determinations of the root mean square local strain, $\epsilon$) amounted to only about 3% of the stored energy measured calorimetrically. Stibitz (1936) has related $\epsilon$ to $U$ (the strain energy per unit volume):

$$U=\frac{3}{2}\frac{E}{2(1+2\nu^2)}\epsilon^2 \tag{4.8}$$

where $E$ is Young's modulus, and $\nu$ is Poisson's ratio.

In the nickel, the strain energy calculated from X-ray determinations was 12% of the stored energy, so these data show that the elastic energy contribution is usually very small, and more recent data on other systems support this general conclusion. The implication of these results is, therefore, that the major part of the energy storage must be ascribed to the presence of lattice imperfections.

### Point defects

When a crystal is deformed its resistivity ($\pi_{el}$) generally increases. The part of this extra resistivity that anneals out at temperatures below which appreciable changes in mechanical properties take place is believed to result from the resistivity due to point defects ($\pi_p$), and the other part to the resistivity of dislocations ($\pi_d$), that is,

$$\pi_{el}=\pi_p+\pi_d) \tag{4.9}$$

Experiments of this type have shown that vacancies are produced by cold work and that the concentration $c$ depends on the amount of plastic strain $\epsilon$ according to the approximate relation

$$c=10^{-4}\epsilon \tag{4.10}$$

These vacancies are presumed to form at jogs produced on dislocation lines by intersecting dislocations, since it is observed that when the deformation is restricted to single slip the vacancy concentration (as shown by the change in electrical resistivity or density) is relatively small, while during double slip the resistance increases and the density decreases. It should be pointed out that no single point defect (that is, vacancy or interstitial atom) has been positively identified as responsible for any particular part of the annealing spectrum of a cold-worked metal. Detailed interpretation of annealing experiments is made very difficult due to the complexities of impurity effects, dislocation effects and point defect interactions. Vacancies must, however, be considered responsible for at least some energy storage. In cold-worked metals this energy is likely to be only a small fraction of the total stored energy.

### Dislocations

The dislocation energy per unit length, $E$, can be obtained from the equation:

$$E=(\mu b^2/4\pi K)\ln(r/r_0)+\text{core energy} \tag{4.11}$$

where $\mu$ is the shear modulus, $r$ is the outer cut-off radius (which could be the specimen dimension), $r_0$ is the inner cut-off radius, $b$ is the magnitude of the Burgers vector, and $K$ is a constant (equal to unity for a screw dislocation and equal to $1-\nu$, where $\nu$ is Poisson's ratio, for an edge dislocation). The core energy has been estimated as about equal to $\mu b^2/10$ (Bailey and Hirsch 1960).

For a dislocation density $\rho$, and neglecting the core energy, the elastic stored energy per unit volume, $E_D$, is then

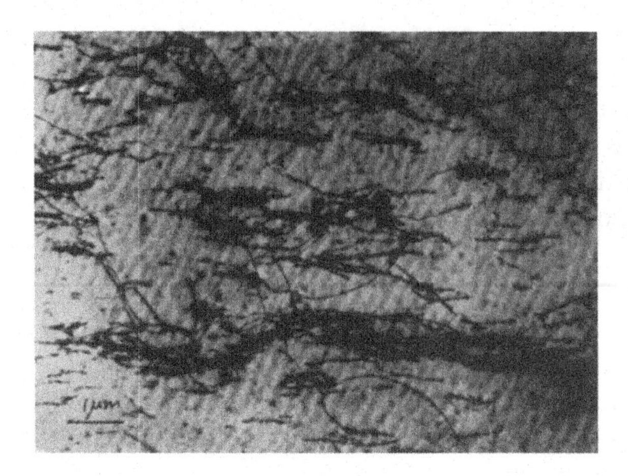


**Figure 4.4** Copper in stage I, (111) section. Shear strain $\gamma$=0.10. (After Humphreys and Martin 1967.)

$$E_D = \rho\,(\mu b^2/4\pi K)\ln(r/r_0) \tag{4.12}$$

It follows that a uniform distribution of dislocations is unstable since the value of $r$, and hence the stored energy, can be reduced by clustering into stress screening arrays. As dislocations become arranged in dense tangles, various stress components can cancel, which leads to an effective lowering of the strain energy. Any attempt to estimate the value of $E_D$ thus requires information on their density and distribution. The direct observations of these parameters come mainly from electron microscopy, and they have been reviewed in the case of single crystal copper by Basinski and Basinski (1966), and polycrystalline materials of medium to high stacking fault energy by Hansen (1990).

In crystals deformed within stage I of the stress–strain curve, fig. 4.4, the dislocation structure is mainly composed of isolated bundles or braids of long, relatively straight, primary edge dislocations lying on closely neighbouring primary glide planes. The lack of contrast across the bundles and also the absence of X-ray asterism (see below) indicates that the Burgers vectors of the dislocations in a bundle add approximately to zero. The dislocations are arranged approximately in pairs of opposite sign.

Stage II is marked by increasing numbers of secondary dislocations. Because most of the dislocations are confined to (111) planes, at the higher dislocation densities occurring in stage II the structure on the slip planes becomes very complicated. Fig. 4.5 is an electron micrograph of a crystal of a copper alloy deformed into stage II sectioned on the $(\bar{1}01)$ plane, which is perpendicular to the (primary) (111) plane. Traces of the slip plane are well defined, and are approximately 0.75 $\mu$m apart and the alternating light and dark contrast across the slip planes indicates lattice misorientations. These misorientations have been measured and are found to be typically of the order 1°. In crystals deformed into stage III of the curve, the dislocation networks become even more dense and the lattice tilts produced across them more pronounced. The dislocation distribution resembles that

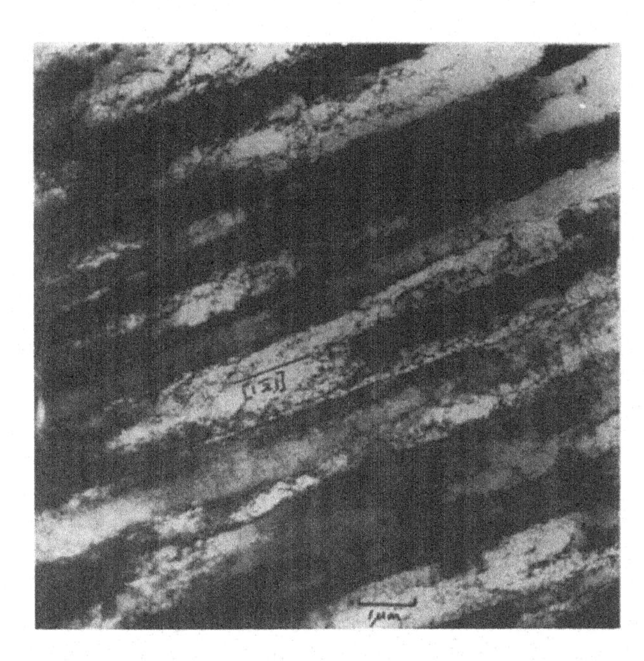

**Figure 4.5** Cu – 0.7 wt% Co, $\gamma=0.6$ ($\bar{1}01$) section. (After Humphreys and Martin 1967.)

in stage II, especially for high deformations at low temperatures. The configuration at this stage can be described as dislocation sheets lying approximately parallel to the main glide plane, their separation decreasing with deformation.

It is generally agreed that the dislocation distribution is highly non-uniform on the electron microscopical scale, the bundles or networks separating regions of the crystal almost free of dislocations. In polycrystals the dislocation structure produced by the deformation generally consists of irregular subgrains separated by tangled dislocation walls and with increasing deformation temperature, the cells become larger and the dislocation walls cleaner (see, for example, Warrington (1961)).

### Planar defects

The X-ray Laue technique provides a simple means of studying, in single crystals, disorientations produced by cold work. Bending of the lattice planes results in the sharp diffraction spots being smeared out into arcs – this phenomenon is known as asterism. The asterisms frequently show intensity maxima which indicate that a type of deformation substructure is being formed, of a grosser nature than the dislocation subboundaries such as those illustrated in fig. 4.5.

These inhomogeneities, which occur with more complex types of deformation can also be revealed by etching, when the changes in orientation show up as differently etched bands known as *deformation bands*. There is considerable confusion in the literature regarding the terminology used to describe these heterogeneities, but Hatherly and Malin (1979) have provided a glossary of the terms used which may provide a basis for a common nomenclature.

Originally a technique of X-ray microscopy (Barrett 1945; Honeycombe 1951) was used to study these inhomogeneities which occur both in the deformation of single crystals and of polycrystalline aggregates. Various techniques of electron microscopy and diffraction are more commonly used today, and Hansen (1990) has summarised the changes in structures found with increasing strain in fcc metals.

The original dislocation 'cell' structure first becomes subdivided by dense dislocation walls (DDWs) into regions taking up a large fraction of the grain. Microbands (MBs), of greater wall thickness, evolve from the DDWs, and groups of MBs (called 'transition bands') develop, and these have relatively high misorientations across them. At very high degrees of strain, such as in heavily rolled metals, 'shear bands' subdivide the structure: these are non-crystallographic, are not restricted to a single grain, and sharp misorientations occur across them.

Large deformation twins, which are visible under the optical microscope, do not contribute appreciably to the stored energy, although twins on a fine scale, and stacking faults formed during deformation, can account for a significant part of the stored energy. A twin fault has about one-half the energy of a stacking fault since only half as many next-near neighbour bond violations are involved, and its value and that of the stacking fault energy depend on the metal itself. Typical estimates of the stacking fault energy in nickel, aluminium, copper and silver are 0.4, 0.2, 0.07 and 0.02 J $m^{-2}$ respectively.

Since X-ray techniques are available for determining both the stacking fault probability and the twin fault probability (see, e.g., Barrett and Massalski (1980)), it is possible to estimate the energy contribution due to twins in materials in which these parameters have been measured.

## 4.1.4 Variables affecting the stored energy

The reported values of the stored energy of cold work can be related to the energy expended during deformation (when this quantity is known), and a problem of special interest is the saturation of a metal with stored energy as the strain increases. The variables which affect the stored energy fall into two classes: those involving the deformation process (particularly the type and extent of deformation and the effect of temperature) and those involving the metal or alloy, such as its composition and metallurgical condition.

### Process variables

The total work done ($W$) in any cold-working process can be described by

$$W = W_{th} + W_e + W_i \tag{4.13}$$

where $W_{th}$ is the theoretical work required to produce the overall shape change, $W_e$ represents the external losses due to friction, and $W_i$ the internal losses due to redundant work – that is, any internal work that is not involved in the first two terms.

Many deformation processes are used in practice, including extension, compression, torsion, wire drawing and rolling, as well as cutting operations such as filing, drilling and orthogonal cutting. Apart from differences in strain and strain rate which may exist between them, more fundamentally the simpler processes like extenson involve negligible stress gradients, no frictional effects and (until necking begins) no redundant work. On the other hand, cutting and forming operations, such as wire drawing and drilling, involve large stress and strain gradients, non-uniform strain rates, large frictional effects and appreciable amounts of redundant work.

It is, of course, very difficult to make a comparison of two of these processes at an equal value of strain, and little effort of this kind is found in the literature. Bever *et al.* (1973) have tabulated a summary of measured values of the stored energy of cold work for many deformation processes, to which reference should be made by the reader for full detail, but their data illustrate the point that the simpler the mode of deformation, the lower the value of the stored energy.

Fig. 4.6 is a plot of the fractional energy stored over energy expended versus energy expended. This shows that the relative amount of energy stored decreases to a constant (saturation) value as the total amount of energy expended increases. The more complex the working process, the higher will be the saturation value of the stored energy.

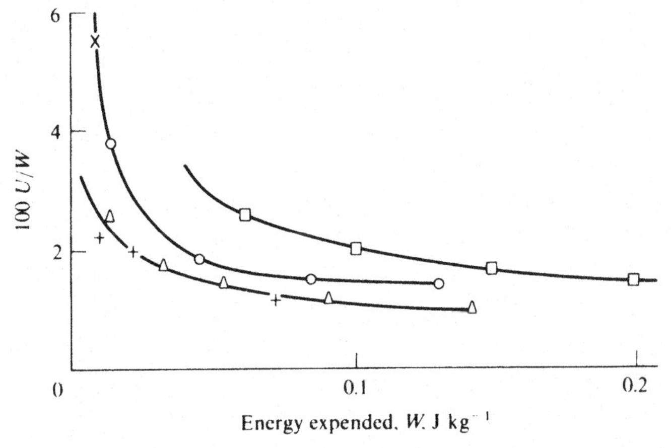


**Figure 4.6** Percentage of energy stored ($U$) as a function of the work of deformation for various metals and methods of working. □=nickel, torsion; ○=arsenical copper, torsion; ×=arsenical copper, tension; △=pure copper, torsion; +=pure copper, compression. (After Clarebrough, Hargreaves and West 1955.)

The effect of the temperature of working is a very important variable affecting the amount of stored energy. Work-hardening rates usually increase with decreasing temperature, and it is also likely that, at lower temperatures, processes of energy release occurring during or immediately after the deformation are suppressed to a greater extent.

### Material variables

**(i) The identity of the metal** The work of Williams (1962) provides a set of comparable values of the stored energy for seven metals which were deformed in the same apparatus. The energy stored at a given strain decreased in the order zirconium, iron, silver, nickel, copper, aluminium and lead: with the exception of silver, this is also the order of decreasing melting temperature.

**(ii) Solute effects** Additions of foreign atoms to a metal invariably allow larger amounts of stored energy of cold work. Thus Clarebrough *et al.* (1955) found that the energy stored in copper of 99.55% purity (arsenical copper) was about 30% greater than that stored in copper of high purity. The effect is also observed in more concentrated solutions; thus Sato (1931) demonstrated that, in general, for a given expenditure of energy, for a series of Cu–Zn alloys, an increase in zinc content was attended by an increase in the stored energy, and the results of Tizhnova (1946) for Cu–Ni alloys show a similar trend.

Fig. 4.7 shows the energy stored as a function of composition of Au–Ag alloys in the form of chips produced by drilling at room temperature and at −195 °C. It can be seen that an alloy containing 75% gold and cold-worked at room temperature stored about four times as much energy as one containing 99%. These data also illustrate well the influence of the temperature of deformation upon the magnitude of the energy stored.

**(iii) Grain size and orientation** One would expect finer grain sizes to lead to higher values of stored energy because of the higher dislocation densities expected in fine grained material than in coarse grained material at the same strain due to the 'complexity effect'. This requires multiple slip in the grains to allow changes in shape while maintaining continuity across grain boundaries; lattice deformation will thus be more complex in material with fine grains, because of, in a given grain, the close proximity of material in other orientations in adjacent deforming small grains. The limited amount of experimental data available are in general agreement with this expectation (see Bever *et al.* (1973)).

A number of workers have investigated the variation in stored energy as a function of shear strain in single crystals of various orientations. Thus Nakada (1965) has studied both aluminium and silver crystals, and fig. 4.8 shows his data for the latter material. It is seen that the curve for the crystal with a centre-of triangle orientation is concave upward and that, at the same strain, the corner ori-

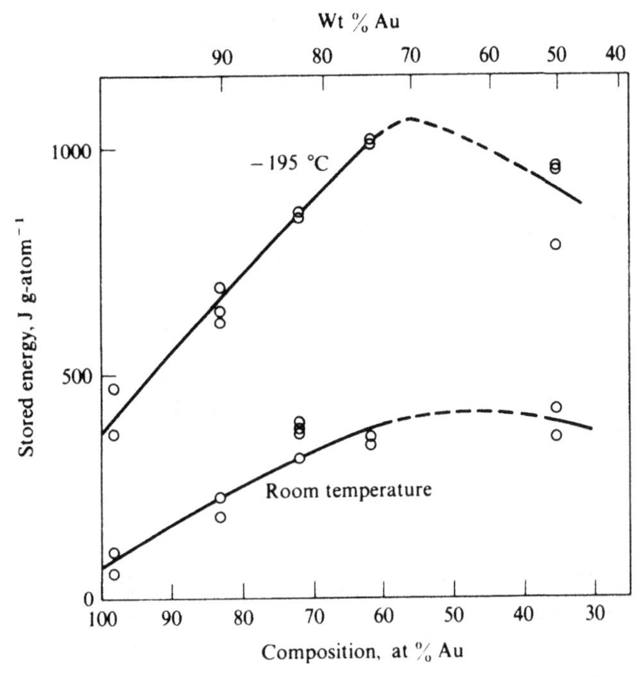


**Figure 4.7** Energy stored as function of compositions of Au–Ag chips formed by drilling at room temperature and at −195 °C. (After Greenfield and Bever 1957.)

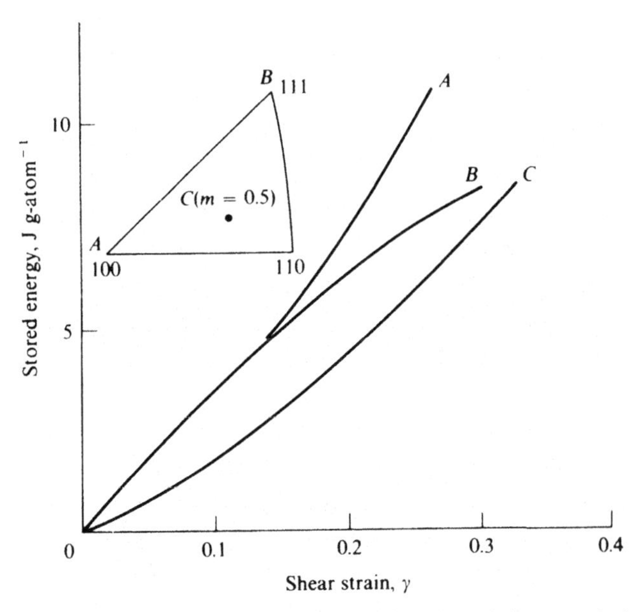


**Figure 4.8** Stored energy as a function of shear strain in single crystals of 99.99% silver compressed at various orientations. Data of Nakada (1965). (After Bever, Holt and Titchener 1973.)

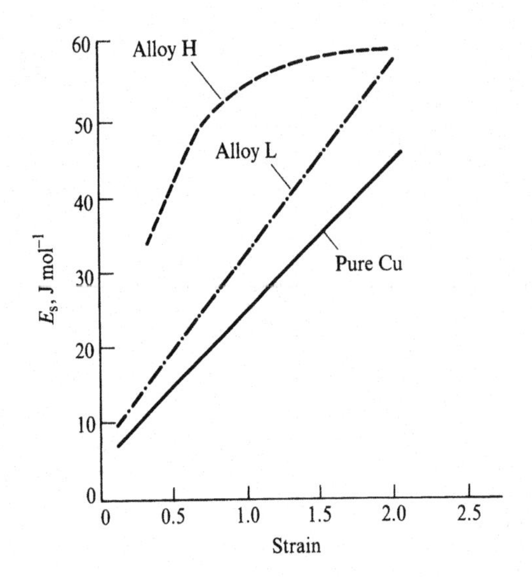


**Figure 4.9** Stored energy $E_s$ as a function of strain in pure Cu and in Cu – 0.395 vol. $Al_2O_3$ (L) and Cu – 1.445 vol. $Al_2O_3$ (H). (After Baker and Martin 1983b.)

ented crystals stored more energy than the centre-of-triangle crystals. This reflects a difference in work-hardening rate.

Relatively little work appears to have been done to determine the dislocation distribution in deformed single crystals as a function of their initial orientation. Price and Washburn (1963) studied copper crystals with a ⟨111⟩ tension axis deformed at room temperture and observed dislocation networks that had little correlation with the traces of close-packed planes, and van Drunen and Saimoto (1971) have observed deformed copper crystals of ⟨001⟩ orientation and again found no clusters of slip bands such as those illustrated in fig. 4.5. It is, however, clear that changes in crystal orientation do change the distribution of dislocations in the cold-worked state.

**(iv) The effect of a dispersed second phase** Differential scanning calorimetry (DSC) has been used by Baker and Martin (1983b) to study as a function of rolling reduction the effect of dispersed alumina particles on the stored energy in copper crystals. The presence of the second phase was found to raise the stored energy by an amount proportional to the volume fraction of particles and also to the magnitude of the applied strain. The increase in stored energy due to particles was shown to increase with increasing strain (fig. 4.9).

An understanding of these effects can be obtained from a consideration of the dislocation microstructures which develop in such systems during deformation.

The stress–strain curves of crystals containing a dispersed second phase may be classified according to whether the second phase deforms with the matrix or not. Work on systems containing a deformable second phase has been mainly confined to alloys containing a coherent precipitate, and it is found that the precipitate causes an increase in the yield stress due to coherency strains, and to several possible short range interactions, but it appears that the subsequent rate

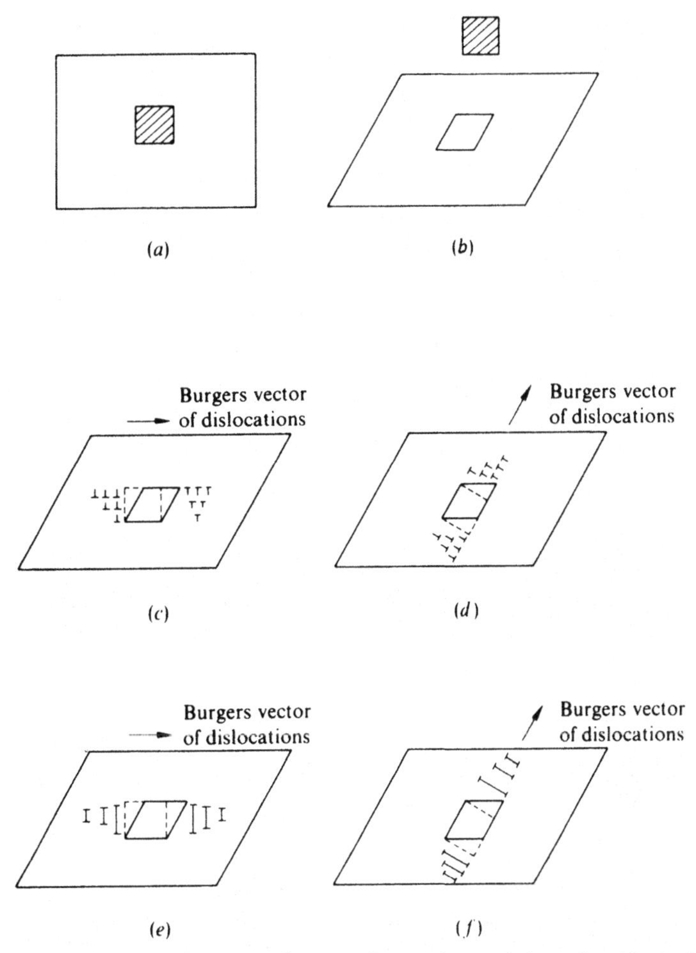


**Figure 4.10** (*a*) shows an element of crystal containing a hard inclusion of a second phase (shown prior to deformation). (*b*) The inclusion has been removed and the crystal sheared uniformly. (*c*) and (*d*) show the hole restored to its original shape by the emission of glide dislocations on (*c*) the primary system and (*d*) a secondary system. (*e*) and (*f*) show the hole restored to its original shape by prismatic glide.

of work-hardening is not significantly different from that of the matrix alone. This behaviour has been reviewed by Kelly and Nicholson (1963), and from the present standpoint we may treat such systems as a special form of solid solution.

Much has been published concerning the deformation of metals containing a non-deforming second phase, and all work has shown that the presence of a second phase of this type leads to a much higher dislocation density at a given strain than in a single-phase metal, and this is discussed and reviewed by Martin (1980).

The problem has been considered by Ashby (1971), whose interpretation may be summarized diagrammatically in fig. 4.10. Fig. 4.10(*a*) shows an element of undeformed crystal containing a hard inclusion of a second phase. In fig. 4.10(*b*) the inclusion has been removed and the crystal sheared uniformly; since the inclusion cannot be deformed, the hole from which it came must be restored to its original shape if the inclusion is to be replaced.

Figs. 4.10(*c*) and (*d*) show the hole restored to its original shape by shear dislocation loops on primary and secondary systems respectively. This process leads to local lattice bending of the matrix in these regions, and is a simplification of the type of process which has taken place in the specimen illustrated in fig. 4.11. This shows a copper crystal containing a dispersion of silica particles (produced by internal oxidation) which has been subjected to a shear strain of 0.2. Complex dislocation interactions have taken place with the larger precipitate particles (in this case those of diameter greater than ~300 nm), there being intense secondary slip activity in the matrix adjacent to them. In figs. 4.10(*e*) and (*f*) the hole is shown restored to its original shape by prismatic glide, which does not involve

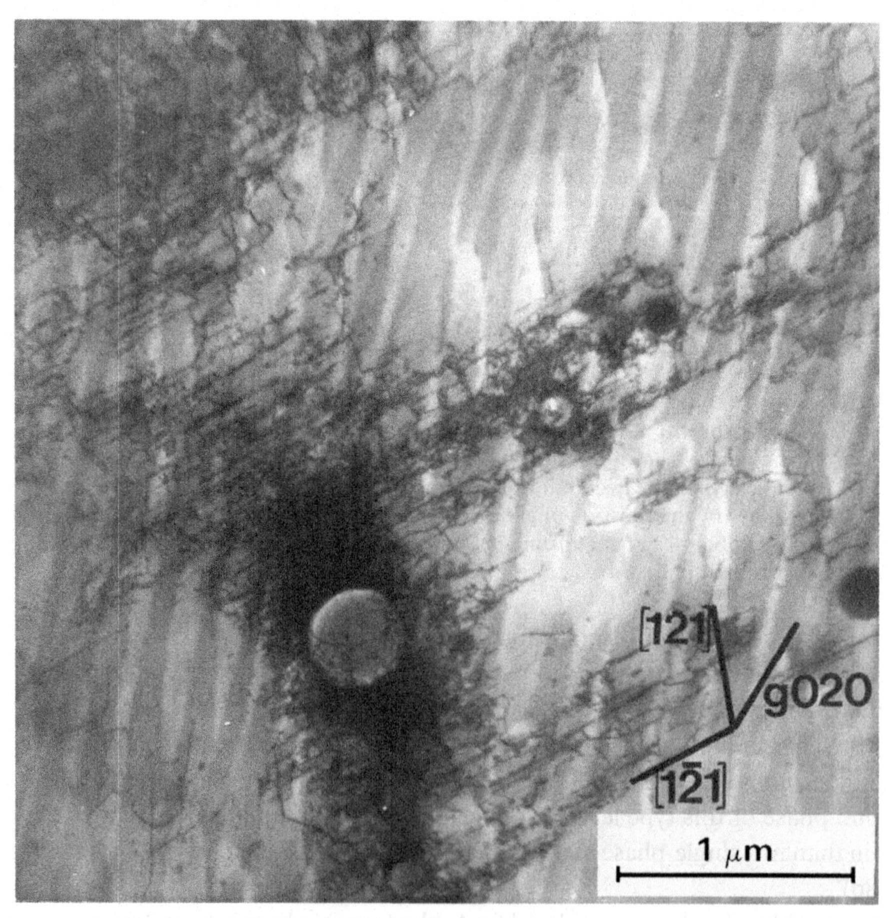


**Figure 4.11** Specimen showing intense local secondary slip activity adjacent to a large silica particle; ($\bar{1}$01) section. (After Rollason and Martin 1970.)

bending the matrix lattice. In fig. 4.10(*c*) loops of primary Burgers vector are formed and in fig. 4.10(*f*) dislocations of secondary Burgers vector are formed. Fig. 4.12 illustrates a dislocation structure corresponding to that of fig. 4.10, and is of a two-phase crystal similar to that illustrated in fig. 4.11 but involves a particle of smaller diameter: prismatic loops of primary Burgers vector are seen to be aligned in a direction parallel to the primary slip direction.

This transition from primary to more complex slip processes in the matrix near undeformable particles appears to be a function of the applied strain as well as the size of the particles. The effect has been studied in detail by Humphreys and Stewart (1972) for the case of silica particles in brass crystals and fig. 4.13 illustrates the conditions under which a simple array of primary dislocations (similar to that illustrated in fig. 4.12) is replaced by more complex interactions as particle size or strain is increased.

Early studies of the effect of a precipitate on the dislocation distribution at

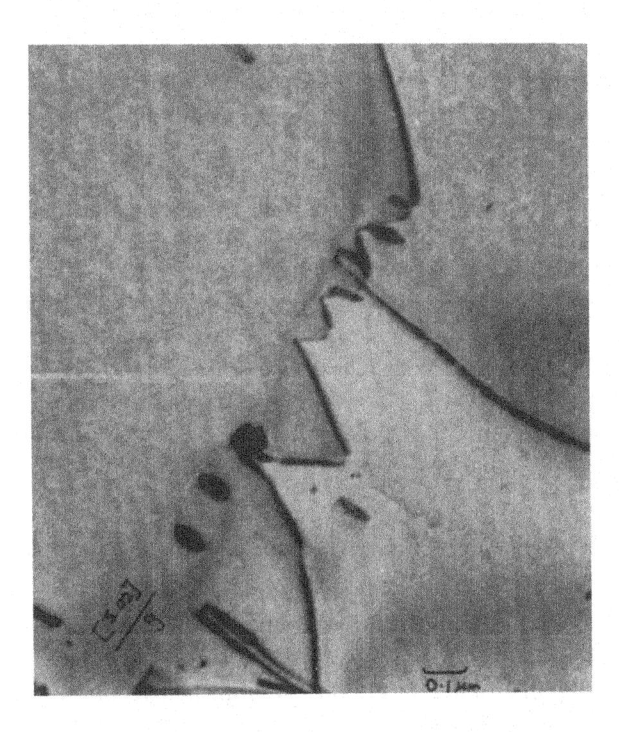


**Figure 4.12** Specimen showing primary prismatic loops aligned with small silica particle; (111) section. (After Humphreys and Martin 1967.)

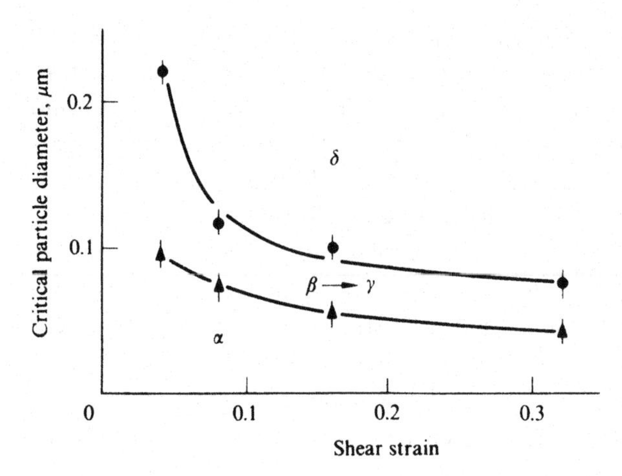


**Figure 4.13** The critical particle size for transitions in structure: $\alpha$ – primary prismatic loops; $\beta$, $\gamma$, $\delta$ – dislocations of secondary Burgers vector. (After Humphreys and Stewart 1972.)

higher deformations were mainly confined to polycrystals, and the results have generally been interpreted in terms of the cell structures reported for single-phase polycrystals. Thus Lewis and Martin (1963) found that at small inter-particle spacings the formation of cells was inhibited at small strains, and at large strains a cell structure of the same scale as the interparticle spacing developed. With coarser dispersions, these authors and Leslie, Michalek and Aul (1963) found the cell boundaries to be associated with the precipitate.

Measurement of lattice misorientations in rolled internally oxidised silver and copper polycrystals (Brimhall and Huggins 1965) has shown that the misorientation across cell boundaries is reduced by the presence of the second phase. X-ray line broadening techniques (Klein and Huggins 1962) have also shown the tendency of the precipitate to produce a more homogeneous distribution of dislocations. These results have been interpreted in terms of slip on many systems, and particles not only acting as barriers to dislocations but also as sources.

More recently, Baker and Martin (1983a) have used transmission electron microscopy to examine the effect of fine (~50 nm) particles on the deformation structure in cold-rolled copper single crystals. They found that dislocation cells formed at lower strains than in pure copper, and although smaller cell sizes occur in two-phase than in single-phase material, *larger* misorientations were found across the cell walls. More prolific microbands were observed in the two-phase crystals, and their density increased with strain.

More detailed quantitative work has been carried out by electron microscopy to estimate the dislocation distributions in two-phase copper crystals containing coarser particles (for example, Humphreys and Martin (1967)). Sectioning of crystals that had been deformed into stage II of the stress–strain curve showed that the precipitate caused lattice misorientations to be decreased, and less inhomogeneous dislocation structures occurred than in single-phase crystals. Fig. 4.14 shows the dislocation distribution in a crystal of Cu – 2.1 wt% Co alloy

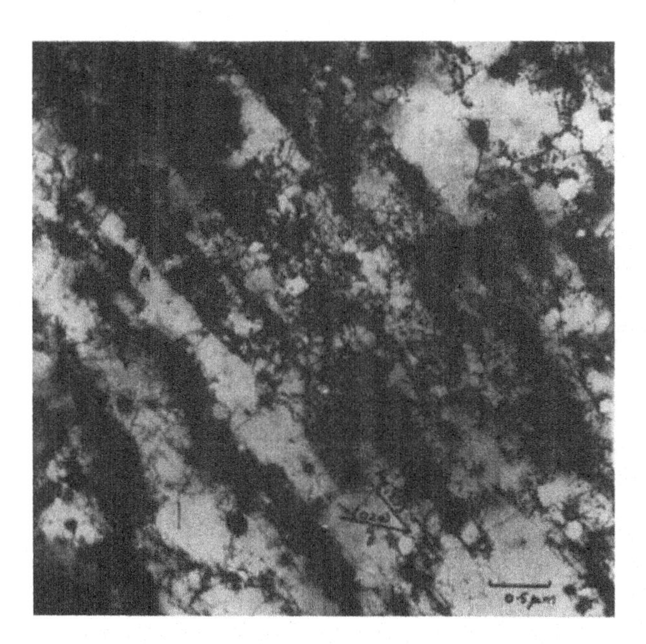

**Figure 4.14** Cu – 2.1 wt% Co, $\gamma$=0.64 ($\bar{1}01$) section. (After Humphreys and Martin 1967.)

aged to form a dispersion of cobalt particles (average diameter 80 nm), and subjected to a shear strain of 0.64. Comparison with fig. 4.5 shows that the dislocation distribution is profoundly changed. In the single-phase crystal, well-defined slip planes are seen and the alternating light and dark contrast across them indicates lattice misorientations which have been measured by means of Kikuchi lines and found in general to be $60' \pm 10'$. In fig. 4.14 the primary slip planes can still be detected, but there is little contrast across them and no misorientations of greater than $10'$ were detected.

In conclusion, we may summarise by saying that slip distances and the number of dislocations per slip line are smaller in two-phase alloys than in single-phase alloys for equivalent stresses, so the structure in stage II is less inhomogeneous in two-phase crystals, and lattice misorientations are lower.

For large particles and/or at high strains, the complexity of the problem precludes a consideration of the movement of individual dislocations. A more macroscopic continuum approach has been made by Humphreys (1983) which can account for the structures observed in the electron microscope in such deformed material. Fig. 4.15(*a*) illustrates a single-crystal matrix deforming on a single slip plane AB intersecting a particle, and fig. 4.15(*b*) shows the particle deforming with the matrix. If the particle is undeformable, then local, plastic deformation must restore the particle to its original shape, and fig. 4.15(*c*) illustrates one way of effecting this. Here a local rotation of a portion of the matrix in the sense shown has occurred: in three dimensions, rotation of a suitable portion of the matrix about *any* axis normal to the slip vector can produce the required shear across CD.

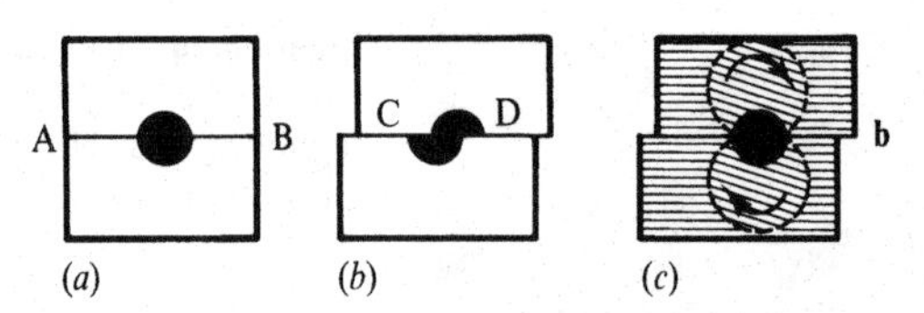


**Figure 4.15** The origin of a deformation zone at a particle. (After Humphreys 1977.)

The size of the deformation zone, and the distribution of orientations within the zone are considered in a model by Humphreys and Kalu (1990), which is based on a modification of Taylor's polycrystalline plasticity theory. They obtain reasonable agreement with measurements of these parameters by means of a transmission electron microscopy microtextures technique.

Humphreys (1983) summarises the effects of applied strain and of particle size upon the local dislocation distribution by means of the deformation mechanism map shown in fig. 4.16. The behaviour summarised by fig. 4.16 can be taken to be representative of an alloy system with hard particles of similar elastic modulus to the matrix, and in which decohesion does not occur at the particle matrix interface.

## 4.2 Recovery

Whenever the kinetic conditions are favourable, a cold-worked metal returns in the direction of the unworked or annealed state, the stored energy providing the driving force for the processes involved. The first distinction to be made is between the processes of recovery and recrystallisation; the term recovery embraces all changes which do not involve the sweeping of the deformed structure by migrating high angle grain boundaries, but in recrystallisation the crystal orientation of any region in the deformed material is altered, because of the passage through the material of high angle grain boundaries.

### 4.2.1 The kinetics of recovery

The kinetics of recovery have been reviewed by Bever (1957) who pointed out that any simple formulation of the rate of recovery is at best an approximation. Since the cold-worked state is complex and non-uniform on a fine scale, a single parameter cannot describe this state adequately, and any rate equation in terms of a single parameter is unlikely to represent the kinetics of recovery for more than a restricted range.

If we call $x$ the instantaneous value of some property in the cold-worked condition, equations intended to express the kinetics of the recovery of cold-worked metals begin with the assumption that, at constant temperature, the rate of decay of $x$ depends on $x$, that is,

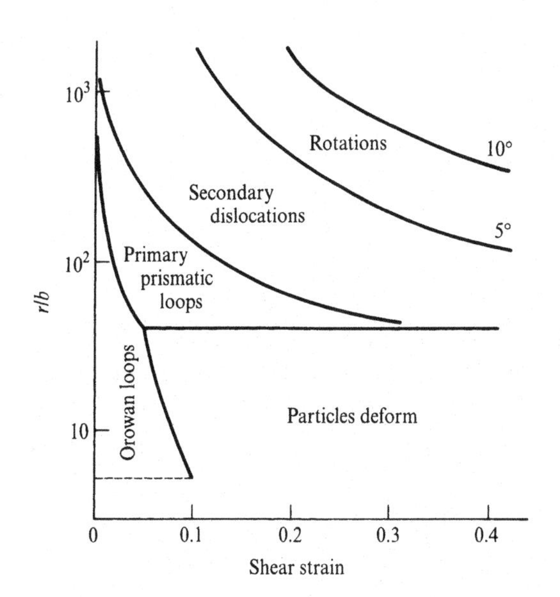


**Figure 4.16** The relationship between the various deformation mechanisms in copper containing silica particles.

$$dx/dt = -\alpha x \tag{4.14}$$

According to this equation, $x$ decreases exponentially with time, and such behaviour has been observed (Åström 1955) for the release of stored energy during one stage of recovery, but it is not the only pattern known.

In eq. (4.14), $x$ has also been considered as the intensity of the lattice imperfections that cause the property changes in the cold-worked metal. In this analysis $\alpha$ (eq. (4.14)) represents the probability that one imperfection disappears in unit time, which is expressed by the relation

$$\alpha = A \exp(-E/kT) \tag{4.15}$$

where $E$ is an activation energy, $k$ is Boltzmann's constant, and $A$ is another constant. Putting eq. (4.15) into eq. (4.14) and integrating gives

$$\ln(x_0/x) = -K \exp(-E/kT)t \tag{4.16}$$

if $E$ is not a function of $x$. At constant temperature, this reduces to

$$\ln(x_0/x) = Ct \tag{4.17}$$

In several investigations, rates of recovery have been observed not to follow the first-order kinetics predicted by eq. (4.17) but they can be expressed approximately by the equation

$$x=b-a\ln t \qquad (4.18)$$

in which $a$ and $b$ are constants for a given temperature. This equation obviously cannot represent the recovery process of very small or very large values of the time, since, physically, $x$ must be finite at zero time and must approach zero as time approaches infinity.

The rate of recovery increases with increasing temperature, as in other thermally activated processes. By finding an activation energy for a given process, one may throw light on the mechanism or mechanisms by which the process occurs. In some instances it has been found experimentally that several activation energies apply within the recovery range of a single specimen, which implies that different processes or mechanisms predominate in different parts of the range.

Kuhlmann (1948) has suggested that the activation energy for recovery is a function of time during isothermal annealing, and that the rate of recovery is then

$$\mathrm{d}x/\mathrm{d}t=-\alpha\exp[-(E_0-bx)/kT] \qquad (4.19)$$

which leads to an equation of the type of eq. (4.18), that is, if the activation energy is a linear fuction of the recovering property, a logarithmic time relation holds. The possibility of $E(E_0-bx)$ increasing during recovery is compatible with the assumption that the most severely deformed regions, where the stored energy $x$ is highest and the activation energy is lowest, recover first and, in general, it is found experimentally that a number of mechanisms operate when a wide enough temperature range is explored.

## 4.2.2 Mechanisms of recovery

We will make a rough classification of the mechanisms by which the stored energy of cold work is gradually annealed out during progressive heating in terms of those mainly involving *point defects* (which is the class with lowest activation energy), those involving *linear defects*, and finally those involving *planar defects* (which is the class with the highest energy of activation).

### Recovery involving point defects

Balluffi, Koehler and Simmons (1963) reviewed the state of knowledge about point defects in deformed fcc metals, and concluded that the presence of dislocations and impurities and the interaction of point defects with one another introduce so many complexities that the detailed interpretation is very difficult. A clearer understanding of the basic properties of point defects in metals has been gained from simpler experiments involving: (*a*) measurement of equilibrium defect concentrations at elevated temperatures; (*b*) quenching in of defects from elevated temperatures; and (*c*) radiation damage. We can here only briefly touch on some of the main mechanisms proposed, and will confine ourselves principally to a consideration of recovery processes in quenched metals, which have also been reviewed by Kimura and Maddin (1971).

The recovery of the quench-hardened state has been investigated in gold, copper, silver and aluminium, but at present gold and aluminium are the only metals in which the defect responsible for the hardening has been identified. In quenched gold the most frequently observed defects are stacking fault tetrahedra, although under certain quenching conditions voids and dislocation loops are also observed. Tetrahedra are observed in gold quenched from above 800 °C, and very small (2.5 nm) defects called 'black spot' defects form in gold foils quenched from below 800 °C and then annealed. Voids have been observed in gold by Yoshida, Kiritani, Shimomura and Yoshinaka (1965) and by Clarebrough *et al.* (1967); void formation is favoured by slow quenching rates (for example, into paraffin oil) and high ageing temperatures (for example, 150 °C). It is, of course, particularly important to understand the conditions which can lead to formation of voids during recovery in situations such as nuclear reactors where point defects arise from radiation damage.

Dislocation loops of various sizes were observed by Clarebrough, Segall, Loretto and Hargreaves (1964) in gold, with very large loops for specimens quenched from 1030 °C at a very slow rate (quenched into carbon tetrachloride).

The quench-hardening characteristics of quenched aluminium are not so sensitive to quenching temperatures as those in copper and gold. Voids, both perfect and prismatic dislocation loops and Frank sessile loops as well as heavily jogged dislocations, are seen in quenched and aged aluminium. The number of loops increases with increasing quenching temperature, while the number of voids decreases. In the case of copper it is difficult to identify the defects responsible for quench-hardening, as it is not yet well known what are the most important factors in controlling the formation of vacancy condensation products – the impurities, the quenching rate or the annealing conditions. Kimura and Maddin (1971) have reviewed the subject in depth and reference should be made to their work for further details.

**Gold** The stacking fault tetrahedra observed in quenched gold exhibit a remarkable stability and a correspondingly high activation energy for their disappearance on annealing. Meshii and Kauffman (1960) showed that quench-hardened gold maintained its high yield stress on annealing at temperatures up to 600 °C, and their observation is consistent with the results of investigations on the recovery of resistivity remaining after low temperature annealing, and on the annealing of stacking fault tetrahedra as observed by transmission electron microscopy (for example, Cotterill (1961)).

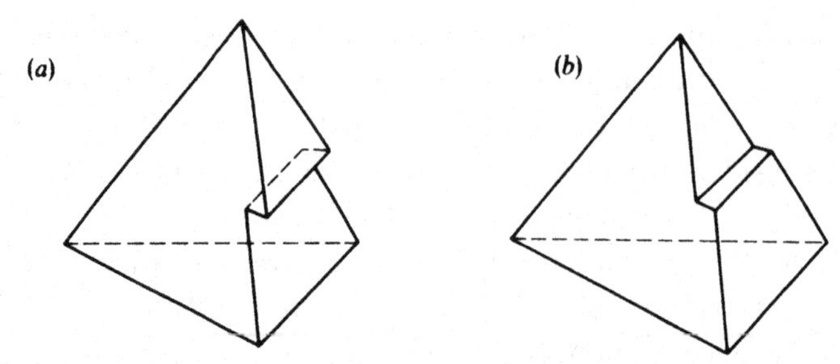


**Figure 4.17** Two types of ledges on a stacking fault tetrahedron: (*a*) a V-ledge, which moves toward the corner to be eliminated; (*b*) an I-ledge, which moves to the edge to be eliminated. The width of the ledges is one atomic distance.

The half-decay time for the recovery of the yield stress at 642 °C was 80 min, and the activation energy for the recovery was found to be 4.8 eV. This large activation energy also implies that the frequency factor in the rate equation for this process should be large, but the reason for this high value is as yet not known. The high stability of these defects and the mechanism of their ultimate disappearance have been discussed in detail by Kuhlmann-Wilsdorf (1965), whose proposal is summarised below.

The growth of stacking fault tetrahedra may proceed by the migration of ledges of stacking fault over their faces, and two types of ledge exist, as indicated in fig. 4.17. If the concentration of point defects (vacancies or interstitialcies) is near equilibrium, the ledges will move towards corners (fig. 4.17(*a*)) or edges (fig. 4.17(*b*)) in order to be eliminated. The nucleation of new ledges is impossible unless there is a super- or sub-saturation of point defects with respect to a critical concentration, $C_{crit}$. The migration of I-ledges by absorption of interstitial atoms is unlikely since no supersaturation of interstitial atoms exists at these annealing temperatures. Again a subsaturation of vacancies below $C_{crit}$ is improbable, since it would be eliminated as the temperature rose to the annealing temperature by the emission of vacancies from other sources.

Kuhlmann-Wildsdorf (1965) suggests that the tetrahedra continue to grow by the absorption of vacancies supplied from climbing network dislocations, voids or loops. Upon reaching a critical size the tetrahedra collapse spontaneously into loops which represent a lower energy configuration than the tetrahedra (the details of the process are discussed in modern textbooks on physical metallurgy). The loop then emits vacancies for further growth of other tetrahedra and eventually all tetrahedra are eliminated. In an alternative theory the nucleation of V-ledges at the edges of the tetrahedra is considered.

**Copper** Kimura and Hasiguti (1962) have investigated the recovery of quench-hardening in copper, and found that their specimens maintained their high yield stress up to an annealing temperature of 700 °C – the activation energy

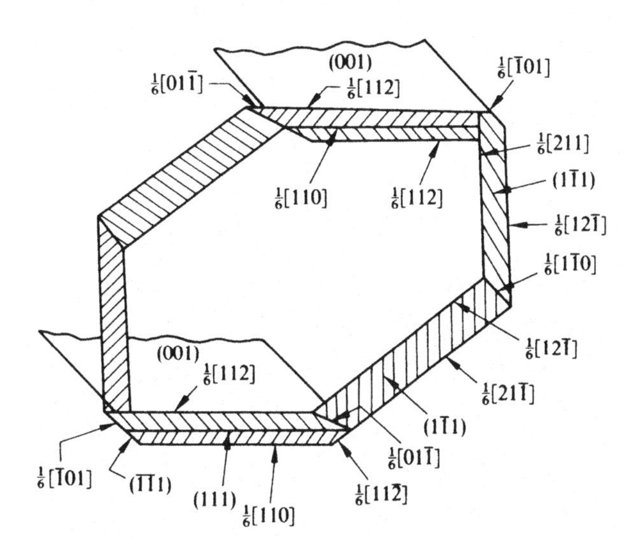


**Figure 4.18** An extended prismatic dislocation loop on a (111) plane with a total Burgers vector of $\frac{1}{2}[110]$. (After Kimura and Hasiguti 1962.)

for the recovery and the frequency factor being quite similar to those in gold. The interpretation is not straightforward, however, since more than two kinds of defects are possible causes of hardening in copper.

Three-dimensional voids could not be stable at temperatures as high as 700 °C (since they disappear at lower temperatures by emitting vacancies), and Frank sessile loops are ruled out by the same reasoning. Stacking fault tetrahedra are a quite likely cause of hardening, in which case the same argument as for gold would apply, but prismatic dislocation loops may also be possible if they extend on the plane parallel to its total Burgers vector, as shown in fig. 4.18. If the degree of extension is appreciable, the energy to form a jog on it would be very high (of the order 3 eV) and the jog would have to be accompanied by vacancies emitted from it for the loop to shrink. The total activation energy for the whole process would then be about 5 eV (3 eV + self diffusion energy of 2 eV), which is in agreement with experimental observation.

**Aluminium** Fig. 4.19 shows the resoftening of quench-hardening in 99.999% pure aluminium and the disappearance of dislocation loops by annealing for 30 min at the temperatures shown. The loop density is seen to decrease rapidly at 150 °C, while resoftening occurs in two stages, one from 100 to 220 °C and the other near 260 °C. Kinked dislocations were observed above 180 °C, after the disappearance of the prismatic loops characteristic of the lower temperature range, and Shin and Meshii considered that the kinked dislocations formed at the expense of the loops. The second stage of softening was thought to be due to the straightening of these kinked dislocations, which impurity segregation would have stabilised.

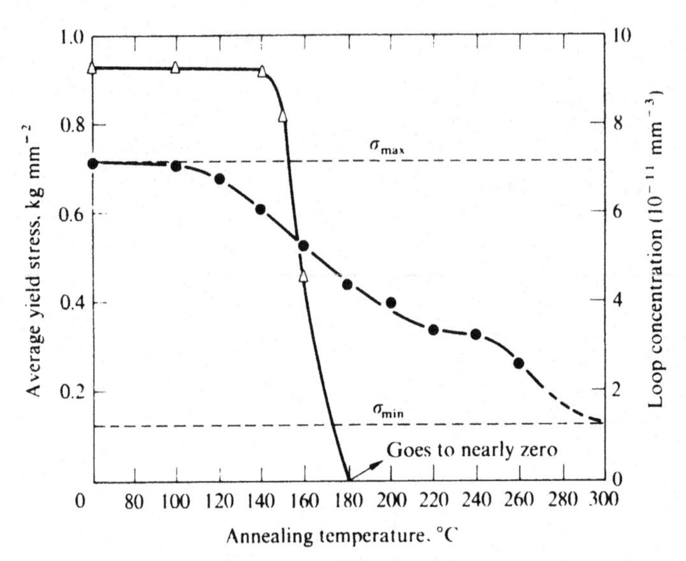


**Figure 4.19** Resoftening of quench-hardening in aluminium and disappearance of dislocation loops by annealing for 30 min at the temperatures shown. ● yield stress; △ loop concentration. (After Shin and Meshii 1963.)

**Silver** Fig. 4.20 shows the resoftening of quench-hardening in 99.999% pure silver by annealing for 1 h at the temperatures shown. A two-stage resoftening is apparent – the first near 350 °C was accompanied by the elimination of 'black spot' contrast, and the second stage near 700 °C was due to the elimination of stacking fault tetrahedra.

### Recovery involving dislocations in cold-worked metals

In discussing recovery mechanisms involving point defects, we have inevitably considered the behaviour of the dislocations produced by vacancy aggregation. We will now turn to a consideration of cold-worked rather than quench-hardened metals, and thus be concerned with changes in the dislocation arrays associated with the deformed state.

In the case of copper crystals, Humphreys and Martin (1967, 1968) report that crystals deformed into stage II and stage II recrystallised rapidly on annealing at 700 °C, and no significant dislocation rearrangement was observed in the crystals prior to their recrystallising. In the case of crystals deformed into stage I, however, it was found that no specimen recrystallised. All crystals developed structures of low dislocation densities by a process of recovery. The edge dipoles, prominent in the deformed structures (fig. 4.4) were absent, and had presumably been annihilated by climb.

There appear to be three main processes whereby the excess stored energy due to dislocations may be released by recovery. Firstly, the elimination of dipoles; secondly, the mutual annihilation of dislocations of opposite sign; and finally,

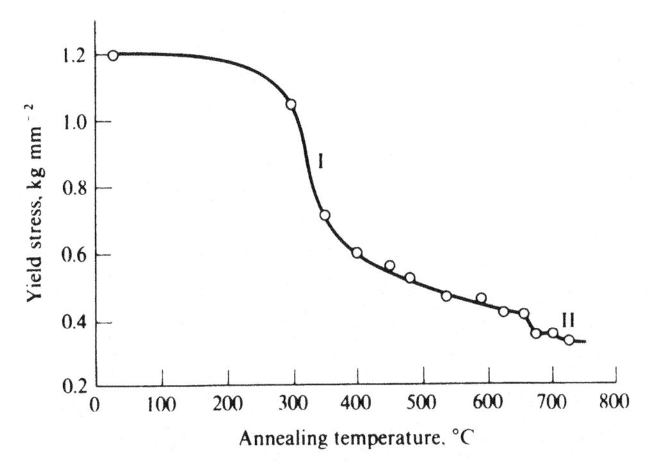


**Figure 4.20** Resoftening of quench-hardening in silver by annealing for 1 h at temperatures shown. (After Mori and Meshii 1963.)

the rearrangement of the remaining dislocations into more stable arrays possessing lower values of strain energy ('polygonisation').

The first of these elementary recovery processes has been treated theoretically by Li (1966), who has considered the annihilation of screw, edge and mixed dislocation dipoles. If the separation of the dipole is $x$, the average velocity of a dislocation may be assumed to be proportional to a power function of the force exerted on it, so that dipole separation will decrease with time as

$$dx/dt = -2\beta f^m \tag{4.20}$$

where $B$ is the mobility (the velocity produced by a unit driving force), $m$ is the dislocation velocity and $f$ is the force of attraction between the pair of dislocations constituting the dipole.

The simplest dipole is a pair of parallel screw dislocations of opposite signs, when $f = \mu b^2/2\pi x$ ($\mu$ being the shear modulus and $b$ the Burgers vector), and if this is substituted in eq. (4.20), we obtain on integrating:

$$x_0^{m+1} - x^{m+1} = \alpha t \tag{4.21}$$

where $\alpha = 2B(m+1)(\mu b^2/2\pi)^m$ and $x_0$ is the separation at $t=0$. Since the dipole disappears when $x = r_0$, the core radius, a dipole of separation $x_0$ has a life $t_f$ given by

$$t_f = (x_0^{m+1} - r_0^{m+1})/\alpha \approx x_0^{m+1}/\alpha \tag{4.22}$$

Eq. (4.22) is plotted in fig. 4.21 to show how the life of a dipole is affected by the mobility $B$, the stress sensitivity of the dislocation velocity, $m$, and the separation, $x_0$.

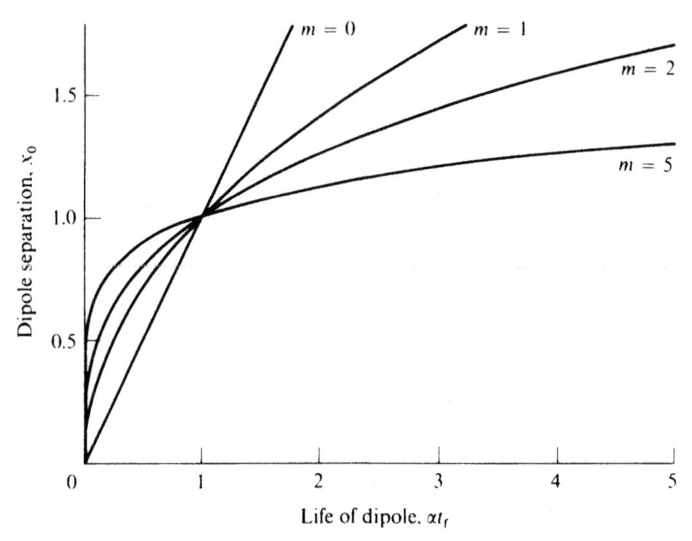


**Figure 4.21** Effect of separation and mobility on the life of screw dislocation dipoles, $\alpha t_f = x_0^{m+1}$; $\alpha = 2B(m+1)(\mu b^2/2\pi)^m$; $V = Bf^m$. (After Li 1966.)

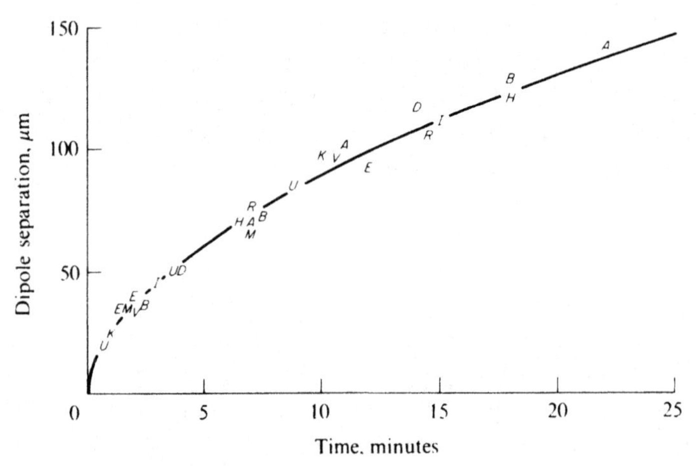


**Figure 4.22** Life of screw dislocation dipoles in LiF crystals annealing temperature 140 °C. (After Li 1966.)

Li (1966) shows the data of Lee, Hu and Li (unpublished results) who measured the separation of screw dislocation dipoles as a function of time during the annealing of lithium fluoride crystals. These results are shown in fig. 4.22 where each letter represents a single measurement; a set of measurements on the same dipole is represented by the same letter. A comparison of fig. 4.22 with fig. 4.21 shows that $m$ is about unity, and this value was found for the five temperatures studied.

The behaviour of edge dislocation dipoles is more complicated than that of

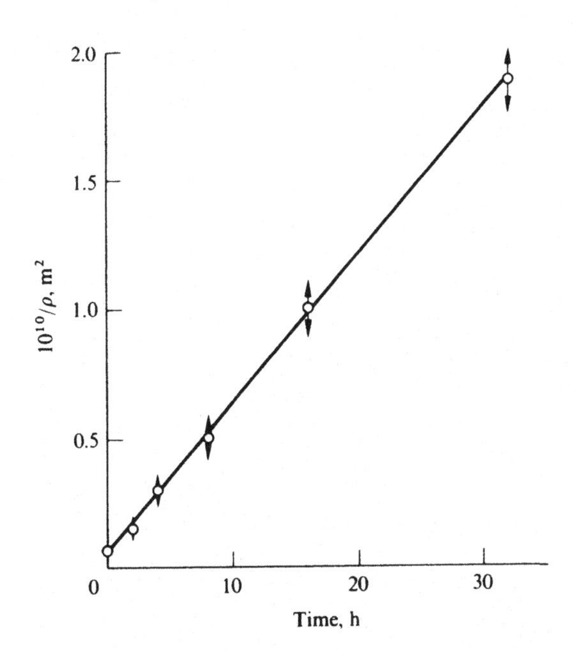


**Figure 4.23** Second-order kinetics of the annealing of dislocation density in LiF crystals. Primary dislocations, 500 °C. $d\rho/dt = -k\rho^2$; $1/\rho - 1/\rho_0 = kt$. (Data of Sankaran and Li, after Li 1966.)

screw dislocation dipoles because of the existence of tangential forces between edge dislocations, but Li's analysis (1966) again predicts a dipole life given by an expression of similar form to eq. (4.22), that is, corresponding to a value of unity for $m$.

The above analysis gives the kinetics for the annihilation of an individual dipole, but to deal with all the dipoles in a deformed specimen a statistical treatment is necessary. By assuming the existence of a set of non-interacting dipoles, Li shows that the kinetics for their annealing corresponds to a simple second-order process

$$d\rho/dt = -k\rho^2 \tag{4.23}$$

where $\rho$ is the dipole density and $k$ is a constant. This idealised model could be a good approximation when the amount of cold work is small, and such kinetics have been observed by Sankaran and Li (Li 1966) in a lithium fluoride crystal after 1% compressive strain and annealing at 500 °C. The reciprocal of the dislocation density is plotted in fig. 4.23 as a function of the annealing time, and from the slope of the straight line obtained a second-order rate constant $k = 1.7 \times 10^{-13}$ mm$^2$ s$^{-1}$ is obtained.

Subsequent to or concurrent with the disappearance of dislocation dipoles during annealing, the excess dislocations of one sign tend to rearrange into lower energy configurations. The simplest process of this kind is *polygonisation*, which has been reviewed by Hibbard and Dunn (1957) and Li (1966).

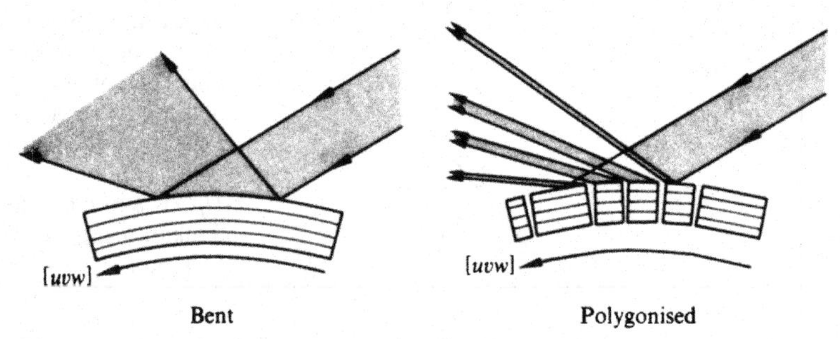


**Figure 4.24** Reflection of white radiation by bent and polygonised lattices.

Cahn (1949) used this term to describe the formation of edge dislocation walls during the annealing of bent single crystals. Crystals of zinc, aluminium, magnesium and sodium chloride were bent about an axis parallel to their active slip planes. Transmission Laue patterns from these crystals taken with the X-ray beam normal to the bent planes showed the expected continuous asterisms, but when a bent crystal was then annealed close to its melting point and a second Laue pattern taken, the asterisms became discontinuous. Cahn explained the break-up of the asterism as due to the formation of walls of dislocations perpendicular to the glide planes as shown schematically in fig. 4.24. The term polygonisation describes the fact that a certain crystallographic direction forms part of an *arc* before annealing and part of a *polygon* afterwards. The elongated Laue streak is thus replaced by a row of small sharp spots – one spot from each of the individual blocks which exist between the walls of dislocations forming perpendicular to the glide planes that were active during the bending.

Cahn (1949) succeeded in obtaining metallographic evidence for this rearrangement of dislocations by the use of etch pits which formed traces in continuous rows on surfaces perpendicular to the active slip direction, as seen in fig. 4.25. To understand the process, it must be that the effect of *bending* the crystal is to introduce a number of excess dislocations of one sign into the crystal to accommodate the curvature. Thus the large difference in dimension between the outer and inner surfaces of a bent crystal implies the presence of many extra lattice planes near the outer surface which terminate at edge dislocations within the crystal (fig. 4.24). Upon annealing, these dislocations rearrange themselves into walls which have a lower total elastic strain energy than the more random dislocation arrangement, and this lowering of strain energy provides the driving force for the process.

Both climb and glide of edge dislocations are required to form the walls, and these processes require thermal activation and determine the rate of polygonisation. Gilman (1955) made the first quantitative study of polygonisation in zinc and found the average angle of polygon walls to vary linearly with the logarithm of time at each temperature. These and similar subsequent data have been

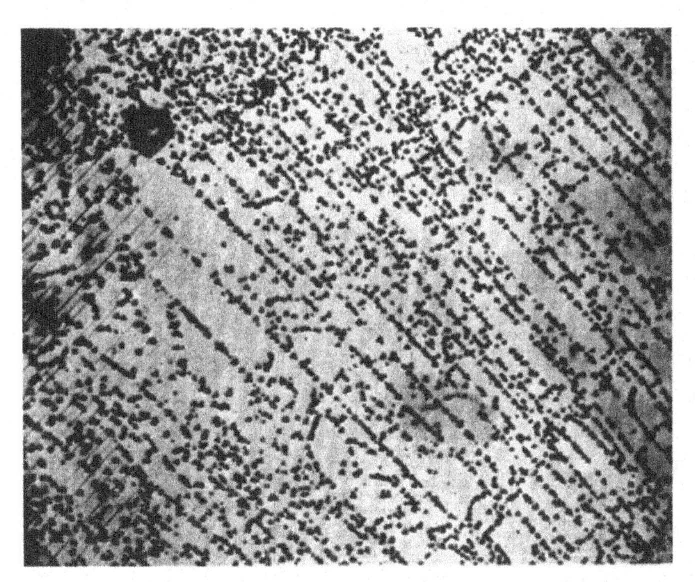

**Figure 4.25** Dislocation etch pits delineating polygonised walls normal to slip planes which are apparent from slip traces induced by further bending after sectioning and etching. Aluminium crystal initially bent and then annealed at 625 °C. (After Cahn 1949.)

accounted for by Li (1960) by assuming that the rate-controlling process is the nucleation of the 'Y' junction, whose growth transforms two walls into one (fig. 4.26). Numerous Y junctions are observed in all polygonisation experiments, and Li predicted that the activation energy for the polygonisation process should increase with the angle of misfit of the wall, and also showed that this result is a direct mathematical consequence of the fact that the angle of the wall is linear with the logarithm of time at each temperature. The predicted variation of activation energy with angle of misfit of the wall has been observed by Feltner and Laughhunn (1962) from the data of Gilman (1955) in zinc, and they also found that the data of Hibbard and Dunn (1956) on Si–Fe and those of Sinha and Beck (1961) on zinc both showed the effect.

Owing to application of electron microscopy to the study of dislocation distributions in metals, it is now recognised that polygonisation is a special form of subboundary formation. Most subboundaries are formed even during deformation, as already mentioned, and these are called 'cell walls'. During annealing, the dislocation distributions within the walls become more regular. A detailed analysis of the distribution of dislocations in recovered structures of two-phase copper crystals has been made by Humphreys and Martin (1967) by studying deformed and annealed specimens in the electron microscope from crystals sectioned in known relation to active slip systems.

Two copper based systems have been studied in this context – Cu–Co and Cu–$SiO_2$ (where the dispersion of silica is produced by the internal oxidation of

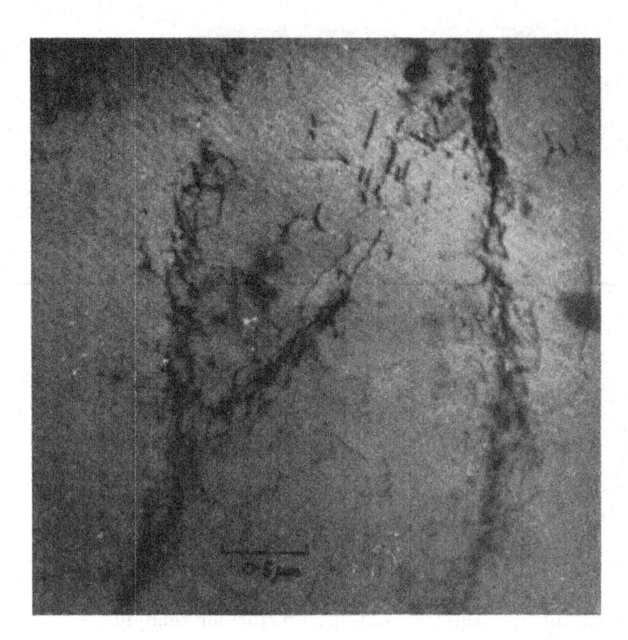

**Figure 4.26** Copper single crystal annealed at 950 °C for 15 h after deforming; (101) section showing 'Y' junction. (After Rollason and Martin 1970.)

a dilute Cu–Si alloy). In both systems a dispersed phase was produced which was of equiaxed form and of high dimensional stability over extended annealing periods, and the results of annealing at 700 °C were dependent upon the stage of deformation of the crystals.

As already discussed (p. 172), crystals deformed into stage I of the stress–strain curve recovered to low dislocation densities on annealing by a process of dipole annihilation by climb. Single-phase crystals deformed into stage II or stage III recrystallised rapidly on annealing, and no significant dislocation rearrangement was observed in these crystals prior to their recrystallising.

None of the two-phase crystals recrystallised, however, but they all recovered within a few minutes to structures consisting of dislocation networks. Sections of annealed crystals cut parallel to the primary slip plane (111) showed (fig. 4.27) extensive network formation on planes close to (111). The networks were complicated and could best be described as small regions of relatively simple networks extending 1–2 $\mu$m, but containing many singular dislocations.

A Burgers vector analysis of the nets in fig. 4.27 showed that this structure consisted of primary dislocations, $a/2[\bar{1}01]$, in an approximately screw orientation, running from $C$ to $B$, which had reacted with two sets of dislocations. At $A$ they have reacted with $a/2[101]$ dislocations which results in square nodes, and at $B$ they have interacted with either $a/2[01\bar{1}]$ or $a/2[1\bar{1}0]$ to form a hexagonal network with Burgers vectors coplanar with (111).

The effect of the precipitate on the network is seen to cause short range disturbances of the regular net, as at $C$, and as the precipitates are frequently asso-

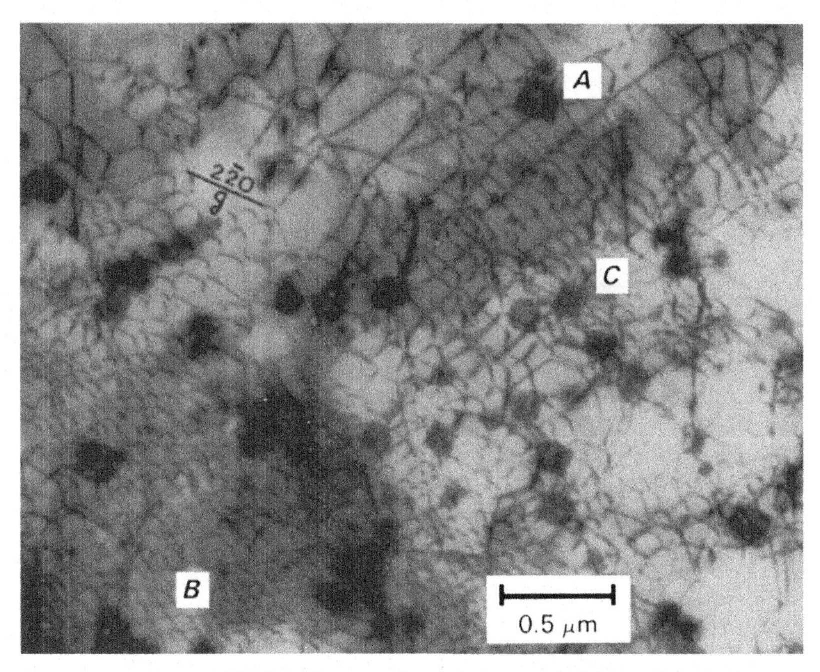


**Figure 4.27** Cu – 2.1% Co. Shear strain $\gamma$=0.64, anealed 2 h at 700 °C; (111) section. (After Humphreys and Martin 1967.)

ciated with stray dislocations, this results in further distortions of the net in the vicinity of the particle.

Examinations of sections perpendicular to the primary slip plane, $(\bar{1}01)$, showed the occurrence of boundaries consisting mainly of straight dislocations parallel to the $[1\bar{2}1]$ direction, as in fig. 4.28. These were determined to be primary $[\bar{1}01]$ dislocations, and from their orientation it can be seen that they are in an edge configuration, and thus constitute a tilt boundary. Many secondary dislocations also thread this boundary, and the plane of these boundaries was usually nearer to the conjugate plane $(\bar{1}\bar{1}1)$ than $(\bar{1}01)$ – which would be the plane of an unstressed pure tilt boundary. Several of the other networks were analysed in detail by Humphreys and Martin (1967), and their formation and distribution could be accounted for in terms of a modification of the structures formed during deformation.

With regard to the behaviour of bcc metals, in the case of iron of very high purity (produced by zone refining), Talbot (1963) has found that recrystallisation on annealing was never observed after straining in tension. The original small grains were retained, and they exhibited a polygonised structure, the ease with which polygonisation occurs being a basic characteristic of very high purity iron. This is presumably due to the high value of the stacking fault energy in the case of iron and to the absence of solute pinning of dislocations – both of which will permit ready cross-slip and climb of dislocations.

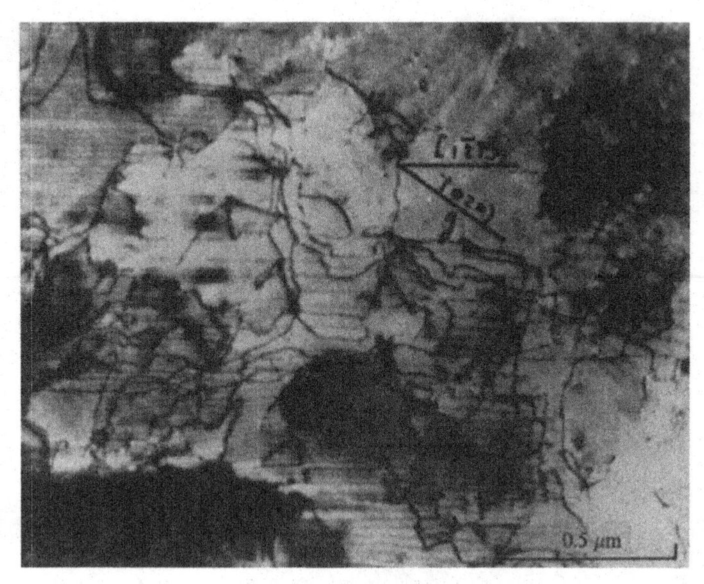


**Figure 4.28** Complex tilt boundary in Cu – 0.05% Si. $\gamma$=0.52, annealed 2 h at 700 °C; ($\bar{1}$01) section. (After Humphreys and Martin 1967.)

Talbot's experiments showed that recrystallisation occurs in iron which has been strained to fracture in tension only when the iron is insufficiently purified. Thus single crystals of zone-melted iron cannot be prepared by the strain–anneal process: the necessary critical strain is greater than the elongation at the point of fracture in the tensile test. The resulting polygonised structure was found to be very stable, and could not be eliminated either by prolonged annealing at high temperature or by thermal cycling between 850 °C and room temperature, and Talbot concluded that the polygonised structure constituted a stable state of low energy in this material.

### Recovery involving subgrain reorientation

There is strong evidence from the study of thin foils by transmission electron microscopy (for example, Hu (1962), Fujita (1961)) that when deformed metals are annealed, the subgrain size gradually increases prior to the appearance of recrystallisation.

Li (1962) suggests that this size increase is due to a coalescence process in which some of the subgrain boundaries gradually disappear while at the same time the adjoining subgrains merge into the same orientation. Fig. 4.29 represents schematically how the process is envisaged; dislocations must gradually move out of the disappearing subgrain boundary by a process involving climb, and the subgrain itself must rotate. In fig. 4.29 the boundary $CH$ is shown being eliminated with the rotation of one subgrain, which involves atom diffusion along boundaries so that matter is transferred from the shaded areas to the cor-

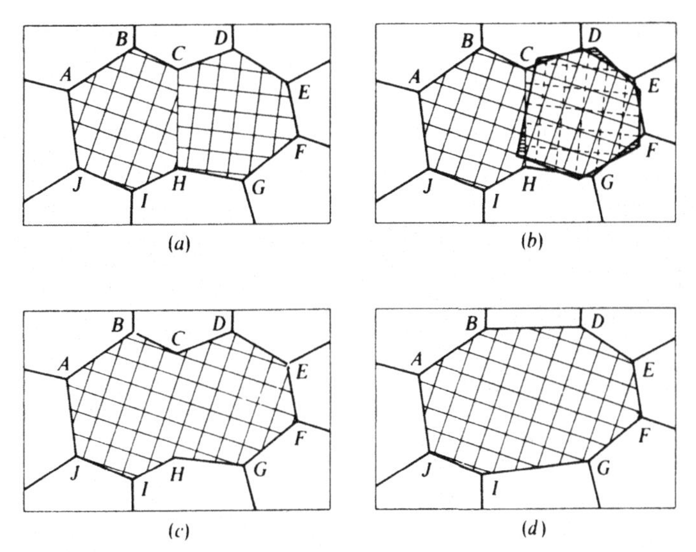


**Figure 4.29** Schematic representation of subgrain coalescence by subgrain rotation. (*a*) Original subgrain structure before coalescence; (*b*) one subgrain is undergoing a rotation; (*c*) subgrain structure just after coalescence; (*d*) final subgrain structure after some subboundary migration. (After Li 1962.)

responding open areas. Hu (1969) has observed a similar process in subgrains in heavily rolled copper, while Bay (1970) has observed it in cold-rolled aluminium.

Li has examined the energetics and kinetics of subgrain reorientation in some detail since this is a necessary mechanism of subgrain coalescence. In terms of a dislocation model (see §1.3.3) the free energy of a subboundary can be expressed as a function of angle, $\theta$, thus:

$$E = E_0\theta(A - \ln\theta) \text{ per unit area} \tag{4.24}$$

where $E_0A$ is the free energy at $\theta=1$ and $A$ is the logarithm of the angle at which the free energy is a maximum. The values of $E_0$ and $A$ will depend upon the type of boundary, but if two boundaries have the same $E_0$ and $A$ values and angles $\theta_1$ and $\theta_2$, they would coalesce into one boundary of angle $(\theta_1+\theta_2)$ with a driving force

$$\Delta F = E_0 \ln[(\theta_1+\theta_2)/\theta_1]^{\theta_1}[(\theta_1+\theta_2)/\theta_2]^{\theta_2} \text{ per unit area} \tag{4.25}$$

which is always positive; it is always thus possible to coalesce boundaries of the same type.

Any subgrain can rotate relative to all its neighbours with a driving force that depends on the axis of rotation and the orientations of all surrounding subgrains. This force can be calculated from the energy changes of all the surrounding subboundaries due to the rotation, and the subboundary that supplied most of the driving force is the one with the largest area, smallest angle, largest value of $E_0$, and a misfit axis nearest to the rotation axis in question. Li (1962) analysed in detail the disintegration of a vertical wall of edge dislocations – the disintegration process being considered to proceed either by the cooperative climb of edge dislocations such that the spacing between them remained uniform at all times, or by cooperative vacancy diffusion such that the subboundary maintained its low energy form at any angle of misfit. The first process yielded for the rate of rotation:

$$d\theta/dt=(3E_0\theta Bb/L^2)\ln(\theta/\theta_m) \tag{4.26}$$

where $b$ is the Burgers vector of edge dislocations in the wall, $\theta_m = \exp(A)$ (as in eq. (4.23)), $L$ is the half size of the wall and $B$ is the mobility of edge dislocations given by

$$B=Db^2j/kT \tag{4.27}$$

where $D$ is the self diffusivity, $j$ the jog density (number per unit length) along the dislocation, and $k$ is Boltzmann's constant. Eq. (4.26) can be integrated to give

$$\ln \ln (\theta_m/\theta)=(3BbE_0/L^2)t+\text{constant} \tag{4.28}$$

The second process (that of cooperative vacancy diffusion) predicts a rate of rotation:

$$d\theta/dt=(4DE_0/kT)(b/L)^3 \ln(\theta/\theta_m) \tag{4.29}$$

which can be integrated to give

$$\text{li}\,(\theta/\theta_m)=(4DE_0/\theta_m kT)(b/L)^3t+\text{constant} \tag{4.30}$$

where li is the logarithmic integral.

Li suggests that the first of the above processes is rate controlling when the jog density is low, the subgrain size or angle small, or there are many vacancy sources and sinks nearby, and the second process is controlling otherwise. When the first process is controlling, the time required for a coalescence is proportional to $L^2$, that is, $L^2$ is linear with time if subgrains coalesce through dislocation climb. $L^3$ would be linear with time if they coalesced through vacancy diffusion. The latter growth law has been observed by McGeary and Lustman (1953) in the annealing of cold-rolled zirconium, and their data are shown in fig. 4.30. On the other hand Dillamore, Smith and Watson (1967) measured the rate of subgrain growth

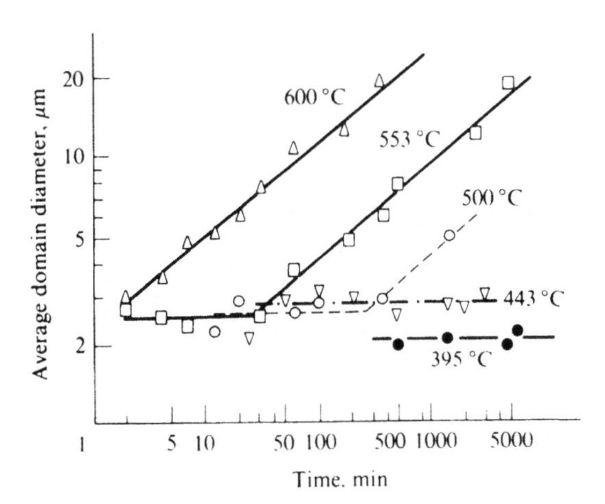


**Figure 4.30** Isothermal domain (subgrain) growth curves for 97% cold-rolled zirconium. Slope=0.36. (After McGeary and Lustman 1953.)

in deformed polycrystalline iron and found it to be much faster than that predicted by Li. This disagreement does not prove that coalescence cannot occur but does suggest that general subgrain growth by boundary migration is, in some cases at least, the faster process. An understanding of growth mechanisms of subgrains is important when considering nucleation of recrystallisation, and Doherty (1974) suggests that since the energy for subgrain coalescence comes from the difference between the high energy of a dislocation in a low angle boundary and the low energy in a higher angle boundary, coalescence of subgrains should be most likely to occur close to grain boundaries or in transition bands.

## 4.3 Recrystallisation

Recrystallisation results from the passage through the material of high angle grain boundaries by a process involving the nucleation of new, strain-free grains in the deformed material which then grow until all the latter is consumed. The formal theory of recrystallisation kinetics has been considered by Burke and Turnbull (1952), and we will review this first.

### 4.3.1 Kinetics of recrystallisation

Fig. 4.31 illustrates the sigmoidal form of the graph obtained of the fraction recrystallised versus the time of isothermal anneal. After an incubation time ($\tau$), recrystallisation can be described in terms of a nucleation frequency ($\dot{N}=dN/dt$, where $N$ is the number of nuclei per unit volume), and of $g$, the linear rate of growth of the new grains. The problem of expressing the volume fraction recrystallised ($X$) as a function of time ($t$) under these conditions is complicated by the

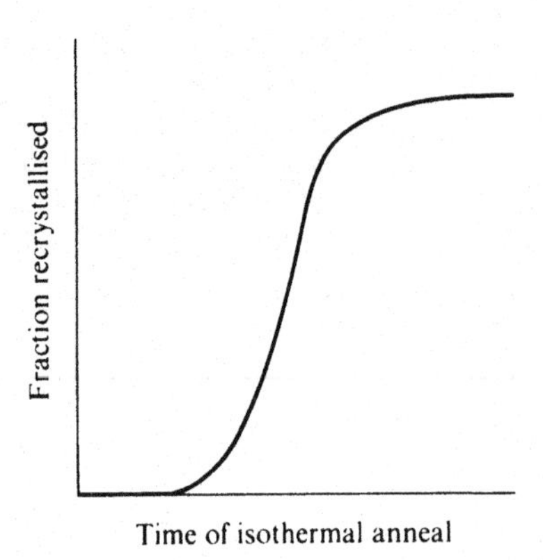


**Figure 4.31** Fraction recrystallised versus time of isothermal anneal.

situation that, as recrystallisation proceeds, the volume available in which new nuclei can form is continuously decreasing. Johnson and Mehl (1939) treated the problem by considering the number $dn'$ of nuclei originating in time interval $dt$ so as to include the number of 'phantom' grains that would have appeared in the volume $X$ had this volume not consisted of recrystallised grains, that is, they wrote:

$$dn' = dn + \dot{N}X\,dt \tag{4.31}$$

where $dn$ is the actual number of nuclei formed in time $dt$ (when the fraction recrystallised is $X$) and $\dot{N}Xdt$ is the number of these 'phantom' grains. If the volume of a recrystallising grain is $v$ at time $t$, we can now calculate what has been called the 'extended volume' of recrystallisation $X_X$:

$$X_X = \int_0^t v\,dn' \tag{4.32}$$

But we can write

$$v = fg^3\,(t-\tau)^3 \tag{4.33}$$

where $f$ is a shape factor and grains are growing linearly (a distance $g(t-\tau)$) in three dimensions, so we can write:

$$X_X = fg^3 \int_0^t (t-\tau)^3 \dot{N}\,dt \tag{4.34}$$

It is now necessary to relate $X$ and $X_X$. This can be done by considering at time $t$, the fraction unchanged $=1-X$. During a further time interval $dt$, the extended volume is increased by $dX_X$, and the true volume is increased by $dX$. Of $dX_X$, a fraction $(1-X)dX_X$ will lie in untransformed material, that is, $dX=(1-X)dX_X$, so we can write:

$$dX/dX_X = 1 = X \tag{4.35}$$

therefore

$$X_X\left(=\int_0^X dX_X\right)=\int_0^X \frac{dX}{1-X}=-\ln(1-X) \tag{4.36}$$

that is,

$$-\ln(1-X)=fg^3\int_0^t (t-\tau)^3\dot{N}d\tau$$

So if $\dot{N}$ is constant we have

$$X=1-\exp(-\tfrac{1}{4}fg^3\dot{N}t^4) \tag{4.37}$$

for three-dimensional recrystallisation.

If a thin sheet is recrystallising and the lateral dimensions of a recrystallising grain become much larger than the sheet thickness, we have a situation of 'two-dimensional' recrystallisation, and eq. (4.33) becomes

$$v=fg^2(t-\tau)^2\delta \tag{4.38}$$

where $\delta$ is the thickness of the sheet, so that eq. (4.37) becomes

$$X=1-\exp(-\tfrac{1}{3}fg^2\delta\dot{N}t^3) \tag{4.39}$$

Similarly, in the case of recrystallising fine wires of diameter $\delta$, if the length of the recrystallising grains becomes much longer than $\delta$ we have 'one-dimensional' recrystallisation with

$$v=fg\,\delta^2(t-\tau) \tag{4.40}$$

and

$$X=1-\exp(-\tfrac{1}{2}fg\delta^2\dot{N}t^2) \tag{4.41}$$

In fact, of course, $\dot{N}$ is not constant, but is some function of time, and in a classical series of experiments by Mehl and his coworkers $\dot{N}$ was measured by a statistical procedure and was found to be an exponential function of time:

$$\dot{N}=a\exp(bt) \tag{4.42}$$

where $\exp(bt)\gg a$ for all times observed.

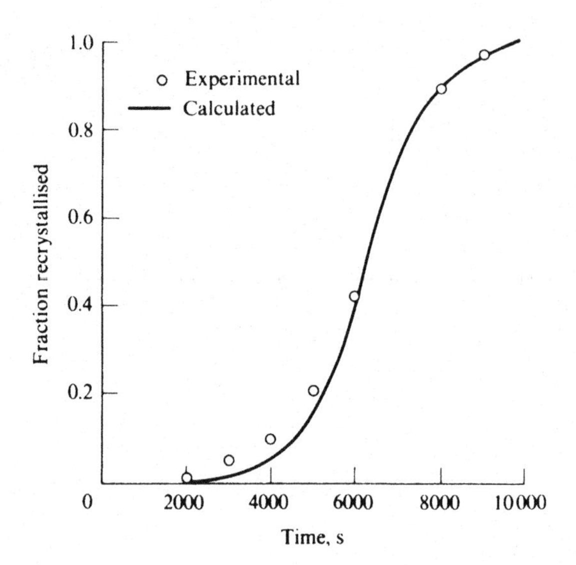


**Figure 4.32** Comparison of experimental data with curve calculated from impingement theory for the recrystallisation of aluminium; temperature=350 °C, $\epsilon$=0.051. (After Anderson and Mehl 1945.)

Anderson and Mehl (1945) assumed two-dimensional recrystallisation, with nuclei growing as perfect circles, with $\dot{N}$ an exponential function of time, and $g$ constant with time. They measured $g$ and $\dot{N}$, using aluminium strained 5.1% in tension annealed at 350 °C to calculate a curve of $X$ versus time from the relation

$$X=1-\exp\left\{\frac{2\pi g^2 a}{b^3}\left[\exp(bt)-\frac{b^2t^2}{2}-bt-1\right]\right\} \tag{4.43}$$

The experimental points and the calculated curve are shown in fig. 4.32 to be in fairly good agreement.

Anderson and Mehl also found that for aluminium reduced 15% by rolling, the growth rate during annealing at about 310 °C was essentially the same as for material reduced 90%. A steady decrease in the final recrystallised grain size was found, but this must have been due to the increase of $\dot{N}$ with higher degrees of deformation. This increase in the magnitude of the stored energy would produce a greater driving force both for recovery and for recrystallisation and would thus produce more potential nucleation sites than material with a lower stored energy.

A more recent attempt to model recrystallisation has been made by Furu, Marthinsen and Ness (1990), which modifies the Johnson–Mehl (JM) approach by considering the factors that may lead to a non-constant growth rate. In general terms the JM equations may be expressed:

$$X=1-\exp(-kt^n) \tag{4.44}$$

where $k$ and $n$ are constants, and $n$ is usually referred to as the Avrami exponent.

For three-dimensional nucleation $n$ is equal to 3 in the limiting case in which

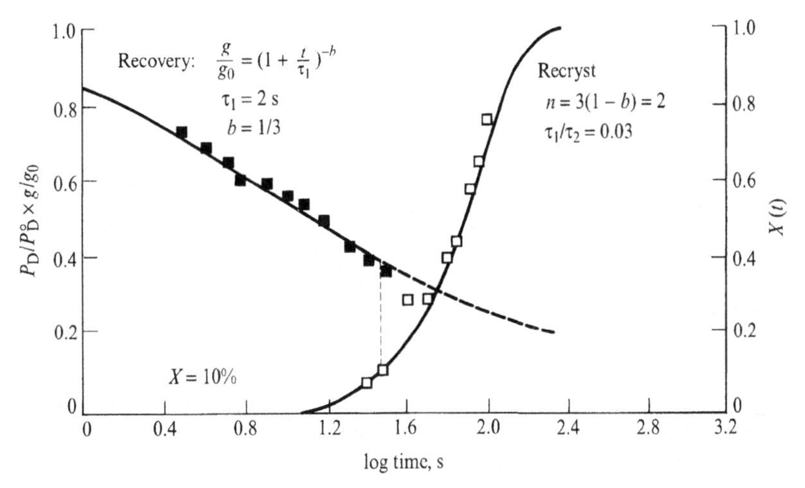


**Figure 4.33** Recovery and recrystallisation reactions modelled according to eqs. (4.45) and (4.44) for commercial purity aluminium alloy (for details, see text).

all nuclei are activated at a very early stage of the transformation (that is, for 'site saturation'), and, as we have shown above, is equal to 4 when $n$ is constant. Avrami exponents larger than 3 are rarely observed; experimentally $n$ is typically around 2 and sometimes even less than 1.

Furu *et al.* (1990) consider two mechanisms that might lead to a non-constant growth rate, namely a spatial variation in the net driving force for recrystallisation (that is, local strain gradients), or simultaneous recovery during the reaction. They demonstrate that either may account for the considerable negative deviations from classical JM kinetics.

Local strain gradients will arise if coarse particles are surrounded by a deformation zone approximately equal to the particle size, and they predict that $n$ decreases with increasing volume fraction of particles.

Recovery will cause a gradual decrease in the amount of stored energy and accordingly a decrease in growth rate which they express in the form:

$$g=g_0[1+(t/\tau)]^{-b}, \quad 0\leq b<1 \tag{4.45}$$

where $\tau$ is a relaxation time parameter which is constant at constant temperature. Furu *et al.* apply their model to the annealing behaviour of commercial purity aluminium, whose softening behaviour at 325 °C has been 'split' into two concurrent recovery and recrystallisation reactions, fig. 4.33. Here it is seen that the recovery reaction is well expressed by eq. (4.45), with $b=1/3$ and $\tau=2$ s. The $X$–$t$ curve was obtained by metallographic analysis, and this is seen to be in accord with eq. (4.44) with $n=2$.

## 4.3.2 Nucleation of recrystallisation

It is now recognised that an analysis of the kinetics of recrystallisation such as that presented above (§4.3.1) is not an approach which can give an insight into nucleation mechanisms. Models of nucleation are based essentially upon observational evidence, and detailed reviews of the various mechanisms have been made by Cahn (1966, 1983) and Doherty (1978). Three principal models have been advanced to account for nucleation; these are 'classical' nucleation, subgrain growth and strain-induced boundary migration.

### Classical nucleation

Burke and Turnbull (1952) first examined whether the classical theory of nucleation could be extended to the situation of recrystallisation. Here a critical nucleus will be stable if the necessary interfacial energy is balanced by the difference in strain energy per unit volume between the cold-worked state and the fully recrystallised condition. The model can account for the existence of substantial incubation periods, since during this period a sequence of thermal fluctuations occurs until finally a sufficiently powerful activated event occurs which leads to the creation of a nucleus of stable size. Burke and Turnbull conclude that the model is energetically feasible since most of the measured stored energy (which, as we have seen, is relatively small) is concentrated in a few compact sites, although Byrne (1965) in a quantitative model that he proposed concludes that an enormous local strain of approximately 0.2 would be required to produce a critical nucleus size of approximately $6 \times 10^{-10}$ m.

Nuclei would be expected on this model always to be close in orientation to that of the matrix, since this would involve the creation only of low angle boundaries of low interfacial energy and thus would imply a smaller critical nucleus size. This is, in general, not observed, but no experimental method of disproving the model appears to exist and it is excluded principally upon the grounds that alternative models fit the facts without the necessity for assuming the existence of very large local strains.

### Subgrain growth

This model, due to Cahn (1966, 1971), is illustrated schematically in fig. 4.34: a small region of high dislocation density (and therefore of high strain gradient and thus presumably of high local misorientation) evolves into a small strain-free cell by a process of dislocation climb and rearrangement. Such a subgrain, once formed, can grow by sweeping up the dislocations in the volume it absorbs. Some of these dislocations will be annihilated by reacting with other dislocations in the advancing subboundary, but most will enter the subboundary, thus increasing the dislocation density there and hence the specific energy of this interface.

**Figure 4.34** Nucleation by subgrain growth (schematic). Subgrain boundaries thickly populated by dislocations (dots) have a high misorientation angle, and are the most likely to migrate.

Cahn (1971) points out that the main reason why a subgrain grows must be that the swept-up dislocations will have a smaller total energy when captive in the subboundary which has just swept them up than they had either as isolated dislocations or as constituents of the low angle subboundaries which have been consumed. The growing subgrain will thus, as its boundary acquires more dislocations by this sweeping-up process, become progressively more misoriented with respect to its neighbours. Eventually the boundary angle of the subgrain will approach the value at which the individual dislocations within it begin to lose their character, and the subboundary may be regarded as having undergone a transition to a conventional boundary. When this occurs it represents a successful nucleation event, and the expanding boundary now destroys dislocations as the grain grows (rather than sweeping up dislocations), so that the rate of release of stored energy will now rise sharply; the rate of growth of the nucleus may therefore be expected to accelerate.

Li's subgrain coalescence process (p. 180 *et seq.*) may be regarded as a variant of the subgrain-growth model of nucleation. Doherty (1978) employed transmission electron microscopy to obtain experimental evidence for this mechanism (in deformed aluminium). It was found to occur in regions where there were already large local misorientations, such as close to a grain boundary or at a deformation band. Hu (1981) observed that grain boundaries act more efficiently as sinks than sources for dislocations, which is an essential requirement for the process of coalescence.

### Strain-induced boundary migration (SIBM)

This process was first reported in aluminium by Beck and Sperry (1950), and the energetics of the process have been analysed by Bailey and Hirsch (1962). This might be termed the 'bulge nucleation' model, and can be understood by reference to fig. 4.35. Here the upper grain is assumed to contain a higher dislocation density than the bottom one so that the boundary will tend to bulge as shown if certain energy relations are satisfied.

Let $\Delta E$ be the stored energy difference per unit volume between the two

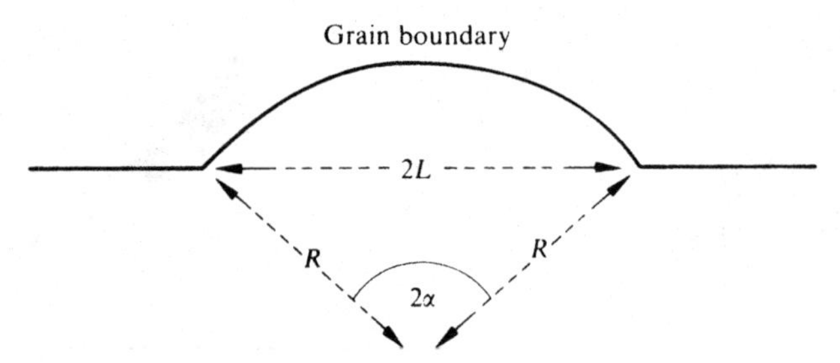


**Figure 4.35** Bulge nucleation.

grains, $\sigma$ the specific grain boundary energy and $A$ the surface area of the bulge. The number of atomic jumps required for the bulge volume to increase by $\mathrm{d}V$ is $\mathrm{d}V/b^3$, where $b^3$ is the atomic volume, and the consequential change in free energy is

$$[\Delta E-\sigma(\mathrm{d}A/\mathrm{d}V)]\mathrm{d}V$$

If $\Delta G^*$ is the activation energy required by an atom to jump and if $\nu$ is the jump frequency, then the rate of transfer of atoms across the boundary (per atom site) is

$$(\nu k/RT)[\Delta E-\sigma(\mathrm{d}A/\mathrm{d}V)]\exp(-\Delta G^*/RT) \tag{4.46}$$

where $k$ equals the atomic weight over the density. Since the number of atom sites on the boundary is $A/b^2$, the total number of jumps ($\mathrm{d}V/b^3$) in time $\mathrm{d}t$ is given by

$$\mathrm{d}V/b^3=(A/b^2)[\Delta E-\sigma(\mathrm{d}A/\mathrm{d}V)](\nu k/RT)\exp(-\Delta G^*/RT)\mathrm{d}t \tag{4.47}$$

whence

$$\mathrm{d}V/\mathrm{d}t=Abf[\Delta E-\sigma(\mathrm{d}A/\mathrm{d}V)] \tag{4.48}$$

where $f=(\nu k/RT)\exp(-\Delta G^*/RT)$. As the bulge grows, the angle $\alpha$ increases, and the rate equation can be rewritten in the form

$$\mathrm{d}\alpha/\mathrm{d}t=(2bf/L)[\Delta E-(2\sigma/L)\sin\alpha](1+\cos\alpha) \tag{4.49}$$

and it follows that for the bulge to grow $\Delta E-(2\sigma/L)\sin\alpha$ must be positive for all values of $\alpha$, that is, for all configurations of the bulge. This calculation follows the classical Volmer–Becker calculation for a critical nucleus radius, since $L$ must exceed a critical value for a given $\Delta E$.

For the method to operate, therefore, the dislocation distribution must be

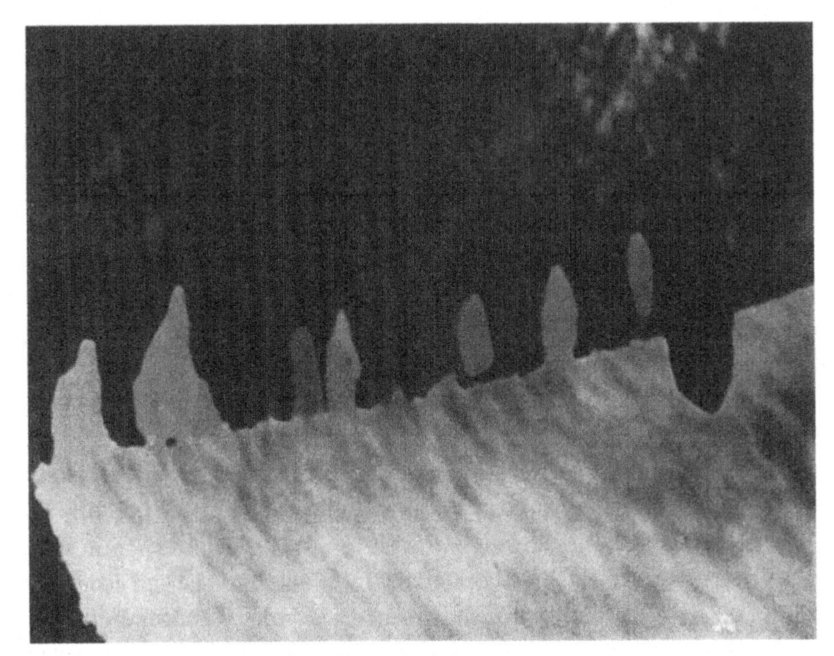

**Figure 4.36** Aluminium deformed 40% by compression. Annealed 1 h at 328 °C. (After Bellier 1971.)

inhomogeneous on a sufficiently large scale – which means effectively that a particularly large subgrain abutting on to the grain boundary will cause that boundary to bulge across the entire width of the subgrain. The Bailey and Hirsch condition is thus equivalent to the formulation of a critical subgrain size in a grain boundary region. Fig. 4.36 illustrates bulge nuclei forming during the annealing of aluminium deformed 40% by compression and annealed 1 h at 328 °C.

It is generally observed experimentally that recrystallisation nuclei form in regions of large strain gradient, such as at grain boundaries or within transition bands (see §4.1.3). Bellier and Doherty (1973) found in their study of deformed and partially recrystallised aluminium that following 20% deformation almost all the new grains originated by SIBM of pre-existing grain boundaries with only a minority of new grains originating from the transition bands. However, after 40% deformation, most of the new grains developed by SIBM of material on either side of a transition band. The implications of these observations are discussed by Doherty and Cahn (1972), and it is clear that the process of SIBM is more complex than has previously been supposed; they consider that a process of subgrain coalescence at boundaries may be a necessary precursor of SIBM.

Local lattice misorientations are thus a necessary and sufficient condition for nucleation of recrystallisation – the larger the magnitude of the strain gradient, the greater is the number of nuclei capable of being formed. High macroscopic strains which do not involve local lattice curvature may thus be expected not to give rise to recrystallisation nuclei on annealing: single crystals of zinc or cadmium may indeed be deformed by single glide to strains of 100% or more without subsequently recrystallising when annealed. Again, van Drunen and Saimoto (1971) strained [001] oriented copper crystals into stages II and III at room temperature. The very uniform dislocation distribution obtained (compared with those in crystals of other orientations) made it possible to anneal the crystals at temperatures up to 1000 °C without recrystallisation occurring.

The basis of recrystallisation is that, when high angle grain boundaries migrate through a deformed matrix, the defects generated during plastic deformation are annihilated because of the ability of these boundaries to act as sinks for such defects. It is ordinarily assumed that high angle boundaries act as perfect sinks in this way, but work by Grabski and Korski (1970) on copper has shown that the relative grain boundary energy (after a recrystallisation anneal at 600 °C) is increased from an equilibrium value of 0.25 to about 0.6 after recrystallisation following 15% strain. (The relative grain boundary energy is the ratio of the grain boundary free energy to the surface free energy, and is determined by measuring the dihedral angle of thermal grooves.) It is suggested that this effect arises because the incorporation of lattice defects into the grain boundary during recrystallisation causes a change in the grain boundary structure which increases the energy of the boundary. Grabski and Korski further showed that subsequent high temperature (1000 °C) annealing led to a return of the grain boundary energy towards the equilibrium value. These observations have revolutionary implications for the structure of grain boundaries and also for recrystallisation itself, and clearly the phenomena deserve greater attention.

### 4.3.3 Growth of new grains in recrystallisation

As we have seen in the previous section, the process of nucleation requires that the potential nucleus is capable of growth into the deformed microstructure and so no complete separation of the two stages is possible. However, we can discuss the factors that are known to affect the rate of grain boundary migration while recognising that these factors can affect the initial nucleation as well as the growth of new grains.

#### Grain boundary mobility in pure metals

The boundary separating the two grains in fig. 4.37(*a*) is moving to the right under the influence of the difference in free energy per atom ($\Delta G$) between the two grains. This difference will, in the case of recrystallisation, be the difference in stored strain energy, but the same arguments would apply if it were due to a chemical free energy difference (chapter 2) or to a reduction in grain boundary surface energy (chapter 5). Straightforward use of simple reaction rate theory

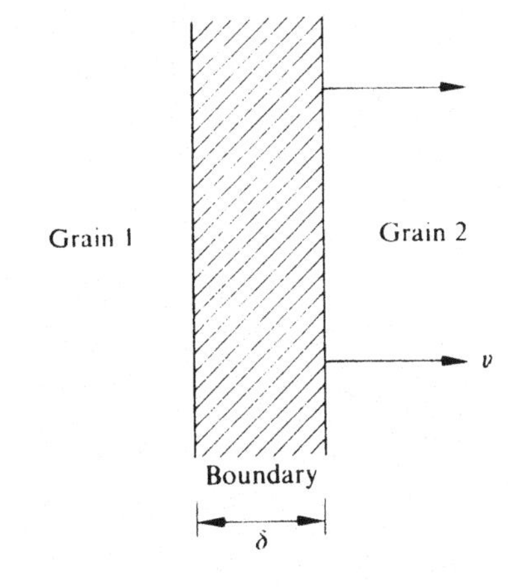


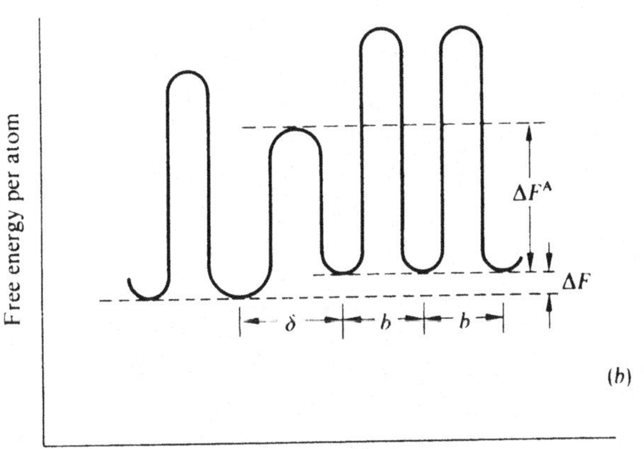


**Figure 4.37** (*a*) The boundary separating grains 1 and 2 is moving to the right with a velocity $v$. (*b*) represents the free energy per atom along a direction normal to the boundary plane; the minima represent stable lattice sites in either grain which are spaced a distance $b$ apart in grain 2.

can be made in this case by application of the energy–distance relationship shown in fig. 4.37(*b*).

The frequency of atomic transfer† from grain 2 to 1 is

$$\nu \exp(-\Delta G^A/kT)\ \mathrm{s}^{-1}$$

while the reverse rate is

$$\nu \exp[-(\Delta G^A+\Delta F)/kT\ \mathrm{s}^{-1}$$

† This assumes that every atom leaving one grain can find a resting site in its neighbour. If this is not the case, an accommodation coefficient $A$ must be introduced into the expression for atom transfer (Chalmers 1964) and if $A \rightarrow 0$, as may be the case for the immobile coherent twin boundary, then the rate of migration falls accordingly. Values of $A$ have been considered for migrating solid–liquid interfaces (Chalmers 1964; Jackson and Hunt 1965), but not for grain boundaries, though ideas of Aaronson and his colleagues (Aaronson 1962; Aaronson *et al* 1970) on the immobility of some interphase boundaries are based on possible accommodation difficulties.

($\nu$ is the vibration frequency of atoms belonging to either grain but facing the grain boundary – this will be of the order $10^{13}\,s^{-1}$.) For each net atom transferred from 2 to 1 the boundary moves forward a distance $b$. Its velocity $v$ is therefore given by

$$v = b\nu \exp(-\Delta G^A/kT)[1-\exp(-\Delta G/kT)] \tag{4.50}$$

Expanding $\exp(-\Delta G/kT)$ for the usual situation in which $\Delta G \ll kT$ and noting that $\Delta G^A = \Delta U^A - T\Delta S^A$ gives:

$$v = b\nu(\Delta G/kT)\exp(\Delta S^A/k)\exp(-\Delta U^A/kT) \tag{4.51}$$

Eq. (4.51) is commonly abbreviated to

$$v = M\Delta G \tag{4.52}$$

where $M$ is the mobility of the boundary – the velocity for unit driving force. The expression for $M$ is, of course,

$$M = b\nu/kT \exp(\Delta S^A/k)\exp(-\Delta U^A/kT)$$

and this is studied by the conventional ln versus $1/T$ plot to determine the activation energy $\Delta U^A$ and the 'pre-exponential' parameter $b\nu/kT \exp(\Delta S^A/k)$.

In a review of grain boundary migration, Gordon and Vandermeer (1966) confirmed the validity of this type of expression for high purity (zone-refined) metals. They also showed that the measured activation energies were all much less than the energies of lattice diffusion and closer to the value of grain boundary diffusion. Surprisingly, the experimental activation energies for lead and tin were found to be even smaller than the activation energy for grain boundary diffusion. This conclusion was, however, based on an early value for the activation energy of boundary diffusion in lead of 65.9 kJ $mole^{-1}$ compared to the energy of boundary migration of 25–30 kJ $mole^{-1}$. A more recent investigation of boundary diffusion in high purity lead by Stark and Upthegrove (1966) produced a value of about 21 kJ $mole^{21}$, however. It would therefore seem that the same expected result of similar activated processes is found for atom transfer across a grain boundary (migration) as for atom transfer down a grain boundary (diffusion).

There is, however, growing evidence for a crystallographic structure of grain boundaries (for example, Bishop and Chalmers (1971), Schober and Balluffi (1969, 1970)). A theoretical investigation of grain boundary diffusion on the basis of this idea has been able to duplicate the experimental results of Turnbull and Hoffman (1954) on the anisotropy of grain boundary diffusion rates in tilt boundaries in which the diffusion parallel to the tilt axis was faster than the diffusion in the boundary plane normal to this axis. No theory is yet available for the expected migration rates of boundaries with crystallographic atomic arrangement.

**Mobility of low and medium angle grain boundaries** As has been described by Doherty and Cahn (1972), there is considerable evidence for the lower mobility of boundaries separating grains whose orientations differ by less than about 10–15°. This phenomenon hinders many of the nucleation processes discussed earlier since, for a nucleus to develop, it must not only be energetically capable of growth (for example, be larger than its surrounding subgrains), but it must also have a boundary with a sufficient misorientation to be mobile.

Many experimental investigations have been carried out to analyse the effect of orientation between the grains on the growth process (Aust and Rutter 1959). Fredriksson (1990) has analysed the results of one such experiment, assuming that the variation in growth rate arises from a variation in the surface free energy with the grain orientation. Assuming the surface free energy is a sinusoidal function of the orientation, Fredriksson made a suitable choice of adjustable parameters and applied absolute reaction rate theory for the transfer of atoms from the grains to the boundary. Fig. 4.38 shows that this approach can describe the experimental results for lead fairly well.

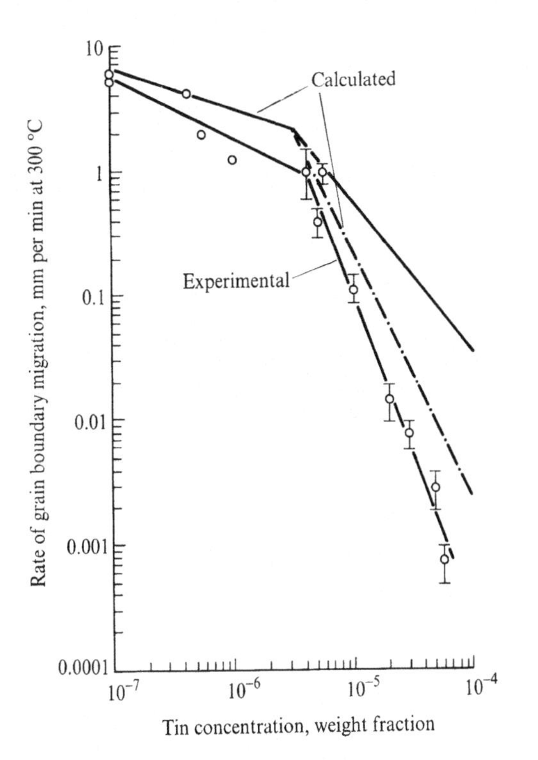


**Figure 4.38** Comparison between theory and experimental results for grain growth in Pb–Sn alloy.

**The effect of solute atoms on grain boundary migration** Experimental results suggest that the main effect of impurities is on the growth process rather than upon the nucleation of recrystallisation, although it is not easy to separate the two processes in practice. The growth of recrystallising grains in the presence of impurity atoms has certainly received the greater attention, both theoretically and experimentally, and we shall confine ourselves principally to consideration of this aspect.

*The impurity drag theory* Lücke and Detert (1957) were the first to attempt to formulate a quantitative atomistic theory of the effect of impurity atoms upon grain boundary migration. Their basic hypothesis was that solute atoms have a lower energy in the neighbourhood of grain boundaries than in a perfect lattice. This implies the existence of an attractive force between a solute atom and a grain boundary which can be expressed by

$$G(x) = -\mathrm{d}U(x)/\mathrm{d}x \tag{4.53}$$

where $x$ is the distance of the solute atom from the grain boundary and $U(x)$ is the free energy of interaction, shown schematically in fig. 4.39($a$).

Due to this interaction, impurities concentrate near the interface separating

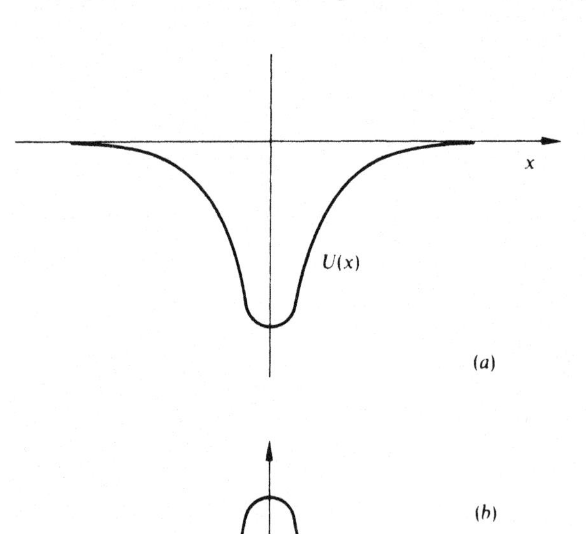


**Figure 4.39** Schematic profiles of interaction energy $U(x)$ and diffusion coefficient $D(x)$ as a function of the distance to the grain boundary plane.

a new recrystallised grain from the cold-worked matrix, and this atmosphere must be dragged by the moving boundary, which thus decreases the velocity of the boundary. The problem has been treated more rigorously by Lücke and Stüwe (1963) and Cahn (1962).

The rate of migration $v$ is proportional to the net driving force on the boundary, that is,

$$v=(1/\lambda)(\Delta G-P_i) \tag{4.54}$$

where $\lambda$ is the reciprocal of the boundary mobility in pure specimens, $\Delta G$ is the driving force resulting from the decrease of free energy of the crystal when the boundary moves, and $P_i$ is the impurity drag force on the boundary.

The magnitude of $P_i$ depends on impurity concentration and on boundary velocity. Cahn calculated its value as a function of $U(x)$ (eq. (4.53)) and of the diffusion coefficient of the impurity in a direction perpendicular to the grain boundary, $D$. As shown in fig. 4.39(*b*), $D$ is a function of the distance ($x$) to the boundary plane. By summing all the forces acting on the grain boundary, Cahn obtained the following expression relating the velocity of boundary migration to the driving force, composition and indirectly to the temperature:

$$\Delta G=\lambda v+\alpha Cv/(1+\beta^2v^2) \tag{4.55}$$

where $C$ is the bulk impurity concentration, and $\alpha$ and $\beta$ are parameters depending on the exact shape of the $U(x)$ and $D$ profiles. The relationship between velocity and driving force is shown schematically in fig. 4.40 for three different solute concentrations, $C_3>C_2>C_1$.

For low concentrations the curve is continuous, and there is only a small deviation from the straight line corresponding to an ideally pure metal. This implies that the foreign atoms cannot follow the migrating interface after it has broken away from its impurity atmosphere. For high concentrations the curve is S-shaped and its central part is unstable; in such a case the velocity will change discontinuously from one branch of the curve to the other at some critical driving force, as is indicated by the dashed lines in fig. 4.40. The relationship is complex, but becomes less so if certain extreme cases are examined.

If, for example, the velocity of the boundary is appreciably less than the drift velocity of the impurity atoms in the boundary, that is, $v\ll 1/\beta$, then eq. (4.55) simplifies to

$$\Delta G=\lambda v+\alpha Cv \quad \text{or} \quad v=\Delta G/(\lambda+\alpha C) \tag{4.56}$$

This low velocity is attainable either by a low driving force, $\Delta G/A<1/\beta$, or a high impurity content, $\Delta G/\alpha C<1/\beta$, and eq. (4.56) shows that $1/v$ should be a linear

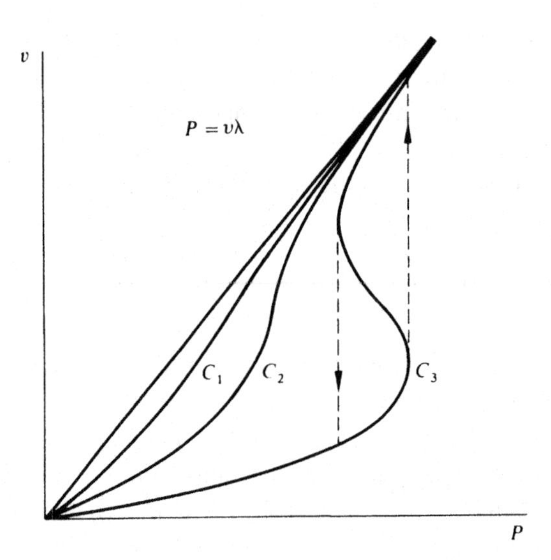


**Figure 4.40** Grain boundary velocity $v$ as a function of the driving force. $P=v\lambda$, for different solute concentrations $C_3>C_2>C_1$ (schematic).

function of $C$ and that furthermore the apparent activation energy will be close to a mean value of the activation energy for impurity diffusion in the boundary region.

Another limiting case occurs when the velocity of the boundary is much greater than the drift velocity of the impurity atoms, that is, $v \gg 1/\beta$, when

$$v=\Delta G/\lambda-(\alpha/\beta^2)C/\delta F \tag{4.57}$$

For low $C$ or high $\Delta G$ the second term is negligible, so that the influence of impurities is not large. The activation energy is mainly determined by the intrinsic mobility $1/\lambda$ and should be of the order of the activation energy for grain boundary self diffusion, which will be smaller than that for the low velocity situation.

We will now examine the extent to which the predictions of the above theory are supported by experiment. There are, however, severe limitations to such comparisons because of the assumptions which have to be made about the values of the different adjustable parameters.

**Experimental observations** There is only a limited amount of data available with which to compare the theory outlined above. Ideally, an experiment aimed at testing the theory should provide a constant and known driving force for recrystallisation in a system consisting of an extremely pure metal containing only one solute in a known amount. These requirements are seldom fulfilled, and we will select data which approach these conditions as far as possible.

The low velocity conditions represented by eq. (4.56) will prevail at relatively high impurity levels, and in such systems the reciprocal of velocity should be a linear function of solute composition at constant driving force and temperature.

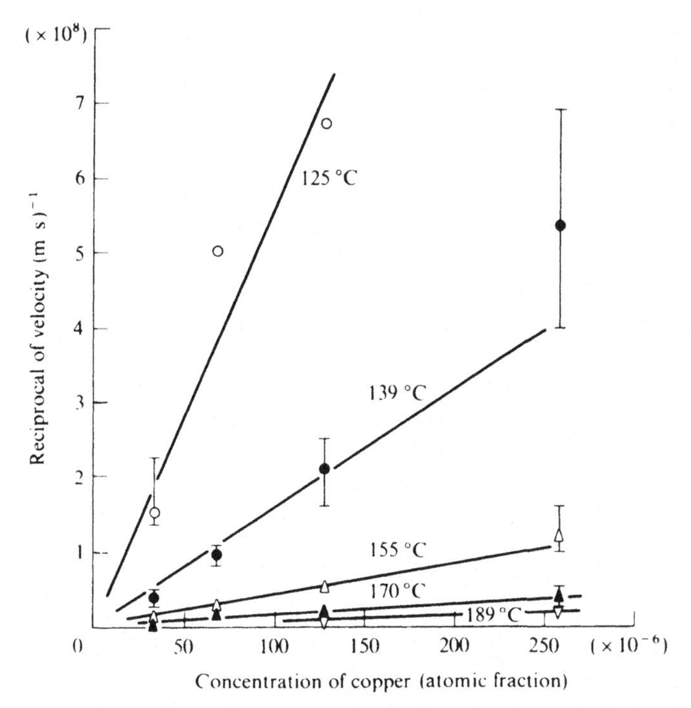


**Figure 4.41** Experimental variation of reciprocal of migration rate with concentration at several temperatures for copper in zone refined aluminium. (After Gordon and Vandermeer 1966.)

Fig. 4.41 shows that such a dependence does exist for zone-refined aluminium alloys containing copper in excess of $17\times10^{-6}$ atomic fraction. Straight lines appear to describe the data fairly well, within the range of estimated uncertainties in the measured velocities represented by the vertical lines through some of the data points.

Dilute aluminium alloys have also been studied by Frois and Dimitrov (1961, 1966), and fig. 4.42 shows typical results of their work for additions of copper or magnesium, which have been reviewed by Dimitrov, Fromageau and Dimitrov (1978). In fig. 4.42 three domains of concentration can be distinguished:

(*a*) A high concentration region ($50\times10^{26}$ for copper, or $390\times10^{-6}$ for magnesium) where the growth rate decreases with increasing impurity content. The linear relation between inverse velocity and concentration predicted by eq. (4.56) and shown in the data of Gordon and Vandermeer is not apparently observed in the data of fig. 4.42: the velocity varies more rapidly with concentration than predicted theoretically. It is concluded that, within the frame of the theory, the concentrated alloys are not entirely in the low velocity limit to which eq. (4.56) applies, but they are still in a transition region.

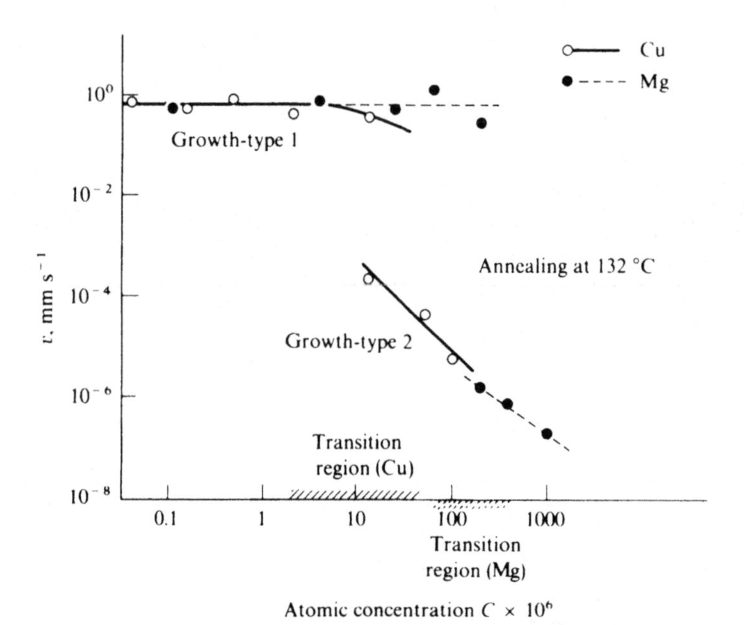


**Figure 4.42** Growth rate of new crystals at 132 °C in cold-rolled aluminium, in the presence of copper or magnesium additions. (After Dimitrov, Fromageau and Dimitrov 1978.)

(*b*) A low concentration region exists (below $2\times10^{-6}$ for copper, or $65\times10^{-6}$ for magnesium) where there is very little influence of the solute element. This situation is appropriate to the application of eq. (4.57) which indeed predicts no large impurity effect.

(*c*) A transition region also exists where both types of growth are observed on one given sample – fig. 4.42 exhibits a discontinuous transition, with the velocity decreasing by a factor of $10^4$ (Cu) or $10^6$ (Mg) when going from one branch of the curve to the other. Whether a transition is continuous or discontinuous depends on the particular solute element. From results of the type given in fig. 4.42 one can define a critical concentration below which no significant effect of the foreign element is detected, and also (in the low velocity region) the solute-independent growth rate for the unspecified particular conditions of temperature and driving force. Theoretically, a particular solute element should be characterised by the values of the interaction energy and of the diffusion coefficient (fig. 4.39) and by their variation in the region of the grain boundary.

With regard to the influence of temperature upon the velocity of migration under these conditions, an apparent activation energy can be derived from the slope of an Arrhenius plot of ln $v$ versus $1/T$. This does not necessarily have a physical significance, but it can be compared with the theoretical predictions.

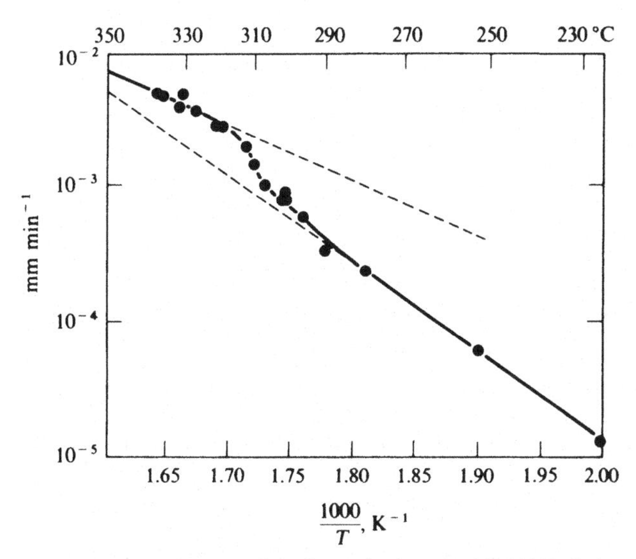


**Figure 4.43** Migration velocity of a 30° [111] tilt boundary in rolled gold, as a function of temperature. (After Grünwald and Haessner 1970.)

Gordon and Vandermeer (1966) explored the temperature sensitivity of migration rate in their experiments on copper additions to aluminium, and Grünwald and Haessner (1970) measured the migration velocity of a recrystallisation boundary in rolled gold single crystals containing 20 parts per million iron as the main impurity. Fig. 4.43 shows the variation of migration velocity as a function of temperature for a 30° [111] tilt boundary. The apparent activation energy is seen to vary as a function of temperature, starting from a value of 1.32 eV at low temperatures, increasing through a maximum at approximately 310 °C, and decreasing to a low constant value of 0.83 eV at high temperatures. This type of behaviour can be understood in terms of the theory previously discussed, for the situation intermediate between the extreme domains described by eqs. (4.56) and (4.57). Here there is a transition region in which the variation of activation energy will depend upon the driving force. For low driving forces, it will go continuously from the low value to the high one, whereas for very high driving forces there will be a discontinuity in the transition region.

When the driving force is of intermediate value, the apparent activation energy will go through a maximum between the values corresponding to the two limiting ranges, and it is this situation to which the data of fig. 4.43 apparently correspond. We may thus conclude that the impurity drag theory mechanism is able to provide an acceptable interpretation of the effects of foreign elements on the recrystallisation process. Although the experimental data are consistent with the theory, it is obvious that strict comparisons are in fact difficult because of the large number of experimental variables involved. Fredriksson (1990) has attempted to apply the absolute reaction rate theory to account for the effect of solute atoms on grain growth during recrystallisation. He treats the boundary as a separate phase with its own free energy, although assumptions have to made regarding the grain boundary equilibrium composition and the diffusion coefficient of an alloying element across a grain boundary. Making reasonable assumptions, good agreement was obtained between the calculated and experimental results for aluminium alloyed with copper, using the data of Gordon and Vandermeer.

A number of questions are, of course, still unanswered, such as how the solute drag is influenced by the relative orientation of the two grains constituting the boundary, and how such effects are related to the *structure* of the boundary.

**Orientation dependence of boundary mobility in the presence of solute atoms** Following the work of Aust and Rutter (1963), it is known that so-called 'special' boundaries, which are close to a coincidence site relationship, are much less affected by impurity than are general boundaries. (Coincidence site relationships are described by Shewmon (1966).) The effect is illustrated in fig. 4.44 taken from the study by Aust and Rutter of boundary migration in zone refined lead doped with impurity. It is clear that in the absence of impurity the general boundaries are no less mobile than the special ones (Gordon and Vandermeer (1966) suggest that, in fact, they may be even more mobile), but that the impurity has a large effect on the general boundaries – presumably by virtue of a greater adsorption at the more open structure of the general boundary. This idea is supported by the experimental studies of Schober and Balluffi (1969, 1970) who demonstrated the existence of arrays of dislocations at boundaries close to coincident site relationships which act to restore the local atomic fit of the perfect coincident site boundary (in the same way as dislocations of low angle boundaries restore the local atomic fit).

These observations of Aust and Rutter go a long way towards explaining the development of annealing textures during the recrystallisation of deformed metals with a well-developed deformation texture; this large area is reviewed by, for example, Beck and Hu (1966).

It should perhaps be noted that Aust and Rutter used very high purity materials and low driving forces (obtained from a solidification substructure) in their work. Some caution should therefore be exercised in applying their conclusions to the normal recrystallisation situation in useful materials, which involves a higher driving force, but also a higher impurity content.

### 4.3.4 The recrystallisation of two-phase alloys

There are three situations we should consider in the context of the recrystallisation of two-phase alloys: firstly, that in which the material contains a finely

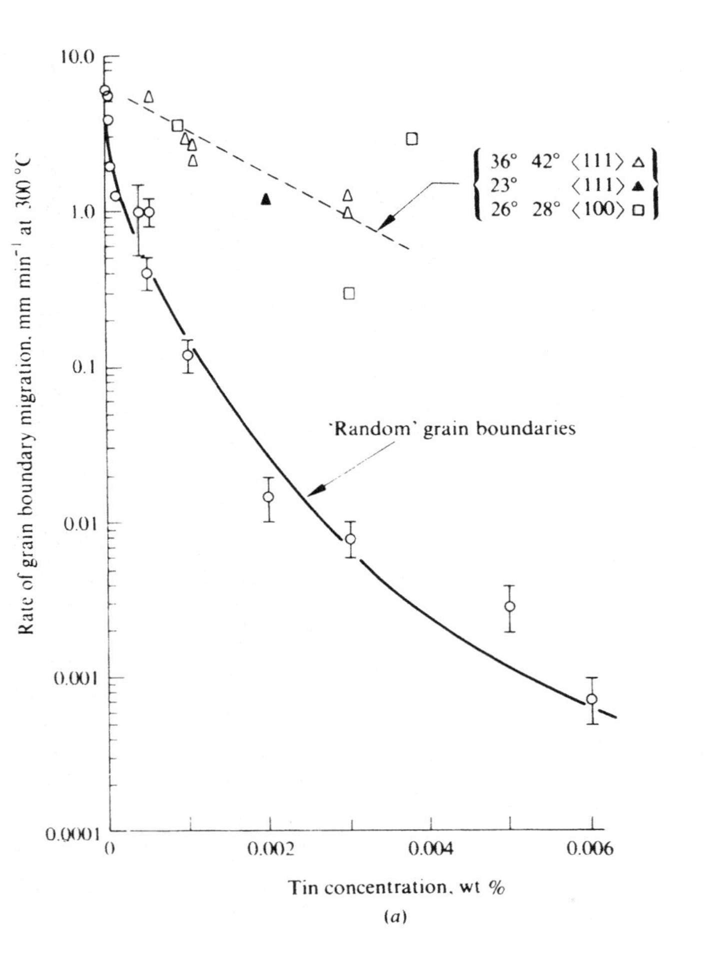


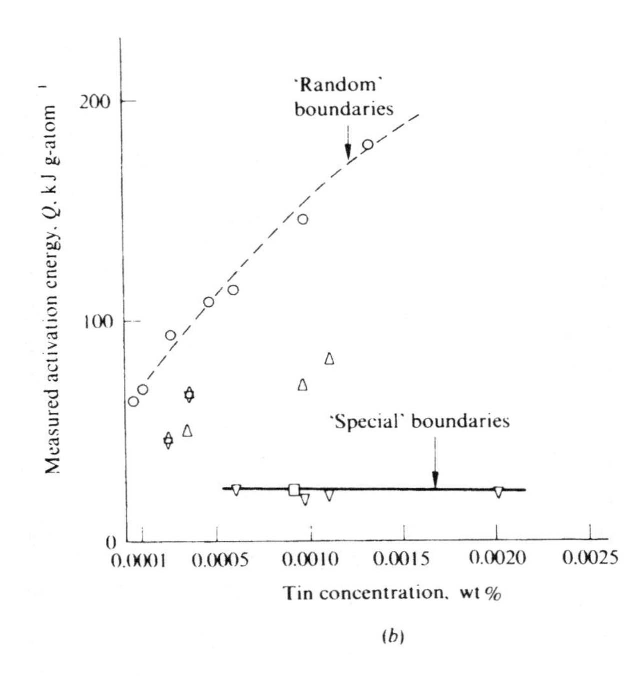


**Figure 4.44** (*a*) Rate of grain boundary migration at 300 °C versus wt% Sn for 'random' grain boundaries and 'special' grain boundaries in zone refined lead. (From Aust and Rutter 1959.) (*b*) Apparent activation energies of grain boundary migration versus wt% tin for 'random' and 'special' boundaries in zone-refined lead. (From Aust and Rutter 1959.)

dispersed second phase prior to being plastically deformed and annealed; secondly, that in which the material consists of a coarse two-phase alloy (such as $\alpha/\beta$ brass) in which the phases effectively form separate grains; and finally, the more complicated situation in which the alloy consists of a supersaturated single-phase solid solution when it is plastically deformed, but which precipitates second-phase particles during recrystallisation.

### Effect of a dispersed phase on recrystallisation

**Recrystallisation kinetics** Martin (1957) using internally oxidised single crystals of Cu–$SiO_2$ alloy found that recrystallisation was *accelerated* by the presence of silica, and that the effect was greatest with the largest volume fraction and the coarsest particles. Other alloys, notably those containing large particles, have shown this effect. Leslie *et al.* (1963) and Gladman, McIvor and Pickering (1971), using transmission electron microscopy, have found that preferential nucleation occurs at large inclusions in iron, and Mould and Cotterill (1967) also ascribe the acceleration of recrystallisation in a two-phase iron–aluminium alloy to the nucleating effect of the particles.

Doherty and Martin (1962/3, 1964) analysed the recrystallisation of cold-rolled two-phase Al–Cu alloys both in polycrystalline and monocrystalline form. They found that the $\theta$ phase could *either accelerate or retard* primary recrystallisation. An approximate assessment of the nucleation rate and of the growth rate was obtained from the fully recrystallised specimens by calculating:

$N'$ = number of grains per unit volume/time for complete recrystallisation, and

$G'$ = average radius of final grains/time for complete recrystallisation.

Although these values are not the same as the conventional values of $\dot{N}$ and $G$, they will not be seriously in error for comparative purposes. They conclude from their data that, for the state of dispersion investigated, the major inhibition was not on growth but on nucleation and, as shown in fig. 4.45, it emerged that the mean particle spacing (in the range 1–4 $\mu$m) was an important parameter in determining recrystallisation kinetics. The same effect was found in internally oxidised Cu–Si alloys by Humphreys and Martin (1966), where the particle diameters were in the range 0.05–0.2 $\mu$m and the spacing 0.5–1.0 $\mu$m.

The situation is summarised in fig. 4.46. Acceleration of recrystallisation is only observed in specimens containing relatively *coarse* particles (that is, of diameter greater than 0.1 $\mu$m) when widely spaced. Very fine particles appear to give rise only to retardation, as their spacing is reduced.

*Retardation* of recrystallisation has been commonly observed, and was first reported by van Arkel and Burgers (1930) working on tungsten containing thoria. Meijering and Druyvesteyn (1947), using internally oxidised copper

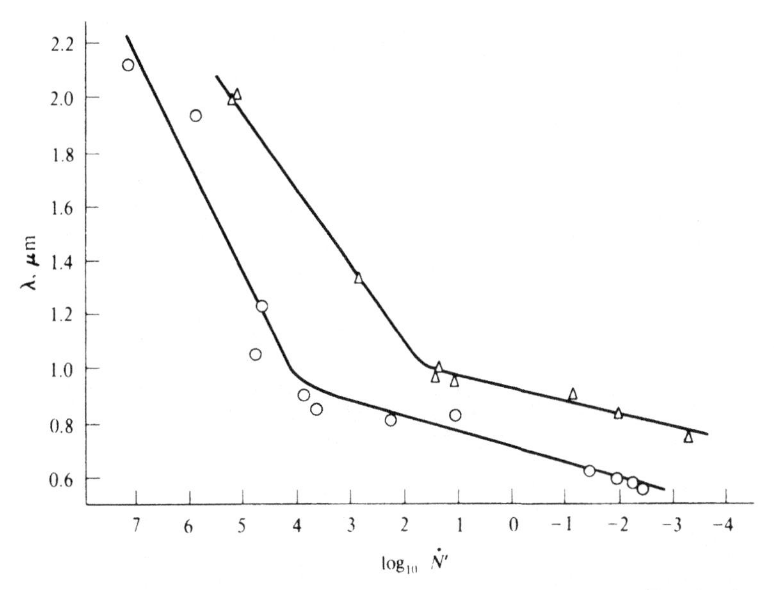


**Figure 4.45** The variation of $\log_{10}$ of the apparent rate of nucleation ($\dot{N}'$) with the interparticle spacing $\lambda$ (in microns) for the single crystals (triangles) and the polycrystalline alloys (circles). (After Doherty and Martin 1964.)

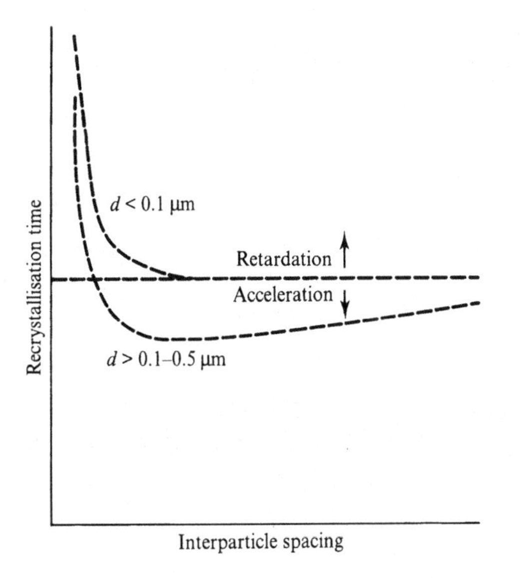


**Figure 4.46** The influence of particle size and particle spacing upon time of recrystallisation of two-phase alloys. (After Hansen 1975.)

alloys, found that in their specimens the central regions which did not contain oxide particles could be made to recrystallise readily, but the oxidised regions resisted recrystallisation up to a temperature near the melting point of the metal. Quantitative work has shown that the finest oxide dispersions are the most resistant to recrystallisation as seen, for example, in the data of Preston and Grant (1961), illustrated in fig. 4.47.

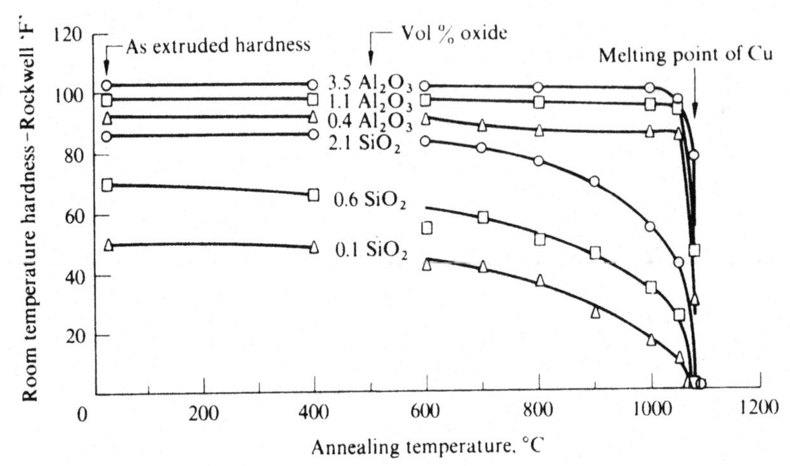


**Figure 4.47** Room temperature hardness after 1 h anneal, for copper powder alloys internally oxidised at 950 °C and extruded. (After Preston and Grant 1961.)

It should be mentioned that small bubbles can act like inclusions in inhibiting recrystallisation. For example, Thompson-Russell (1974) has shown that in doped tungsten wires, as the strain is increased the bubble dispersion is progressively refined, leading to an increase in recrystallisation temperature (fig. 4.48). The general problem of interactions between pores and grain boundaries has been reviewed by Cahn (1980).

## Micromechanisms

**Retardation effects** Retardation of recrystallisation can arise through two effects: firstly the second-phase particles may cause an inherently stable dislocation structure, and secondly they may hinder dislocation (and network) rearrangements thus hindering the formation of grain boundaries and their migration.

(*a*) *Inherent stability* The application of transmission electron microscopy has enabled the initial dislocation configurations in deformed two-phase alloys (see pp. 160–6), as well as their subsequent rearrangement after annealing to be observed. Thus Brimhall, Klein and Huggins (1966), working with silver containing a dispersion of magnesium oxide of radius of 5 nm and interparticle spacing of 50 nm, and with copper containing alumina of radius of 30 nm and interparticle spacing of 0.25 $\mu$m, reported that after cold-rolling a cell structure of dislocations was formed. From a statistical analysis of the misorientations (calculated from rotations in the electron diffraction patterns imaged from different cells) found in the as-deformed pure metals and two-phase alloys, they concluded that lower angle subgrain boundaries existed in the two-phase alloys and that therefore the deformation structure of the two-phase alloys did not

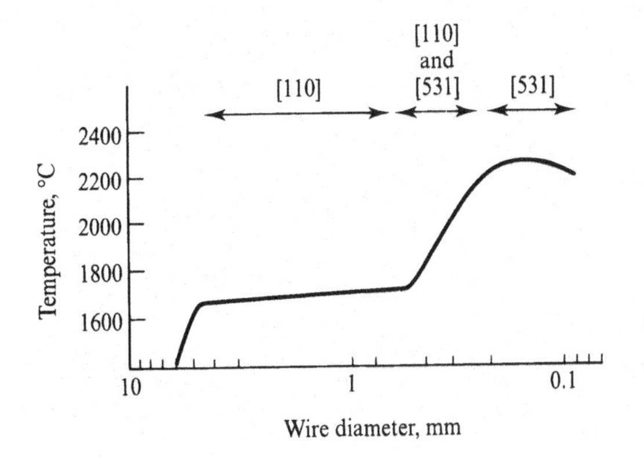


**Figure 4.48**
Development of increasing recrystallisation temperature and changes in texture of recrystallised W wire with decreasing wire diameter. Specimens annealed 10 min. (Thompson-Russell 1974.)

contain regions of severe plastic curvature; consequently the accumulation of misorientation to allow the formation of mobile interfaces is difficult.

Humphreys and Martin (1967, 1968) measured the misorientations developed in Cu–$SiO_2$ and Cu–Co single crystals after deformation (see p. 164–6), and showed that when a suitably oriented crystal was deformed into stage II of the shear stress/shear strain curve, the presence of the particles caused a greater number of slip planes to be active. The number of dislocations accumulated in these planes is thus reduced (in comparison with a similarly strained single-phase crystal), so that the lattice misorientation across the primary slip planes is reduced (see fig. 4.14). These dislocation configurations were shown to be highly resistant to an annealing heat-treatment and their inherent stability was demonstrated by annealing the crystals above the solvus temperature of the cobalt particles. Rollason and Martin (1970) explored this effect in greater detail, and were able to identify those strains leading to inherent stability and those in which particle pinning is important. They confirmed that a crystal deformed in tension into a region in which a high rate of work-hardening occurs (for example, stage II in the case of a single-glide oriented crystal) develops inherently stable dislocation arrays: fig. 4.49 illustrates a Cu – 2% Co crystal which had been aged to contain a dispersion of incoherent cobalt particles and then deformed into stage II. Its structure would then be similar to that illustrated in fig. 4.14; it was then annealed for 98 h at 950 °C at which temperature all the cobalt particles would dissolve within a few seconds, yet it is seen that no recrystallisation has occurred. Walls of dislocations aligned approximately along the (111) plane traces delineate bands across which misorientations of less than 1° were measured. Tensile deformation of crystals of either single- or double-glide initial orientation to high strains did not show the above spectacular recrystallisation resistance, and neither did crystals deformed by rolling. It was concluded that the loss of inherent stability is associated with the presence of unequally stressed slip systems involving appreciable slip on the cross-slip plane; this, combined with primary and conjugate slip, produces non-laminar slip and thus high local inhomogeneities of deformation.

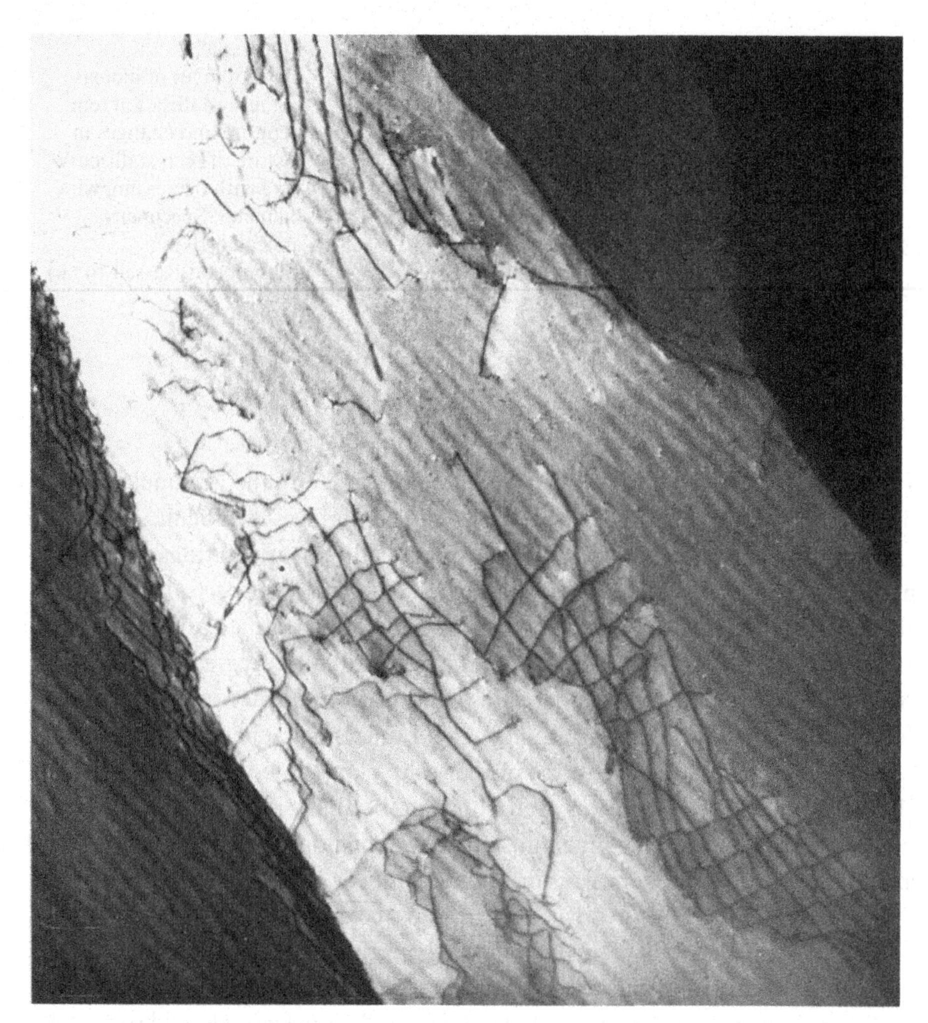

**Figure 4.49** Specimen annealed for 98 h at 950 °C, (112) section, $g$=020. (After Rollason and Martin 1970.)

In summary, therefore, two types of dislocation–particle interaction are found, depending on the size of the particles and the applied plastic strain. At large strains and/or particle sizes, dislocations form as secondary glide loops which involve local lattice curvature adjacent to each particle; when the dispersed particle size and/or the strain is small, only primary prismatic dislocations exist, which do not involve local curvature of the matrix lattice, and in this case no recrystallisation embryos are observed after annealing.

When polycrystalline dispersion-hardened materials in technological use are subjected to mechanical and thermal treatment, from the above considerations it is improbable that retardation of recrystallisation will arise from an enhanced homogenisation of dislocation structures. Such homogeneity is seen only under

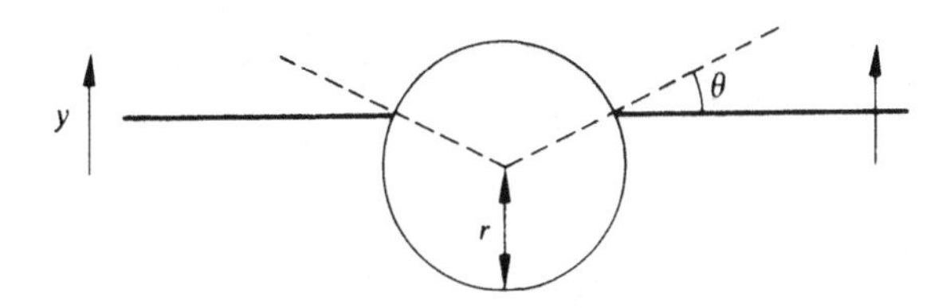


**Figure 4.50** Interaction of migrating grain boundary with an inclusion.

certain special conditions of particle size and applied strain, so that in the majority of practical situations particle pinning effects will be the principal source of recrystallisation resistance.

(*b*) *Pinning effects* Zener (1948) was the first to point out that a dispersed second phase will retard a grain boundary migrating under a fixed driving force. Zener's approach will be considered in more detail in the following chapter, where the effect is considered of a dispersion limiting the grain size attainable during grain growth – where the driving force continuously diminishes as growth continues. We can, however, use Zener's analysis to derive a relation between the drag per unit area of boundary and the characteristics of the dispersion itself.

Fig. 4.50 represents a grain boundary migrating upwards and intersecting a spherical inclusion. The drag exerted by the inclusion, resolved in the $y$ direction will be given by $\pi r \sigma_B \sin 2\theta$ where $\sigma_B$ is the specific boundary surface energy (which is effectively equivalent to a surface tension). The maximum value of this force will thus be exerted where $\theta=45^\circ$, when its value will be $\pi r \sigma_B$.

If there are $N$ of these particles per unit volume, their volume fraction ($f$) is $(4/3)\pi r^3 N$. A planar boundary of unit area will intersect all particles within a volume $2r$, that is, $2rN$ particles, so the number of particles intersecting unit area of a boundary ($n$) is given by $3f/2\pi r^2$. Thus the retarding force per unit area of boundary ($p_r$) is given by

$$p_r = -3f\sigma_B/2r \tag{4.58}$$

This equation is based on the assumptions that the boundary itself is inflexible, that the interaction between the grain boundary and the particle is independent of the nature of the particle, and also that the particles are distributed uniformly throughout the matrix. It follows from this expression that, for a given volume fraction, smaller particles hinder grain boundary migration more than large ones. For high degrees of deformation, therefore, and coarse dispersions, the effect of this retarding force can be neglected. Thus, for a volume fraction $f$ of 0.04 of dispersed phase with $r=1$ $\mu$m, and a boundary surface energy of 0.5 $\mathrm{J m^{-2}}$, the retarding force would be only $3\times10^4$ N $\mathrm{m^{-2}}$. On the other hand, the driving force ($p_d$) can be estimated by considering a recrystallisation front migrating into a deformed region being associated with a change in dislocation density of $\delta\rho$, so that $p_d$ will be given by the gain in energy per unit volume traversed by the recrystallising boundary, that is,

$$p_d = \mu b^2 \, \delta\rho \qquad (4.59)$$

where $\mu$ is the shear modulus and $b$ the Burgers vector. The order of magnitude of $p_d$ in the case of heavily deformed metals is thus 10 MPa, so the retarding force due to the particles in the situation discussed above can be neglected in comparison with the driving force, and only under conditions of low driving force are such dispersions likely to exert an effective retardation. For high driving forces, therefore, a particle diameter of the order of 0.1 $\mu$m at a volume fraction of about 0.05 will be required in order to have an appreciable effect on a migrating grain boundary. Since $r$, $f$ and $\lambda$ (the mean spacing between particles) are connected by the equation

$$\lambda = r(4\pi/3f)^{1/3} \qquad (4.60)$$

relatively close particle spacing will be associated with effects of retardation, as shown in fig. 4.46.

**Acceleration effects** It is well established from metallographic evidence that nucleation of recrystallisation may occur in the vicinity of large second-phase particles. Humphreys (1977, 1979) has studied the conditions under which this *particle-stimulated nucleation* (PSN) is observed, and fig. 4.51 shows the conditions of deformation and particle size for which nucleation is observed to occur at particles of Si in rolled aluminium. Although the occurrence of PSN is associated with the existence of lattice misorientations at particles, Humphreys shows that there is a factor of about 10 between the particle size at which rotations occur and that at which PSN occurs. The occurrence of rotation is thus not a sufficient criterion of PSN.

Humphreys (1977) found that recrystallisation originates at a pre-existing subgrain within the deformation zone at the particle (fig. 4.15) which grows by rapid subboundary migration until the deformation zone is consumed. The small grain thus formed will not continue to grow into the matrix unless the driving force due to the stored energy $E$ in the matrix exceeds the force due to the curvature $R$ of the boundary, that is,

$$R > 2\sigma_b/E \qquad (4.61)$$

where $\sigma_b$ is the grain boundary energy. If the radius of curvature of the nucleus is taken to be equal to that of the particle $r$, then the critical particle radius for PSN is when $R = r$. If reasonable values for the material parameters are assumed,

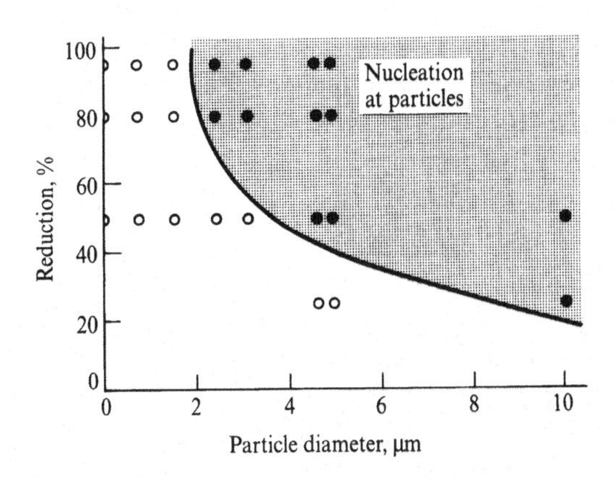


**Figure 4.51** The conditions of deformation and particle size for which nucleation is observed to occur at particles. Closed circles indicate observations of nucleation at particles. (After Humphreys 1977.)

then eq. (4.61) leads to critical particle diameters of 0.5–2 $\mu$m which are in accord with experimental observation and which will vary with strain in a qualitatively similar manner to that shown in fig. 4.51.

**The effect of deformable particles** If an alloy contains a dispersion which deforms with the matrix, then the recrystallisation behaviour is different. Coherent phases will not generate 'geometrically necessary' dislocations, so no local lattice curvature is expected in association with the particles. Again, the associated change of orientation with a migrating, high angle grain boundary implies that coherency with the dispersed phase must be lost, unless the particle were to rotate by the appropriate angle, or the migrating boundary were to pass through the particle, or re-solution and reprecipitation of the particle were to take place.

An important family of alloys strengthened by coherent dispersions is the nickel based superalloys. Relatively simple systems have been found to imitate certain microstructural features of superalloys, and Haessner, Hornbogen and Mukherjee (1966) studied the Ni–Cr–Al ternary system. Haessner *et al.* found that $\gamma'$ usually dissolves at the recrystallisation front and reprecipitates behind it. Ringer, Li and Easterling (1992) have put forward a physical and kinetic model for particle rotation, and show that particle rotation is controlled by interfacial diffusion and depends on alloy composition, time, temperature and particle size and shape. Experimental results from a Ti stabilised austenitic stainless steel containing a dispersion of coherent TiC precipitates support the particle rotation model.

In commercial superalloys the microstructures are very complex, and their behaviour has been reviewed by Ralph and his coworkers (Ralph, Barlow, Cooke and Porter 1980). A range of recrystallisation behaviour is found in different superalloys: nucleation sites vary between alloys, and recrystallisation has been found to nucleate at grain boundaries, intragranular carbides, and $\gamma'$-denuded regions near grain boundaries. The recrystallisation interface causes major disruption to the $\gamma'$ distribution.

Even hard second-phase particles are likely to deform if they are sufficiently small or thin: this may lead at intermediate strain to *inhomogeneous slip distribution* in the alloy, which can promote recrystallisation and at higher strains to a refinement of the dispersion, which will retard recrystallisation. This effect has been demonstrated by Kamma and Hornbogen (1976) in a hypoeutectoid steel containing small carbide plates, *F* in fig. 4.52. Here three dispersions are compared, namely coarse (2700 nm diameter) spheres (*C*), medium-sized (200 nm diameter) spheres (*M*) and fine plates (*F*). *C* recrystallises faster than *M* at all strains due to PSN, but *F* lies to the left of *M* in fig. 4.52 due to the microscopic heterogeneity of strain in this microstructure.

**The effect of bimodal dispersions on recrystallisation** Many technologically important alloys contain a mixed particle structure, consisting of a coarse distribution of large inclusions due to the casting, and a fine particle dispersion introduced during subsequent processing. When these materials are recrystallised, the number of potential nuclei depends upon the number of large particles, and the rate of nucleation upon the distribution of the small particles.

Nes (1976a) has considered the kinetics of recrystallisation of such alloys in terms of the parameter $f/r$ for the fine dispersion (proportional to the grain boundary pinning force per unit area, eq. (4.58)) and assuming site-saturated nucleation of new grains at the large inclusions. This approach has been shown (Nes 1976b) to account for the recrystallisation data of an Al–Mg–Si alloy obtained by Scharf and Gruhl (1969).

**Recrystallisation of coarse two-phase alloys** Duplex structures are of interest because, for example, they may provide superplastic properties at elevated temperatures. Their recrystallisation behaviour has been reviewed by Hornbogen and Köster (1978). Honeycombe and Boas (1948) studied the deformation and recrystallisation of an $\alpha/\beta$ brass containing 60% copper and 40% zinc, using microscopic and X-ray methods, and Clarebrough (1950) made

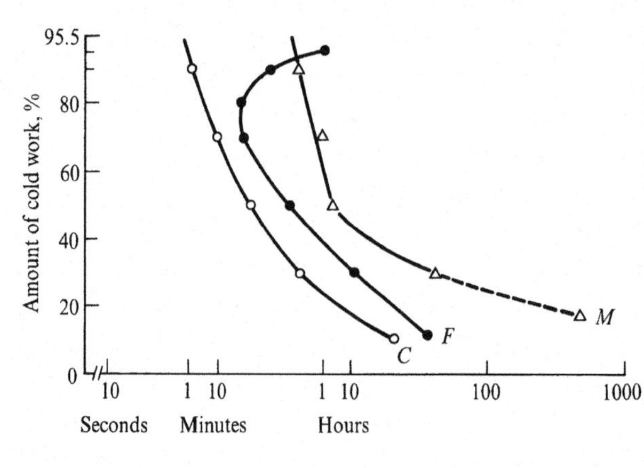


**Figure 4.52** The time to start recrystallisation as a function of the amount of cold work and pre-treatment of a hypoeutectoid steel recrystallised at 550 °C. (After Kamma and Hornbogen 1976.)

a similar study on $\alpha/\beta$ Ag–Mn alloys in which both phases are ductile. A more detailed study was made by Mäder and Hornbogen (1974), who felt that this alloy might be expected to show all the phenomena which *can* occur in such alloys since the two phases differ strongly in crystal structure, diffusivity, stacking-fault energy and atomic order. Their results are summarised in fig. 4.53.

For strains less than 40%, and annealing at the temperature of original precipitation of the $\alpha$ phase (fig. 4.55(*a*)), the $\alpha$ phase recrystallises discontinuously from grain boundary sites while the $\beta$ phase shows 'continuous recrystallisation' by subgrain growth. For annealing temperatures different from the original precipitating temperature, the volume fractions of the $\alpha$ and $\beta$ phases must change so that $\alpha$ nuclei may form within the $\beta$ grains (fig. 4.53(*b*)) which, in turn, can act as recrystallisation nuclei for the original deformed $\alpha$ grains (fig. 4.53(*c*)).

For strains of about 40%, the $\beta$ phase starts to transform martensitically into the $\alpha$ phase ($\alpha_M$), and this process is complete for strains above 70% (fig. 4.53(*d*)). On annealing (fig. 4.53(*e*)), cross-diffusion tends to equalise the compositions of both fcc phases; a fine grained duplex structure forms (fig. 4.53(*f*)) via recrystallisation nuclei of the $\alpha$ phase and retransformation nuclei of the $\beta$ phase.

A contrasting approach has been made by Vasudevan, Petrovic and Robertson (1974), who studied Ag–Ni as a model system in which both phases are ductile and whose compositions and volume fractions do not vary with temperature. Isochronal annealing experiments were conducted on powder metallurgically prepared samples of 57 vol% Ag phase, 43 vol% Ni phase. Recrystallisation was followed by macrohardness measurements and by microhardness measurements on the individual phases of the alloy. The results are shown in fig. 4.54: the dashed line represents the volume fraction hardness calculated from the measurements on the individual phases, and it is seen that this provides a good description of both shape and value of the experimental Ag–Ni macrohardness curve. The overall recrystallisation is thus related to that of the individual phase recrystallisation on a volume fraction basis.

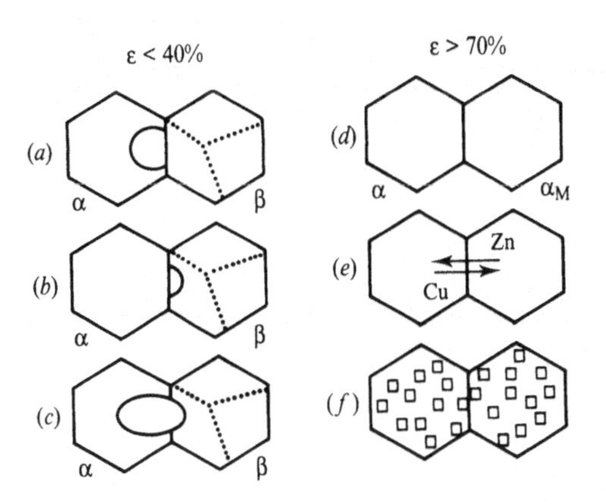


**Figure 4.53** Schematic representation of the micromechanisms of recrystallisation in $\alpha/\beta$ brass for two different deformation stains. (After Mäder and Hornbogen 1974.)

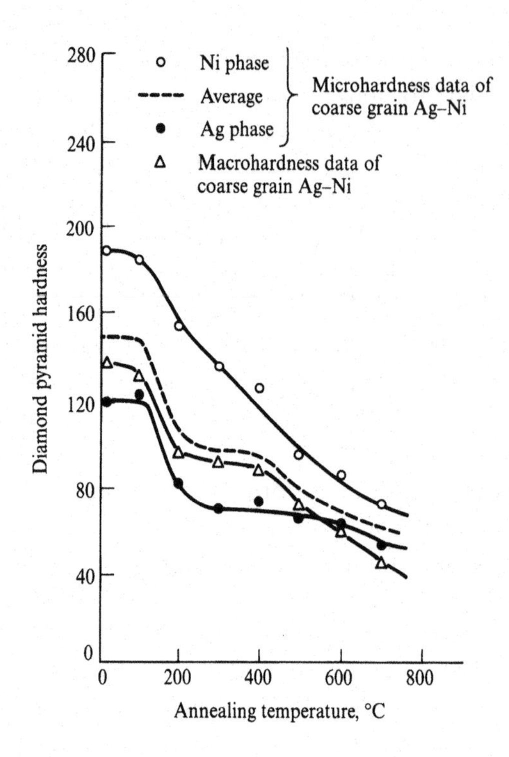


**Figure 4.54** Comparison of Ag–Ni macrohardness to microhardness of the individual phases. (After Vasudevan, Petrovic and Roberson 1974.)

**Precipitation during recrystallisation** The recrystallisation of supersaturated solid solutions has been reviewed by Hornbogen and Köster (1978). Except when recrystallisation is complete before the precipitate is nucleated, precipitation and recrystallisation exert a mutual influence upon each other. Thus precipitate particles hinder the formation and migration of 'recrystallisation fronts', while the lattice defects themselves promote the nucleation of precipitates.

The temperature of dependence of recrystallisation and of precipitation processes can be expressed by considering $t_r$, the time for the start of recrystallisation (assuming a constant size for the recrystallisation nuclei):

$$t_r = K_r \exp(Q_r/kT) \tag{4.62}$$

where $K_r$ is a factor containing the driving force for the reaction, an entropy term and geometric factors, and $Q_r$ is the activation energy for the formation of a recrystallisation front. The time $t_p$ for the start of precipitation of a single stable phase can be written:

$$t_p = K_p \exp[(Q_n + Q_D)/kT] \tag{4.63}$$

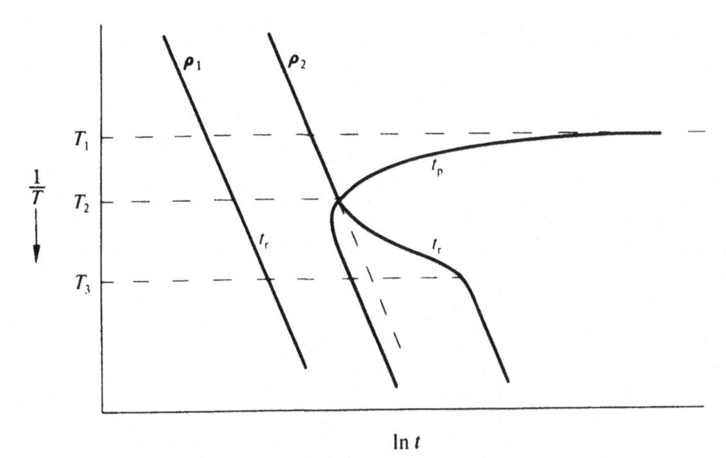


**Figure 4.55** Temperature dependence of the incubation time for precipitation, $t_p$, and for recrystallisation, $t_r$, as a function of dislocation density $\rho$. It is assumed here that the dislocations have little influence on the start of precipitation. Dislocation density $\rho_1$: recrystallisation is always complete before precipitation. Dislocation density $\rho_2$ ($\rho_2 < \rho_1$): when $T > T_1$, single-phase stable: $T_1 > T > T_2$, recrystallisation complete before precipitation; $T_2 > T > T_3$, discontinuous recrystallisation but slowed by pinning effect of precipitates; $T_3 > T$, continuous recrystallisation only. Schematic diagram which is representative for Cu–Co alloys. (After Kreye and Hornbogen 1970.)

where $K_p$ is a factor containing the driving force, an entropy term and geometric factors, and $Q_n$ is the activation energy for nucleation of the second phase whose value will depend strongly on the degree of supercooling below the equilibrium temperature ($T_1$). Thus $t_p$ only becomes dependent solely upon the activation energy for diffusion ($Q_D$) when $Q_D \gg Q_n$, that is, under conditions of high supercooling, giving rise to the familiar '$C$' curve when $1/T$ is plotted against ln $t$. Such a graph is illustrated in fig. 4.55, and this enables us to define a temperature $T_2$ at which $t_r = t_p$.

Below $T_2$ there will be an interaction between recrystallisation and precipitation for an alloy of a given dislocation density and composition. The effect of increasing $\rho$, the dislocation density, will cause the time $t_r$ to be displaced to smaller values, and temperature $T_2$ will be displaced to a lower value. Above a certain critical dislocation density, recrystallisation will always be complete before the start of precipitation: this is illustrated in fig. 4.55, the assumption being made that the dislocations have little effect on the beginning of precipitation.

The growth rate ($v$) of the recrystallisation front will depend upon the grain boundary mobility ($m$) and the driving force ($p$), according to $v = mp$. The value of $m$ will be strongly dependent upon the temperature, the structure of the grain boundary and the impurity content. The net driving force, $p$, is the resultant of three forces: that due to the elimination of dislocations ($p_\rho$), whose value will be

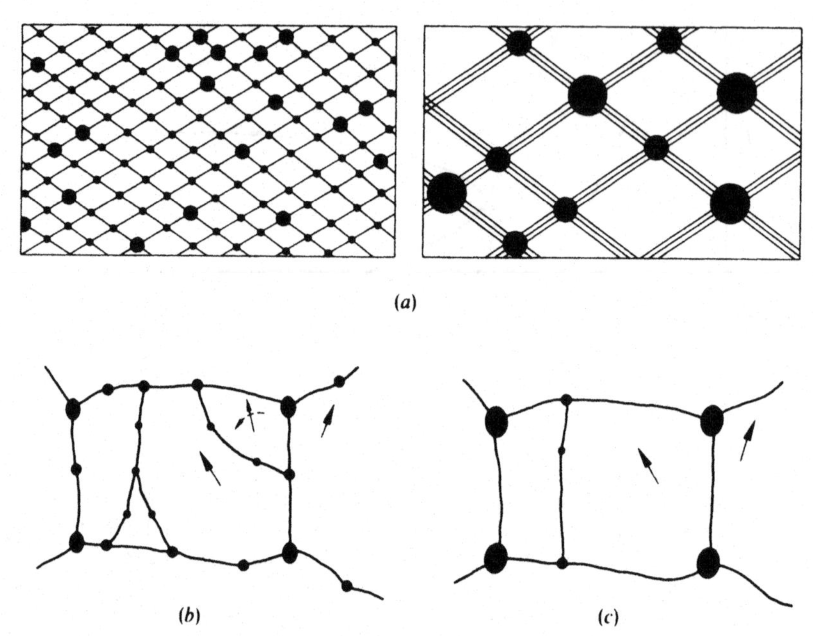


**Figure 4.56** Schematic sketch of continuous recrystallisation. (After Hornborgen 1970.) (*a*) No grain boundaries are able to move as recrystallisation front. Annihilation and rearrangement of the dislocations are controlled by the growth of the particles. Subgrains form and their size and orientation differences (indicated by the thickness of the lines) increase. (*b*) Subboundaries are pinned by particles. (*c*) After dissolution of the smallest particles, a subgrain can anneal out by Y-node motion (left) or by subgrain rotation (right).

given approximately by eq. (4.59), a chemical driving force due to the process of discontinuous precipitation ($p_c$), and a retarding force ($p_f$) due to precipitated particles, whose value given by eq. (4.58) will increase strongly with decreasing temperature since $f$, the volume fraction of precipitated phase, will increase, and $r$, their average radius, will decrease.

In the temperature range below $T_2$ (fig. 4.55), in which the precipitation and recrystallisation processes mutually interact, *discontinuous* or *continuous* recrystallisation can occur.

In discontinuous recrystallisation, a grain boundary serves as a reaction front and reduces the dislocation density, and it will be observed if a grain boundary capable of migration exists and if the sum of the driving forces exceeds the retarding force, that is,

$$p_\rho + p_c > p_f$$

If no grain boundaries are present, or the existing boundaries are pinned by preferential precipitation, no discontinuous recrystallisation can take place and the

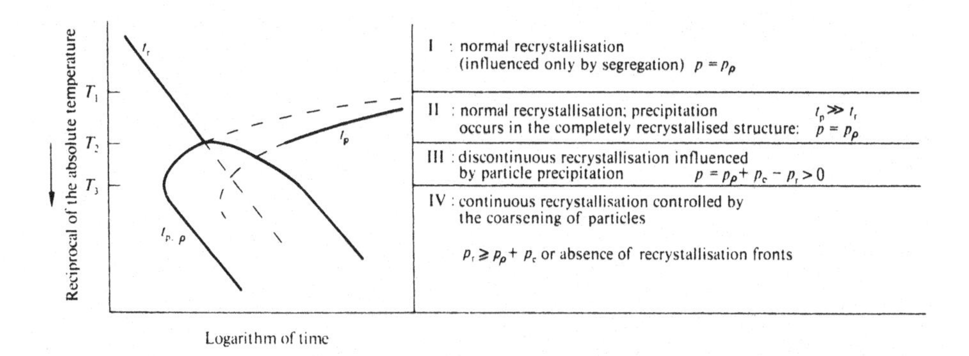


**Figure 4.57** Schematic diagram showing the possibilities for interaction between recrystallisation and precipitation processes. $t_r$: start of recrystallisation; $t_p$: start of precipitation in undeformed crystal; $t_{p,\rho}$: start of continuous precipitation on dislocations. (After Köster 1971.)

annealing process is controlled by particle growth. This is termed *continuous, or in-situ recrystallisation*, and is illustrated schematically in fig. 4.56. Dislocation rearrangement is controlled by the dissolution of small particles which exert a pinning effect, and subboundaries can also anneal out as particles dissolve. A completely recrystallised structure can thus eventually form, although no 'recrystallisation front' migrates, and the process has been reported in Al–Cu alloys by Köster and Hornbogen (1968) and Ahlborn, Hornbogen and Köster (1969), who suggest the use of the term 'continuous recrystallisation' for the process, rather than 'recrystallisation *in situ*', which had been used previously. Hornbogen and Köster (1978) have summarised the situation diagrammatically, as shown in fig. 4.57, demonstrating the possibilities for interaction between recrystallisation and precipitation processes. $T_1$ is the solvus temperature for the precipitate, so that if $T<T_1$ normal recrystallisation will proceed. If $T_1>T>T_2$, normal recrystallisation again occurs, but precipitation will take place in the completely recrystallised structure: $t_p$ is the time for precipitation in undeformed crystal (eq. (4.59)). If $T_2>T>T_3$, $Q_n$ in eq. (4.62) is influenced by the lattice defects present, so that the precipitation curve is accelerated due to the precipitation on dislocations to the times represented by the curve $t_{p,\rho}$. Discontinuous precipitation will occur, but the net driving force will decrease as the temperature is lowered until, below a certain temperature ($T_3$), $p_p+p_c \leq p_f$ and the discontinuous process will cease. If $T<T_3$, only continuous recrystallisation can proceed, at a rate controlled by the rate of coarsening of particles of precipitate.

The principles outlined above are valid for all supersaturated solid solutions. Hornbogen and Köster (1978) pointed out that the wide range of behaviour observed in different alloys arises from the wide variety of nucleation behaviour of stable and metastable phases in perfect or defect matrix lattices. They pointed out that if the defect dependence of the nucleation behaviour of a particular alloy is known, conditions can be predicted under which the different reactions will occur.

# Chapter 5

## Microstructural instability due to interfaces

### 5.1 Introduction

The three main interfaces which are important in metallic systems are the solid–gas interface (the external surface), the interface between two crystals of the same phase which differ only in orientation (the grain boundary) and the interface between two different phases (the interphase boundary). The interphase boundary provides an almost infinite range of possibilities since in addition to the possible difference of orientation of the two crystals, the crystals can also differ in crystal structure, lattice parameter and in composition. In this almost infinite array of possible structures and thus properties two limiting conditions can be recognised. In one case where the interface is formed, as it often is in metallic systems, by precipitation of a second phase within a primary crystal structure then a particular orientation relationship develops between the phases. This produces an interface with a close atomic fit which minimises the interfacial energy (see §2.2.2 and Doherty (1982)). This, in turn, can introduce difficulties for interfacial mobility in that growth ledges may be needed, see §1.3. The opposite extreme occurs when the two phases have no orientation relationship with each other. As a result their interface will be a high energy, incoherent one that usually provides no particular crystallographic barrier to mobility. Examples of this type of incoherent interface arise when the two phases come in contact by growth processes rather than by nucleation. A common way that this may occur is when a high angle matrix grain boundary sweeps past a precipitated second phase so changing the orientation of the matrix but not the particle. This phenomenon has been reviewed and discussed in some detail by Doherty (1982).

There are several important properties of interfaces which affect the

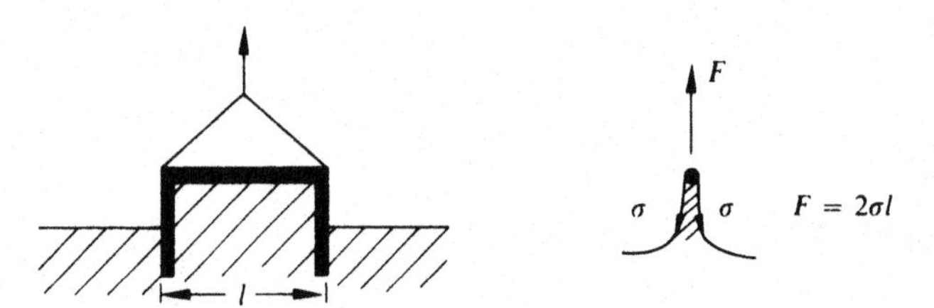


**Figure 5.1** Determination of surface tension $\sigma$ of a liquid–air interface by differential weighing of a wire attached to, and then subsequently free of, a thin film of water.

development of microstructure – examples include the mobility of the interface and interfacial diffusion. From the standpoint of this chapter the most important is the *energy* of unit area of the interface. As we shall see, this energy is a free energy. The reduction of interfacial area and thus energy provides the driving force for many important changes in structure, such as the achievement of equilibrium shape of crystals, precipitate coarsening, spheroidisation of two-phase lamellar structures, grain growth etc. All of these changes are ones that reduce the excess surface area, and thus the energy, of the system. The ultimate stage of these processes is, for a single-phase material, a single crystal having the 'equilibrium shape'. For a two-phase alloy, the final stage would be a complex equilibrium shape which would have to contain a single interphase interface. This interface would be one of low energy, where the two crystal structures would fit together well. What makes the subject both difficult and interesting is that this stage is almost never reached in times of interest to users of practical materials, so the kinetics of the changes, and the resulting intermediate structures produced, are much more important than the final equilibrium structure.

## 5.2 Surface energy and surface tension – $\sigma$

The surface of a liquid such as water behaves as if it had a surface skin exerting a tension. This is readily shown using a piece of bent wire (fig. 5.1), suspended from a chemical balance, which is lowered into a beaker of water and partially withdrawn. A film of water is formed inside the wire frame and this film exerts a readily measurable *force* of $2\sigma l$. This experiment in surface physics, which is used to determine the surface tension of the liquid–vapour interface, will be familiar to many scientists. The equivalent demonstration and determination of the solid–vapour surface tension is much less well known – but it is equally important. The experiment consists of measuring the specimen shape changes that occur on holding a fine drawn metal wire at a temperature just below the melting point of the metal, when the mobility of the atoms by diffusion is significant, typically where the diffusion coefficient is about $10^{-13}$ m$^2$ s$^{-1}$. The specimen is seen to shrink in length in order to reduce its surface area. This process was known for many years and is readily illustrated in any filament of syrup pulled from a jar by a spoon and broken. It was first used by Udin, Shaler and Wulff (1949) to measure the surface tension of fine wires by the 'zero creep' method. In this method, the surface tension of the wire is balanced by the gravitational force exerted on a small blob of metal on the end of the wire, usually formed by local melting and freezing (fig. 5.2). The significant distinction between solids and liquids, even viscous liquids, in respect of their response to the small surface tension forces is in their viscosities which differ by many orders of magnitude.[†]

The zero creep experiment, which is perhaps the most reliable method of measuring the surface tension of solids, will be described in some detail since it most clearly demonstrates the essential property of the surface energy or tension of solids. The values of surface tension of solids determined by this method are the fundamental measurements from which most other solid state interfacial energies have been determined. Other methods of measuring the energy of the solid–gas interfacial tension have been reviewed, for example, by Inman and Tipler (1963) and Blakely (1973).

In the length ($l$) of the wire, shown in fig. 5.2($a$), there are $n+1$ grains. Each grain takes up the entire cross-section of the wire so producing a 'bamboo' structure. There are then $n$ grain boundaries along this length of the wire. The bamboo structure is readily established by annealing a high purity wire at temperatures close to the melting point of the metal ($0.9T_m$ or higher) to give recrystallisation (if the wire has been cold drawn) followed by grain growth. Grain growth ceases when each grain boundary lies normal to the surface of the wire. The wire has a radius $r$ and on the end of the wire is the melted blob of metal of mass $m$. The *average* surface tension of the solid–gas interface is $\sigma$ and that of the grain boundaries is $\sigma_b$. As is discussed in the next section, the interfacial tensions, $\sigma$ and $\sigma_b$, are numerically equal to the interfacial free energies. That is, $\sigma$ and $\sigma_b$ are the free energies per unit area of the surface and the grain boundaries respectively.

The total surface free energy, $\sigma_T$, is $2\pi r l\sigma + n\pi r^2\sigma_b$. If the wire increases in length by $\partial l$ at constant volume (that is, plastically) and at constant pressure:

$$\partial\sigma_T/\partial l = 2\pi r\sigma + 2\pi l\sigma(\partial r/\partial l) + 2\pi n r\sigma_b(\partial r/\partial l) \tag{5.1}$$

[†] Bikerman (1965) attacked the whole idea of surface energy of solids and criticised many of the techniques used for measuring it. His principal objection was that solid surfaces are rough and appear to stay rough, while liquid surfaces rapidly become smooth. However, he failed to recognise that at temperatures where atom motion in solids becomes significant, solid surface roughness does, indeed, decay and this phenomenon has also been used to measure the surface energy (Mullins 1959; Blakely and Mykura 1962). A major distinction between liquids and solids is in their viscosities. So whenever a liquid has an extremely high viscosity, for example, window glass at room temperature, scratches in the 'liquid' do not then disappear. Argon (1965) very adequately answered Bikerman's criticisms of the currently accepted concepts of the surface tension of solids.

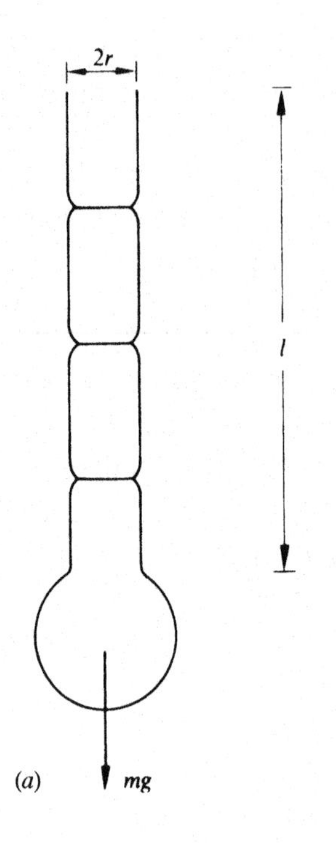


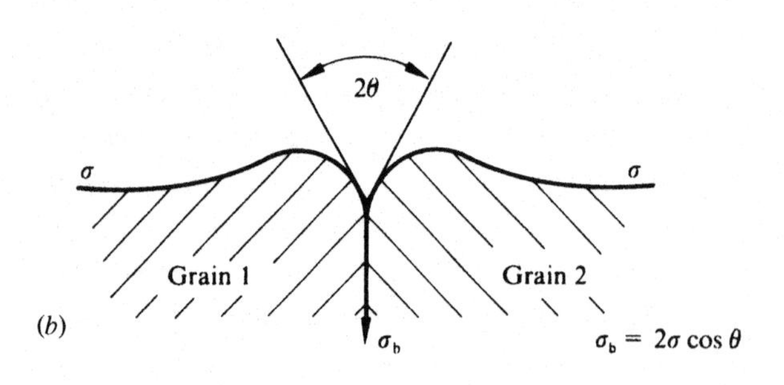


**Figure 5.2** (*a*) Zero creep method of measuring the surface energy $\sigma$ of a solid. This uses wire of radius $r$ containing only transverse grain boundaries ('bamboo' structure) with a large melted blob on one end. The grain boundary energy is $\sigma_b$.

(*b*) Section through a grain boundary groove formed at the surface of a solid where a grain boundary meets the external surface. The effect is not to scale as $2\theta$ should be much closer to an angle of 180° than is shown. The 'bumps' on the surface are only expected if the groove forms by surface or volume diffusion, and should not be present if evaporation/condensation is the main transport process.

If the mass $m$ has the critical value $m^*$ at which the surface forces are in equilibrium with the gravitational force on the blob ($m^*g$), where $g$ is the acceleration due to gravity, we obtain:

$$\delta\sigma_T/\partial l = m^*g$$

It is assumed here that the mass of the wire is negligible compared to the mass of the blob. Since the volume, $V$, remains constant:

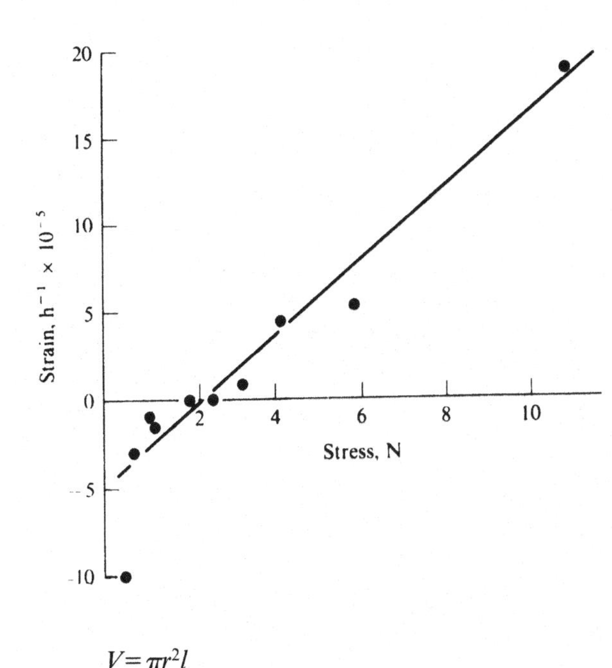


**Figure 5.3** Variation in strain rate in a wire subjected to long term loading by the gravitational force on a melted blob. (Typical experimental result due to Hayward and Greenough (1960), courtesy of the Metals Society.)

$$V = \pi r^2 l$$

and so

$$\partial V/\partial l = \pi r^2 + 2\pi r l(\partial r/\partial l) = 0$$

Then $\partial r/\partial l = -r/2l$ and, from eq. (5.1), we obtain:

$$m^*g = \pi r\sigma - \pi r^2 n\sigma_b/l \qquad (5.2)$$

The usual experimental arrangement is to have a series of wire samples each with a different mass of the melted blob. An initial anneal is given to establish the bamboo structure, the sample lengths are measured accurately and the wires are then reannealed, at the same temperature, for several days. To prevent possible evaporation and/or surface contamination, the annealing is carried out in a vacuum in a chamber made of the same metal as the samples. At the end of the experiment, the length changes of each wire are measured and these changes are plotted against the mass of the blob. A typical experimental plot is shown in fig. 5.3 where the creep strain rate has been plotted against the full applied stress ($mg/\pi r^2$). The first important result seen in this figure is that the value of zero strain (*zero creep*) is found at a finite stress, so giving the critical value of $m^*g$ that satisfies eq. (5.2).

A second important result of this experiment is that there is a *linear* relationship between the net applied stress, $\tau$, and the resultant strain, $\epsilon$, or more precisely the strain rate, $\mathrm{d}\epsilon/\mathrm{d}t$: $\tau$ is given by:

$$\tau = \{mg - [\pi r\sigma - (\pi r^2 n\sigma_b/l)]\}/\pi r^2$$

This linear dependence of strain rate on stress, at a very low stresses, provides a clear example of what is called 'diffusion creep'. This is creep arising from the diffusive motion of atoms changing places with vacancies. For the lengthening reaction in a wire with a bamboo structure, there is a vacancy flux from vacancy *sources* at the grain boundaries to the vacancy *sink* at the surface. The coupling mechanism between the net stress and the vacancy flux is the change of the formation energy of the vacancies at the grain boundaries transverse to the stress produced by the net tensile stress. The energy needed to form a vacancy at the transverse grain boundary under a net tensile stress is *reduced* by a factor, $V_a\tau$ (where $V_a$ is the volume of one atom), that is proportional to the stress. Under the high temperature conditions of the experiment this gives rise to an increase in the equilibrium vacancy concentration at the grain boundary. The increase of vacancy concentration is also proportional to the net stress. As a result of diffusion of vacancies down the concentration gradient from the boundaries to the surface, a linear stress–strain rate relationship is established. Such strain by diffusion creep is seen at stresses much less than that which would cause strain by dislocation motion. A detailed discussion of diffusion creep has been given by Ashby (1972) and by Frost and Ashby (1982). For polycrystalline materials two major mechanisms occur. The first is by volume diffusion of vacancies. This gives a strain rate that varies as the inverse square of grain size, $d\epsilon/dt = \kappa D/d^2$ (where $D$ is the volume self-diffusion coefficient, $d$ is the grain size and $\kappa = 42V_a\tau/kT$), usually called Nabarro–Herring creep. The second mechanism is transport by grain boundary diffusion. Here the strain rate varies as the inverse cube of grain size, $d\epsilon/dt = \kappa' wD_b/d^3$ (where $D_b$ is the grain boundary self-diffusion coefficient, $w$ is the width of the grain boundary and $\kappa' = 42\pi V_a\tau/kT$), a process known as Coble creep.

A very interesting negative result is the observation that surface energy induced creep of single crystal wires does not appear to occur – even though edge dislocations within the wire could act as *alternative* vacancy sources and sinks. Edge dislocation climb that produced extra atomic planes, normal to the tensile stress, should act as the vacancy source while climb of edge dislocations whose Burgers vectors are parallel to the stress should act as the vacancy sink, fig. 5.4. A flux of vacancies between such dislocation sources and sinks should, at first glance, also give continued diffusion creep. However, this does not happen. The difficulty seems to be that continued creep under these conditions would require *curvature* in the climbing dislocations from their nodes (exactly as in the glide operation of Frank–Reed sources for a network of dislocations, see, for example, Hull and Bacon (1984)). Such curvature in the climbing edge dislocations, if it were to lead to continued production and destruction of vacancies needed for continuing strain, would require stresses of the order of the yield stress. This

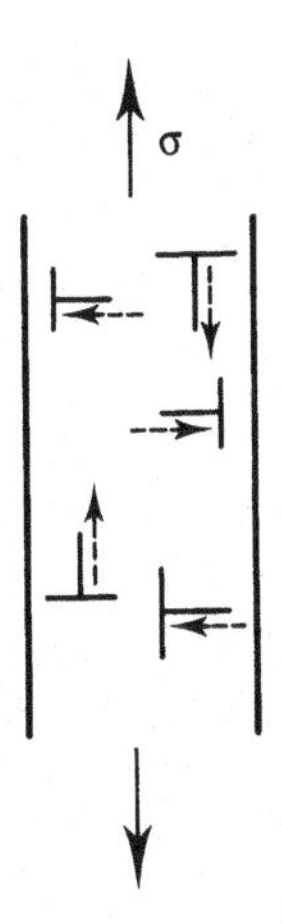


**Figure 5.4** Creep of single crystal by edge dislocation climb under very low net tensile stress. Direction of dislocation movement under a net tensile stress is indicated by the dashed arrows.

deformation process is well known and described for climbing dislocation by Bardeen and Herring (1952). The yield stress is determined by the usual equation for dislocation generation:

$$\tau = Gb/2R_c = aGb(\rho)^{0.5}$$

Here $G$ is the shear modulus, $b$ the Burgers vector and $R_c$ the critical radius of curvature that is related to the dislocation density, $\rho$, by:

$$R_c = 1/a(\rho)^{0.5}$$

The factor $a$ is an algebraic constant of the order 0.5–1; its value depends on the detailed dislocation arrangement. There is clear evidence that small transient creep strains do, in fact, occur under small stresses by the operation of climbing edge dislocation sources but that these strains are very limited; see, for example, Hayward and Greenough (1960). The strain due to edge dislocations climbing under very small stresses would be expected to end as soon as the pinned dislocations achieve a radius of curvature $R_c$ given by, $\tau' = Gb/2R_c$, where $\tau'$ is the net stress due to unbalanced surface tensions.

Low stress–low strain rate creep deformation has been studied in detail; see, for example, Ardell and Przystrupa (1984) and Ardell and Lee (1986). This type of deformation, following the classic studies of Harper and Dorn (1957), is usually called Harper–Dorn (H–D) creep. The characteristics of H–D creep, according to Ardell and Lee (1986), are that it occurs at high temperatures, it has a primary creep strain and it shows a steady state creep rate that increases linearly with stress, for stresses less than about $5\times10^{-6}$ of the shear modulus. The microstructure has a steady state dislocation density that is independent of stress and the strain rate is found to be independent of grain size. H–D creep is usually studied in single crystals to avoid the complication of Nabarro–Herring or Coble creep occurring as additional mechanisms. This interesting topic, however, is not relevant to the subject of this chapter and will not be discussed further.

We now return to the topic of major interest here – the use of the zero creep experiment to determine interfacial energies. A further relationship between $\sigma$ and $\sigma_b$ is required, in addition to eq. (5.2), to yield the individual values of $\sigma$ and $\sigma_b$. Such a relationship is provided by measurements of the equilibrium *groove* angle $2\theta$, formed between a surface and a grain boundary, see fig. 5.2(*b*). It should be noted that equilibrium requires the grain boundary to lie normal to the surface. This normal incidence of grain boundaries is what gives rise to the bamboo structure of wires annealed at high temperature.

Typical values for $\sigma$ and $\sigma_b$ are given in table 5.1. The value of the surface free energy, $\sigma$, has been shown, for example, by Jones (1971), to increase steadily with the melting temperature, $T_m$, of the metal (in K). The linear dependence of $\sigma$ with $T_m$ arises from the dependence of both on the interatomic bond energy of the metal. A similar linear relationship with $T_m$ has been established for the surface energy of solid–liquid interfaces, $\sigma_{sl}$ (Turnbull 1950), for the liquid–vapour interfacial energies, $\sigma_{lv}$ (Skapski 1948), and the grain boundary energies, $\sigma_b$ (Hondros 1969). Jones and Leak (1967) have tabulated values for the rate of change of interfacial energy with temperature, $d\sigma/dT$. For pure metals in a vacuum, where surface adsorption of impurities is not significant, $d\sigma/dT$ is the *negative surface entropy* (see below). Typically the experimental value of $d\sigma/dT$ is $\sim -1\ \text{mJ m}^{-2}\text{K}^{-1}$.

**Table 5.1** *Surface and grain boundary energies of pure metals.*

| | Surface energy (Jones 1971) $\text{J m}^{-2}$ | Grain boundary energy (Hondros 1969) $\text{J m}^{-2}$ |
|---|---|---|
| Aluminium | 1.1±0.2 | 0.6[a] |
| Gold | 1.4±0.1 | 0.4 |
| Copper | 1.75±0.1 | 0.53 |
| Iron (bcc) | 2.1±0.3 | 0.8 |
| Iron (fcc) | 2.2±0.3 | 0.79 |
| Platinum | 2.1±0.3 | 0.78 |
| Tungsten | 2.8±0.4 | 1.07 |

[a]Determined by calorimetry, so this is the internal enthalpy, $\sigma(H)$, which will be larger than the free energy by virtue of the entropy contribution to the latter.

The identification of the interfacial tension, $\sigma$ (in N m$^{-1}$), with the interfacial free energy per unit area (in J m$^{-2}$) is readily demonstrated. A surface of width $l'$ with a tension $\sigma$ parallel to the surface† can be expanded reversibly by a small length $\mathrm{d}l$ giving an increase in area $\mathrm{d}A$, where $\mathrm{d}A = l'\,\mathrm{d}l$, at constant temperature, volume and pressure. The work done on the sample, $\mathrm{d}w$, is given by:

$$\mathrm{d}w = \sigma\,\mathrm{d}A$$

From the first law of thermodynamics, the change in internal energy $\mathrm{d}U$ is given by:

$$\mathrm{d}U = \mathrm{d}q + \mathrm{d}w$$

For any reversible change, the increment of heat added to the sample $\mathrm{d}q_{rev}$ is given by an increase of entropy $\mathrm{d}S$:

$$\mathrm{d}q_{rev} = T\,\mathrm{d}S$$

So provided that the area change is accomplished reversibly,

$$\mathrm{d}U = T\,\mathrm{d}S + \sigma\,\mathrm{d}A$$

The definition of Gibbs free energy $G$ is $G = H - TS = U + PV - TS$. So, at constant pressure, constant volume and constant temperature, and provided that the change is reversible, we obtain:

$$\mathrm{d}G = \mathrm{d}U - T\,\mathrm{d}S = \mathrm{d}q_{rev} + \sigma\,\mathrm{d}A - T\mathrm{d}S = \sigma\,\mathrm{d}A \qquad (5.3)$$

If the atomic structure of the interface maintains its equilibrium arrangement of the surface atoms during the change, then the area change will indeed be reversible. We then find (eq. (5.3a)) that the surface tension (in N m$^{-1}$) and the specific surface free energy (in J m$^{-2}$) are numerically identical.

$$\sigma = (\mathrm{d}G/\mathrm{d}A)_{T,P,V} \qquad (5.3a)$$

The atomic structure at the interface will be expected to remain constant if the change of shape occurs at a sufficiently high temperature for atomic mobility to maintain the equilibrium interfacial structure.

† For anisotropic interfacial free energies, see §4.4, there can be additional forces (torque terms) *normal* to the interface trying to rotate the interface to a lower energy orientation (see Mykura (1961)).

Since $\sigma$ is a Gibbs free energy, it will be made up of two terms. The first term is the excess interfacial energy, strictly *interfacial enthalpy*, per unit area of interface, $\sigma(H)$, and the second term arises from the excess *interfacial entropy* per unit area of interface, $\sigma(S)$. That is:

$$\sigma(G)=\sigma(H)-T\sigma(S) \tag{5.3b}$$

From (5.3b) using the standard methods of thermodynamics, see, for example, Gaskell (1984), it follows that this excess surface entropy, $\sigma(S)$, is given by:

$$\sigma(S)=-\partial G/\partial T=-\partial\sigma/\partial T \tag{5.3c}$$

The negative values found experimentally, with pure metals, for $\partial\sigma/\partial T$ then indicate that their surface entropies are indeed small and positive.

## 5.3 Atomic origin of the interfacial free energy

Taking, as our example, the {100} surface of an fcc crystal, fig. 5.5, we can see that each atom will have four nearest neighbours in the surface plane and four more in the plane beneath the surface. There are four 'missing' atoms in the atomic plane that has been removed from above the surface. These missing atoms would have been in positions directly above the atoms, shown as broken circles. So, on the simple bonding model using only the nearest atoms, a surface atom on {100} has only eight out of the twelve possible bonds – the other four will be 'broken bonds'. Since each bond lowers the energy of an atom by $H_b/2$, each surface atom will have its energy raised by $2H_b$ by virtue of these broken bonds. ($H_b$ is the energy of a bond between two atoms.) The product of $2H_b$ and the

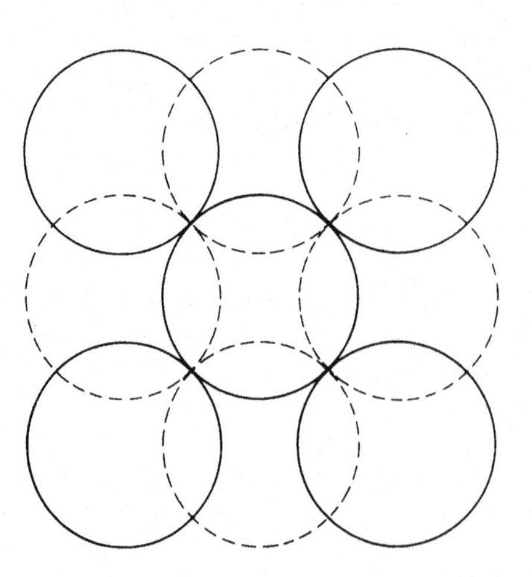

**Figure 5.5** Arrangement of atoms on a {100} surface of an fcc crystal (circles) showing the atom positions in the plane below the surface and the equivalent four missing atoms in the plane that would lie above the surface plane if the surface were not present (broken circles).

number of atoms per unit area will be the excess interfacial *enthalpy*, $\sigma(H)$, of the interface. It is, however, only part of the excess interfacial free energy, $\sigma=\sigma(G)$.

At any non-zero temperature, there is a small but finite vapour pressure of evaporated gas atoms in equilibrium with the crystal – these all have an extra *enthalpy* of $6H_b$ but this increased enthalpy is balanced by the extra configurational *entropy* of the gas atoms to give the same free energy per atom in each phase, crystal and gas. The necessarily large value of the configurational entropy of the gas atoms, especially at low temperatures, is achieved by each gas atom occuping a very large volume, that is, by the gas exerting a sufficiently small partial pressure. Atoms on a crystal surface also have a higher configurational and vibrational entropy than bulk atoms. However, such an excess interfacial entropy, $\sigma(S)$, is limited in magnitude and it thus cannot offset the enthalpy, even at high temperatures. That is, the surface will always have a finite excess surface free energy, $\sigma(G)$. So the only route towards equilibrium for a pure material is the reduction of the surface area. For an impure system, however, adsorption of impurities on the surface can itself reduce the surface free enthalpy, see, for example, the experimental review by Hondros (1965) and the thermodynamic review by Gaskell (1984). According to the Gibbs adsorption relationship (1878) the surface free energy $\sigma$ is reduced by impurity adsorption† by

$$d\sigma=-\Gamma_B RT d\ln a_B \tag{5.4}$$

$\Gamma_B$ is the surface excess of the impurity component, B, and $a_B$ the activity of the impurity. In this case for a gaseous impurity such as oxygen, the activity is given from the partial pressure $p_B$ of the gas ($a_B=p_B/p_B^\circ$ where $p_B^\circ$ is the pressure of the standard state, typically 1 atm)

$$d\sigma=-\Gamma_B RT d\ln p_B \tag{5.4a}$$

Several investigators have shown that as expected from the adsorption isotherm above, measured values of $\sigma$ from zero creep experiments fall steadily with, for example, oxygen partial pressure (for example, Hondros (1965, 1968)) and, from the data, values of surface excess of the impurity, $\Gamma_B$, can be determined. The values of this excess are usually a few tenths of a monolayer.

The origin of the excess surface entropy is found in both vibrational and configurational effects. The reduced binding energy of the surface atoms causes

† A clear demonstration of the effect of adsorption on surface free energy can be made with soap, salt and water. The soap molecule with its polar and non-polar ends is adsorbed strongly and positively ($\Gamma_B \gg 0$) at the water–air interface, so *lowering* the surface tension. This allows the easy creation of a soap lather in fresh water. In salt water, however, the Na and Cl ions avoid the surface, so their adsorption $\Gamma_B$ is here $<0$ and so the surface energy is increased. This makes it much more difficult for soap to form a lather in salt water than in fresh water.

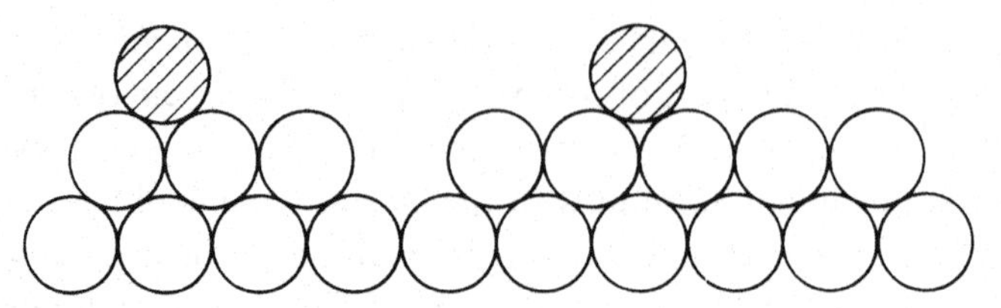

**Figure 5.6** Section through the surface of a crystal showing both 'missing' atoms (surface vacancies) and 'adatoms' on the surface, defects that are expected at any temperature significantly above 0 K.

a lower vibration frequency of the surface atoms and this raises the vibrational entropy (for example, Swalin (1972)). In addition, as the temperature rises, surface vacancies can form, fig. 5.6, by atoms leaving the surface plane and becoming 'adatoms' on the surface for a similar reason to that which causes vacancies to form inside the crystal. The formation of both surface and bulk vacancies raises the energy and entropy but it does so to give a minimum in the free energy at the equilibrium concentration of each type of vacancy. For both these reasons there is expected to be a positive surface entropy (Jones 1971). For copper, silver and gold, $\sigma(S)$ is about 0.5 mJ m$^{-2}$ K$^{-1}$.

Chalmers (1959) pointed out that for these three metals, the ratio of the latent heat of vaporisation per atom to the surface free energy per atom, near the melting point, varies from 7 (copper ) to 4 (silver) with an average of 5. In order to vaporise a close-packed fcc crystal all the bonds to the nearest neighbours must be broken, so the surface free energy apparently corresponds to 12/5, that is, 2.4, broken bonds. (Twelve broken bonds give two vapour atoms and five broken bonds two surface atoms.) However, the latent heat is an *enthalpy*, change not a free energy change, so the calculation is more usefully made for the surface energy $\sigma(H)$, not the surface free energy $\sigma$. Assuming $\sigma(H)$ and $\sigma(S)$ are approximately constant, then we obtain:

$$\sigma(H)=\sigma+T\sigma(S)$$

For copper at 1300 K, $\sigma$ is 1.65 J m$^{-2}$ and $\sigma(S)$ is about 0.5 mJ m$^{-2}$ K$^{-1}$, so $\sigma(H)$ should be about 2.3 J m$^{-2}$. This change has increased the value of the surface energy to be used by 50%, from $\sigma$, used in Chalmers's calculation, to $\sigma(H)$. The use of this larger value of the surface ethalpy increases the predicted number of broken bonds, per surface atom, from 2.4 to 3.6. This is a more reasonable value since for fcc metals on a close-packed {111} surface there will be three broken bonds, on {100} there will be, as shown above, four broken bonds and for all other surface planes the number of broken bonds will be even higher. That is, the simple Chalmers calculation is improved when the surface enthalpy is used in the model.

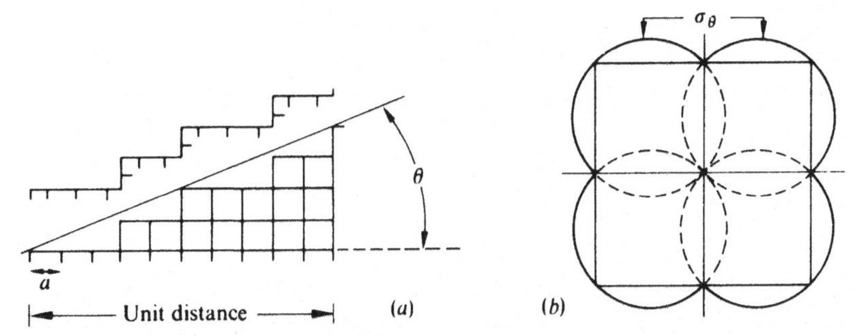


**Figure 5.7** (*a*) Showing a surface plane at an overall angle $\theta$ to a close-packed {01} plane in a simple two-dimensional cubic crystal together with the 'matching' surface. (*b*) The expected variation of surface energy $\sigma$ as a function of $\theta$ bonds formed to nearest neighbours only. (After Mullins 1963, courtesy of ASM.)

## 5.4 The anisotropy of surface free energy – the $\sigma$ plot and the Gibbs–Wulff theorem

For crystalline solids there is, as described above, a variation in the number of 'broken bonds' at different surfaces. Since the surface free energy is due to the 'broken bonds' it would be expected that the measured values of the interfacial free energy $\sigma$ should be anisotropic, that is, different for different crystal surfaces. This is commonly found; Shewmon and Robertson (1963) gave a review of the major results up to 1963 and Dunn and Walter (1966) extended this by a couple of years.

The main theoretical ideas are readily demonstrated using a simple two-dimensional square crystal with a chemical bond only to the nearest neighbour atoms (fig. 5.7). This model calculation is due to Mullins (1963). If we have a crystal surface at an angle $\theta$ to the two-dimensional (01) plane, this will be made up of a series of terraces and steps – having a readily determined number of broken bonds. This surface has been created from a region of perfect crystal by separation from its mating surface, which is also shown. The energy required for creation of these two surfaces is $H_b$ times the number of broken bonds $f(\theta)$, where $f(\theta)$ is the number of broken bonds on each surface corresponding to a unit distance along the (01) base.

$$f(\theta)=1/a+(\tan\theta)/a$$

where $a$ is the interatomic spacing. The first term gives the vertical bonds on the (01) plane and the second the horizontal bonds, on the (10) plane. The surface free energy per unit area, $\sigma_\theta$, is then:

$$\sigma_\theta=[H_b f(\theta)\cos\theta]/2$$

Note that $2/\cos\theta$ is the surface area created. Combining the last two equations we obtain:

$$\sigma_\theta=H_b(\cos\theta+\sin\theta)/2a=H_b\cos(\theta-\pi/4)/2a \qquad (5.5)$$

This equation describes a circle passing through the origin with a diameter of $H_b/a\sqrt{2}$. The diameter is found at an angle of $\pi/4$ to the horizontal axis. The part of this circle shown as a solid curve is for the actual surface seen, with the physically unreal rest of the circle shown as a dashed line in fig. 5.7(*b*). From symmetry, the same arguments can be applied to the other three {10} surfaces, restricting the validity of each of the relationships in (5.4) for $\sigma_\theta$ to one quadrant only. This diagram gives, for the angular variation of $\sigma_\theta$, the solid line in fig. 5.7(*b*) – the original circle and the other three circles required by symmetry. This type of polar representation of $\sigma$ as a function of the angular variation of the surface from an initial plane is usually called the '$\sigma$-plot'.

For a two-dimensional crystal the two-dimensional $\sigma$-plot is a complete description, but for a three-dimensional crystal a three-dimensional plot is needed. The same calculation for a three-dimensional simple cubic crystal, again with only bonds to the nearest neighbour being considered, gives a $\sigma$-plot consisting of eight *spheres* passing through the origin with their diameters lying along the eight ⟨111⟩ directions and producing cusps along the six ⟨100⟩ directions.

Similar bonding calculations for real fcc and bcc crystal structures have been made, for example, by Mackenzie, Moore and Nicholas (1962) and Sundquist (1964). Sundquist considered different values of the bond energy to the first and second nearest neighbours (with $\rho$ the ratio of the energy of bond to the second nearest neighbour to that to the first neighbour), and calculated the surface free energy as a function of orientation for different values of $\rho$. The results of these calculations are shown in fig. 5.8, where contours of equal surface free energy $\sigma$ are plotted in a standard triangle of the stereographic projection. Cusps are produced at low index crystallographic planes. For the bcc structure, for either $\rho=0$ or $\rho=0.5$, the {110} planes had the lowest energy, and for the fcc structure, with either $\rho=0$ or $\rho=0.1$, the close-packed {111} planes had the lowest energy.

The conventional representation of the variation of $\sigma$ with the orientation of the surface plane as a section through the polar $\sigma$-plot allows a clear demonstration of the Gibbs–Wulff theorem. This theorem, due to Wulff (1901), is the solution to the problem raised by Gibbs in 1878: For a fixed amount of matter in a single particle, what shape gives the lowest value of the total surface free energy $\int\sigma' \mathrm{d}A'$? Here $\mathrm{d}A'$ is the area of an element of the interface whose interfacial energy is $\sigma'$.

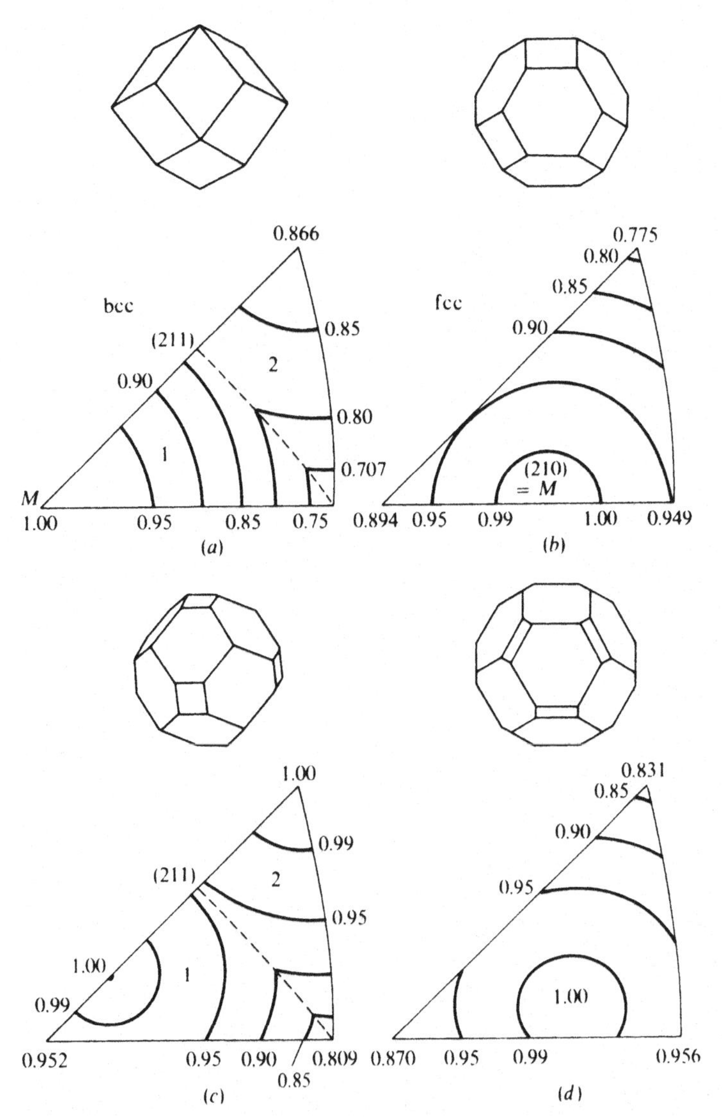


**Figure 5.8** $\sigma$-plots in a stereographic triangle ((100), (110) and (111)) and the corresponding equilibrium shapes for (*a*) bcc with $\rho=0$, (*b*) fcc with $\rho=0$, (*c*) bcc $\rho=0.5$, and (*d*) fcc with $\rho=0.1$; $\rho$ is the relative energy of the bond to the second nearest neighbour to that to the nearest neighbour. (After Sundquist 1964, courtesy of *Acta Met.*)

The solution to this problem gives the *equilibrium shape* of the particle. For a liquid droplet where $\sigma$ is constant all round the surface, the minimum value of $\int\sigma' \mathrm{d}A'$ is given by $\sigma\int \mathrm{d}A$. Since the minimum surface area of a given volume is found for a sphere, the equilibrium shape for a liquid drop is correctly predicted to be spherical. However, when $\sigma$ is not constant, the value of $\int\sigma' \mathrm{d}A'$ can be reduced by allowing planes with low values of $\sigma$ to occupy

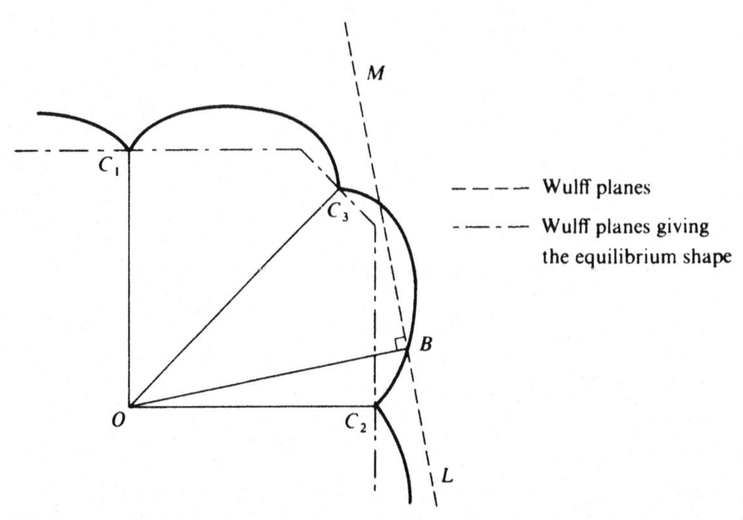


**Figure 5.9** Wulff construction to give the equilibrium shape of a crystal. The Wulff planes are those such as *MBL*, which lies normal to the vector *OB* where *B* lies on the σ-plot. The Wulff planes at the cusps $C_1$, $C_2$, $C_3$, give the inner envelope of Wulff planes, and thus the equilibrium shape.

more of the surface. Wulff proposed the construction shown in fig. 5.9. From the origin, *O*, of the σ-plot, a line is drawn to cut the σ-plot at a position *B*, the plane *LBM* is then drawn normal to *OB* to cut the σ-plot, at *B*. Such a plane is now called a *Wulff plane*. The Wulff plane is the plane of the crystal whose interfacial energy is the value of σ at the point where the plane cuts the σ-plot. The Wulff *theorem* is that the equilibrium shape is bounded by those parts of Wulff planes that can be reached from the origin without crossing any other Wulff planes. In other words the equilibrium shape is made up by the 'inner envelope' of Wulff planes. For example, in fig. 5.7(*b*) the σ-plot shown would give rise to an equilibrium shape that is the square formed by the four Wulff planes at the four cusps. In three dimensions our simple cubic crystal discussed previously will be bounded by the six Wulff planes corresponding to the six cusps in ⟨100⟩ directions. In this case they will be {100} planes and the equilibrium shape will be that of a cube. Sodium chloride crystals formed from aqueous solutions do indeed have this shape. More complete discussions of the Wulff theorem and the equilibrium shape are available, see, for example, Christian (1975).

The Wulff construction clearly allows the low σ planes to be extensively represented in the equilibrium shape. It seems, according to Mullins (1963), that the theorem has only been proved if it is first assumed that the equilibrium shape is convex. However, since no observations have ever been reported of non-convex crystal shapes, it appears that the Gibbs–Wulff theorem has been proved. We then have the important result that for any observed equilibrium shape, relative

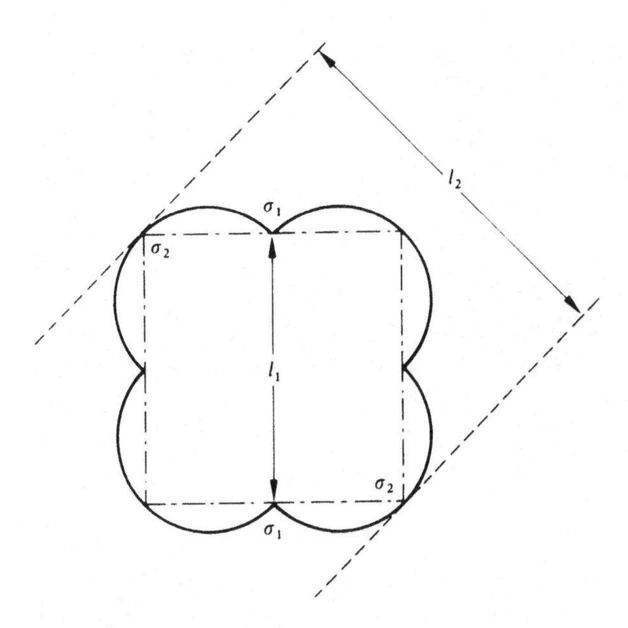


**Figure 5.10** Showing a possible section through the $\sigma$-plot for a cubic crystal and the expected equilibrium shape. Measurements on the equilibrium shape can give the relative surface energies of $\sigma_1$ and $\sigma_2$.

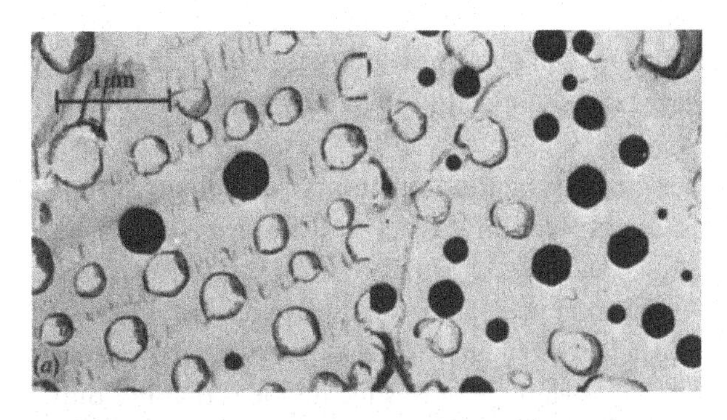


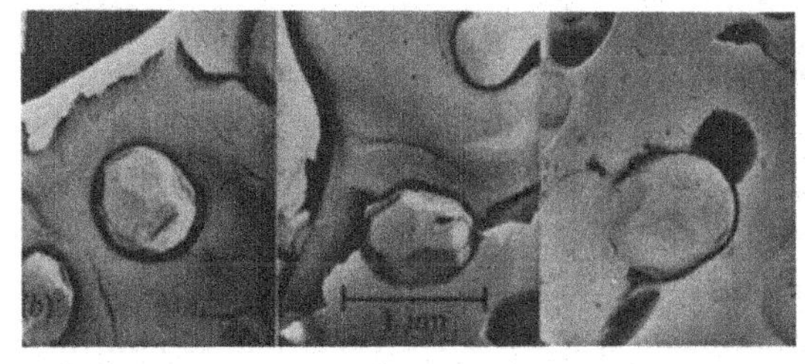


**Figure 5.11** Examples of (*a*) profiles and (*b*) typical replicas of small fcc nickel particles annealed in dry hydrogen at 1273 K for 100 h. The equilibrium shape is faceted, but with rounded corners and edges. (After Sundquist 1964, courtesy of *Acta Met.*)

values of $\sigma$ can be determined from measurement of the perpendicular distances from the centre of the particle, see fig. 5.10, where $\sigma_1/l_1 = \sigma_2/l_2$.

Sundquist observed the equilibrium shapes of small (1 $\mu$m) crystals of various fcc and bcc metals, and as expected, facets of {100} and {111} planes are seen on fcc metal crystals and {110} and {100} 'facets' on bcc metals. Some of Sundquist's equilibrium shapes are shown in figs. 5.11 and 5.12. A comparison between fig. 5.12 and the predicted $\sigma$-plot for fcc metals with $\rho$=0.1 (fig. 5.8)

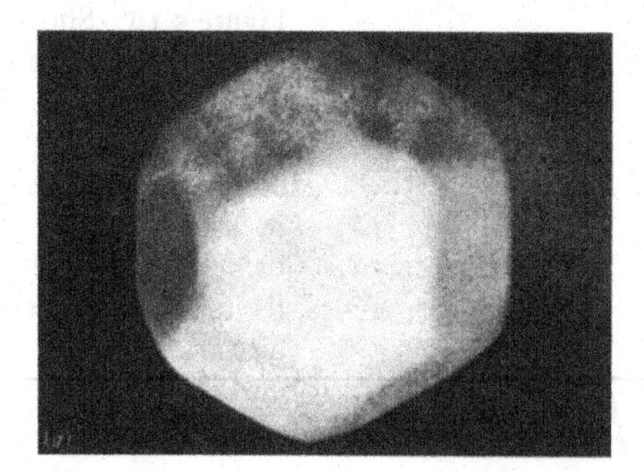

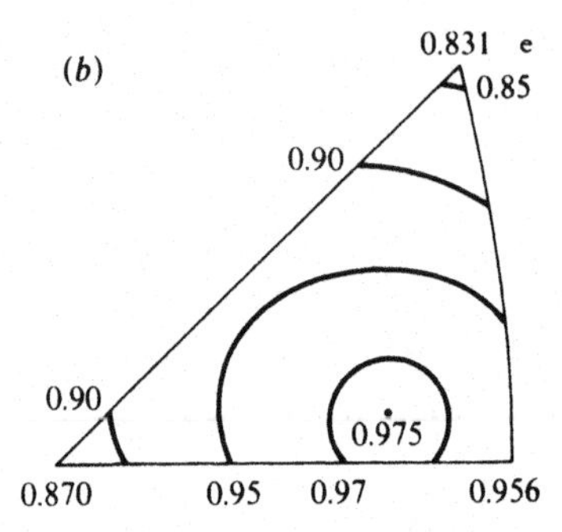


**Figure 5.12** (*a*) Model of the experimentally determined typical equilibrium shape of fcc crystals. (*b*) The $\sigma$-plot that would give rise to this experimentally determined shape, scaled for comparison with Sundquist's calculated model illustrated in fig. 5.8(*d*). The hexagonal faces in (*a*) are the $\{111\}$ planes and square faces $\{100\}$. (After Sundquist 1964, courtesy of *Acta Met.*)

shows excellent agreement (the anisotropy ratio $\sigma_{max}/\sigma_{min}$ from fig. 5.12(*b*) is 0.975/0.831, that is, 1.17). The agreement between theory and experiment is much less good for bcc iron, but it is not clear whether this disagreement comes from use of an oversimplified theory. The other possibility is that the measured shapes of iron crystals might have been significantly affected by impurity adsorption.

There is another, and rather elegant, method of observing the equilibrium shapes of clean metal crystals, due to Nelson, Mazey and Barnes (1965). This consists of making holes in crystals, 'negative crystals', by condensation of excess vacancies or by precipitation of injected inert gases, helium, neon, etc., in metal crystals. The holes produced by such experiments are best observed by transmission electron microscopy. The great advantage of this technique is that not only are very small voids produced, so needing very short equilibrium times, but more importantly, there is much less possibility of any impurity adsorption, at least from the gas phase. The results from the study by Nelson *et al.* support the general ideas so far discussed in that polyhedral shapes are found with rounded corners, but the detailed shapes were different from Sundquist's results, especially for the fcc metal, copper. For copper the (110) planes had, surprisingly, the lowest energy; for aluminium, however, the expected (111) planes had the lowest energy. For bcc metals, the results given by Nelson *et al.* (1965) were in agreement with Sundquist's *calculation* in which the $\{110\}$ planes had the lowest energy, but disagreed with Sundquist's results where the $\{100\}$ plane had the lowest energy.

An alternative technique for measuring the anisotropy of surface energy has been introduced by McLean and Gale (1969). This consists of measuring the variation of grain boundary groove angle in wire bicrystals with the boundary transverse to the wire axis. They showed that for copper in a pure atmosphere of

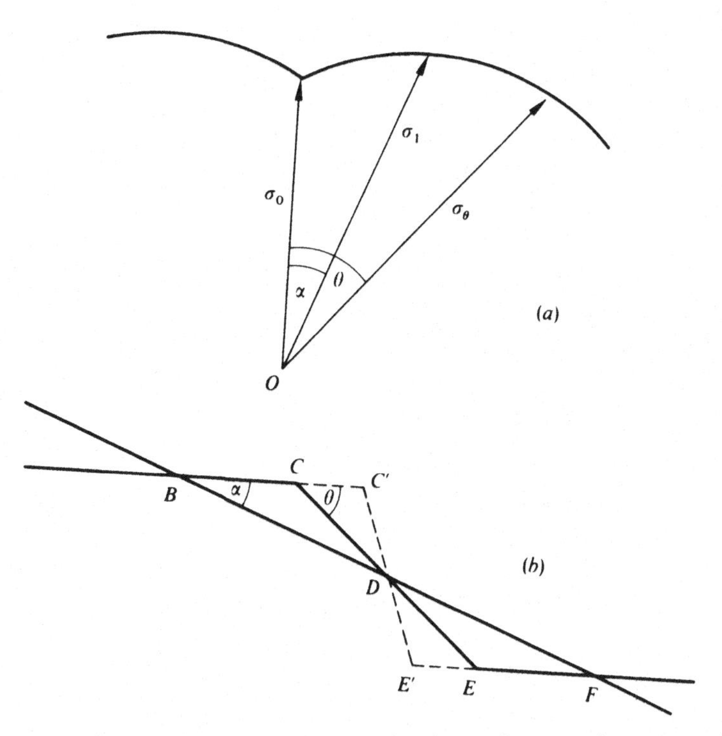


**Figure 5.13** (*a*) Portion of a $\sigma$-plot showing surface energies $\sigma_0$, $\sigma_1$ and $\sigma_a$. (*b*) The original section in *BDF* ($\sigma_1$) which has faceted to give low energy 'facet' planes *BC* and *EF* ($\sigma_0$) and 'continuation' planes such as *CDE* ($\sigma_\theta$). *CDE* can rotate to *CDE'* to achieve lowest possible energy. (After Mykura 1966, courtesy of Routledge & Kegan Paul.)

dry hydrogen having an oxygen partial pressure of only $10^{-25}$ atm, the order of the surface energies was the expected: $\sigma_{111}<\sigma_{100}<\sigma_{110}$. This was changed by the presence of a trace of oxygen, at $10^{-15}$ atm, to $\sigma_{100}<\sigma_{110}<\sigma_{111}$. Other results using this technique and illustrating the important role of surface adsorption include those of McLean (1971) and Mills, McLeall and Hondros (1973).

Although there may be some doubts as to whether either the theoretical models or the current experiments in metals are as yet completely reliable, the general result of an *anisotropic* surface energy of crystals and the resulting polyhedral equilibrium shape is firmly established as is the validity of the Gibbs–Wulff theorem for the equilibrium shape of crystals.

### 5.4.1 Faceting

There is another problem equivalent to that of equilibrium shape; this concerns the possible lowering of the energy of an initially flat crystal surface by virtue of the creation of *facets* on the surface. Fig. 5.13(*a*) shows a $\sigma$-plot with a cusp at a low index plane. A flat surface at an angle $\alpha$ to the low index plane is exposed by sectioning a crystal; fig. 5.13(*b*) shows this exposed plane *BDF*. The surface free energy of this plane is $\sigma_1$ and facets *BC* and *EF* of the low energy plane may develop if the total energy of the faceted structure per unit area of the original plane, $\sigma_f$, is less than $\sigma_1$. The energy of the low index plane is $\sigma_0$ and the 'continuation plane' *CDE* is $\sigma_\theta$. By consideration of the triangle *BCD* it is readily shown that

$$\sigma_f = \sigma_0 \sin(\theta - \alpha)/\sin\theta + \sigma_\theta \sin\alpha/\sin\theta \tag{5.6}$$

So facets develop if

$$\sigma_1 > \sigma_0 \sin(\theta - \alpha)/\sin\theta + \sigma_\theta \sin\alpha/\sin\theta \tag{5.6a}$$

Further reduction in energy may be possible by rotation of the continuation plane *CDE* to a new position *C′DE′* by increasing the angle $\theta$ to $\theta'$. The equilibrium angle, $\theta'$, of lowest energy is found by differentiation of $\sigma_f$ with respect to $\theta$ and equating $d\sigma_f/d\theta$ to zero. After manipulation this then yields

$$\sigma_0 = \sigma_\theta \cos\theta' - (d\sigma/d\theta)\sin\theta' \tag{5.7}$$

The equilibrium value of $\theta'$ is that which satisfies this equation.

These equations are most clearly illustrated by use of the *reciprocal* $\sigma$-plot. This plot is one in which the reciprocal of the surface free energy ($1/\sigma$) is plotted as the polar coordinate (Frank 1963). This plot has vertices where the $\sigma$-plot has cusps and straight lines where the $\sigma$-plot has circles through the origin. Mullins (1963) noted that with this reciprocal construction the $\sigma$-plot of fig. 5.7(*a*) becomes a square with its corners in the $\langle 10 \rangle$ directions.

Use of the reciprocal $\sigma$-plot for the faceting problems above is illustrated in fig. 5.14. The low index *BC* plane is at *B*, the original surface *BDF* is at *S*, and the continuation plane *CDE* ($\sigma_\theta$) at *D*. With the origin at *O*

$$OB = 1/\sigma_0,\ OS = 1/\sigma_1 \text{ and } OD = 1/\sigma_\theta$$

By constructing the tangent from *B* to touch the reciprocal plot at *D*, it can be shown that $\theta$ has the equilibrium value, $\theta'$, of eq. (5.7). It can also be shown that, by extrapolation of *OS* to *S′* on this tangent, the value of $1/OS'$ is that of the minimal value of $\sigma_f$, given by eq. (5.6), with $\theta = \theta'$ from eq. (5.7). This means that faceting can occur to reduce the surface energy if *OS′* is greater than *OS*, so that construction of the reciprocal plot with tangents from all the vertices such as *B* to the curve will indicate the orientations expected to facet as those which lie between the maximums in the reciprocal $\sigma$-plot, and their tangent points, for example, *D* to *D′* in fig. 5.14. The $\sigma$-plot of fig. 5.7(*b*) which gives a square reci-

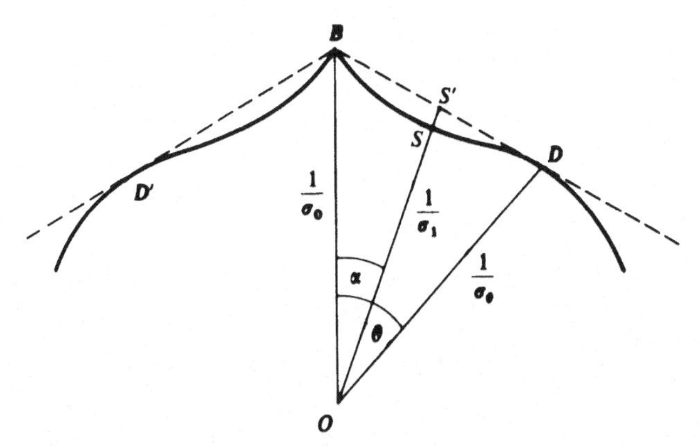


**Figure 5.14** Reciprocal $\sigma$-plot for the situation of fig. 5.13; $BD$ is tangent to the $1/\sigma$-plot at $B$. Only regions that lie underneath such tangents, for example, $D'$ to $B$ and $B$ to $D$, are expected to facet.

procal $\sigma$-plot is just stable against faceting – any slight increase in anisotropy would allow facets to develop.

Although there appears to be little doubt that the Wulff–Gibbs theory of the equilibrium shape is confirmed by experiment and by analysis, there is some doubt whether the faceting process is due *solely* to these surface energy considerations (Moore 1963) since surface evaporation may play a role in faceting in some occasions. Nevertheless, faceting occurs on planes close in orientations to known low energy planes such as the (111) planes in silver, and many other predictions of the surface energy theory of faceting are supported by experiments from the early work of Chalmers, King and Shuttleworth (1948) up to the review by Moore in 1963. Since this is not a review of faceting, all that concerns us is that the expected results of an anisotropic surface free energy are found for the simplest crystal surface, the crystal–vapour interface. It will be assumed for the purposes of the remainder of this chapter that all the ideas and relationships developed here apply equally to all other crystal interfaces.

## 5.5 Precipitate coarsening: 'Ostwald ripening'

One of the most important metallic microstructures is that of a fine dispersion of hard precipitate particles in a metallic matrix. As discussed in many articles on strengthening processes (for example, Kelly and Nicholson (1971) and Martin (1980)), at small interparticle spacings the precipitates are usually cut by dislocations as they glide on the slip plane. Spacings at which the particles are cut are typically of the order of 25 Burgers vectors ($b$), and this situation occurs during the early or underaged stage of precipitation-hardening where the alloy yield stress is increasing. At longer ageing times the precipitates become larger

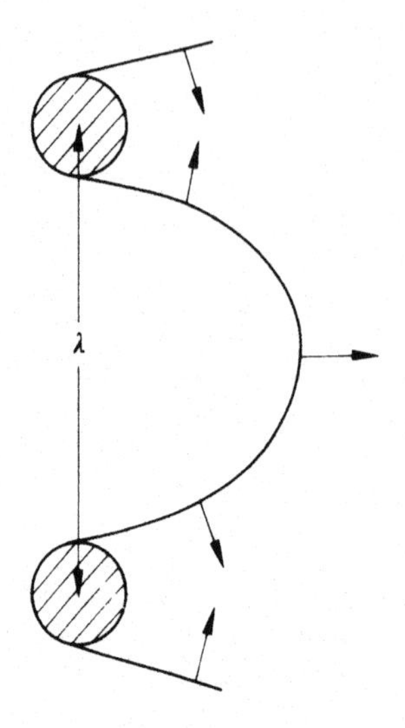


**Figure 5.15** Orowan looping of a dislocation past an array of precipitates of spacing λ.

and thus stronger obstacles. At these later stages of the precipitation process, when the interparticle spacing may be of the order of $100b$, the dislocations are able to bypass the precipitates by looping between them (fig. 5.15). The increase in shear stress ($\Delta\tau$) over the value for the metal without precipitates, due to bowing of dislocations of line tension $T_l$ between particles of mean planar separation, $\lambda$, is given, approximately, by the well-known Orowan equation:

$$\Delta\tau = 2T_l/\lambda b \tag{5.8}$$

The significance of this relationship is that any increase in the interparticle spacing, $\lambda$, will *reduce* the strength of the heat-treated alloy. Such a fall in strength with longer times of ageing, known as 'overageing' is characteristic of precipitation hardening at least for artificial ageing.[†] The most important mechanisms which cause overageing are: (i) the formation of more stable precipitates as discussed in chapter 2, since the more stable precipitates are usually more widely spaced; and (ii) the coarsening of the precipitate due to the interfacial free energy between the precipitate and the matrix. This second process, first identified by Ostwald in the physical chemistry of salt precipitates in aqueous solutions, is often called 'Ostwald ripening'. The earliest application of quantitative ideas to this process for metallic precipitates was by Greenwood (1956) whose initial treatment forms the basis of the description given here. Greenwood (1969) also provided an early but still relevant review of the topic.

[†] Ageing at room temperature (natural ageing) in aluminium alloys usually ends when the artificially high atomic mobility due to quenched-in vacancies disappears with the loss of the excess vacancies. This causes the ageing process to continue at such a slow rate that the microstructure appears unchanging over the lifetime of the manufactured part.

## 5.5.1 Interfacial free energy

The specific interfacial free energy of metallic precipitates in a metallic matrix is found to vary from about 0.02 J m$^{-2}$ to 0.6 J m$^{-2}$ (20–600 mJ m$^{-2}$) – see table 5.2. The smallest values are for fully coherent precipitates with a small lattice misfit and with a similar chemistry to the matrix, for example, for ordered $Ni_3Al$ ($\gamma'$) precipitates in a nickel matrix or the similar $Al_3Li$ ($\delta'$) precipitates in an aluminium matrix. The energies rise to intermediate values for coherent precipitates with large differences in chemistry, for example, cobalt rich zones in a copper rich matrix. Even larger values of interfacial energy are found for fully incoherent precipitates where the energy is comparable to, but usually somewhat less than, high angle matrix grain boundary (Smith 1952). The largest values have been reported for metal–oxide interfaces, such as that between nickel and thoria where not only is the chemistry and the atomic structure different across the interface, but the *type* of bonding (metallic to ionic) also changes.

An alloy containing a volume fraction, $f$, of 0.05 of precipitates with an interparticle spacing of only 30 nm (100$b$) in a simple cubic array, will have approximately $10^{22}$ precipitates per cubic metre with a total interfacial area of $10^7$ m$^2$ m$^{-3}$. If the interfacial free energy $\sigma$ is 0.2 J m$^{-2}$, as is found for the coherent cobalt precipitates in copper (Servi and Turnbull 1966), the total interfacial free energy is 2 MJ m$^{-3}$ of alloy. This can be compared to the free energy change for the initial precipitation of such an alloy, given by eq. (5.9) (a relationship derived, for example, by Doherty (1984)) see §1.2:

$$\Delta G_V = -(RT/V_m)\ln N_0/N_\alpha \text{ (for dilute solutions)} \tag{5.9}$$

where $\Delta G_V$ is the free energy change per unit volume *of the $\beta$ precipitate*, $R$ is the gas constant ($R$=8.31 J g-mole$^{-1}$), $T$ is the absolute temperature, $V_m$ the volume of 1 g-mole of the precipitate and $N_0/N_\alpha$ is the ratio of the alloy solute content before precipitation to the equilibrium solubility in the $\alpha$ matrix at the precipitation temperature. If precipitation occurs at a temperature of 600 K with a solubility ratio of $10^3$ and the gram-molar volume has the usual value of about $10^{-5}$ m$^3$ g-mole$^{-1}$, then the energy released on complete precipitation will be 125 MJ m$^{-3}$ of alloy. (The volume fraction is still assumed to be 5%.) It can therefore be seen that this is 60 times the interfacial free energy that remains if the precipitation is on the fine scale of an interparticle spacing of 100$b$. In other words, at the end of a successful precipitation-hardening heat-treatment, about 2% of the original driving force remains in the form of the interfacial free energy. If the interparticle spacing is 10 times larger (1000$b$) then this energy ratio falls to only 0.2% and if the spacing is 100 times larger (3 $\mu$m) with a particle radius of 1 $\mu$m, then the interfacial energy is only $2\times10^4$ J m$^{-3}$ – 100 times less than for the fine dispersion. This simple calculation shows that for the very fine precipitates seen by electron

**Table 5.2** *Values of the specific interfacial free energy for a number of metallic systems.*

| System | Interface type | Misfit % | Energy J $m^{-2}$ | Method of measurement |
|---|---|---|---|---|
| Ni–Si | Coherent | 0.3 | 0.011 | Precipitate ripening (Rastogi and Ardell 1971) |
| Ni–Al | Coherent | 0.5 | 0.014 | Ardell (1968) |
| Ni–Ti | Coherent | 0.9 | 0.021 | Ardell (1970) |
| Cu–Co | Coherent | 1.8 | 0.20 | Nucleation (Servi and Turnbull 1966) |
| Cu–Co | Coherent | 1.8 | 0.23 | Calculation (Servi and Turnbull 1966) |
| Cu–Co | Coherent | 1.8 | 0.18 | Precipitate ripening (Ardell 1972*a*) |
| Cu–Zn–Sn ($\beta/\gamma$) | Coherent | 0.4 | 0.06–0.12 | Precipitate morphology development (Malcolm and Purdy 1967) |
| Cu–Zn ($\beta/\gamma$) | | 0.1 | 0.13 | Precipitate growth kinetics (Bainbridge 1972) |
| Cu–Zn ($\alpha/\beta$) | Incoherent | | 0.5 | Precipitate growth kinetics (Purdy 1971) |
| Cu–Zn ($\alpha/\beta$) | Incoherent | | 0.42 | Equilibrium with grain boundaries (Smith 1952) |
| $\alpha$ Fe–$Fe_3C$ | Incoherent | | 0.74 | |
| $\alpha$ Fe–$\gamma$ Fe | Incoherent | | 0.56 | |
| $\alpha$ Fe–Cu(fcc) | Incoherent | | 0.50 | Equilibrium with grain boundaries |
| $\alpha$ Fe–Cu(fcc) $\langle 111\rangle_{\alpha} \| \langle 110\rangle_{Cu}$ | Coherent | 3.0 | 0.125 | Equilibrium shape (Speich and Oriani 1965) |
| Ni–$ThO_2$ | Incoherent | | 1.5 | Precipitate coarsening (Footner and Alcock 1972) |

**Table 5.2** (*cont.*)

| System | Interface type | Misfit % | Energy $J\ m^{-2}$ | Method of measurement |
|---|---|---|---|---|
| Al–Zn $\langle 111\rangle_{Al} \parallel \langle 0001\rangle_{Zn}$ | Coherent | | 0.07 | Eutectic growth (Chadwick 1963) |
| Cd–Zn (both cph) | Coherent | | 0.11 | Eutectic growth (Moore and Elliott 1968) |
| $Al–Al_2Cu$ | Coherent | | 0.34 | Chadwick (1963) |
| | | | 0.09–0.11 | Whelan and Howarth (1964) |
| Ag–Pb (both fcc) | Coherent | | 0.09–0.11[a] | Moore and Elliott (1968) |
| | | | 0.11–0.19 | Moore and Elliott (1968) |

[a]Eutectic growth.

microscopy and which give precipitation-hardening there exists a considerable residual driving force for precipitate-coarsening. However, if the dispersion is coarse enough to be resolved by optical microscopy then there will be a very much smaller amount of interfacial energy driving the coarsening process.

## 5.5.2 The mechanism and kinetics of Ostwald ripening

The physical mechanism by which the microstructure coarsens and releases its excess surface energy arises from the increased solubility of small particles. These small particles have a larger ratio of surface area to volume. The increase of solubility can be readily demonstrated for the typical phase diagram and free energy–composition relationships shown in fig. 5.16. The equilibrium solubilities, $N_\alpha$ and $N_\beta$, are changed to $N_\alpha(r)$ and $N_\beta(r)$ for a fine dispersion of $\beta$ precipitates of radius $r$, in an $\alpha$ matrix. (The solute contents are here considered in atomic or molar fractions.) This 'Gibbs–Thomson' change of solubility arises because of the increase in chemical potential $\Delta\mu$ due to the energy $\sigma$ of the interface between the matrix and the precipitate. The intercepts of the common tangents with the pure B axis are the partial molar free energies, $\bar{G}_B(r)$ and $\bar{G}_B(\infty)$ (for radii of $r$ and $r=\infty$) as discussed in any textbook on solution thermodynamics (for example, Swalin (1972)).

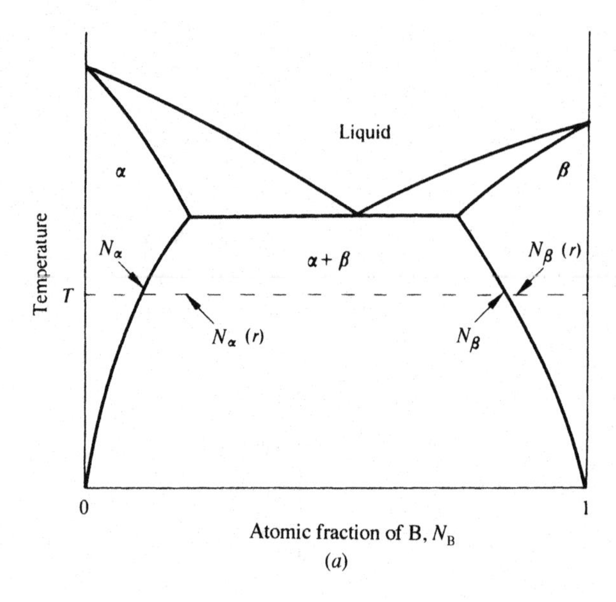


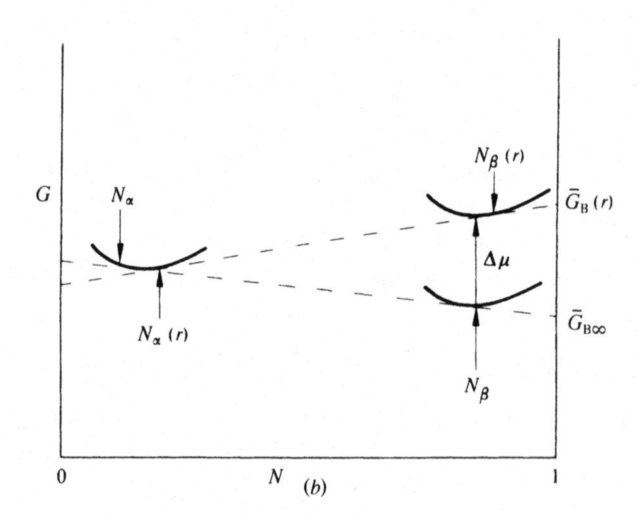


**Figure 5.16** (*a*) Typical phase diagram showing limited solid solubility $N_\alpha$ and $N_\beta$ at a temperature $T$. (*b*) Schematic representation of the free energy per mole ($G$) as a function of the atomic fraction of B, $N$, for both the case when the $\beta$ precipitates have a flat interface ($r=\infty$) and when they have a finite radius $r$. The curvature of the interface raises the free energy of the $\beta$ precipitates by $\Delta\mu$, so increasing both the partial molar free energy of B from $\overline{G}_B(r)$ to $\overline{G}_B(\infty)$ and the solubility of B in $\alpha$ from $N_\alpha$ to $N_\alpha(r)$.

The increase in the free energy per mole of precipitate, the chemical potential $\Delta\mu$, under a curved interface, is obtained by consideration of the transfer of d$n$ moles of atomic fraction $N_\beta$ from a precipitate with a plane interface ($r=\infty$) to a precipitate of radius $r$. This transfer produces an increase in free energy d$G$ given by:

$$\mathrm{d}G=\Delta\mu\mathrm{d}n=\sigma\mathrm{d}A$$

where $A$ is the area of the precipitate and the interfacial energy $\sigma$ is assumed to be isotropic. Therefore:

$$\Delta\mu = \sigma \mathrm{d}A/\mathrm{d}n$$

But

$$A = 4\pi r^2 \text{ so } \mathrm{d}A/\mathrm{d}r = 8\pi r$$

The number, $n$, of moles of precipitate in a sphere of radius $r$ is

$$n = 4\pi r^3/3V_\mathrm{m} \text{ so } \mathrm{d}n/\mathrm{d}r = 4\pi r^2/V_\mathrm{m}$$

where $V_\mathrm{m}$ is the molar volume of the precipitate. So

$$\mathrm{d}A/\mathrm{d}n = (\mathrm{d}A/\mathrm{d}r)(\mathrm{d}n/\mathrm{d}r) = 2V_\mathrm{m}/r$$

and

$$\Delta\mu = \sigma V_\mathrm{m} 2/r \tag{5.10}$$

The factor $2/r$ is the *curvature* of a spherical surface. For a surface of more general shape, this becomes $(1/r_1 + 1/r_2)$ where $r_1$ and $r_2$ are any two *perpendicular* radii of curvature. For anisotropic $\sigma$ Herring (1949) has shown that

$$\Delta\mu = V_\mathrm{m}\{(1/r_1)[\sigma + (\partial^2\sigma/\partial\mathbf{n}^2)_{r_1}] + (1/r_2)[\sigma + (\partial^2\sigma/\partial\mathbf{n}^2)_{r_2}]\} \tag{5.10a}$$

where $\sigma$ is a function of the unit normal vector $\mathbf{n}$. This has been applied to solid state precipitates by, for example, Speich and Oriani (1965).

From eq. (5.10) and the thermodynamic relationships:

$$\bar{G}_\mathrm{B}(r) - G^\circ_\mathrm{B} = RT \ln a_\mathrm{B}(r) \text{ and } \bar{G}_\mathrm{B}(\infty) - G^\circ_\mathrm{B} = RT \ln a_\mathrm{B}(\infty)$$

eq. (5.11) can be derived using the information of fig. 5.16; $a_\mathrm{B}(r)$ and $a_\mathrm{B}(\infty)$ are the activities of B in the $\alpha$ phase for precipitate radii of $r$ and infinity respectively, and $G^\circ_\mathrm{B}$ is the molar free energy of pure B in its standard state. For the limiting case when

$$[N_\alpha(r) - N_\alpha]/(1 - N_\alpha) \ll 1$$

we have from fig. 5.16(*b*),

$$[G_\mathrm{B}(r) - G_\mathrm{B}(\infty)]/(1 - N_\alpha) = \Delta\mu/(N_\beta - N_\alpha)$$

So

$$G_B(r)-G_B(\infty)=(2\sigma V_m/r)[(1-N_\alpha)/(N_\beta-N_\alpha)]$$

But

$$G_B(r)-G_B(\infty)=RT\ln[a_B(r)/a_B(\infty)]$$

$$=RT\ln[N_\alpha(r)/N_\alpha(\infty)]-RT\ln\{\gamma_\alpha[N_\alpha(r)]/\gamma_\alpha[N_\alpha(\infty)]\}$$

where $\gamma_\alpha[N_\alpha(r)]$ and $\gamma_\alpha[N_\alpha(\infty)]$ are the activity coefficients of B in the $\alpha$ phase at concentrations $N_\alpha(r)$ and $N_\alpha(\infty)$ respectively. ($a_B=\gamma_\alpha N_\alpha$.) Therefore

$$\ln[N_\alpha(r)/N_\alpha(\infty)]=(2\sigma V_m/RTr)[(1-N_\alpha)/N_\beta-N_\alpha)]-\ln\{\gamma_\alpha[N_\alpha(r)]/\gamma_\alpha[N_\alpha]\} \quad (5.11)$$

In the limiting case of the two phases being almost pure A, $N_\alpha\rightarrow0$, and pure B, $N_\beta\rightarrow1$, and with $N_\alpha\rightarrow0$, the activity coefficient, $\gamma_\alpha$, is constant in the dilute solution of the $\alpha$ matrix. So eq. (5.11) reduces to the more commonly quoted form of the Gibbs–Thomson equation:

$$\ln[N_\alpha(r)/N_\alpha]=2\sigma V_m/RTr \quad (5.12)$$

If typical values such as those used in §5.5.1 are substituted into the right hand side of the equation, we obtain:

$$\ln[N_\alpha(r)/N_\alpha]=10^{-9}/r$$

So, for any precipitate whose radius is larger than a few nanometres, we can further simplify eq. (5.12) to:

$$N_\alpha(r)=N_\alpha[1+(2\sigma V_m/RTr)] \quad (5.12a)$$

If the radius is larger than a few nanometres but the two phases are not limiting solid solutions then, as Purdy (1971) has pointed out, the equation becomes

$$N_\alpha(r)=N_\alpha\{1+[(1-N_\alpha)/(N_\beta-N_\alpha)](2\sigma V_m/RT\epsilon_\alpha r)\} \quad (5.12b)$$

where $\epsilon_\alpha$ is the Darken factor (§2.1.1).

$$\epsilon_\alpha=1+(\partial\ln\gamma_\alpha/\partial\ln N_\alpha)$$

Use of this non-ideality correction, $\epsilon_\alpha$, may be very significant since values of $\epsilon_\alpha$ as large as 10 have been reported by Purdy (1971) for precipitates forming from a matrix which is an intermediate phase. An important example is the precipitation of the fcc $\alpha$ phase from the $\beta$ phase in Cu–Zn brass.

Of much more frequent occurrence is the strong enhancement of the Gibbs–Thomson effect for intermediate phase *precipitates* by virtue of the composition term, $(N_\beta - N_\alpha)/(1-N_\alpha)$. This composition term increases from being close to 1 for precipitates that are terminal solid solutions such as Al–Si, where $N_\beta = 1$, to a value of 3 for precipitates of $Al_2Cu$ in Al and to a value of 7 for $Al_6Mn$ in Al. This is, when precipitates are close in composition to the matrix the effect of this term is to increase, significantly, the rise in solubility of the phase under the curved interface.† One way that this effect can be introduced into the Gibbs–Thomson equation while retaining the form given in eq. (5.12a) is to define the molar volume, $V_m$, as the volume occupied by an Avogadro's number ($6\times10^{23}$) of *solute* atoms. Then, in an precipitate of atomic composition, $A_nB$, $V_m$ will be increased from the typical metallic value of $10^{-5}$ m$^3$ mole$^{-1}$ to about $(n+1)\times10^{-5}$ m$^3$ mole$^{-1}$. Ardell (1988, 1990) used this form of the Gibbs–Thomson equation for his discussions of the Al–$Al_3Li$ and Ni–$Ni_3Al$ systems.

For the purpose of the rest of this discussion we shall use the simplified expression of eq. (5.12a), and maintain the definition of $V_m$ used throughout this volume that $V_m$ is the volume of an Avogadro's number of all atoms, ($V_m \approx 10^{-5}$ m$^3$ mole$^{-1}$). The correct expression for the Gibbs–Thomson relationship can be readily obtained for any equation by multiplying the interface energy $\sigma$ by $f(N)$ where

$$f(N) = (1-N_\alpha)/[(N_\beta - N_\alpha)\epsilon_\alpha]$$

A final point should be made about the Gibbs–Thomson equation. The solubility expression contains the term $\sigma/r$, where $\sigma$ is the isotropic interfacial energy and $r$ the radius of the spherical precipitate, that is, the distance of the curved interface from the centre of the precipitate. Correction for *anisotropic* interfacial energy is easily made on the basis of the Wulff theorem. The theorem requires, for precipitates with the equilibrium shape, that the ratio $\sigma_i/x_i$ is a constant. Here $\sigma_i$ is the energy of the $i$th portion of the interface and $x_i$ is the distance of that portion of the interface from the centre of the precipitate (fig. 5.10). The solubility of a precipitate with an equilibrium shape must be constant all round that precipitate. Examples of non-spherical equilibrium shapes are extremely

† The opposite effect, but coming from the same physical origin, is seen in precipitate nucleation. This is the reduced driving force per unit volume of precipitate at the same supersaturation – ln $(N_0/N_\alpha)$, when precipitate and matrix approach each other in composition, $N_\beta \rightarrow N_\alpha$ – eq. (1.11), §1.2.

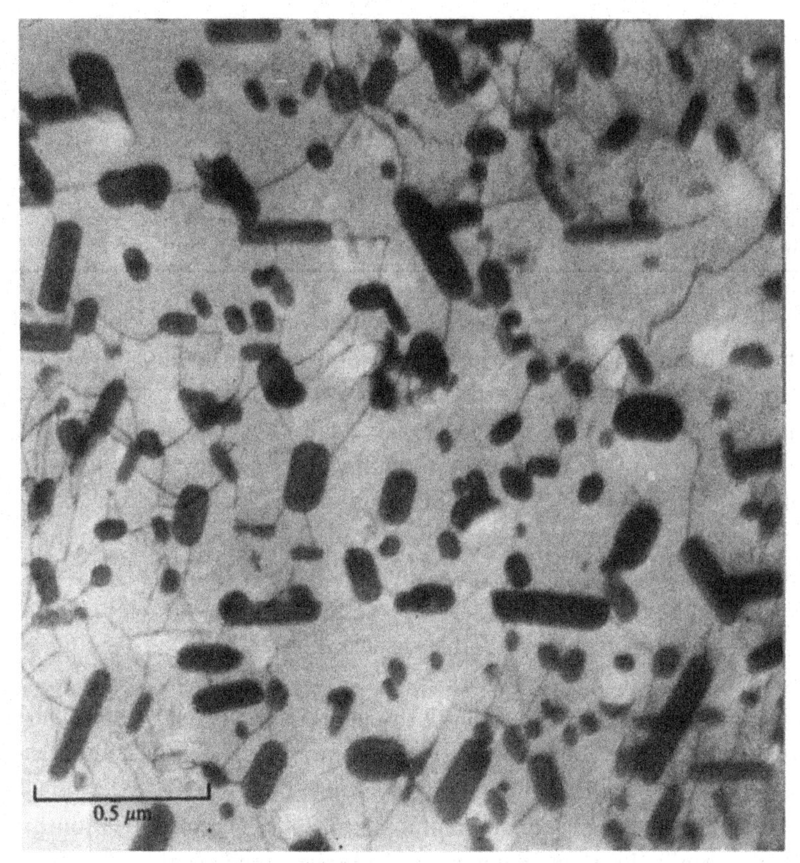


**Figure 5.17** Rod-shaped precipitates of copper in an iron matrix obtained by ageing an Fe – 4 wt% Cu alloy for 300 h at 953 K. Transmission electron micrograph showing precipitates of different orientations. Rods that are nearly normal to the foil appear as circles, and only those parallel to the foil are seen as rods with spherical caps. (After Speich and Oriani 1965, courtesy of AIME.)

common in precipitation reactions, §§1.3, 2.2.4, particularly precipitation strengthening reactions since particles with plate-like or rod-like morphologies provide many more barriers for dislocation glide than precipitates of equivalent volume fraction but with a spherical shape. Due to their importance for precipitation hardening, precipitate coarsening studies have been frequently reported for non-spherical precipitates. For fcc copper precipitates in a bcc iron matrix, Speich and Oriani (1965) found that the equilibrium shape was that of a rod with curved ends (fig. 5.17). The axis of the rod was parallel to the close-packed direction, $\langle 110 \rangle$ in the fcc precipitates and $\langle 111 \rangle$ in the bcc matrix, which allows a good fit for planes containing these directions while the ends of the rod did not allow any crystallographic fit. The diameter/length 'aspect' ratio was 4, indicating the low value of $\sigma$ for the surfaces containing the close-packed direction. It should be recognised that this type of anisotropy can only apply for a

**Figure 5.18** Thin disc-like precipitates of $\theta'$ in Al – 4% Cu aged for 24 h at 513 K seen edge on with dark field illumination. The high aspect ratio of diameter to thickness ($\approx$40) stayed constant during anneals of up to 40 h. (After Boyd and Nicholson 1971, courtesy of *Acta. Met.*)

fixed orientation relationship between the matrix and the precipitate, and would be destroyed if the orientation of one of the phases was changed – for example, if a high angle grain boundary passed through the matrix but not through the precipitate. (Smith 1952, Doherty 1982). Of all the possible relative orientations that could occur between a matrix and a precipitate, the special one that gives rise to the low values of $\sigma$ for certain parts of the interface arises from the nucleation process (§2.2.2), since only precipitates with low interfacial energies will nucleate at a significant rate.

The opposite, plate-like, morphology was initially studied by Boyd and Nicholson (1971a,b) for $\theta'$ and $\theta''$ precipitates in Al–Cu alloys (fig. 5.18). As discussed in chapter 2 (see fig. 2.11), for only one plane is it possible to have a coherent interface between $\theta'$ or $\theta''$ and the matrix, so the precipitates were seen to have the shape of flat discs where the aspect ratio of diameter to height was 25 for the $\theta''$ precipitate and 40 for the $\theta'$ precipitate. Even larger values of the aspect ratio have been found for hexagonal $Ag_2Al$ on {111} planes in aluminium (§1.3), see, for example, Ferrante and Doherty (1979) and Rajab and Doherty (1989). These extreme values of shape anisotropy are usually not due solely to the anisotropy of the interfacial *energy* but may be non-equilibrium shapes produced by differences in interfacial *mobility* as noted, for example, by Frank(1958). The good-fit interface between a precipitate and the matrix with different crystal structures needs growth ledges to migrate, §1.3, and if the spacing of these ledges is too large then the thickening reaction of the plates will be slowed relative to the lengthening reaction. As a result, a much larger aspect ratio than that set by the Wulff theorem can then develop, see Ferrante and Doherty (1979), Doherty and Cantor (1982, 1988) and Enomoto (1987). The aspect ratio, $A$, defined as the radius of the plate, $a_y$, divided by the half thickness, $a_x$, in these circumstances is larger than the equilibrium aspect ratio, $A_{eq}$, given by the Wulff theorem. The estimated value of the equilibrium aspect ratio of $\theta'$ in Al–Cu is likely to be about

20 (Merle and Merlin 1981; Doherty and Purdy 1982) and that of $Ag_2Al$ in Al, with a fully coherent habit plane interface, fig 1.11(*b*), is likely to be about 10 which falls to about 3 if the precipitate adopts its equilibrium lattice parameter (Ferrante and Doherty 1979):

$$A_{eq}=\sigma_r/\sigma_f$$

where $\sigma_r$ is the high interfacial energy rim of the precipitate and $\sigma_f$ is low energy of the good-fit facet interface. Ferrante and Doherty (1979) developed the Gibbs–Thomson equations for the $\alpha$ matrix solubility in equilibrium with the facet plane, $N_f$, and in equilibrium with the rim, $N_r$, for a cylindrical plate of aspect ratio $A$ and radius, $a_y$. These equations are

$$N_f=N_\alpha[1+f(N)2\sigma_r V_m/RTa_y] \tag{5.12c}$$

$$N_r=N_\alpha[1+f(N)(1+A/A_{eq})\sigma_r V_m/RTa_y] \tag{5.12d}$$

These equations show two important results. First, when the precipitate has the equilibrium shape, $A=A_{eq}$, the precipitate has, as is obviously required for an equilibrium, a *constant solubility around the precipitate*. The second result is that eq. (5.12d) shows that for precipitates such as those seen in Al–Cu and Al–Ag, where $A \gg A_{eq}$, these non-equilibrium shapes will have a greatly enhanced solubility increase at their *rims* and this enhanced solubility promotes the kinetics of coarsening, at least the coarsening kinetics measured by the increase in mean *length* of the precipitates. In the Al–Cu system it has been found that the precipitates can rapidly shrink back to allow the aspect ratios to achieve near equilibrium values (Merle and Merlin 1981; Merle and Doherty 1982). In the Al–Ag system, however, the hexagonal precipitates maintained their non-equilibrium shapes for long times (Ferrante and Doherty 1979; Rajab and Doherty 1989).

In other cases nearly spherical precipitates are found. This occurs for fcc cobalt precipitates in an fcc copper matrix (Livingston 1959; Haasen and Wagner 1992). In this system, since the two crystals have the same structure, the interface will have the same coherence all around the particle and so show an almost isotropic $\sigma$, but one of low energy. Alternatively, if the precipitate and matrix have no plane or direction of good fit, e.g. $\alpha$ manganese in magnesium (Smith 1967), oxides in copper (Ashby and Centamore 1968) or silicon in aluminium (Humphreys 1977), then the interface will be incoherent at all orientations, and although $\sigma$ will be larger, it will again be nearly isotropic and will thus give an almost spherical equilibrium shape.

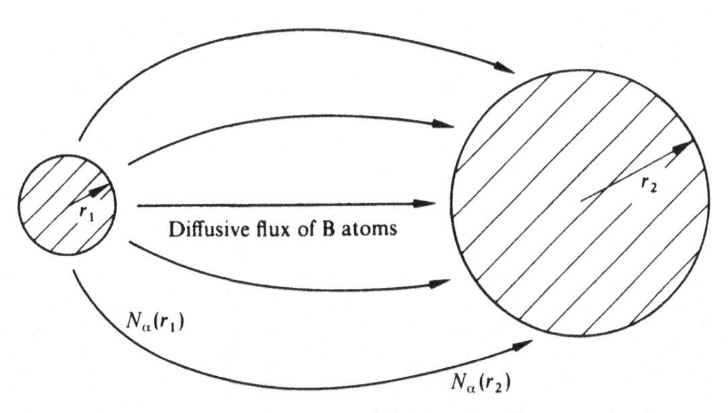


**Figure 5.19** Two spherical $\beta$ precipitates of radius $r_1$ and $r_2$ in an $\alpha$ matrix. Since $N_\alpha(r_1) > N_\alpha(r_2)$ there is a diffusive flux from 1 to 2.

## 5.5.3 Lattice diffusion – the rate controlling process

The application of the Gibbs–Thomson equation to precipitate coarsening is qualitatively obvious even if quantitatively complex. We will start with the simplest system consisting of two spherical precipitates of different radii, $r_1$ and $r_2$ (fig. 5.19). In this treatment, we assume that the rate controlling process is that of lattice diffusion. This means that the solute concentrations in the matrix at each precipitate will be the equilibrium Gibbs–Thomson values. On this basis, the solute content in equilibrium with precipitate 1 will be greater than that with precipitate 2, and so there will be a diffusive flux of B atoms, through the matrix, from 1 to 2. This flux results in the shrinkage of the smaller precipitate and the growth of the larger. This exercise in power politics will continue, but at an accelerating rate as $r_1$ becomes smaller, and $N_\alpha(r_1)$ greater, leading to the inevitable disappearance of the small precipitate, so leaving the system with the one, somewhat larger, precipitate.

The process occurring in a real system, with many particles, has been analysed at various levels of sophistication. It is accepted that the theoretical results reported independently by Lifshitz and Slyozov (1961) and Wagner (1961) – 'LSW' – are rigorously derived at least for the conditions that they considered. This was of a negligible volume fraction of particles ($f \rightarrow 0$) and with no elastic interaction between particles. We shall, however, follow Greenwood (1956, 1969) in developing equivalent equations less rigorously. We will then compare the results of this simplified treatment to the rigorously derived results. This approach has the advantage of simplicity but it also recognises the important contribution to the theory made by Greenwood (1956). His contribution appears to be often overlooked in recent literature.

If we consider the array of particles shown in fig. 5.20($a$) where the average particle size is $\bar{r}$ and concentrate on the particle shown whose radius is $r$, the radial solute distribution around this precipitate is shown as being $N_\beta$ (assumed

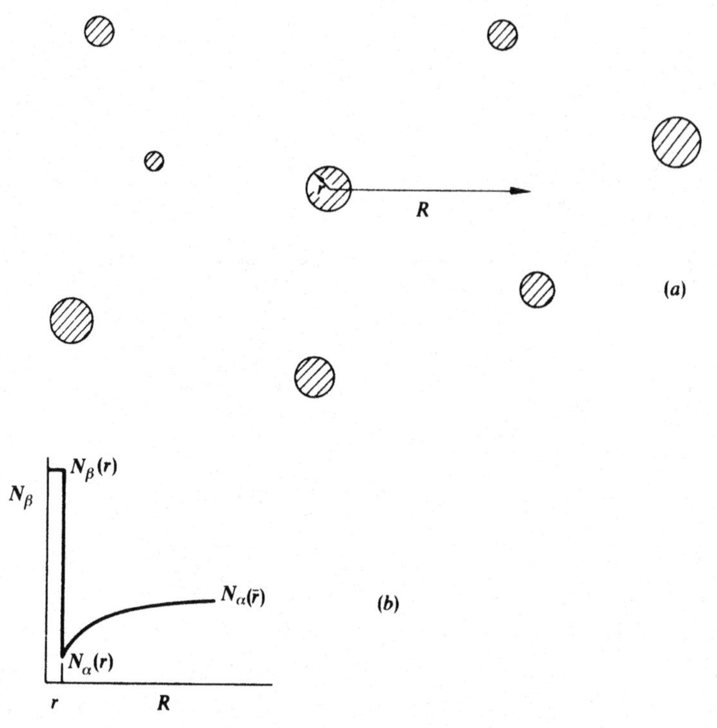


**Figure 5.20** (*a*) An array of precipitates around a particular precipitate of radius *r*. (*b*) Variation in concentration away from the centre of this precipitate. It is assumed that the concentration rises to the average value $N_\alpha(\bar{r})$ before the nearest neighbours and their diffusive fields are approached.

$\sim$1) inside the precipitate, $N_\alpha(r)$ at $R=r$, and as rising, at $R \gg r$, to $N_\alpha(\bar{r})$ – the solute concentration corresponding to the average particle radius. For this to be valid it is clear that the spacing, $\lambda$, between the precipitates must itself be much larger than the mean radius so this is a model appropriate for a very small volume fraction of particles, $f_v$.

$$(r/\lambda)^3 \approx f_v \ll 1$$

The value of $N_\alpha(r)$ is taken as the equilibrium Gibbs–Thomson value (5.12) which applies if the two phases are in local equilibrium, that is, if there is no significant barrier to atom transfer across the precipitate–matrix interface (§1.3). The only departure from equilibrium will be the non-uniform solute concentration in different regions of the matrix so causing lattice diffusion – in other words, the slowest, and thus rate-determining, step will be that of lattice diffusion. This model then is that of volume diffusion control.

The diffusive flux, in moles per second, to the precipitate, at a distance $R$ from the precipitate is flux=area×diffusion coefficient×concentration gradient at $R$

$$=-4\pi R^2 D(\mathrm{d}C/\mathrm{d}R) \tag{5.13}$$

where $C$ is *concentration* in atoms per unit volume. That is,

$$C=N/V_m=\text{molar (atomic) fraction/molar volume}$$

$$\text{flux}=-4\pi R^2 D(\mathrm{d}N/\mathrm{d}R)/V_m \tag{5.13a}$$

This negative flux, in the $-R$ direction, gives the positive flux that feeds the growth of the precipitate, $\mathrm{d}r/\mathrm{d}t$. This can be readily determined since the volume of the precipitate, $V$, is given by

$$V=4\pi r^3/3$$

and so the extra number of g-moles, $n$, of solute inside the precipitate, $N_\beta-N_\alpha$, is given by

$$n=V/V_m=4\pi r^3(N_\beta-N_\alpha)/3V_m$$

Therefore

$$\mathrm{d}n/\mathrm{d}r=4\pi r^2(N_\beta-N_\alpha)/V_m$$

and

$$\mathrm{d}n/\mathrm{d}t=(4\pi r^2/V_m)(N_\beta-N_\alpha)\mathrm{d}r/\mathrm{d}t \tag{5.14}$$

For simplicity in the analysis of Greenwood and LSW it was assumed that the $\beta$ precipitates were almost pure B, so $N_\beta\approx1$ and the matrix was almost pure A, $N_\alpha\approx0$. Under this simplifying assumption, $(N_\beta-N_\alpha)\approx1$, so that eq. (5.14) is usually given as

$$\mathrm{d}n/\mathrm{d}t=(4\pi r^2/V_m)\mathrm{d}r/\mathrm{d}t \tag{5.14a}$$

Eq. (5.14a) is the flux required for growth of the precipitate, so equating the fluxes, eqs. (5.13a) and (5.14a), we obtain:

$$(4\pi r^2/V_m)\mathrm{d}r/\mathrm{d}t=4\pi R^2 D(\mathrm{d}N/\mathrm{d}R)/V_m \tag{5.15}$$

or

$$dR/R^2 = D\,dN/r^2(dr/dt) \tag{5.16}$$

Integrating this from the precipitate radius, $r$, where $N=N_\alpha(r)$, out to large values of the radial distance $R$, we obtain:

$$\int_{R=r}^{R=\infty} dR/R^2 = \int_{N_\alpha(r)}^{N_\alpha(\bar{r})} D\,dN/r^2(dr/dt) \tag{5.16a}$$

$N$ at $R=\infty$ is taken as $N_\alpha(\bar{r})$, the composition of the solution in equilibrium with a precipitate of average radius $\bar{r}$. To integrate this equation we will assume that the instantaneous growth rate, $dr/dt$, and the radius, $r$, are each effectively constant, *for the time that it takes for a solute atom to diffuse upto the precipitate*, so that the composition profile does not change. This is valid for very small values of $[N_\alpha(\bar{r})-N_\alpha(r)]/(N_\beta-N_\alpha)$ which is always true in coarsening as was discussed for diffusional precipitate growth at small supersaturations (§2.3.1). By integration of eq. (5.16a) we then obtain

$$1/r = D[N_\alpha(\bar{r})-N_\alpha(r)]/r^2(dr/dt) \tag{5.17}$$

Rearrangement of eq. (5.17) yields:

$$dr/dt = (D/r)[N_\alpha(\bar{r})-N_\alpha(r)] \tag{5.17a}$$

and substitution of the Gibbs–Thomson equation (5.12a) for spherical particles into eq. (5.17a) gives the expected growth rate as:

$$dr/dt = (2D\sigma N_\alpha V_m/RTr)(1/r-1/\bar{r}) \tag{5.17b}$$

This is shown in fig. 5.21 for two values of the average particle radius $\bar{r}$. From this relationship several factors can be seen:

(i) all particles smaller than $\bar{r}$ are shrinking. The rate of shrinkage increases rapidly as $r/\bar{r}$ approaches zero;

(ii) all particles larger than $\bar{r}$ are growing, but the rate of growth in terms of $dr/dt$ increases from zero at $r=\bar{r}$ to a maximum value at $r=2\bar{r}$ and then decreases;

(iii) as $\bar{r}$ increases, the rate of growth of precipitates larger than $\bar{r}$, falls.

These factors will cause an initially fine dispersion with a range of particle sizes to change, with the small particles initially becoming smaller and eventually disappearing while the larger particles grow. The dispersion becomes coarser and the mean particle radius $\bar{r}$ increases with time. This has fatal results for even those particles which were initially just growing, with $r\approx\bar{r}$ since their slow growth will be less than $d\bar{r}/dt$ and so they will be overtaken by $\bar{r}$ and they will

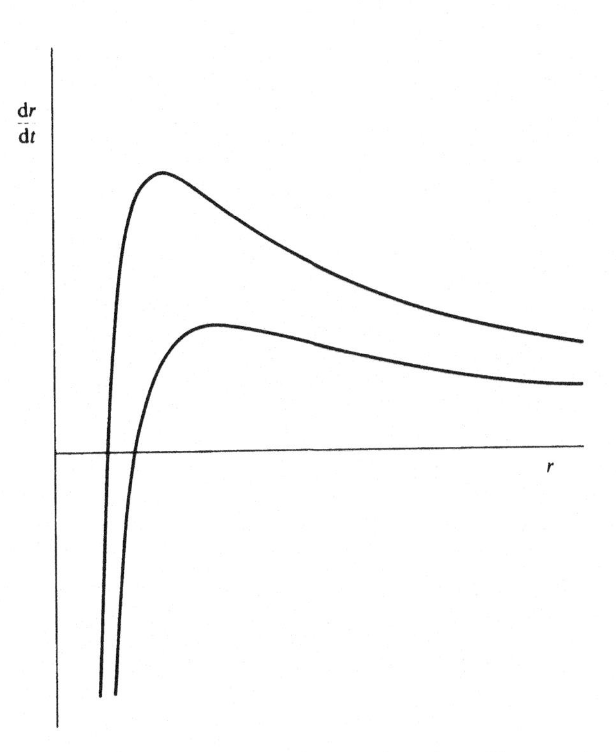


**Figure 5.21** Variation in precipitate growth rate $dr/dt$ with precipitate radius $r$ for diffusion controlled growth (eq. (5.17)) for two different values of the mean radius $\bar{r}$ – for the lower curve $\bar{r}$ is increased by 50%. (After Greenwood 1969, courtesy of the Metals Society.)

then shrink. Precipitates larger than $2\bar{r}$ will grow, but only very slowly, as they require much more solute for growth than particles with $r \approx 2\bar{r}$, and the very large particles will be caught up by the particles whose radius $r$ is about $2\bar{r}$. The result of these actions is that no particles larger than $2\bar{r}$ will be expected after a significant amount of coarsening has occurred.

An approximate solution for the variation of $\bar{r}$ with time can be achieved by assuming that this will be the same as the maximum value of $dr/dt$ at $r=2\bar{r}$, that is,

$$d\bar{r}/dt \approx (dr/dt)_{max} = D\sigma N_\alpha V_m / 2RT\bar{r}^2$$

So

$$\bar{r}^2 d\bar{r} = D\sigma N_\alpha V_m dt/2RT$$

Integration of this equation from $\bar{r}=\bar{r}_0$, at time $t=0$, to $\bar{r}=\bar{r}_t$, at time $t$, then yields

$$\bar{r}_t^3 - \bar{r}_0^3 = 3D\sigma N_\alpha V_m t/2RT \qquad (5.18)$$

The rigorous LSW analysis gives a very similar result. The LSW analysis showed that an initially narrow Gaussian distribution of precipitate sizes relaxes to a steady distribution of precipitate sizes (fig. 5.22). The size distribution is usually

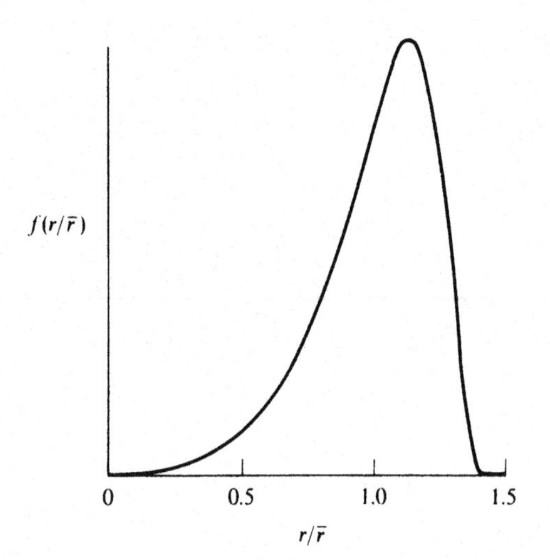


**Figure 5.22** Expected distribution of particle size shown as frequency of occurrence of reduced radius ($r/\bar{r}$) as a function of the reduced radius for diffusion controlled coarsening. (After Greenwood 1969, courtesy of the Metals Society.)

plotted against the reduced radius ($r/\bar{r}$). Unlike the simple treatment leading to eq. (5.18) which would predict a cut-off at $r/\bar{r}=2$, the LSW distribution has a cut-off at $r/\bar{r}$ of 1.5. The LSW equation for the coarsening rate, after this steady distribution has been achieved, is

$$\bar{r}_t^3-\bar{r}_0^3=8D\sigma N_\alpha V_m t/9RT \tag{5.18a}$$

This is the same as given by the earlier elementary derivation, apart from the difference in numerical constant of 8/9 rather than 3/2. It should be remembered that the approximate and the LSW equations were derived for the assumption that $N_\beta \approx 1$ and $N_\alpha \approx 0$. Correcting for the general case involves two modifications. Firstly the Gibbs–Thomson correction, $[(1-N_\alpha)/N_\beta-N_\alpha)](1/\epsilon_\alpha)$ from eq. (5.12b) needs to be included. Secondly the amount of solute needed for precipitate growth $(N_\beta-N_\alpha)$ also needs to be included so the correct general form of the LSW equation is:

$$\bar{r}_t^3-\bar{r}_0^3=8D\sigma N_\alpha(1-N_\alpha)V_m t/9\epsilon_\alpha(N_\beta-N_\alpha)^2RT \tag{5.18b}$$

In most cases the matrix is a dilute solution, $N_\alpha \rightarrow 0$, so that the Darken correction $(1/\epsilon_\alpha)$ is 1, so the non-constancy of the activity coefficient is rarely a concern unless the matrix is an intermediate phase, for example, $\beta$ brass. In many real coarsening situations, for example, $Al_2Cu$ or $Al_6Mn$, the *precipitate* is an intermediate phase so that $(N_\beta-N_\alpha)<1$. The effect of the composition term, $(1-N_\alpha)/(N_\beta-N_\alpha)^2$, will be to accelerate the expected coarsening rate rather significantly. This important acceleration of coarsening by factors as large as 10 ($Al_2Cu$) and, in one case ($Al_6Mn$), of nearly 50 times must not be forgotten.

A more trivial point, but one that sometimes leads to confusion, is when the concentration term in eq. (5.18a) is given as $C_\alpha$, the number of moles of solute in unit volume of the matrix. This leads immediately (since $C_\alpha = N_\alpha / V_m$) to

$$\bar{r}_t^3 - \bar{r}_0^3 = 8 D \sigma C_\alpha V_m^2 t / 9RT \quad (5.18c)$$

The LSW size distribution is the only distribution that can, mathematically, maintain a steady state form (see, for example, Hillert, Hunderi, Ryum and Saetre (1989) and Hoyt (1990)). Such a steady state distribution allows what is usually called 'self similar' coarsening to occur (Mullins and Vinals 1989). A self similar microstructure is one that evolves under coarsening in a way that only requires a reduction in magnification to remain apparently unchanging. It should be noted that the reasons for the lack of particles between $r = 1.5\bar{r}$ (LSW) and $r = 2\bar{r}$, discussed by Greenwood (1956) are not clear. The conclusion that only the LSW distribution can maintain a stable or self similar form has been challenged by Brown (1989, 1990a,b, 1992a,b). Brown (1989) found, by numerical simulation of coarsening, that given a wide size distribution, as is very likely to occur by the prior process of nucleation and growth, particles can maintain a few particles in the size range above $1.5\bar{r}$ for extended times. Hillert *et al.* (1989) found a similar result for one of the distributions used by Brown but it is interesting to note, that when this distribution was truncated at $r \geq 2\bar{r}$ the distribution then returned to the LSW distribution. In further discussions Hillert, Hunderi and Ryum (1992a,b) argued that the LSW would always be produced *eventually*. Achieving the LSW distribution may, however, take a very long time – so that experimental results may well almost always occur with the particle size distributions not having achieved the stationary state.

Fang and Patterson (1993) have experimentally studied the effect of different initial particle size distributions on coarsening. They used tungsten powders in liquid Ni–Fe matrix, at a rather high volume fraction of solid particles ($f_s = 0.6$). They found, as might be expected, that the rate of coarsening was controlled by the size distribution, fig. 5.23. Narrow distributions coarsened slowly and wide distributions coarsened quickly. The distributions changed rapidly at the earliest times as measured by the natural logarithm of the standard deviation, fig. 5.24. During the time interval when the size distributions were changing rapidly, the coarsening equation did not show the cube law expected for diffusion limited coarsening, fig. 5.23(*b*). However, the distributions, particularly the wide distribution had not reached a really unchanging form even after extensive coarsening. During the period when the distributions were changing slowly, however, the cube law still fitted the data to a very high precision. This result is important in showing that a good experimental fit to the cube law does not guarantee that the kinetic coefficient will have the value expected for true steady state coarsening. In addition it can be noted that the ratio of maximum size particle radius to the mean particle radius, $r_{max}/\bar{r}$, *remained above* 2 for all the distributions. Of course,

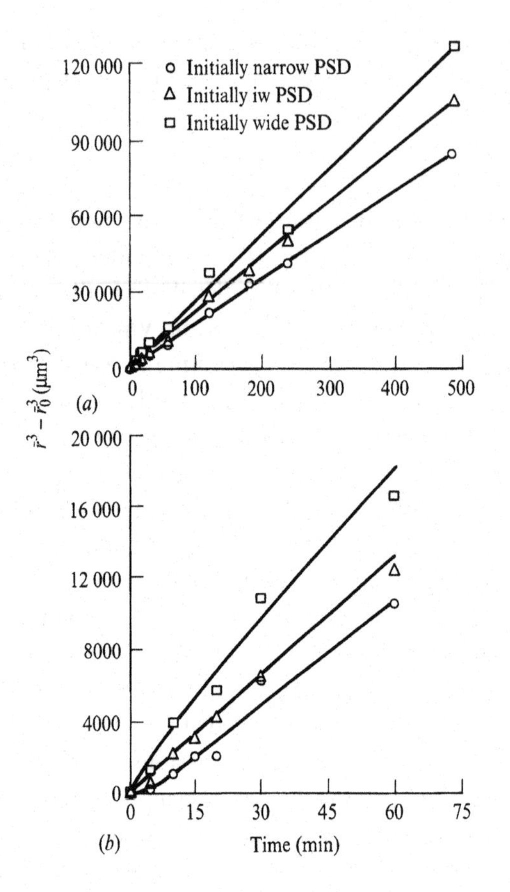


**Figure 5.23**
Experimental values of the coarsening as $r^3 - r_0^3$ (in $\mu$m) for solid, tungsten rich, particles in a liquid matrix in W–14Ni–6Fe for different initial particle size distributions: (*a*) for the entire coarsening period and (*b*) for the early times when the distribution was changing rapidly. (After Fang and Paterson 1993. Courtesy of *Acta Metall. et Mater.*)

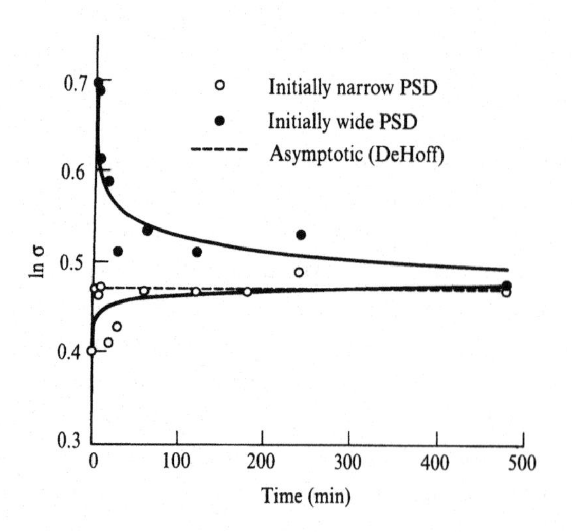


**Figure 5.24**
Experimental values of the width of the particle size distributions, given as the natural logarithm of the standard deviation $\sigma$, during the coarsening shown in figure 5.23. (After Fang and Patterson 1993. Courtesy of *Acta Metall. et Mater.*)

even the narrow size distribution started with $r_{max}/\bar{r}>2$. If the results shown by Fang and Patterson can be duplicated for *low* volume fractions of second phase particles, §5.5.7, then the problem of the particle size distributions first raised by Greenwood (1969) will have been solved. The steady state particle size distribution of the classic LSW model may only be experimentally accessible in real experiments from initially *very narrow distributions* where particles of a size greater than $1.5\bar{r}$ are not present. Most two-phase metallic microstructures arise through precipitation that always gives rise to wide size distributions, §5.5.8. So it is quite likely that the LSW distribution will not often be found, §5.6.

### 5.5.4 Interface kinetics: atom transfer across the interface as the rate controlling step

Under the conditions in which the slowest step in the coarsening process is the transfer of atoms across the precipitate matrix interface, lattice diffusion will then be able to make the matrix solute concentration, seen to have a diffusion profile in fig. 5.20, almost constant at the value where $r\rightarrow\infty$, that is, $N=N_\alpha(\bar{r})$. The local rate of precipitate growth will then be determined by the rate at which solute atoms can join (or leave) the precipitate across the interface, a process driven by the local departure from equilibrium. This departure from equilibrium can be calculated from the equilibrium solute content at a precipitate of radius $r$, $N_\alpha(r)$, given, as before, by the Gibbs–Thomson equation. The equilibrium partial molar free energy (chemical potential) of solute atoms at a precipitate of radius $r$ is

$$G_\alpha(r)=G^\circ+RT\ln a(r)=G^\circ+RT\ln N_\alpha(r)+RT\ln\gamma_\alpha(r)$$

The difference between the actual molar free energy of $G_\alpha(\bar{r})$ and the equilibrium value of $G_\alpha(r)$ is therefore (for the situation of a constant activity coefficient in the matrix, $\gamma_\alpha(r)$, found in a dilute solution)

$$G_\alpha(\bar{r})-G_\alpha(r)=RT\ln[N_\alpha(\bar{r})/N_\alpha(r)]\approx RT[N_\alpha(\bar{r})-N_\alpha(r)]/N_\alpha(r) \qquad (5.19)$$

The approximation in eq. (5.19) is valid since $N_\alpha(\bar{r})/N_\alpha(r)\approx 1$. The driving force for atom transfer across the interface, the difference in chemical potential (5.19), is then proportional to the solute difference: $N_\alpha(\bar{r})-N_\alpha(r)$. So, if the velocity, $dr/dt$, is linearly dependent on the driving force (as for the discussion of grain boundary migration of §3.3.3), then we obtain the equation used in eq. (2.30)

$$dr/dt=K[N_\alpha(\bar{r})-N_\alpha(r)] \qquad (5.20)$$

where $K$ is the proportionality constant that includes the unknown interfacial mobility. (In the diffusion controlled case, $K\rightarrow\infty$.) For the case under discussion where $N_\beta \approx 1$, $K = M'_i$ (eq. (1.16d)). Eq. (5.20) may be compared with the corresponding expression for $dr/dt$ in the case of diffusion control (eq. (5.17a)). The critical dimensionless number, $D/rK = D/rM'_i$, was shown in §1.3 to determine if there was diffusion control when $D/rK \ll 1$, mixed control when $D/rK \approx 1$, or interface control when $D/rK \gg 1$.

For interface control, we can substitute the Gibbs–Thomson values (eq. (5.12a)) into eq. (5.20) and obtain

$$dt/dt = (2K\sigma N_\alpha V_m/RT)[(1/\bar{r}) - (1/r)] \qquad (5.20a)$$

On the basis of this relationship and the previous assumption that

$$d\bar{r}/dt \approx dr/dt \text{ for } r = 2\bar{r}$$

we obtain

$$\bar{r}\,d\bar{r} = K\sigma N_\alpha V_m dt/RT$$

Integration of this expression to give $\bar{r}$ as a function of time yields the relationship:

$$\bar{r}_t^2 - \bar{r}_0^2 = 2K\sigma N_\alpha V_m t/RT \qquad (5.21)$$

The expression derived by the rigorous treatment is given by Wagner (1961) as

$$\bar{r}_t^2 - \bar{r}_0^2 = (64/81)K\sigma N_\alpha V_m t/RT \qquad (5.21a)$$

## 5.5.5 Ripening of precipitates on grain boundaries

Another type of precipitate ripening process has been considered theoretically by Speight (1968), Kirchner (1971) and Ardell (1972a). This occurs when all the precipitates are situated on grain boundaries and when the dominant diffusion path is that along the grain boundaries. This situation requires a very fine grain size, of the order of the interparticle spacing, and also a moderately low temperature, so that the following relationship holds:

$$wD_{gb} \gg dD_{lattice}$$

where $D_{gb}$ and $D_{lattice}$ are the diffusion coefficients in the grain boundary and lattice respectively, $w$ is the grain boundary thickness, usually considered to be 1–2 atom spacings, §1.3, and $d$ is here the mean grain size.

Table 5.3 shows the temperatures below which $wD_{gb}/dD_{lattice}>1$ for two metals, fcc silver and bcc $\alpha$ iron, for a range of grain sizes. The temperatures shown are the ones below which grain boundary diffusion will be the dominant diffusion path. This change in diffusion path occurs as the temperature falls since the activation energy for grain boundary diffusion ($H_{gb}$) is smaller than the activation energy of lattice diffusion ($H_{lattice}$). This distinction between two structures is likely to represent a general tendency, since James and Leak (1965) have noted that the activation energy ratio is usually about 0.5 for fcc and 0.7 for bcc metals. The figures in table 5.3 suggest that for the finest microstructures likely to be commonly encountered, where the grain size is limited by an interparticle spacing which is 100 nm–1 $\mu$m, grain boundary diffusion will predominate at all temperatures; for coarser microstructures, more particularly for the bcc metals, the grain boundary mechanism may not operate unless the temperature is quite low.

The equation that has been derived by Kirchner (1971) for the coarsening of particles situated on grain boundaries where the grain boundary paths dominate the diffusion process is

$$\bar{r}_t^4-\bar{r}_0^4=(9/32)(wD_{gb}\sigma N_\alpha(\mathrm{gb})V_m t/ABRT) \tag{5.22}$$

where $N_\alpha$(gb) is the solute content at a grain boundary in equilibrium with an infinitely large precipitate, and $A$ and $B$ are parameters defined as

$$A=2/3+(\sigma_b/2\sigma)+(1/3)(\sigma_b/2\sigma)^3$$

$$B=(1/2)\ln(1/f_b)$$

where where $\sigma_b$ is the grain boundary energy and $f_b$ the fraction of the grain boundary covered by precipitates.

The value of $r$ is the radius of curvature of the spherical caps of the lens shaped

**Table 5.3** *Temperatures below which boundary diffusion will be faster than volume diffusion.*

| Metal | $H_{gb}/H_{lattice}$ | $T_m$ (K) | $d$: $10^{-7}$ m | $10^{-6}$ m | $10^{-5}$ m | $10^{-4}$ m |
|---|---|---|---|---|---|---|
| Silver (fcc) | 0.48 | 1234 | $T_m$ | $T_m$ | $0.8T_m$ | $0.6T_m$ |
| Iron (bcc) | 0.69 | 1808 | $T_m$ | $T_m$ | $0.7T_m$ | $0.5T_m$ |

These values were calculated from the tabulated diffusion data given for silver by Shewmon (1963) and iron by Buffington, Hirano and Cohen (1961).

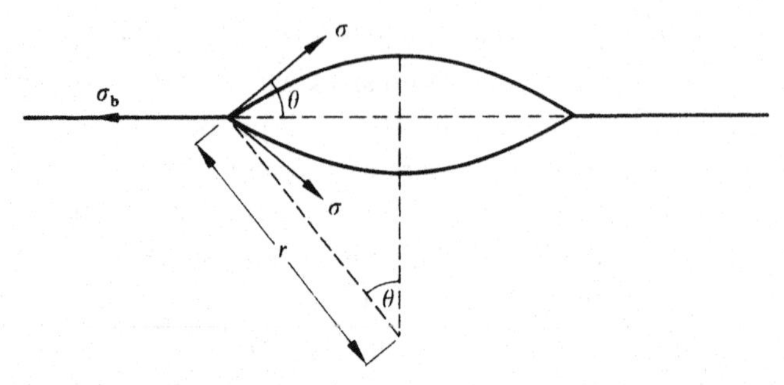


**Figure 5.25** Equilibrium shape of a precipitate on a grain boundary if the interfacial energy with both the matrix grains is constant. It consists of two spherical caps each of radius $r$. (After Greenwood 1969, courtesy of the Metals Society.)

precipitate on the grain boundary as is illustrated in fig. 5.25. This radius will be significantly larger than the radius of a sphere containing the same volume of material. The predicted distribution of particle sizes for the grain boundary situation was determined by Kirchner (1971) who found that the distribution was somewhat narrower than in the case of ripening controlled by volume diffusion.

## 5.5.6 Ripening of precipitates on low angle (dislocation) boundaries

Kreye (1970) and Ardell (1972b) discussed the coarsening behaviour of precipitates on low angle subboundaries where pipe diffusion down the dislocations is the rate controlling process. If $b$ is the Burgers vector of the dislocation, then the dislocation spacing, $h$, is related to the boundary misorientation $\theta$ by the standard equation, see, for example, Hull and Bacon (1984)

$$h=b/\theta$$

When the value of $h$ is larger than the maximum particle radius (only one dislocation per particle) a new power law is derived in which

$$\bar{r}_t^5-\bar{r}_0^5=Ct \tag{5.23}$$

When the subboundary misorientation is sufficiently large, the spacing $h$ between the dislocations becomes much smaller than the average precipitate radius. In this situation many dislocations, in the form of a low angle, dislocation boundary, cut each precipitate and Ardell (1972b) derived an $r^4$ relationship as found for the geometrically similar grain boundary process:

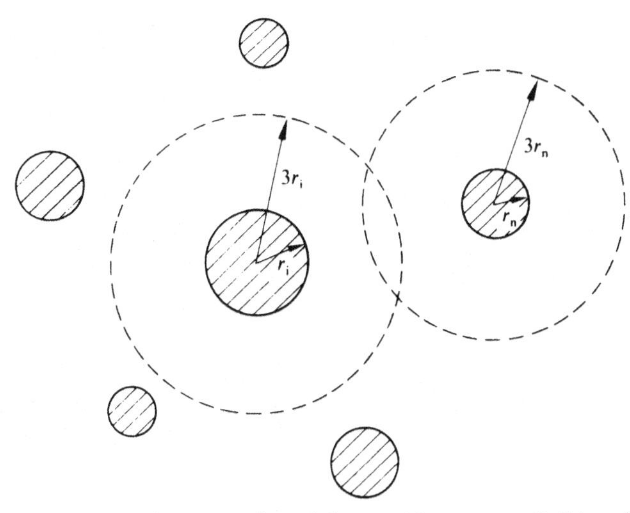


**Figure 5.26** An array of precipitates with a non-negligible volume fraction ($f_v$) such that the distances between precipitates is only somewhat larger than the mean precipitate radius. Under these conditions the solute distribution will be distorted from that shown in fig. 5.20(*b*). (After Ardell 1972*a*.)

$$\bar{r}_t^4 - \bar{r}_0^4 = (64/27)(qD_d\eta\sigma N_\alpha(\mathrm{d})V_m t/\pi hRT) \tag{5.22a}$$

Here $q$ is the cross-section of the dislocation that carries the flux of solute, $D_d$ is the dislocation core diffusion coefficient, $N_\alpha$(d) is the solute content in dislocations in equilibrium with a precipitate of infinite radius, $\eta$ is a function of the precipitate volume fraction given by Ardell (see §5.5.7), and $h$ is the dislocation spacing.

Ardell (1972b) indicated that where an intermediate value of $h$ between the two extremes occurs, an intermediate value of the radius exponent between 4 and 5 would be expected.

## 5.5.7 Effect of volume fraction on coarsening rates

### When lattice diffusion is controlling

Asimov (1963) and then Ardell (1972a) initially discussed the role of the volume fraction of the precipitate. Ardell pointed out that the original treatment of Lifshitz and Slyozov (1961) and Wagner (1961), as in the simplified discussion presented here, assumed that eq. (5.16) could be integrated out from the precipitate radius to distances that were very large compared with this radius, with no disturbances from other precipitates. However, if the volume fraction, $f_v$ is not negligibly small, the precipitate radius cannot be considered as a *negligible fraction* of the interparticle spacing, as seen in fig. 5.26, since for any particle such as $r_i$, at any distance $<3r_i$ from this precipitate we find ourselves at a distance $<3r_n$

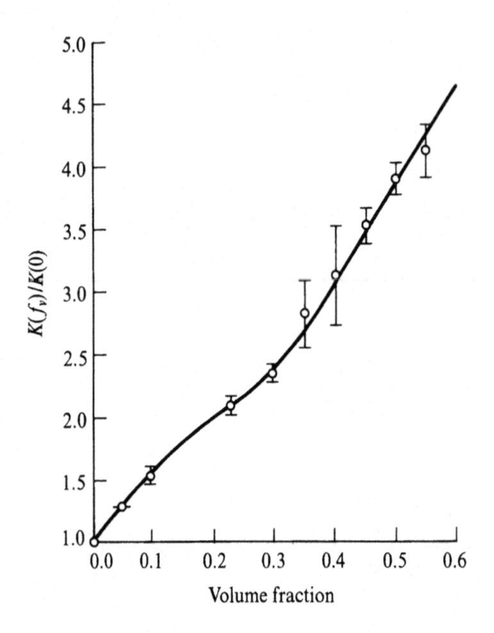


**Figure 5.27** The volume fraction enhanced rate constant, $K(f_v)$, of the coarsening reaction, from computer simulation, normalised by the LSW value, $K(0)$, as $f \to 0$, as a function of the volume fraction of particles. The error bars come from 3 sets of $\langle r \rangle^3$ against time runs for each volume fraction. (After Voorhees and Glicksman 1984c. Courtesy *Acta Met.*)

from its neighbouring precipitate whose radius is $r_n$. Ardell analysed this problem after making some simplifying assumptions and obtained a result in which the growth equation for the average precipitate radius (eq. (5.18a)) is modified by a parameter, $k(f_v)$: $k(f_v)$ is solely a function of the precipitate volume fraction, to give:

$$\bar{r}_t^3 - \bar{r}_0^3 = 8k(f_v)D\sigma N_\alpha V_m t/9RT \tag{5.18d}$$

This modification should be highly significant. $k(f_v)$ is unity, giving the LSW relationship (5.18a) for a zero volume fraction of precipitate. However, $k(f_v)$ was found, by Ardell, to increase to much larger values at larger volume fractions – as high as $k(f_v)=10$ for a volume fraction of 0.25 (25%).

Ardell also found that as the volume fraction of precipitates increases, the dispersion of precipitate sizes was predicted to broaden. Various other authors including Brailsford and Wynblatt (1979), Voorhees and Glicksman (1984a,b,c) and Enomoto, Tokuyama and Kawasaki (1986) have reanalysed the problem both analytically and numerically. Although there are significant differences between the various models, all agree that the coarsening rate increases with volume fraction but at a slower rate than first suggested by Ardell. Typically $k(f_v)\approx 4$ at volume fractions of about 0.5 (50%), fig. 5.27. It is also found that the steady state size distribution is broadened and made more symmetrical by increased volume fraction, see fig. 5.28. Insight into the faster coarsening with the higher volume fractions was shown by measurement of the mean flux to or from a particle. As the volume fraction increased

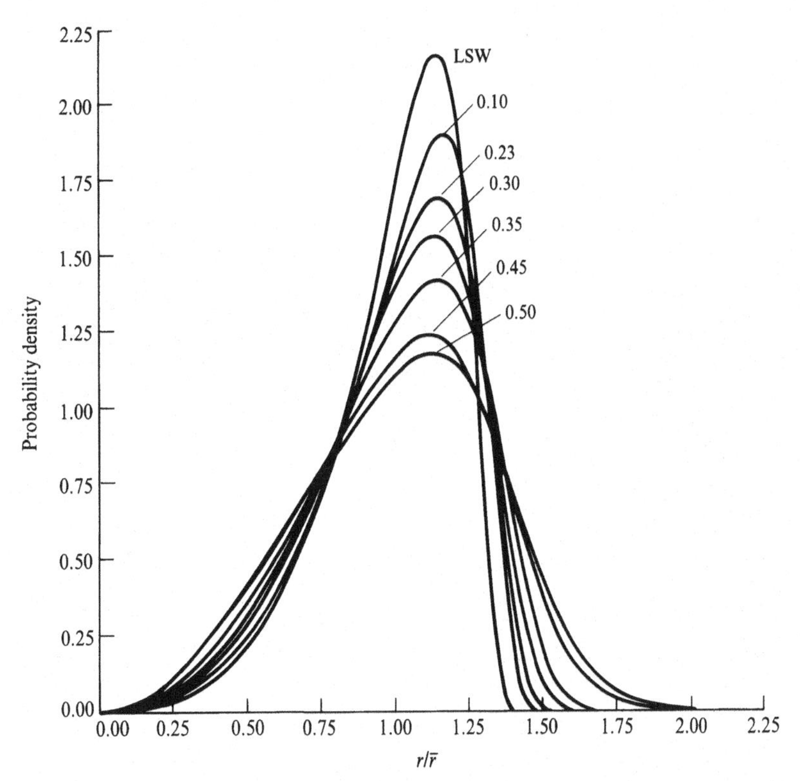


**Figure 5.28** The computer simulated, steady state particle size distributions in bulk diffusion limited coarsening for different values of the volume fraction of the particles. (After Vorhees and Glicksman 1984c. Courtesy of *Acta Met.*)

the flux from small particles was larger than expected for the zero volume fraction (LSW) while that to large particles was also larger than that expected. Voorhees and Glicksman simulated the coarsening numerically using various imposed initial size distributions. They found in their simulations that, at each volume fraction, initial distributions, narrower than the steady state distributions, evolved quickly towards the steady state. During the early stage transient while the standard deviation was increasing the mean particle size actually *fell* due to the shrinkage of small particles. Once the small particles started to vanish, however, the mean particle size rose. For the distributions that were initially wider than the steady state distribution, the usual experimental situation, it was found that the distribution took much longer to evolve towards the steady state size distribution.

### When grain boundary diffusion is controlling

Ardell also modified Kirchner's (1971) result for a non-zero volume fraction of precipitate and has obtained similar changes to that discussed for lattice diffusion. The rate of coarsening increases from that predicted for a small volume

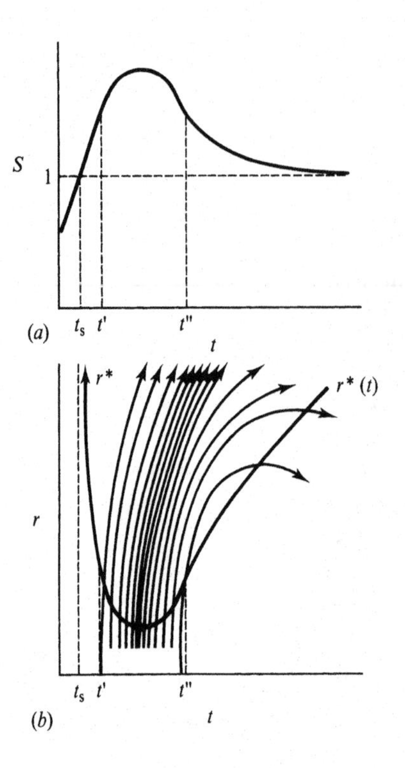


**Figure 5.29** Schematic competitive coarsening during precipitation. An alloy is cooled becoming supersaturated at time $t=t_s$. The first nucleation occurs at $t'$ and the last at $t''$. As the supersaturation, $S=N_\alpha/N_\alpha(T)$, falls with precipitation, and as the cooling ends, the critical radius $r^*(t)$ rises and the smallest precipitates redissolve. (After Kampmann and Kahlweit 1970).

fraction, and the distribution of precipitate sizes becomes larger, as the volume fraction increases.

However, the proportional increase in coarsening rate was much less than that for lattice diffusion – the maximum predicted increase was only just over 20%. This change is likely to be much less than the uncertainties in the values of $\sigma$, $\delta D_{gb}$, $N_\alpha$(gb), etc.

## 5.5.8 Onset of the coarsening reaction

It is commonly assumed, and occasionally stated, that the coarsening reaction starts or becomes significant after the completion of the precipitation reaction – that is, when the solute content of the matrix approaches the Gibbs–Thomson solubility, $N_\alpha(\bar{r})$. This might appear reasonable in view of the much larger driving force for precipitation than for coarsening (see §5.5.1) but this simple viewpoint misses a critical factor first identified by Kampmann and Kahlweit (1967, 1970) and illustrated in fig. 5.29. Here a numerical simulation was given of a binary alloy system cooled at a steady rate from the single phase field where the supersaturation $S<1$, past the solvus temperature where $S<1$ into the two phase field where $S=1$. The supersaturation, $S$, is defined as

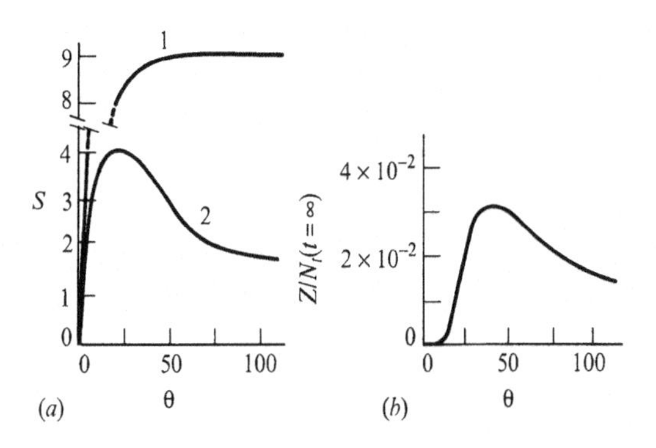


**Figure 5.30** Numerical simulation of the combined precipitation and coarsening process of fig. 5.29: (*a*) the variation in supersaturation $S$ with time $\theta$, curve 1 with no precipitation, curve 2 with precipitation; (*b*) the variation in the density of the precipitates, $Z$, scaled by the equilibrium solute constant $N_t\,(t=\infty)$. (After Kampmann and Kahlweit, 1970.)

$$S=\bar{N}_\alpha/N_\alpha(T)$$

where $\bar{N}_\alpha$ is the average solute content, which is initially the alloy content, and $N_\alpha(T)$ is the equilibrium solubility that varies as a function of temperature, $T$, and thus time $t$. The critical radius for nucleation is $r^*$, §2.2.2, and so $r^*$ falls as $S$ increases on cooling.

$$r^*=-2\sigma/\Delta G_v=2\sigma/RT\ln S$$

The critical radius is that below which any particle will try to dissolve. So, during precipitation the critical radius functions exactly like the mean radius $\bar{r}$ during the coarsening process in a 'fully precipitated' alloy at a constant temperature, §5.5.2. When precipitation starts to occur at a significant rate, $S$ stops decreasing and starts increasing as soon as $\bar{N}_\alpha$ falls faster than $N_\alpha(T)$. The critical radius, $r^*$, will then start to rise. Once any increase in critical radius occurs, particles just nucleated that are small and growing slowly, fig. 5.29, will be overtaken by the rise in $r^*$ and these small precipitates will then redissolve. As a consequence, the characteristic feature of Ostwald ripening, that the density of particles, $Z$, decreases with time ($\mathrm{d}Z/\mathrm{d}t<0$) is predicted to start to occur *during the precipitation reaction*, see fig. 5.30. That is, interfacial energy driven Ostwald ripening can be expected to begin even while $\bar{N}_\alpha$ is still significantly larger than $N_\alpha(\bar{r})$ and the volume fraction of precipitates is still growing significantly. For the reaction to be just coarsening

$$\bar{N}_\alpha\approx N_\alpha(\bar{r}) \tag{5.24}$$

Even after this condition has been reached there will be an increase in the volume fraction of precipitate but here it is due only to the change of Gibbs–Thomson solubility. For a very finely dispersed system of particles there will be a significant rise in solubility as the mean radius grows. The near full equilibrium atomic or volume fraction of the precipitates, $f_e$, will not be achieved until a great deal of coarsening has occured, that is, when $\bar{r}\rightarrow\infty$.

$$f_e=(N_0-N_\alpha)(N_\beta-N_\alpha) \tag{5.25}$$

while

$$f_v(t)=[N_0-N_\alpha(\bar{r})]/(N_\beta-N_\alpha) \tag{5.25a}$$

where $N_0$ is the solute content of the alloy. The simple, but incorrect, assumption that the volume fraction does not change during coarsening leads immediately to eq. (5.26) for the time dependence of the precipitate density, $Z(t)$ (number of particles per unit volume) in a precipitation reaction, where supersaturation driven growth, with $\bar{N}_\alpha>N_\alpha(\bar{r})$, had lasted for a time, $t_1$. It is reasonable to define, as is done here, that the distinction between combined precipitation and coarsening reaction and a coarsening only reaction, described as 'late stage coarsening', occurs at the time, $t_1$, after which $\bar{N}_\alpha\approx N_\alpha(\bar{r})$.

$$Z(t)=k_2(t-t_1)^{-1} \tag{5.26}$$

with

$$k_2=3f_e/4\pi k_1 \tag{5.26a}$$

where $k_1$ is the proportionality constant in the standard LSW equation (5.18a)

$$k_1=8D\sigma N_\alpha V_m/9RT$$

Eq. (5.26) arises since

$$f_v(t)=Z(t)(4\pi\bar{r}^3/3)$$

Ardell (1988, 1990) has given an exact relationship for the case in which the volume fraction is increasing to satisfy the change of Gibbs–Thomson solubility with increase in mean particle radius with time. Ardell's improved model leads to the following predicted variation of particle density with time

$$Z(t)=k_2(t-t_1)^{-1}-k_3(t-t_1)^{-4/3} \tag{5.27}$$

where $k_3$ is given by

$$k_3 = 3\alpha(1-f_e)/4\pi(N_\beta - N_\alpha)k_1^{4/3} \quad (5.27a)$$

with

$$\alpha = 2\sigma N_\alpha V_m / RT \quad (5.27b)$$

By multiplying both sides of eq. (5.27) by $(t-t_1)$ Ardell produced (5.27c)

$$Z(t)(t-t_1) = k_2 - k_3(t-t_1)^{-1/3} \quad (5.27c)$$

The change of mean concentration in the matrix $\bar{N}_\alpha(t)$ with time during the coarsening regime, $t>t_1$ is given by Ardell (1988) in the form used by Xiao and Haasen (1991) as

$$\bar{N}_\alpha(t) - N_\alpha = k_4(t-t_1)^{-1/3} \quad (5.28)$$

where

$$k_4 = (3N_\alpha V_m \sigma / 2RTD^{1/2})^{2/3} \quad (5.28a)$$

This equation can be used to give the change of volume fraction, $f(t)$ with coarsening time, due to the change in Gibbs–Thomson solubility, which following Xiao and Haasen (1991), is

$$f_e - f_v(t) = k_5(t-t_1)^{-1/3} \quad (5.28b)$$

where

$$k_5 = [N_\alpha(N_\beta - N_0)/(N_\beta - N_\alpha)^2]\,(9V_m\sigma/4R^2T^2N_\alpha D)^{1/3} \quad (5.28c)$$

An important feature of these equations is that studies of the rate of coarsening by observation only of the change in mean radius (eq. (5.18)) measure the product, $\sigma D$, of the interfacial energy and the diffusion coefficient. However, parallel measurements of the rate of change of the mean radius together with either the rate of change of mean solute content, which yields $k_4$, eq. (5.28a), or of particle density, using the full Ardell equation, which yields $k_3$, eq. (5.27a), provide two different relationships for $\sigma$ and $D$, so that both parameters can then be determined independently.

An important result of the idea from the initial analysis of Kampmann and Kahlweit (1967, 1970), as illustrated by figs. 5.29 and 5.30, is that during precipitation a wide spread of the times of nucleation will lead to a wide spread in particle sizes during and at the end of precipitation. This experimental spread will then influence the kinetics of coarsening as discussed in §5.5.3.

### 5.5.9 The coarsening reaction in ternary and higher order alloys

Bhattacharya and Russell (1972) and Bjorklund, Donaghey and Hillert (1972) initially analysed the coarsening rate of a dispersed precipitate phase in a ternary alloy. The first analysis considered stoichiometric oxides such as $SiO_2$ in Cu or $Al_2O_3$ in Ni. Bhattacharya and Russell (1972) suggested that the coarsening rate was determined by the component with the smallest product of the diffusivity and concentration in the matrix, as in eqs. (5.17b) and (5.18). As a consequence of this, the coarsening rate will be modified by adding an excess of the faster diffusing component. Such an excess of the faster diffusing component would depress the solubility of the other component and slow the coarsening rate. Bjorklund *et al.* (1972) considered the situation where the diffusion coefficients of the various components are very different. An important example is the coarsening of carbides in steel where the interstial diffusion of carbon is very much faster than that of the substitutional elements. Bjorklund *et al.* (1972) considered the coarsening of the binary iron carbide, $Fe_3C$, but with different levels of partitioning of the third element in the carbide. For a $\beta$ precipitate in an $\alpha$ matrix, and introducing a new concentration parameter $u_i$ for $i$=Fe, M and C, where

$$u_i=N_i/(1-N_C)=N_i(1+u_C)$$

and for a modified partition coefficient for the ternary metallic element M, $k^*_M$ is given by

$$k^*_M=u_M(\beta)u_{Fe}(\alpha)/u_M(\alpha)u_{Fe}(\beta)\approx N_M(\beta)/N_M(\alpha)=k_M \text{ (for low } N_M(\alpha))$$

where $k_M$ is the usual definition of the partition coefficient. The coarsening equation for spherical $\beta$ precipitates was found to be that given by eq. (5.18e), at least after long times,

$$\bar{r}_t^3-\bar{r}_0^3=8D_M(\alpha)\sigma V_m t/[27RT(1-k^*_M)^2u_M(\beta)] \quad (5.18e)$$

The ratio of the coarsening rates for the carbide phase $\beta$ between that in the ternary alloy and that in the binary is then

$$\text{coarsening rate ratio}=D_M(\alpha)/3(k_M-1)^2D_C(\alpha)u_M(\beta)N_C(\alpha)$$

where $N_C(\alpha)$ is the carbon content in the $\alpha$ phase and $D_M(\alpha)$ and $D_C(\alpha)$ are the diffusion coefficients of the two elements in the $\alpha$ phase. That is, coarsening will be significantly slowed for ternary elements with a low diffusion coefficient in the matrix, a large partition coefficient ($k_M \gg 1$) and when there is a high concentration of the ternary element in solution in the matrix.

The coarsening resistance of rod-like $M_2C$ precipitates in $\alpha$ iron, containing a range of metallic elements $m$, was analysed by Lee, Allen and Grunjicic (1991a). The precipitate shape is similar to that of the copper particles in iron, fig. 5.17, with an aspect ratio, $A$, defined as the length to diameter ratio so $A>1$. The increase of the radius of the rods, assuming an equilibrium shape is maintained, was found to be given by eq. (5.18d).

$$\bar{r}_t^3 - \bar{r}_0^3 = 8\sigma V_m Kt/9RT$$

with $K$ being given by

$$K = \left[\sum_{m=M} (k_m - k_{Fe})(k_m - 1)N_m(\alpha) / D_M(\alpha)\right]^{-1} / \ln(2A) \qquad (5.18f)$$

As in the previous analyses the rate of coarsening falls with increasing amounts of slow diffusing metallic elements particularly those that partition most strongly to the carbide phase, $k_m \gg 1$. The influence of multiple elements, $m = i, j, k, \ldots$, is shown by eq. (5.18f) and has the form equivalent to adding electrical resistances in series:

$$1/K = 1/K_i + 1/K_j + 1/K_k + \ldots$$

So multiple element addition will, in general, further retard coarsening even though one element may provide the dominant resistance. Examples of use of this analysis in designing alloys for coarsening resistance have been given by Lee (1990) and Lee, Allen and Grunjicic (1991b).

## 5.6 Experimental investigations of precipitate coarsening

### Kinetic laws of particle coarsening: diffusion or interface controlled kinetics

Early experimental studies on coarsening were reviewed by Greenwood (1969). He reported that precipitates of various shapes: spherical, rod-like or cubic, which could be coherent, partially coherent or fully incoherent, all gave the same kinetic result, in which $\bar{r}^3$ was linear with coarsening time, fig. 5.31. This is the

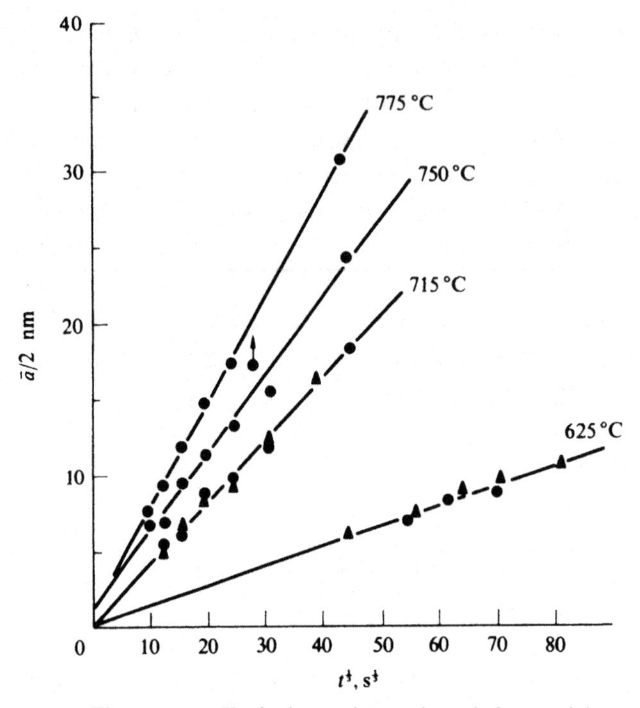


**Figure 5.31** Typical experimental result for particle coarsening showing a linear variation of particle size ($a/2$ for cube shaped particles) against the cube root of time for $\gamma'$ precipitates in Ni–Al alloys: ▲=6.35 wt% Al, ●=6.71 wt% Al. (After Ardell 1969, courtesy of the Metals Society.)

prediction for diffusion controlled coarsening under near steady state, precipitate size distributions. From the slopes of the plots, the product of $D\sigma N_\alpha$ could be obtained (eq. (5.18a)), and if the diffusion coefficient and the equilibrium solubility were known then the interfacial energy $\sigma$ could be determined and compared to the expected value for the microstructure (table 5.2).

## 5.6.1 Coarsening of spherical or near spherical precipitates

In many examples excepting the nickel alloys (§5.6.3), unusual values of the interfacial energy $\sigma$ were initially deduced by this method (that is, unusual with respect to values of $\sigma$ measured by other methods – see table 5.2). For cobalt in copper and copper in iron precipitates the value of $\sigma$ was somewhat larger than expected but for manganese precipitates in magnesium (Smith 1967) it was much smaller than expected. Smith suggested that some measure of interfacial inhibition (with $\bar{r}^2=kt$) might be present, but his experimental plots do not support this idea. However, unless data are both accurate and very extensive, distinction between different power laws based on simple inspection of various $\bar{r}^n$ against $t$ plots is difficult. Some small inhibition of the interfacial process might account for his results, that is, $D/rK$ might have been $\approx 1$, so that both diffusion and interface processes might be contributing to the rate of coarsening.

Other early data that supported the same power law, for diffusion controlled precipitate growth included $Fe_3C$ in $\alpha$ iron at 700 °C (Bannyh, Modin and Modin 1962), $Al_2O_3$ in nickel (Dromsky, Lenel and Ansell 1962, as discussed by Oriani 1964), and $UAl_2$ in $\beta$ uranium (Smith 1965). However, only for the $Fe_3C$ were sufficient data available to allow an estimation of the interfacial energy. To do this Oriani (1964, 1966) modified the analysis to take account of coupled diffusion of the two components. The diffusion coefficient of carbon in $\alpha$ iron at 973 K is large ($9 \times 10^{-11}$ m$^2$ s$^{-1}$), that of iron itself is small ($6 \times 10^{-18}$ m$^2$ s$^{-1}$) but the growth of the precipitate requires that the change in volume of the reaction, $Fe_3C \Leftrightarrow 3Fe + C$, must be accommodated at both the growing and the shrinking precipitates and to do this some flux of iron atoms is required to accompany the carbon flux.

Oriani calculated an effective diffusion coefficient, $D_e = 6 \times 10^{-14}$ m$^2$ s$^{-1}$, and use of this suggested a value of $\sigma$ for the $\alpha$ Fe–$Fe_3C$ interface of about 1.0 J m$^{-2}$, which was about right, if a little high (table 5.2). Li, Blakely and Feingold (1966) adopted a different technique for calculating the coupled diffusion coefficient and as a result further reduced $\sigma$ to 0.7 J m$^{-2}$. If the unmodified diffusion coefficient for carbon in $\alpha$ iron were used, then the calculated coarsening rate would be much larger than the observed rate by a factor of about $10^3$ at 700 °C. These ideas on coupled diffusion were, however, challenged by Bjorklund *et al.* (1972) who suggested that the effective reduction in diffusion coefficient may be caused by the presence of slowly diffusing third elements that dissolve preferentially in $Fe_3C$ (see §5.6.4). Das, Biswas and Ghosh (1993) investigated coarsening in a range of steels with no carbide stabilising element. They reported a significant volume fraction effect, discussed in §5.6.4, and after extrapolation to a eliminate the volume fraction effect, found values of the product $\sigma D_e$ that matched closely the coupled diffusion calculations of Oriani – with $\sigma$ of 0.6 Jm$^{-2}$ and $D_e$ of $5 \times 10^{-14}$ m$^2$ s$^{-1}$ at 973 K. The measured activation energy of the product $\sigma D_e$ was 170 kJ mole$^{-1}$, intermediate between the values for diffusion of carbon, 85 kJ mole$^{-1}$, and self diffusion of iron, 280 kJ mole$^{-1}$, in $\alpha$ iron.

Other early investigators found that the coarsening kinetics are described by the power law expected for diffusion controlled coarsening for precipitates of $Fe_3Si$ in $\alpha$ iron (Bower and Whiteman 1969; Schwartz and Ralph 1969b). There were two early cases reported of the growth kinetics giving the $r^2$ power law expected for interface controlled coarsening. Schwartz and Ralph (1969a) studied, by field ion microscopy (FIM), the coarsening of the very fine precipitates of a few nanometres in size of $V_4C_3$ formed in an iron matrix by the 'interphase' precipitation process, see Berry, Davenport and Honeycombe (1969) and Davenport, Berry and Honeycombe (1968). Interphase precipitation involves the formation of precipitates on the moving interphase boundary as $\gamma$

iron transforms to $\alpha$ iron. The results were clearly in good agreement with $r^2$ kinetics and gave an apparent activation energy of 196 kJ mole$^{-1}$, which is less than that expected for lattice diffusion of the slow-diffusing solute (vanadium) in $\alpha$ iron, and similar to that expected for grain boundary diffusion of the solute in the same phase. The assumption was that activated atom transfer across the precipitate matrix interface should be similar to activated diffusion down a grain boundary. The activation energy determination, however, did not take into account the possibility of the change of solubility of $V_4C_3$ with temperature – this solubility is likely to increase with temperature, and so the corrected activation energy for interface mobility will be even less than 196 kJ mole$^{-1}$.

Youle, Schwartz and Ralph (1971) also used FIM to observe the coarsening of fine gold precipitates in a quenched and aged Fe–Au alloy. They again found the $r^2$ kinetics of interface control and an activation energy for the whole coarsening process that was about half of that expected for diffusion of gold in iron. In both these studies, the very high resolution of FIM enabled the coarsening rates to be studied for very small precipitates (1–40 nm), a condition that favours interfacial control ($d/rK \gg 1$) if $K$ has an appropriate value. But not all FIM observations of coarsening of very fine precipitates in iron have this result. Schwartz and Ralph (1969b) found that in an Fe–Ti–Si system the fine precipitates of $Fe_3Si$ type grew with the kinetics such that $r^3$ was linear with time – the diffusion controlled result. Faulkner and Ralph (1972) also found by FIM diffusion controlled kinetics for the coarsening of fine coherent precipitates in a nickel based alloy, a result duplicated in many studies both by FIM and other techniques discussed below.

The extensive coarsening studies in three systems giving near spherical coherent precipitates, Ni–$Ni_3Al$, Al–$Al_3Li$ and Cu–Co, have been reviewed in detail by Ardell (1988). Studies on these systems used a range of techniques that included FIM supplemented by atom probe measurements of composition, by conventional TEM and high resolution TEM, 'HREM' (see Xiao and Haasen (1991)) and by magnetic measurements. Ardell used the results of this large range of studies to test not only the original LSW model for diffusion controlled coarsening but also the modified equations that predict the change of solubility and the density of precipitates during coarsening, eqs. (5.26) and (5.27), and particularly the various models for the increase of coarsening rate with increased volume fraction of precipitate, eq. (5.18b). Ardell also considered the possible effects of a change of misfit strain, $\delta_m$, that increases from the small value of 0.0008 in Al–$Al_3Li$ to 0.0045 in Ni–$Ni_3Al$ to the higher value of $\approx$0.018 in Cu–Co.

$$\delta_m = 2\,|a_\alpha - a_\beta| / (a_\alpha - a_\beta) \tag{5.29}$$

The large value the misfit parameter, $\delta_m$, for Cu–Co means that during coarsening when the precipitates become larger than about 40 nm in diameter they lost

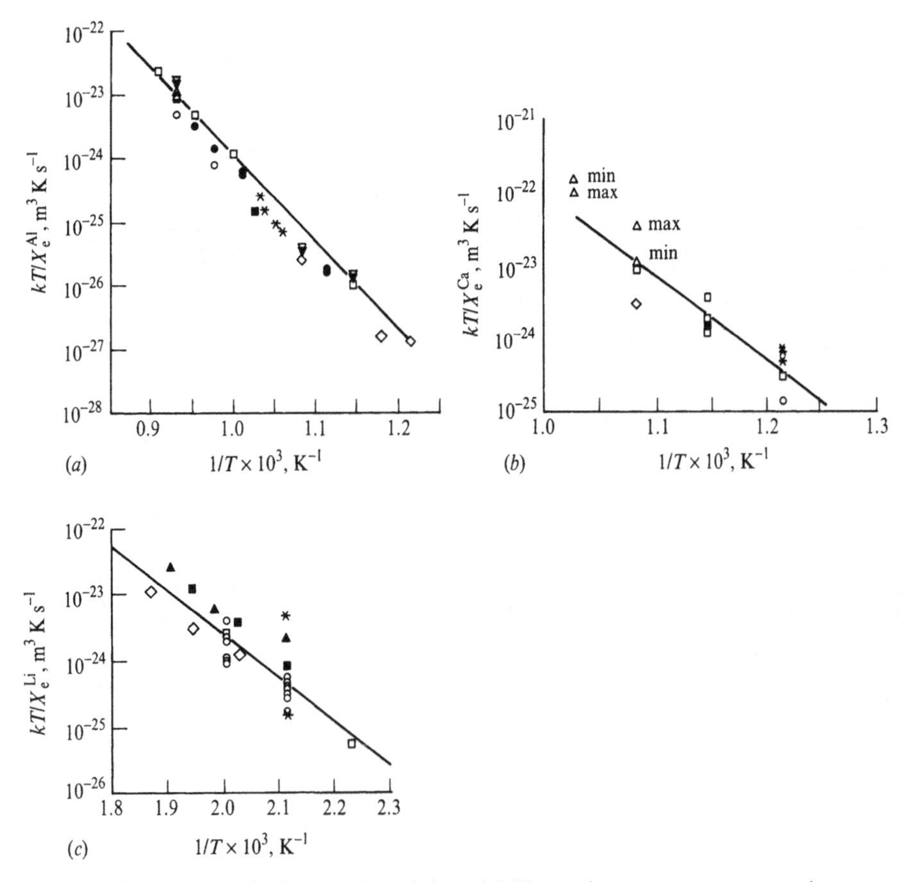


**Figure 5.32** Arrhenius plot of the solubility and temperature corrected coarsening rate constants, $k_1T/N_\alpha(T)$, (from $\bar{r}^3$–$t$-plots) for three alloys: (*a*) $\gamma'(Ni_3Al)$ in Ni–Al, (*b*) Co in Cu–Co and (*c*) $\delta'(Al_3Li)$ in Al–Li. The different symbols refer to different experimental data sets reviewed with the references cited, by Ardell. (After Ardell 1988. Courtesy of the Metals Society.)

coherency. The very low value of the misfit for the Al–$Al_3Li$ system means that elastic stress effects should have a minimal impact in that system. In all three systems the expected $r^3$ relationship with coarsening time was accurately obeyed and the observed temperature dependence of the coarsening parameter, corrected for changes in precipitate solubility (eq. (5.18)) were compatible with diffusion controlled coarsening, fig. 5.32. The activation energies measured were close to those reported for diffusion in each system. The activation energies for the transport process were measured from logarithmic plots of $k_1T/N_\alpha(T)$ against $1/T$. Here $k_1$ is the measured slope of an $\bar{r}^3$ plot against time. The inclusion of the reciprocal of $N_\alpha(T)$ eliminates the complication of the change of solubility with temperature from measurement of the activation energy. The large scatter in the data in these plots for all the systems, especially for Al–$Al_3Li$ at the lowest temperature, should be noted. This large scatter indicates that

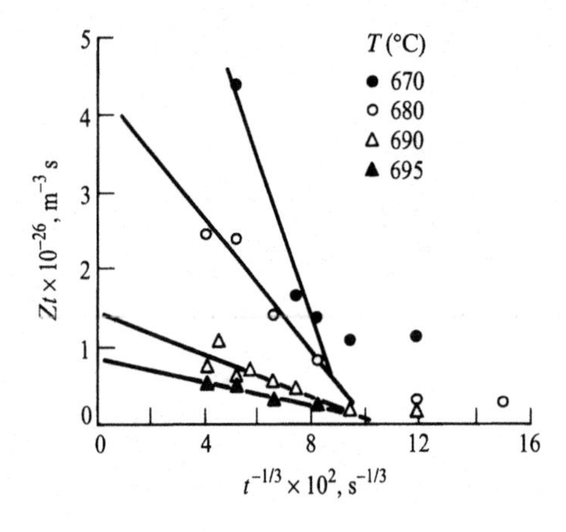


**Figure 5.33** The data of Hirata and Kirkwood (1977) on the density of $\gamma'$ precipitates, $Z$, with ageing time in Ni – 12.3% Al, plotted as the product of $Z$ and $t(Zt)vt^{-1/3}$. (After Ardell, 1988. Courtesy of the Metals Society).

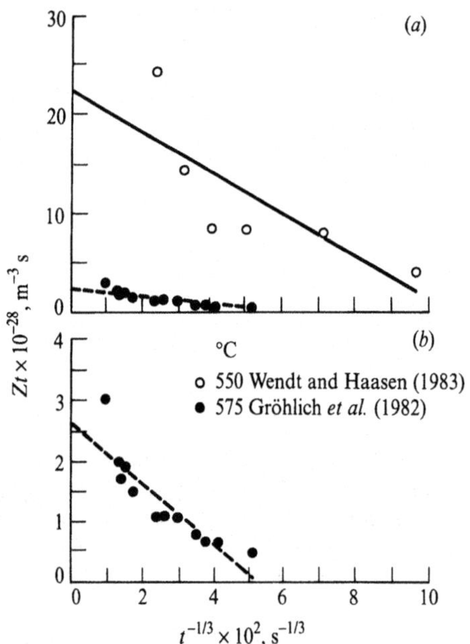


**Figure 5.34** Data of Grohlich *et al.* (1982) in Ni – 15.4% Al and Wendt and Haasen (1983) in Ni – 14% Al plotted as in fig. 5.33. The ordinate scale in (*b*) is expanded with respect to that in (*a*). (After Ardell 1988. Courtesy of the Metals Society.)

simple use of the standard LSW equation will give highly variable values for the interfacial free energy.

Ardell used the data of Hirata and Kirkwood (1977), Grohlich, Haasen and Frommeyer (1982) and Wendt and Haasen (1983) on the change of particle density, $Z$, during precipitation of $Ni_3Al$ from nickel, to test eq. (5.27) under the assumption that the precoarsening time, $t_1$, was $\approx 0$. To test the equation both sides of eq. (5.27) were multiplied by time $t$ so that the product $Z(t)t$ should vary linearly with $t^{-1/3}$, as in eq. (5.27c), see figs. 5.33 and 5.34. The data of Hirata and Kirkwood, at the earliest times, that is, at the largest values of $t^{-1/3}$, did *not* fit a linear plot. Such non-linear behaviour is expected if the values correspond to non-negligable precoarsening times, $t < t_1$ ($t_1$ is the time at which the average solute content falls to the Gibb–Thomson solubility of the average size particle). The overall data shown in figs. 5.33 and 5.34 do, however, provide clear evidence that validates Ardell's equation given here as eqs. (5.27) and (5.27c). Ardell (1988) used eq. (5.27c) to compare the observed slopes and intercepts of plots such as those of figs. 5.32 and 5.33 with the values predicted. In all cases, see table 5.4, the agreement was within half an order of magnitude and in many cases within a factor of 2. Xiao and Haasen (1991) repeated the coarsening studies in the Ni–$Ni_3Al$ system and using the same analysis (eq. (5.27)) also obtained, from the slope and the intercept of a plot of $(t-t_1)Z(t)$ against $(t-t_1)^{1/3}$ (eq. (5.27c)), reasonable values of $\sigma=0.021$ J m$^{-2}$ and $D=1.6\times10^{-21}$ m$^2$ s$^{-1}$.

Ardell (1988) also used a modified form of eq. (5.27) in which $4\pi\bar{r}^3/3$ was multiplied to both sides of (5.27) to yield

$$4\pi\bar{r}^3 Z(t)/3=f_e-\alpha(1-f_e)t^{1/3}/(N_\beta-N_\alpha)k_1^{1/3} \qquad (5.27d)$$

The data from Grohlich *et al.* (1982) and Wendt and Haasen (1983) were tested by Ardell as shown in fig. 5.35 and the results for the intercepts and slopes compared in table 5.5 with the experimental values of the volume fractions of $Ni_3Al$ and the slopes predicted from eq. (5.27d) with very close agreement. This method of plotting avoids the need to use values of the interfacial energy or the diffusion coefficient to calculate the value of the intercepts used in table 5.4.

Ardell (1968) had previously tested the expected variation of solubility in Ni–Al with coarsening by measurement of the change of the magnetic Curie temperature of the nickel rich matrix. This type of study, when combined with

**Table 5.4** *Comparison of the experimental values of the intercepts and slopes of plots of $Z(t)$, t against $t^{-1/3}$ for $Ni_3Al$ in Ni with the values calculated from eq. (5.27) with $t_1\approx0$ (Ardell 1988).*

| | | | | Intercept (m$^{-3}$ s) | | −Slope (m$^{-3}$ s$^{4/3}$) | |
|---|---|---|---|---|---|---|---|
| $T$ (°C) | $k$ (m$^3$ s$^{-1}$) | $f_{\gamma'}$ | $X_e^{Al}$ | Measured | Calculated | Measured | Calculated |
| 670 | $7.75\times10^{-30}$ | 0.031 | 0.119 | $9.72\times10^{26}$ | $9.63\times10^{26}$ | $1.04\times10^{28}$ | $1.58\times10^{28}$ |
| 680 | $9.88\times10^{-30}$ | 0.023 | 0.120 | $4.35\times10^{26}$ | $5.44\times10^{26}$ | $4.29\times10^{27}$ | $1.16\times10^{27}$ |
| 690 | $1.72\times10^{-29}$ | 0.015 | 0.121 | $1.43\times10^{26}$ | $2.02\times10^{26}$ | $1.33\times10^{27}$ | $5.61\times10^{27}$ |
| 695 | $2.98\times10^{-29}$ | 0.010 | 0.122 | $8.68\times10^{25}$ | $8.08\times10^{25}$ | $7.75\times10^{26}$ | $2.73\times10^{27}$ |
| 750 | $1.01\times10^{-28}$ | 0.053 | 0.127 | $6.17\times10^{25}$ | $1.27\times10^{25}$ | $2.68\times10^{26}$ | $5.29\times10^{26}$ |
| 800 | $5.85\times10^{-28}$ | 0.014 | 0.131 | $1.20\times10^{25}$ | $5.71\times10^{25}$ | $5.32\times10^{25}$ | $5.44\times10^{25}$ |
| 550 | $1.71\times10^{-31}$ | 0.266 | 0.107 | $2.20\times10^{29}$ | $3.72\times10^{29}$ | $2.12\times10^{30}$ | $1.79\times10^{30}$ |
| 575 | $2.00\times10^{-31}$ | 0.361 | 0.110 | $2.61\times10^{29}$ | $4.30\times10^{29}$ | $5.08\times10^{30}$ | $1.28\times10^{30}$ |

the studies of the change of mean radius, gave the values of $\sigma$ shown in table 5.2 ($\sigma$=0.014 J m$^{-2}$) and values of $D$ that were within a value of about 2 with independent measurements of the solute diffusion coefficient in Ni–Al. Xiao and Haasen (1991) directly measured the change of volume fraction of $Ni_3Al$ with effective coarsening time $(t-t_1)$, and plotted $f_e-f(t)$ against $(t-t_1)^{-1/3}$ (eq. (5.28b)), see fig. 5.36. From the slope of the $\bar{r}^3$ plot versus time and the slope of fig. 5.36 they also obtained a value of $\sigma$ of 0.014 J m$^{-2}$, a little smaller than the value obtained by them using the particle density method $\sigma$=0.021 J m$^{-2}$, but with an essentially identical value of $D$. The values of $D$ from both methods,

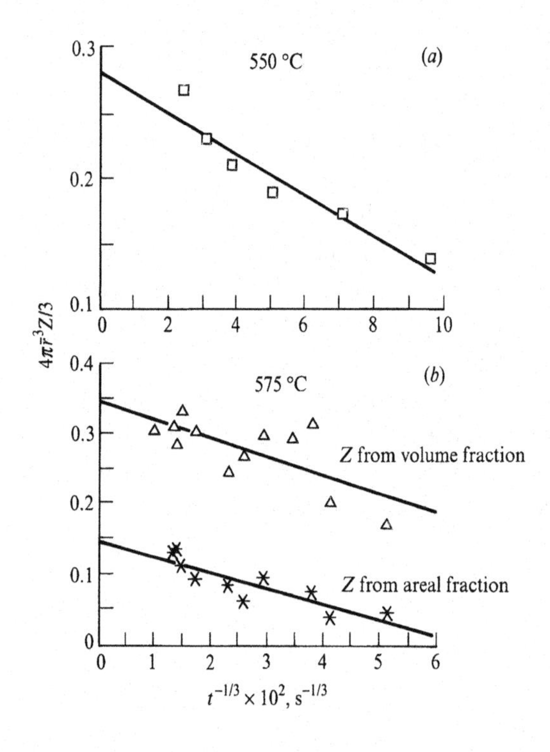


**Figure 5.35** Data of Grohlich *et al.* (1982) in Ni–15.4% Al at 575 °C and Wendt and Haasen (1983) in Ni–14% Al at 550 °C plotted as the volume fraction of precipitates, $f=4\pi\bar{r}^3Z/3$, vs $t^{-1/3}$. The two sets of data in (*b*) came from two different methods used by Grohlich *et al.* to measure $Z$. (After Ardell 1988. Courtesy of the Institute of Materials.)

**Table 5.5** *Comparison of the experimental values of the intercepts and slopes of plots of fig. 5.35 for $Ni_3Al$ in Ni with the values calculated from eq. (5.27d) The intercept $f_e$ is compared with the calculated value $f_{\gamma'}$ and the slope is compared with the calculated slope of $\alpha(1-f_e)/(N_\beta-N_\alpha)k_l^{1/3}$ (Ardell 1988).*

| $T$(°C) | $f_e$ | $f_{\gamma'}$ | −Slope | $\alpha(1-f_e)/\Delta Ck^{1/3}$ |
|---|---|---|---|---|
| 550 | 0.284 | 0.266 | 1.594 | 1.263 |
| 575 | 0.347 | 0.361 | 2.650 | 1.063 |

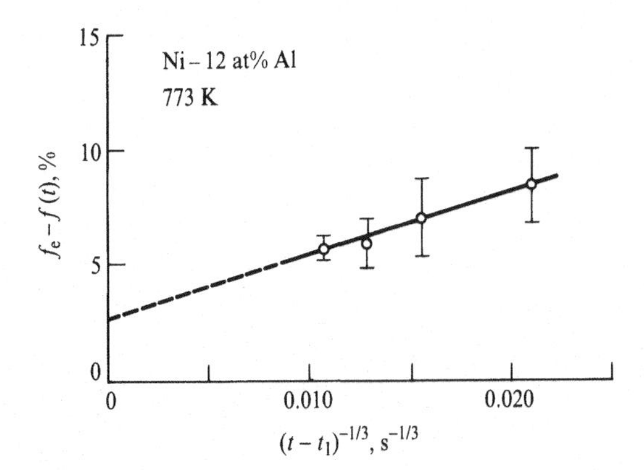


**Figure 5.36** Plot of the measured value of the time dependent volume fraction $f(t)$, as $f_e - f(t)$, against time, as $(t-t_1)^{-1/3}$, showing the expected linear relationship (5.28b) but with an unexplained intercept as $t \to \infty$. ($f_e$ is the lever rule value of the volume fraction and $t_1$ is the time at which the mean solute level loses the supersaturation except that arising from the Gibbs–Thomson effect.) (After Xiao and Haasen 1991. Courtesy of *Acta Met.*)

change of volume fraction and particle density, were four times larger than extrapolated high temperature data. A similar result, with $D$ about twice the expected value, was found for combined magnetic measurements of solute and coarsening studies in Ni–Ti by Ardell (1970). However, for the Ni–Si system (Rastogi and Ardell 1971) the discrepancy between $D$ measured by coarsening and the independent value was much larger (over 20). Use of the relationship from Li *et al.* (1966) for diffusion constrained by volume considerations for the coarsening of these coherent particles reduced the discrepancy in $D$ to a factor of only 2.

This rather satisfactory agreement between theory and experiment for coarsening in these systems with nearly isotropic interfacial energies and thus near spherical shapes is, however, strongly challenged by discrepancies in the volume fraction effect, eq. (5.18d). For cobalt precipitates in copper with a large misfit, a definite volume fraction effect was found in both the increase in rate of coarsening and in the increase of width of the distribution – however, only quite small volume fractions – upto about 5% – have been studied in this system. For the much studied Al–$Al_3Li$ system similar evidence for a clear volume fraction enhancement of coarsening has been found and the results here have explored much larger volume fractions, figs. 5.37 and 5.38. Although there is a large scatter in the data, for any one set of data such as those of Mahalingam, Liedl and Sanders (1987) there is a clear increase of coarsening rate, fig. 5.37, and a clear broadening of the size distribution as the volume fraction of $Al_3Li$ increases. Studies of coarsening of solid particles in liquid matrices allow studies of the volume fraction effect in a situation where no complication due to elastic strain is present. However, the problems of gravity segregation inhibit studies at low volume fractions of solid ($f_e<0.3$) at least in earth gravity. Studies on the

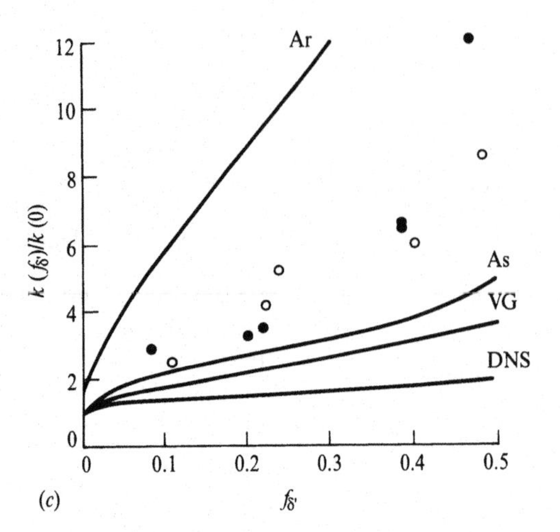


**Figure 5.37** Data of Mahalingam *et al.* (1987) on the normalised coarsening rate constants of $\delta'(Al_3Li)$ in Al–Li at 200 °C (open symbols) and 225 °C (closed symbols) as a function of the $\delta'$ volume fraction, $f_{\delta'}$. The predictions of various models (Ar – Ardell 1972a, As – Asimov 1963, VG – Voorhees and Glicksman 1984a, DNS – Davies *et al.* 1980) are also shown. (After Ardell 1988. Courtesy of the Institute of Materials.)

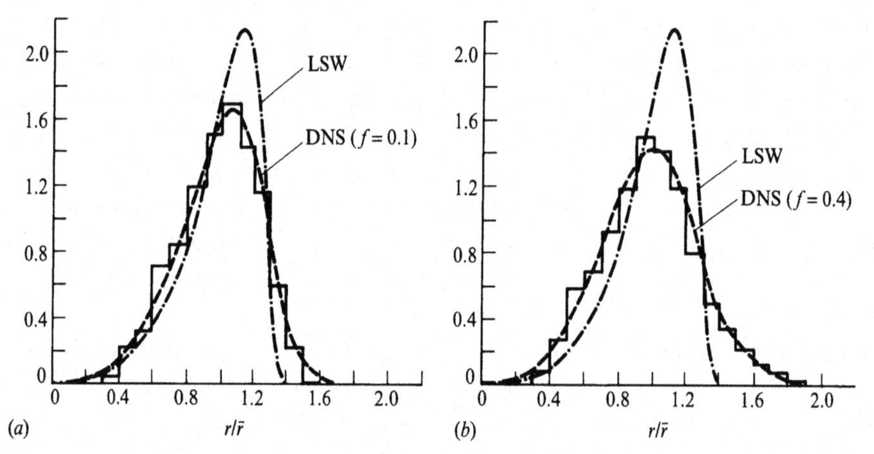


**Figure 5.38** Particle size distributions of the normalized size $u$ ($u=r/\bar{r}$) of the data from Mahalingam *et al.* (1987) for two different volume fractions of $\delta'(Al_3Li)$ in Al–Li coarsened at 200 °C. (*a*) $f$=0.1: Al–8.6% Li, 10 d and (*b*) $f$=0.4: Al–15.6% Li, 48 h. The theoretical distributions of LSW (fig. 5.22) and Davies *et al.* (1980) at the two volume fractions are superimposed. (After Ardell 1988. Courtesy of the Institute of Materials.)

coarsening of iron particles in a copper rich liquid by Courtney, as described by Voorhees and Glicksman (1984a), showed a clear acceleration of coarsening at a rate close to that predicted by the models, particularly that of Voorhees and Glicksman. Fig. 5.39 shows Courtney's data for the observed rate of coarsening at constant temperature as a function of the increasing volume fraction of copper particles. The theoretical curve is that given by Voorhees and Glicksman though with the coarsening rate, at all volume fractions, increased from the theoretical value by 2. This is a not unreasonable correction given the uncertainties

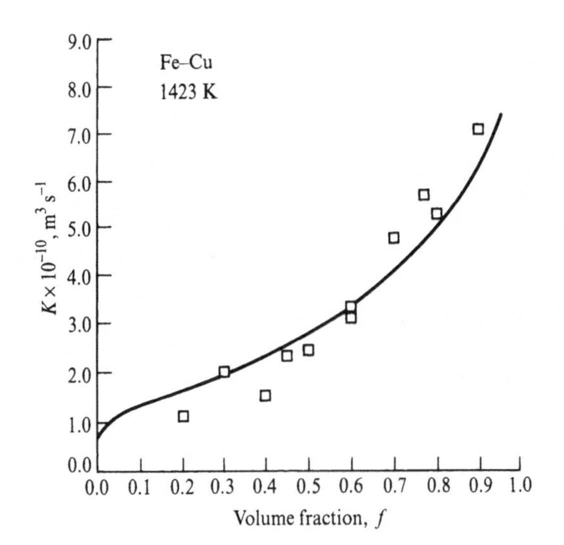


**Figure 5.39**
Experimental data on the coarsening of iron rich solid particles in liquid copper at 1423 K as a function of the volume fraction of the iron particles. The solid line is the theoretical prediction of Voorhees and Glicksman similar to their computer simulated values seen in fig. 2.7. (After Voorhees and Glicksman 1984a.)

in the alloy material constants, $D$, $\sigma$, etc. In this type of liquid matrix coarsening there are, of course, no elastic stresses.

Bender and Ratke (1993) studied the coarsening of cobalt particles in a copper rich liquid and they showed negligible increases of the spread of their experimental particle size distributions, fig. 5.40. Bender and Ratke (1993) mention that the initial sizes after 5 min were 2.5, 3.0 and 3.1 $\mu$m. The theoretical lines in fig. 5.40 are those modified from Voorhees and Glicksman (1984c), previously given as fig. 5.28. It should be noted that the experimental and the theoretical plots are for two-dimensional sections – corresponding to the experimental (scanning electron microscope) metallographic methods used. The theoretical plots show the expected slight increase in the width of the size distribution while the experimental values appear unchanged. In addition, as can be seen from the reported size changes in fig. 5.40 and as shown directly in fig. 5.41 for the three-dimensional radii, there was *no detectable acceleration of the coarsening rate* as the volume fraction increased from 0.3 to 0.5. Although not given by Bender and Ratke (1993), the experimental value of the coarsening rate constant in fig. 5.40 is about 2.2 $\mu m^3\ s^{-1}$ for the three volume fractions. This is about 16 times larger than the expected value for negligible volume fraction (eq. (5.18a)) using the approximate values of 0.2 J $m^{-2}$ for the solid–liquid interfacial energy, $\sigma$, $10^{-5}$ $m^2\ s^{-1}$ for the liquid diffusion coefficient of cobalt in copper and $10^{-5}\ m^3\ mole^{-1}$ for the molar volume of cobalt. Voorhees and Glicksman (1984a) predict increases of about 2.2 and 3.5 of the coarsening rate for the volume fractions of 0.3 and 0.5, compared to that at zero volume fractions. That is, coarsening should have increased by about 60% from $f$=0.3 to $f$=0.5, however, no such increase can be seen in the data of figs. 5.40 and 5.41.

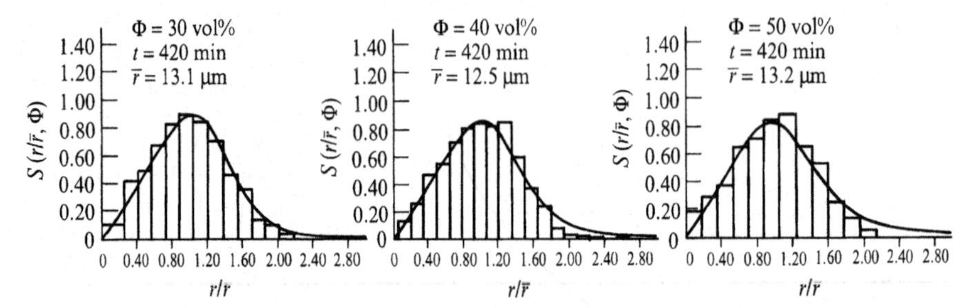


**Figure 5.40** Experimental values (the histograms) of the particle size distributions of solid cobalt particles in liquid copper at different volume fractions. The solid lines are the theoretical curves from Voorhees and Glicksman seen in fig. 5.28. Note the lack of any significant increase in the experimental size distributions and the mean particle size with increased volume fractions. (After Bender and Ratke 1993.)

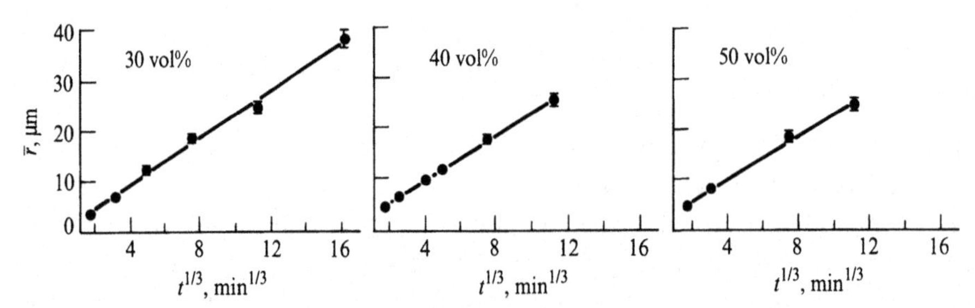


**Figure 5.41** Increase of the mean particle size with the cube root of annealing time for cobalt particles in liquid copper. The coarsening rates did not increase with volume fraction of cobalt – compare with figs. 5.27, 5.37 and 5.39. (After Bender and Ratke 1993. Courtesy of *Scripta Metal. et Mater.*)

Even more striking are the results of Ardell's analysis (Ardell 1988) of the possible volume fraction effect in the much investigated Ni–$Ni_3Al$ system. Fig. 5.42 shows the wide scatter in measured values of the volume fraction parameter, $k(f_v)$, in eq. (5.18b) as a function of the volume fraction of the $\gamma'$ phase, $Ni_3Al$. Although there is a large scatter with many values apparently larger than unity – *for each investigation there was no trend for $k(f_v)$ to increase with the volume fraction, $f_e$*. In addition, the successful analyses for all other aspects of coarsening, reported in tables 5.4 and 5.5, were based on the absence of any volume fraction effect – that is, the LSW values of $k_1$ were used at all volume fractions and these require $k(f_v)/k_0$ to be 1. As Ardell's review clearly indicates there is a very serious problem with the lack of consistent results between different alloy systems for the reality of a volume fraction effect despite the near unanimous theoretical expectation that there should be a significant increase of the coarsening rate as the diffusion distance from shrinking to growing particles falls as

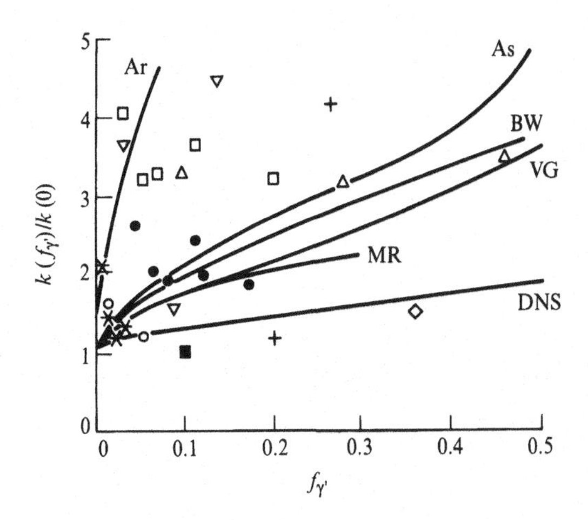


**Figure 5.42** Data from various studies on the normalised coarsening rate constants of $\gamma'(Ni_3Al)$ in Ni–Al (previously seen in fig. 5.32) as a function of the volume fraction of $\gamma'$. Note that for each data set, a constant symbol, there is no increase in coarsening rate with volume fraction. There is, however, a considerable scatter in the measured rates between different experiments. The same symbols were used in fig. 5.32(*a*). The predictions of various theoretical models are also plotted as in fig. 5.37. Here MR is Marquese and Ross (1984). (After Ardell 1988. Courtesy of the Institute of Materials.)

the volume fraction of the coarsening phase rises. Subsequent to Ardell's review, Jayanth and Nash (1990) did find evidence for an increase of coarsening rate with increased volume fraction of precipitate in the Ni–$Ni_3Al$ system. The problem first noted by Asimov (1963) and Ardell (1972a) has not been resolved in the subsequent 30 years of theoretical and experimental study. It is possible that elastic stress effects may have some influence on this problem at least for the case of $Ni_3Al$ and similar precipitates in Ni. Rastogi and Ardell (1971) reported that the observed spread of the particle size distribution increased from the usual LSW form not as the volume fraction increased but as the misfit, $\delta_m$, eq. (5.28), increased.

### Computer simulation of coarsening

Voorhees and Glicksman (1984a,*c*) developed a multi-particle computer model that they used to compare with the original LSW model at negligible volume fraction and to explore the effect of an increasing volume fraction of the dispersed spherical particles. They obtained many interesting and important results. Firstly, they showed that with a negligible volume fraction the system maintained a particle size distribution very close to the LSW distribution of fig. 5.22 – apart from the statistical noise at the limits of the distribution where there are very few particles. They also reproduced, to a very high accuracy, the LSW slope of eq. (5.18a). They found that, as they increased the volume fraction, the particle size distribution widened and they obtained a steadily accelerating rate of coarsening that increased by four times at a volume fraction of 0.5, figs. 5.27

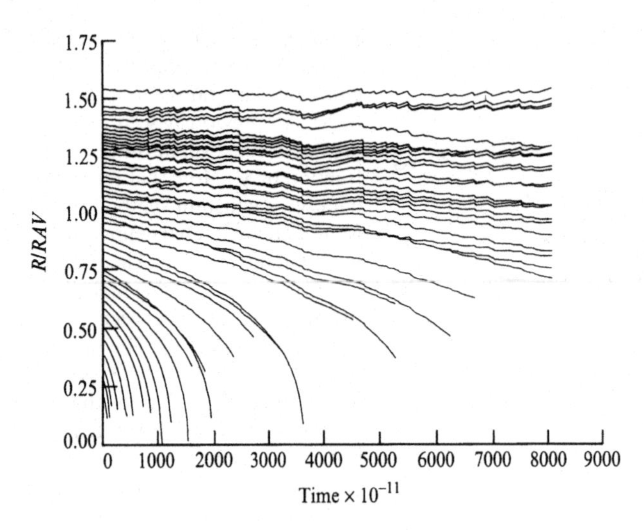


**Figure 5.43** Computer simulated coarsening of a range of particles of different radii, $r$, normalised by the growing mean radius $\bar{r}$ at a volume fraction, $f$, of 0.35. The jagged form of the lines arose from the loss of individual particles. Note the crossing of the lines indicating different coarsening rates of particles of the same size due to the effect of their local environments. (After Voorhees and Glicksman 1984c. Courtesy of *Acta Met.*)

and 5.28. Fig. 5.43 shows the simulated change of the size of individual particles (normalised by the growing mean particle size $\bar{r}$ (*RAV* is the label on the figure)) with coarsening time. The data was obtained after the system had achieved a stationary steady state distribution. As can be seen particles, at or below the mean size, showed an accelerating rate of dissolution – exactly as expected from the simple picture given by fig. 5.21. However, the largest particles showed a steady value of $r/\bar{r}$ (*R*/*RAV*) – indicating that in a multi-particle system the largest particles do, in fact, coarsen at the same rate as the average particles. The constancy of this parameter for the largest particles is what allows the steady state distribution of the LSW model to be maintained. Fig. 5.43 makes clear the important difference between the coarsening rate of a particle of average size, which is on average zero (fig. 5.21) and the rate of growth of the average particle size which is, of coarse, growing due to the steady disappearance of the many small particles. Fig. 5.43 also shows that some of the lines cross – a result indicating that the growth or shrinkage rate of an individual particle depends not only on its relative size ($r/\bar{r}$) but also on its local environment. A particle surrounded by several larger particles will grow slower, or shrink faster, than a particle of the same size whose neighbours are small. The effect is shown more clearly in fig. 5.44 where the flux, $B_i$, to or from each of 2000 particles is plotted. Superimposed on the trend of small particles shrinking ($B_i<0$) and large ones growing ($B_i>0$) is a wide scatter in fluxes for particles of the same size. Voorhees and Glicksman (1984c) also explored the evolution of the particles towards the steady value particle size distribution after extensive coarsening. During transient coarsening with a broadening size distribution they found that the larger particles showed an increase in $r/\bar{r}$ – at least for the earliest times. During this transient period the average particle size actually fell due to many particles shrinking in size before

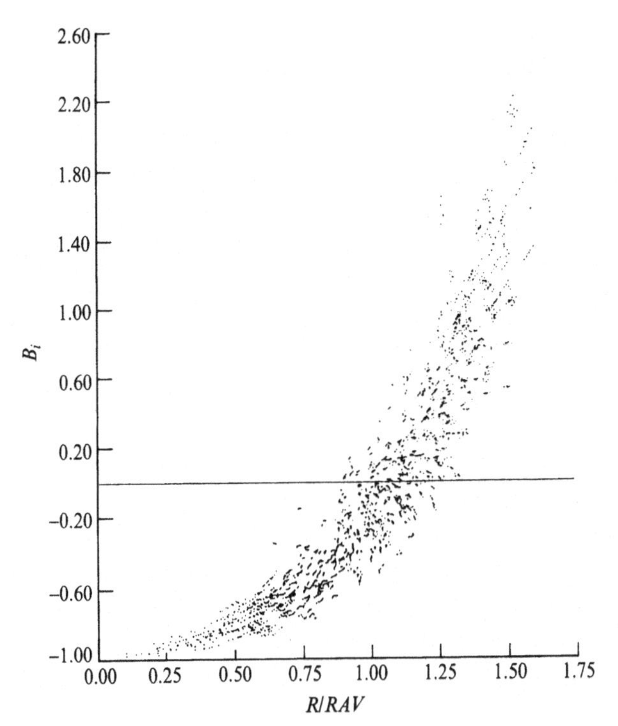


**Figure 5.44** The flux to ($B_i$>0) or from ($B_i$<0) for 2000 particles as a function of reduced particle size, $r/\bar{r}$, a few of these particles were plotted in fig. 5.43. Note the scatter in the value of the flux for particles of the same size – this again arises from the differing local environments of each particle. (After Voorhees and Glicksman 1984c. Courtesy of *Acta Met.*)

any of the smallest particles vanished. This observation is unlikely to be found in experimental precipitate coarsening situations since, as discussed above, interfacial energy driven particle loss begins early in the precipitation process when the average particle size will be growing due to precipitation from the still supersaturated solution. The computer model is able to set up an initially fully precipitated two-phase structure achieved without any prior coarsening during precipitation.

Of major interest to precipitation, however, are the simulations described by Voorhees and Glicksman (1984c) where the initial distributions were made broader than the steady state form. They reported that under these circumstances 'the time interval required for (the) evolution (of the size distribution) to its steady state form was far greater than for distributions narrower than the steady state form.' Brown (1989) also used a simulation model and found that some other size distributions *appeared* stable in the times studied. His ideas were challenged particularly by Hillert *et al.* (1989, 1992a,b). Hillert *et al.* agreed, however, that for Brown's distributions wider than the LSW the rate of change back to the LSW distribution may require very long coarsening times. The computer simulations all seem to indicate that wide particle size distributions relax back rather slowly to the steady state form which gives the LSW coarsening rate. The same result was directly observed in experimental liquid phase coarsening studies by Fang and Patterson (1993) as shown in figs. 5.23 and 5.24. It is thus very likely that in almost all experimental coarsening studies the observations will be influenced by the fact that the precipitate size distribution has not achieved the limiting LSW form. In turn, this means that the kinetics of coarsening will, in general, be somewhat larger than the LSW prediction (eq. (5.18a)) even at negligible volume fractions. It is worth noting that the rate predicted by Greenwood's original (1956) simplified model (eq. (5.18)) might actually turn out to be closer to reality than the apparently mathematically more rigorous model!

### Precipitate movement during coarsening

At non-zero volume fractions of precipitates the fluxes arriving at different parts of the surface of a growing spherical particle can vary. Direct evidence for this was provided by Voorhees and Schaeffer (1987) who observed the coarsening of liquid acetone particles in solid succinonitrile – both materials being transparent. Although some particles were stationary, *many particles were seen to move* relative to each other. Velocities as high as 0.6 $\mu$m h$^{-1}$, averaged over 190 h, for particles of diameters of 100 $\mu$m were seen. That is, many particles moved distances comparable to their diameter. Some particles became slightly ellipsoidal with the long axis in the direction of movement. The authors' analysis of the phenomenon showed that the motion could arise both from coarsening interactions between particles and also from long range solute concentration effects due to non-uniform precipitation. The effect of movement driven by the latter process makes the positions of the particles less clustered. In this study, as in the computer simulation shown in figs. 5.43 and 5.44, particles of the same size were often seen to grow at different rates. Such variable growth rates are due to the different fluxes from adjacent particles. A particle surrounded by many big particles will tend to shrink while a particle of an initially identical size but surrounded by many small particles will tend to grow. A similar effect will also be discussed later for the growth of individual grains in grain growth (§§5.8.2 and 5.8.3).

### Coarsening in ternary systems

Data on coarsening in ternary systems, especially that of cementite in steel, was reviewed by Bjorklund *et al.* (1972) as discussed above. They reported general agreement with their model of coarsening controlled by diffusion of the segregated ternary alloy additions. Bhattacharya and Russell (1972) applied their model to earlier data on oxide coarsening in Cu and Ni and again obtained reasonable agreement. In a subsequent experimental study, Bhattacharya and Russell (1976) investigated the coarsening of $SiO_2$ in Cu as modified by variation in the oxygen activity through control of the furnace atmosphere. The measurements of particle sizes were made on carbon extraction replicas. This allowed the true diameters to be measured at high resolution. The authors found that, despite extensive coarsening in which the mean radius increased by a factor of about 2, the particle size distribution which was much wider than the standard LSW distribution of fig. 5.22, showed only a reduction in the frequency of the small particles in the size distribution. They also found little change in the frequency of particles that were significantly larger than the 1.5 $\bar{r}$ limit of the LSW distribution. Despite the clearly non-stationary particle size distribution, the results fitted the usual $\bar{r}^3$ against time form of eqs. (5.17) and (5.18) but with rate constants 20–30 times faster than expected.

Chen *et al.* (1987, 1990) have studied the coarsening of coherent $Al_3(V,Zr)$ precipitates in melt spun Al–V–Zr alloys. This system is of interest as a potential high temperature aluminium alloy and was discussed by Lee (1990) using eq. (5.18f) which had been developed for ternary alloy coarsening. His model predicts the slowest coarsening for a ternary alloy and qualitatively this was found by Chen *et al.* A further ternary aluminium system that is of interest for strength retention at high temperatures is the Al–Fe–Ce system which has low solubility of the ternary intermetallic phases and also low diffusion coefficients of the alloying element in aluminium. It does appear from these studies that the coarsening rates of binary alloys can be significantly reduced by careful use of ternary solute additions as suggested by Lee (1990). A further example is provided by the analysis and observations of the inhibited coarsening of rod-like $M_2C$ carbides in steel by Lee and Allen (1991) which is discussed below.

## 5.6.2 Coarsening of non-spherical precipitates

In the earliest reported study, that of rod shaped, copper rich precipitates in $\alpha$ iron by Speich and Oriani (1965), see fig. 5.17, an apparently stable equilibrium shape was found and the expected diffusion controlled coarsening rates were found. In an early study of coarsening of the platelet precipitates of $\theta'$ and $\theta''$ in the Al–Cu system and in the same system as modified by a trace of cadmium by Boyd and Nicholson (1971b) values of the interfacial energy were obtained that were much too large. The value for the high energy incoherent rim of the $\theta'$ platelets reached the absurdly high value of 100 J m$^{-2}$. The observed value of the aspect ratio $A \approx 40$ for $\theta'$, fig. 5.18, was also larger than either the theoretically estimated anisotropy of $\sigma$ or the more recently observed values of the equilibrium aspect ratio, $A_{eq}$ (Merle and Merlin 1981; Doherty and Purdy 1982) where $A_{eq}$ appears to be $\approx 17$. It is now widely accepted that the very high values of aspect ratio seen in precipitation of plate-like precipitates in Al–Cu and Al–Ag arise from slow thickening due to an insufficient density of growth ledges on the habit plane of the precipitates (§1.3), see, for example, Ferrante and Doherty (1979), Howe, Aaronson and Gronsky (1985), Enomoto (1987) and Rajab and Doherty (1989). The inhibited thickening reaction during precipitation leads directly to enhanced lengthening and thus large values of $A$ and for any given amount and density of precipitates an increased interfacial area of the precipitate–matrix interface. Such a result has at least two important effects. One result is an increased density of barriers to dislocation motion and thus an enhanced precipitation hardening effect. The other result is that with $A>A_{eq}$, eq. (5.12d) shows that such higher ratios of $A_{eq}/A$ increase the Gibbs–Thomson rise in solubility at the mobile rim of the plate-like precipitates. This rise in solubility leads directly to two effects seen in coarsening. These are:

(i) The relaxation of the shape back to the equilibrium value. This process requires thickening and thus mobility of the good-fit interfaces so there needs to be some method of generating the necessary growth ledges. Studies on the kinetics of $A \rightarrow A_{eq}$ have been given, for example, experimentally by Merle and Merlin (1981) in Al–Cu and Rajab and Doherty (1989) in Al–Ag and using analytical models by Merle and Doherty (1982) and with numerical models by Doherty and Rajab (1989), fig. 5.45. In both sets of experiments, precipitates were formed by ageing and then coarsening *at the same temperature*. It was found by Rajab and Doherty (1989) that the aspect ratio in Al–Ag declined from the high values formed during precipitation but at a rate very much smaller than was predicted on the models that assumed diffusion control. The interpretation offered was that the ledges required for thickening did not form under the low driving forces of coarsening, (Weatherley 1971; Doherty and Rajab 1989). For reasons that are not understood, the observed relaxation towards the equilibrium shape occurred faster in the Al–Cu system than in Al–Ag.

(ii) If a high value of $A_{eq}$ is maintained, with $A \gg A_{eq}$, this can lead to an *acceleration* of the coarsening reaction if measured by the rate of growth

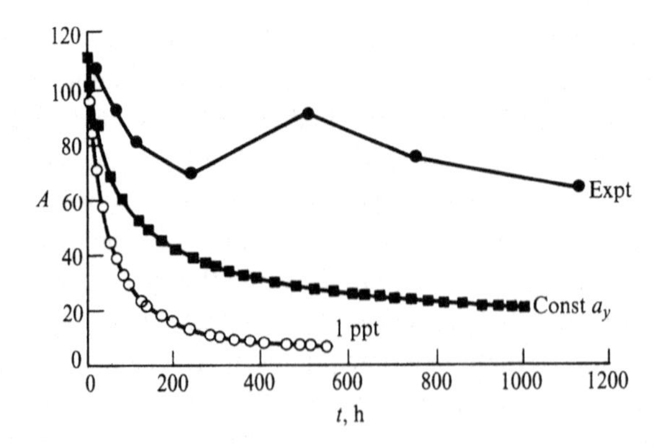


**Figure 5.45** Experimental observations of the mean aspect ratio, $A$, and two numerical simulations as a function of time at 406 °C. In one simulation the precipitate was held at a constant length, $a_y$, as observed experimentally and in the other simulation, '1 ppt', only one precipitate was modelled so its length fell as the thickness grew and $A$ decreased. (After Doherty and Rajab 1989. Courtesy of *Acta Met.*)

of the mean plate length, $\langle \bar{a}_y \rangle$. Such an effect was reported in Al–Ag and analysed in an approximate manner by Ferrante and Doherty (1979), fig. 5.46. In this set of experiments, finely dispersed precipitates were formed by initial precipitation at a low temperature, 200 °C, followed by coarsening at a higher temperature, 410 °C. There was an observed growth of the mean length of the precipitates but with little change in the aspect ratio. A simple model for the accelerated growth of the mean length of the precipitates, with $A \gg A_{eq}$, was developed that appears, as shown by fig. 5.46, to match the observations rather well. The physical origin of the accelerated lengthening is the enhanced Gibbs–Thomson solubility at the rims of precipitates with aspect ratios much larger than the equilibrium value, as described by eq. (5.12).

There are many interesting aspects of coarsening of plate-shaped precipitates that clearly need more detailed experimental and analytical studies in a wider range of systems than have been so far studied. Given that Al–Cu alloys are used in the skin structure of supersonic aircraft where they are exposed to extended periods of time at elevated temperature in service the subject of the stability of plate-like microstructures appears to be not only of scientific but also of real engineering importance.

Lee and Allen (1991) observed, using extraction replica techniques, the coarsening of rods of $M_2C$ formed during the tempering (secondary hardening) of a range of high alloy steels. They observed that for different alloys, nearly constant aspect ratios were found ($A$ from 3 to 8) during extended coarsening indicating that near equilibrium shapes may have been achieved. The results fitted surprisingly well, given the alloy complexity, the expected coarsening condition for

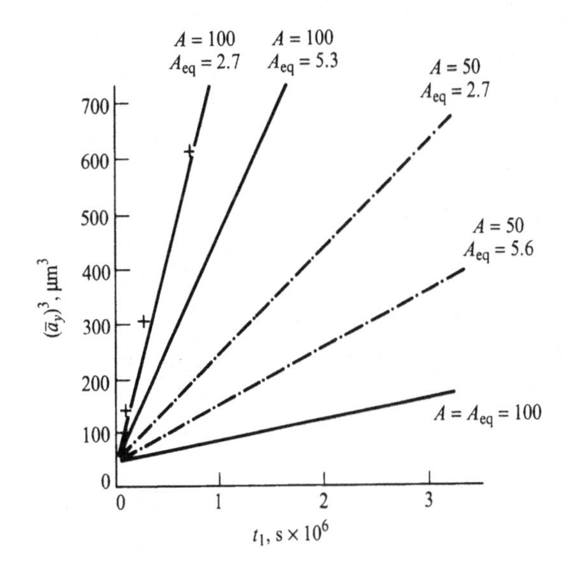


**Figure 5.46**
Experimental values, shown as +, and predicted diffusion limited coarsening for the *lengthening* of plate-like precipitates in Al–Ag. In this experiment, the aspect ratios remained almost constant so the models were also held at the indicated constant aspect ratios for different assumed values of $A_{eq}$. The results indicated accelerated coarsening for the precipitate lengths with $A \gg A_{eq}$. (Ferrante and Doherty 1979. Courtesy of *Acta Met.*)

volume diffusion control as predicted by the model of Lee *et al.* (1991a,b) given as eq. (5.18d). The activation energy for solute transport was found to be close to that of the slowest diffusing element Mo that was strongly partitioning to the $M_2C$ precipitates. The alloy with the slowest coarsening rate was found to maintain its measured hardness most successfully during the extended coarsening anneals at a temperature of 783 K. The agreement between prediction and experiment in this complex system is very satisfactory and it indicates that the multi-element coarsening model is apparently soundly based and should be usable for alloy development purposes.

### 5.6.3 Coarsening kinetics suggesting substructure enhanced diffusion

Early studies of coarsening, for example, by Bannyh *et al.* (1962) and Smith (1965), found evidence at *low temperatures* that the operating power laws during coarsening were

$$\bar{r}^n \approx kt \tag{5.30}$$

where the exponent $n$ was greater than 3. For iron carbide in a quenched and tempered steel, where many low angle boundaries in the matrix are expected from the decomposition of the martensite, Bannyh *et al.* (1962) found that $n$ was approximately 5 at 500 °C, the value for dislocation pipe diffusion rising through $n=4$ (boundary diffusion) to 3 (volume diffusion) at 700 °C. Similarly, for $UAl_2$ in $\alpha$ uranium where the matrix had a subgrain structure of about 3 $\mu$m size, $n$ was about 5 at 590 °C and fell to about 4 at 655 °C and only became 3 at higher temperatures where the matrix had become $\beta$ uranium (Smith 1965). Other early observations that indicate at least some degree of substructural enhancement of diffusion if not its domination were made by Mukherjee, Stumpf, Sellars and McTegart (1969), Stumpf and Sellars (1969) and Mukherjee and Sellers (1969) for carbides in quenched and tempered steels. These studies demonstrated that if the quenched and tempered structure of low angle boundaries linking all the particles (fig. 5.47) were to be eliminated by recrystallisation, the coarsening rate was much reduced (fig. 5.48). Moreover, a creep deformation applied to the recrystallised structure before coarsening to reintroduce a dislocation substructure raised the coarsening rate again. Creep *concurrent* with measured coarsening increased the coarsening rate by about three orders of magnitude (Mukherjee and Sellars 1969).

Doherty and Martin (1964) reported a similar result, namely that low angle boundaries introduced into an Al–Cu alloy by cold work caused a marked increase in the coarsening rate of $\theta(Al_2Cu)$ precipitates in aluminium. In none of these early studies were attempts made to fit the proportionality constants, $k$, to

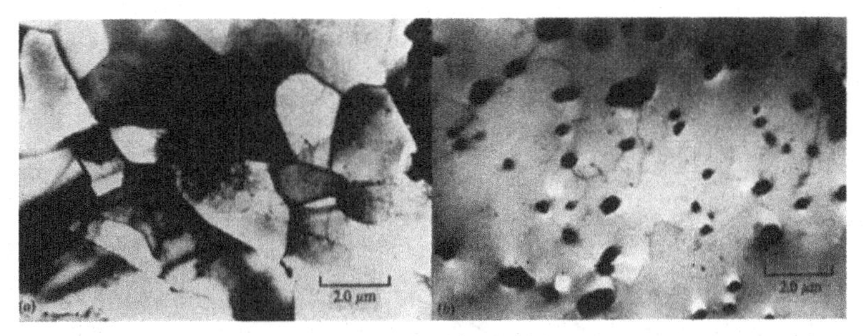

**Figure 5.47** Transmission electron micrographs of 0.21 wt% carbon steel: (*a*) quenched and tempered at 973 K for 10 h, and (*b*) tempered for 1 h and recrystallised by cold swaging 27% and annealing at 973 K for 4 h. The substructure of (*a*) is believed to be responsible for the more rapid coarsening of the carbide particles shown in fig. 5.48. (After Stumpf and Sellars 1969, courtesy of the Metals Society.)

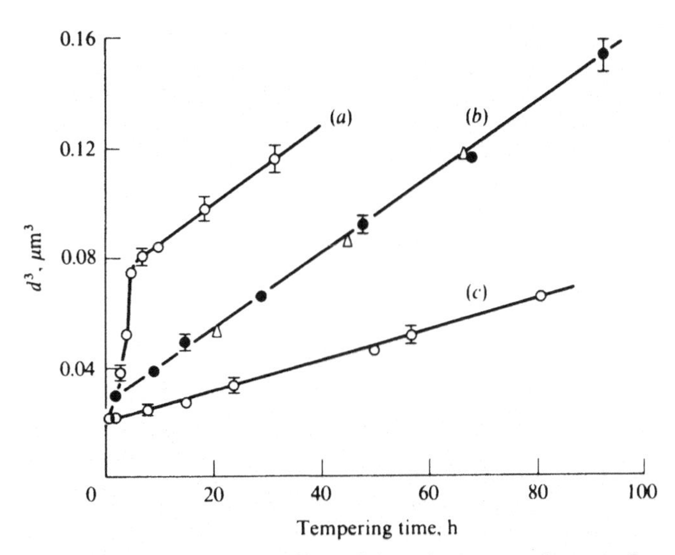


**Figure 5.48** The effect of the substructure shown in fig. 5.47 on the growth of carbides in (*a*) quenched and tempered, (*b*) recrystallised, and (*c*) recrystallised and then deformed 0.21% plain carbon steel by 11% creep strain. The cube of the mean particle diameter is plotted against tempering time at 700 °C. Note the good linear plot for coarsening of the recrystallised material in the absence of substructure. (After Stumpf and Sellars 1969, courtesy of the Metals Society.)

the expected relationship $\bar{r}^n = kt$ when the exponents were greater than 3. Benedek (1974) observed the coarsening of $Fe_2Ti$ precipitates in $\alpha$ Fe in coarse grained materials and again after the same material had been heavily deformed producing a subgrain structure connecting each precipitate, fig. 5.49. The grain boundaries were found by TEM Kikuchi diffraction to be medium angle boundaries of 10–13° misorientation with the angle increasing both with higher rolling strain (70–90% reduction) at the same particle size and spacing, and with

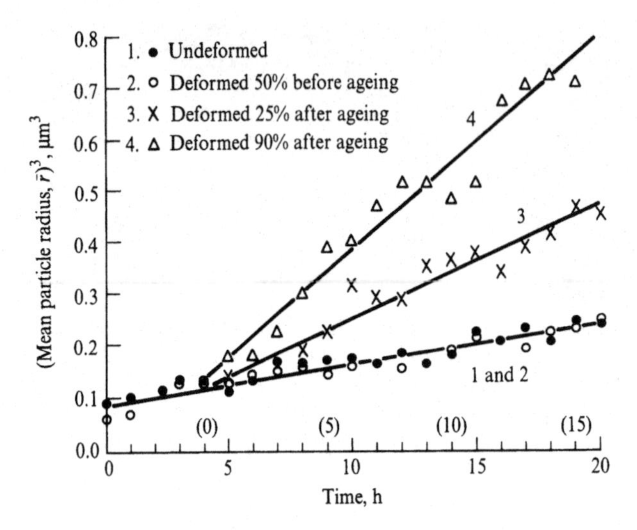


**Figure 5.49** The cube of the mean particle radius in $\mu$m, $\bar{r}^3$, of $Fe_2Ti$ precipitates in an Fe–Ti–Ni alloy as a function of coarsening time, in hours, at 850 °C. Plot 1, open circles, is for the undeformed alloy, plot 2, filled circles, is for an alloy cold rolled 50% before precipitation, plot 3, crosses, is for the undeformed alloy (1) cold rolled 25% after precipitation and holding for 4 h and plot 4, open triangles, is for the undeformed alloy (1) cold rolled 90% after precipitation and holding for 4 h. The numbers in parentheses refer to the times after cold rolling for plots 3 and 4. Clear accelerated coarsening by deformation induced substructure, after precipitation, is seen. (After Benedek 1974.)

increased particle size and spacing at constant reduction. The increased misorientations were also shown to promote superplastic flow, Benedek and Doherty (1974). The coarsening observations were only made over small size increments due to the very slow coarsening of the micrometre sized particles. Benedek (1974) reported that the coarsening was significantly accelerated by about two times by the addition of the substructure after 25% reduction and by four times with 90% reduction – all data measured from an $\bar{r}^3$ annealing time plot, fig. 5.49. More detailed analysis of the data showed that the undeformed material was best fitted to an $\bar{r}^3$ annealing time plot while after 90% deformation the results fitted slightly better to the expected $\bar{r}^4$ annealing time plot. However, in both cases, the experimental coarsening rate constants were about 25 times larger than predicted by the two relevant coarsening equations – eqs. (5.18) and (5.22). It is very likely that these coarsening results like others were significantly modified by the particle size distributions not having achieved the steady state form. It would be good for this type of experiment to be repeated with much finer distributons of stable precipitates to allow the distributions to evolve closer to the steady state values required by theory and to allow more precise measurements. It should also be noted that for the range of data usually obtained, simply plotting the data for different assumed values of the exponent $n$ will not usually distinguish the different results. All the plots will appear nearly linear. It is usually necessary, as was done for the data of Benedek (1974), to test for the least squares fit in a set of plots at different assumed values of the radius exponent $n$ – and to find which value of $n$ gives the highest correlation coefficient.

Higgins, Wiryolukito and Nash (1992) and Yang, Higgins and Nash (1992) observed coarsening of second-phase particles located at grain boundary triple points for two systems: $\alpha/\beta$ brass and Ag–Ni. They found the expected $\bar{r}^4$ annealing time plots (eq. 5.22) but the rates of coarsening were very much faster, by several orders of magnitude, than could be accounted for by conventional coarsening driven by the difference in curvature of the $\alpha/\beta$ interfaces *between different particles*. Differences in solubility between particles give rise to the conventional models of coarsening described in §5.5. Higgins *et al.* (1992) and Yang *et al.* (1992) reanalysed their results in terms of a new model that they developed. This model is of particles being *moved* by the unbalanced grain boundary tensions. The fundamental mechanism analysed by the new model is the movement of a particle, at a triple point, by solute transport *between differently curved surfaces of an individual particle*. The particle movement caused the second-phase $\beta$ particles to grow by particle *coalescence* – that is, by the particles, at the triple points of a shrinking grain, meeting and merging into one single particle. This phenomenon and its analysis are discussed further in §5.8.4 (see figs. 5.85 and 5.86) and again in §5.8.5. The mechanism proposed is likely to be a general and important one for particles in fine grained structures.

## 5.6.4 Some applications of the coarsening theory

In all the different types of coarsening, the parameters of the interfacial energy, $\sigma$, and the solubility, $N_\alpha$, directly control the rate of coarsening, and, for all except the very rare interface controlled situations, the diffusion coefficient $D$ is also vital. We would therefore expect that alloys for high temperature use where coarsening is to be resisted will have low values of one or more of the parameters $\sigma$, $N_\alpha$ and $D$. This expectation is well borne out by various classes of high temperature materials. It is interesting to note that many of the currently used high temperature materials were developed empirically before 1955 when quantitative understanding of precipitate coarsening theory was first developed by Greenwood.

### Low interfacial energy

**Nimonic alloys** As clearly demonstrated, the Ni–$Ni_3X$ systems have exceptionally low values of $\sigma$ ($\approx$0.02 J m$^{-2}$) and therefore would be expected to maintain their fine microstructure at high temperature. This idea is amply confirmed by the success of these materials in high temperature gas turbine applications. The strengthening process is mainly based on the ordered nature of the precipitate requiring dislocations to travel in pairs, but a fine dispersion and a large volume fraction of the ordered $Ni_3X$ phase are required to develop the strength. An interesting development is that optimum creep resistance in this series of alloys has been found at a zero misfit between the two phases (Mirkin and Kancheev 1967). A reduction in the misfit from 0.2% to zero increased the creep-rupture life by 50 times. Since the elastic strains should not greatly affect the coarsening process *directly* – elastic energy being dependent on the volume fraction of the precipitate and independent of the interfacial area (Nicholson 1969) – this is surprising at first but a possible explanation might be that alloys with a large misfit might be expected to acquire interfacial dislocations during creep (Weatherly and Nicholson 1968) and the semicoherent precipitates that result might coarsen much more rapidly (Nicholson 1969). This is discussed in §2.3.4. Davies and Johnston (1970) describe one alloy in which an early onset of creep failure occurred despite otherwise good mechanical strength. This alloy showed a very high thermal stability of the precipitates in the unstressed grip regions but a poor thermal stability in the stressed gauge length – and this alloy had a misfit of 0.2%. They developed a new alloy (Ford 406) in which with a 60% volume fraction of the ordered phase a zero misfit was also achieved. This alloy had very good creep-rupture properties.

**Low solubility oxide dispersed materials** The solubility of oxides such as alumina in metals is exceptionally low and so an alumina dispersion cannot be produced by age hardening. However, other techniques such as powder metallurgy, sometimes combined with internal oxidation (Adachi and Grant 1960), can produce fine dispersions. The low solubility of the oxide then ensures a very considerable resistance to coarsening, and the fine dispersions are indeed maintained together with good mechanical properties to very high temperatures (Gregory and Grant 1954; Bonis and Grant 1962). These alloys resist coarsening and matrix recrystallisation even when considerable substructure has been introduced by cold working, and the mechanical properties reflect a contribution from both the dispersed phase and the substructure. The stability of the whole microstructure is chiefly determined by the resistance to coarsening of the dispersed phase. This principle was originally applied to thoria-dispersed tungsten for light-bulb filaments and to thoria-dispersed nickel (von Heimendahl and Thomas 1964). The coarsening of thoria in nickel based alloys was studied by Footner and Alcock (1972) and was shown to be at the very slow rate predicted by diffusion controlled theory for the highly insoluble oxide. The solubility was only $5\times10^{-2}$ wt% at 1350 °C and the diffusion coefficient of thorium was only $5\times10^{-16}$ m$^2$ s$^{-1}$. The average particle size was still only 60 nm after 200 h at a temperature of $0.94T_m$.

**Low diffusion coefficient** Dispersions of iron carbide in tempered steels coarsen very quickly owing to the high diffusion coefficient of carbon in iron, though as described in §5.6.1, coupled diffusion effects limit the coarsening rate somewhat. However, in a ferrous alloy in which there is a third component which segregates preferentially to the carbide phase (for example, chromium, molybdenum or vanadium) then the coarsening of the carbide requires diffusion of both carbon and also this third element which, being substitutional, has a much lower rate of diffusion. This has been discussed and experimental data reviewed by Bjorklund *et al.* (1972), and it was shown that very considerable reductions in the rate of coarsening are possible by the third element effects. In fact they suggested that the trace impurities present would have been sufficient to render unnecessary the correction for the coupled diffusion effect. The very slow coarsening of thoria in nickel (Footner and Alcock 1972) and of iron and Zr containing precipitates in aluminium discussed earlier in §5.6 arises from the low diffusion coefficients of these solute additions.

Another high temperature alloy based on similar principles is the Al–Fe system which was developed (Towner 1958; Jones 1969; Thursfield *et al.* 1970) for production by high speed splat cooling from the liquid, in which the two metals are completely miscible, to the solid, in which the solubility is extremely limited (§2.5). The resulting very fine dispersion of highly insoluble intermetallic-phase in the aluminium matrix is very resistant to coarsening and excellent medium to high temperature properties are obtained. The rapid solidification approach has been used in many other cases to produce fine dispersions of second-phase particles with very low solubilities and diffusion coefficients. Examples include the Al–Fe–Ce system with incoherent particles and the Al–Zr–V system that has in addition to low solubilities and low diffusion coefficient a coherent interface so giving low interfacial energies as well (Lee 1990; Lee *et al.* 1991a). These more recent alloy development programmes are now being guided by the insights of coarsening theory.

An additional factor arises in steel if the carbide-forming element is present at high concentrations since more stable carbide structures will be produced, which in addition to the advantages due to slower diffusion will also have a lower solubility. For both these reasons the low alloy steels that are used for medium temperature creep resistance invariably have additions of such strong carbide forming elements. Lee *et al.* (1991a) considered the combined effects of lower diffusion coefficients, reduced solubilities and ternary and higher element effects in their analysis of coarsening of $M_2C$ carbides in steel.

### 5.6.5 Systems that resist coarsening

There are several reported cases (Gaudig and Warlimont 1969; Warlimont and Thomas 1970) where a very fine dispersion of a coherent ordered precipitate resists coarsening. This behaviour is shown, for example, in alloys of iron containing 5–9 at% Al in which the ordered $Fe_3Al$ phase has precipitated. Warlimont and Thomas (1970) showed that the particle radius was less than 10 nm, that there was a precipitate density of $10^{23}$ m$^{-3}$ for an alloy with 15 at% Al and that the microstructure did not change detectably even after long annealing times. The electron microscopic observation of fine coherent precipitates which give rise to the so-called apparently stable 'tweed' structure is often reported (Tanner 1966; Fillingham, Leamy and Tanner 1972; Higgins and Wilkes 1972).

Brown, Cook, Ham and Purdy (1973) introduced the idea that microstructural stability might be achieved through elastic interaction between precipitates. They considered coherent elastically strained misfitting $\theta'$ precipitates in Al–Cu having a disc-like shape. They pointed out that the tetragonal distortion of adjacent precipitate discs could be reduced if they were at right angles to each other on different variants of the {100} habit plane. The elastic energy of a structure with two such adjacent discs is a minimum if both precipitates have the same size – a tendency opposed by surface energy which has a *maximum value* if the precipitates have the same size. Using the principle that the elastic energy is independent of precipitate size for a constant volume fraction ($f_v$) while the surface energy becomes less important as the mean precipitate size increases, Brown *et al.* were able to predict that a dispersion of fine disc shaped precipitates resistant to further growth would be achieved once a precipitate radius $r_0$ had reached a limiting size given by:

$$2r_0 = 5\sigma/\mu\delta^2 f_v \tag{5.31}$$

Here $\delta$ is the misfit and $\mu$ the matrix shear modulus. In an fcc structure, where elastic energy favours disc shaped precipitates forming on the cube planes, a regular array of such precipitates on the three {100} planes is then possible and such an array would resist growth. Using appropriate values of the parameters in eq. (5.22), Brown *et al.* were able to predict reasonably correct values of $2r_0$ of 30 nm for Cu–Be and other 'tweed' alloys, 1 $\mu$m for the $\theta'$ plates in Al–Cu and 0.3 $\mu$m for a nickel based alloy containing the plate-like precipitates investigated by Brown *et al.* themselves. The expected decrease of the stable precipitate size with volume fraction was also reported for this alloy. These successes suggested that the elastic energy analysis of Brown *et al.* may be valid. Johnson and Voorhees (1985) have analysed the influence of elastic interactions between *spherical* precipitates on the solute distribution around the particles. They found that significant changes in local concentrations around the particles are possible but no generalised influence on coarsening has so far been identified. However, given the magnitudes of the energies involved some effects appear to be likely, see, for example, Johnson (1987a,b) and Johnson and Voorhees (1987). This is an area in which improved theoretical understanding is clearly needed.

An early example of resistance to coarsening was found by Martin and Humphreys (1974) in Cu–Co, where, soon after the growing spherical precipitates of cobalt become semicoherent, they stopped growing – an effect not predicted by any theory discussed in this section. Livingston's observation (1959) of uninterrupted particle coarsening in this system was for particle sizes smaller than the critical radius (40 nm) found by Martin and Humphreys.

### 5.6.6 General conclusions on particle coarsening

From the preceeding discussion the following points emerge:

(1) In all qualitative respects the current theories of particle coarsening appear to be satisfactory, but there remain quantitative discrepancies in some cases, most notably concerning the effect of volume fraction particularly in the commercially very important system of $Ni_3X$ precipitates in nickel.

(2) In almost every system, diffusion control has been found to operate, exceptions being the case of very fine vanadium carbide precipitates produced by interphase precipitation and fine precipitates of gold in iron. This suggests that most interfaces are sufficiently mobile for diffusion control: $(D/rM'_i) \ll 1$ at least for *spherical* precipitates (where $M'_i$ is the modified interface mobility, see eq. (1.16d)).

(3) Substructure accelerates the coarsening process and can sometimes lead to the predicted higher power laws. It is likely that particle movement caused by the grain boundaries can contribute to accelerated coarsening in many cases especially with fine grain sizes.

(4) For several cases of plate-like precipitates the non-equilbrium shape produced by initial precipitation ($A \gg A_{eq}$) due to immobility of the good-fit interfaces modifies the coarsening reaction. The modifications can involve a rapid change of shape towards the equilibrium shape (as in Al–Cu) or if the out of equilibrium shape is maintained (as in Al–Ag) then this leads to *accelerated coarsening* as measured through the increase in the mean lengths of the precipitates. This acceleration of coarsening despite inhibited mobility arises from an increase in Gibbs–Thomson solubility at the mobile rims of the precipitates when $A \gg A_{eq}$, eq. (5.12d).

(5) For a dispersion to resist coarsening it needs to have low values of one or more of the following parameters: interfacial energy, solubility and diffusion coefficient.

(6) Coarsening is reduced in ternary and higher order alloy precipitates with multiple solutes with different diffusion coefficients.

(7) It has been found that for some alloys with an appreciable elastic misfit a

stable distribution of precipitates can occur that appears to resist coarsening. This phenomenon, however, does not yet appear to be fully understood.

## 5.7 Stability of lamellar or fibrous microstructures

For many years there has been growing interest in both metallic and non-metallic aligned 'fibre' composite materials for strength applications (see, for example, Kelly and Davies (1965) or Kelly (1966)). Amongst the many ways of making metallic fibre composites, that of the unidirectional growth of two-phase microstructures by eutectic growth, has received particular attention (Hertzberg 1965). Early reviews of eutectic growth were provided by Hunt (1968) and Hogan, Kraft and Lemkey (1971) and more recently by Kurz and Fisher (1989). Interest in eutectics for fibre composite applications arises since the two phases are in chemical equilibrium with each other and therefore are not subject to chemical degradation. They also have a strong interfacial bond which allows efficient load transfer between the two phases. In addition, eutectic microstructures have excellent thermal stability (Hertzberg 1965) that is associated with their low energy interfacial structure formed as a result of the eutectic solidification processes (Hogan *et al.* 1971). While some stability of eutectics and eutectoids might be expected, it is not immediately clear how this highly elongated microstructure can resist degradation so successfully as the composite should try to reduce its interfacial area by breaking up towards the equilibrium shape. Eutectic structures are very commonly produced in shaped castings (Flemings 1974) and again here there is much practical interest in the thermal stability of the eutectic microstructure.

### 5.7.1 Rod or fibrous eutectics

Cline (1971) discussed three ways in which a rod-like composite such as the much studied Al–$Al_3Ni$ can degrade. These ways can be described as:

#### (1) Ostwald ripening

Growth of thicker rods can occur at the expense of the thinner ones. Cline (1971) and Ardell (1972d) both showed that the general form of the process is very similar to the Ostwald ripening of precipitates discussed in the previous section. However, the microstructure will be stabilised by the initial, very uniform rod diameter produced by eutectic growth. The time required for the steady state distribution of rod radius to be built up during coarsening may be much longer than that for normal precipitate coarsening since a whole range of particle sizes will be expected at the end of precipitation, see §5.5.8.

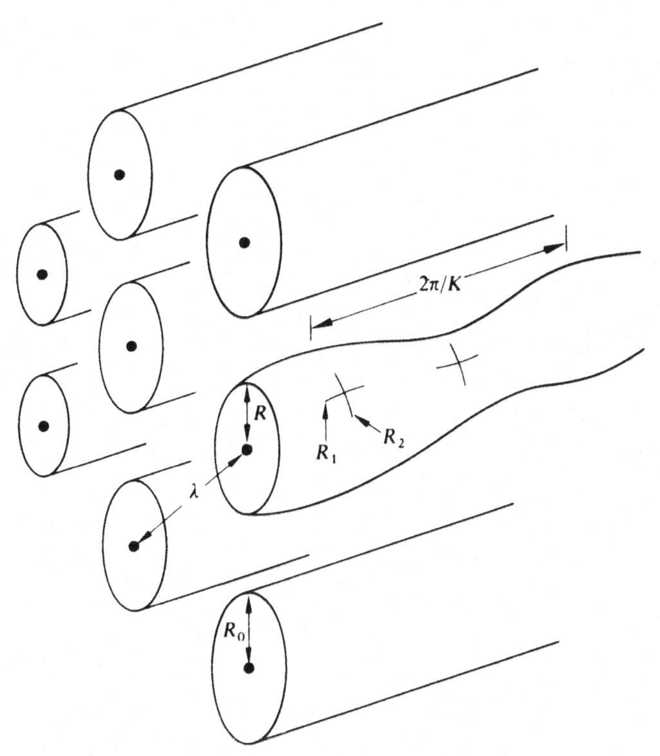


**Figure 5.50** A sinusoidal variation of rod radius $R$ in a single rod in an array of unperturbed rods. The wavelength of the perturbation is $2\pi/K$, $R_1$ and $R_2$ are two perpendicular radii of curvature. (After Cline 1971. Courtesy of *Acta Met.*)

### (2) The Rayleigh instability

Lord Rayleigh (1879) showed theoretically that a cylindrical column of liquid will break down to a row of spherical drops, at intervals of $9R$ along the length of the column ($R$ is the original radius). The origin of this is that the column is unstable with respect to any sinusoidal perturbation of the radius for all wave lengths greater than $2\pi R$. As can be seen in fig. 5.50 at positions of minimum value of the radius, $R_2$, of the cylinder, the total curvature, $1/R_1+1/R_2$, will tend to increase at the minimum radius of $R_2$ since here $1/R_2$ is larger. The instability can, however, only develop at large wavelengths since the other radius $R_1$ becomes *negative* at the minimum in $R_2$, while $R_1$ is *positive* at the maximum in $R_2$. There is a 'saddle shaped' configuration at the developing minimum. For perturbation with long wavelengths the total curvature will indeed be greater at the minimum radius since the increase in curvature from $1/R_2$ offsets the smaller decrease in curvature due to $1/(-R_1)$. Under this increased curvature, material will flow away from the minimum radius to regions where the total curvature is less, that is, where the maximum in $R_2$ is found. The material flow from minimum radius to maximum radius increases the radius differences leading to the break-up of the cylinder into a series of spheres. The wavelength giving the maximum rate of growth of the wavelength is larger, by $\sqrt{2}$, than the minimum wavelength, $2\pi R$, for instability in an isolated cylinder. Such a break-up of a cylinder of fluid into a series of drops can be seen at any water fountain or under any slowly running tap.

Cline showed that the kinetics of this process is complicated in the eutectic situation by virtue of the diffusional interaction between adjacent rods; nevertheless exactly the same result, a break-up of the rods into a series of spheres, is expected. Cline made the important suggestion that cusps in the $\sigma$-plot (producing faceted rods) will stabilise the fibres against the initial perturbation by virtue of the large value of the change of surface energy as the surface plane moves away from that of the cusp orientation (that is, the rate of change of interfacial energy with rotation of the interfacial plane, $d\sigma/d\theta$, is very large close to a cusp). So such cusps in the $\sigma$-plots should tend to prevent Rayleigh instability from operating. Unfortunately, no analysis of this idea was offered by Cline.

#### (3) Fault migration

It is known that the rod phase has faults at which additional rods form, by branching, and existing rods terminate during the initial process of eutectic growth (fig. 5.51). The branches would be expected to fill in (that is, migrate in the growth direction) and the termination to shrink backwards as shown by fig. 5.52. This occurs since the total curvature, $1/R_1+1/R_2$, is a maximum at the termination and a minimum at the branch ($R_2$ is negative). Weatherly and Nakagawa (1971) and Cline (1971) have both produced a theory for the kinetics of this process.

### 5.7.2 Lamellar eutectics

For lamellar eutectics, such as Al–$Al_2Cu$, where the two phases are present as alternating sheets, neither of the first two mechanisms of instability can occur. Both mechanisms are prevented by the absence of any interfacial curvature, so that the chemical potential is everywhere constant (except at the edges of the specimen where the interphase boundaries reach the external surface). Any perturbation on a *flat* interface such as those between the phases in a lamellar eutectic would be expected to decay in the way described by Mullins (1959) for undulations on an external surface of a single crystal. The fault migration mechanism can occur, however. This was originally discussed by Graham and Kraft (1966) for lamelar eutectics, and appears to be the only mechanism for instability in these eutectic structures.

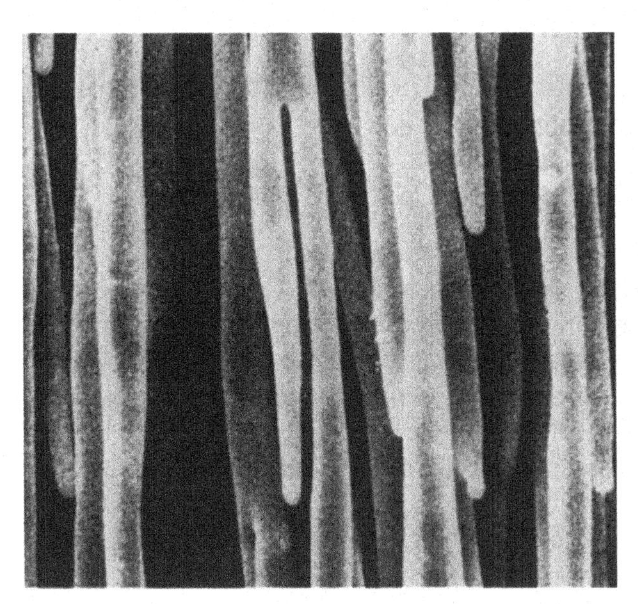

**Figure 5.51** Scanning electron micrographs of a Ni–Al–Cr eutectic in which the matrix has been removed by preferential etching revealing the faulted Cr rods. Both branches and terminations can be seen with all terminations pointing in the direction of growth which was from top to bottom. Magnification ×8600. (After Cline 1971. Courtesy of *Acta Met.*)

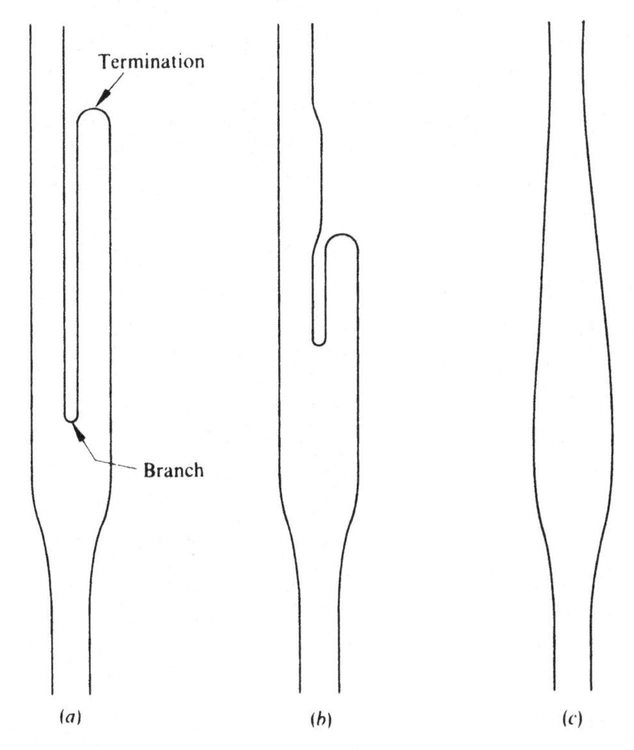


**Figure 5.52** Illustrating the instabilities of faults in a rod eutectic. (*a*) The initial geometry; (*b*) migration of faults towards each other as the termination shrinks and the branch grows; (*c*) annihilation of the two opposite faults probably leaving a bulge. (After Cline 1971. Courtesy of *Acta Met.*)

### 5.7.3 Experimental observations of coarsening of fibrous eutectic microstructures

The rod-like Al–Ni eutectic has been studied by several investigators. Bayles, Ford and Salkind (1967) showed that carefully grown eutectics maintained their morphology and strength after long anneals close to the melting point. They found some coarsening of the rod spacing at the highest temperature, particularly for the morphologies that were initially finer, having been grown from the melt at a faster rate. At the highest temperature there was a 'cross-over' in that samples, initially with the finest spacings, after annealing had the coarsest spacing. The suggestion was made that the stability was partially due to the low energy interface between the aluminium matrix and the intermetallic fibre phase. This idea was challenged by Jaffrey and Chadwick (1970) who grew a curved Al–Ni eutectic ingot in a bent mould and thus destroyed the orientation relationship between the phases, since although the aluminium kept the same orientation – that is, remained a single crystal – the fibre $Al_3Ni$ phase was shown to keep a constant (010) growth axis. This material also resisted coarsening for all interfacial orientations, that is, the stability appears to arise solely from the orientation of the fibre phase. Nakagawa and Weatherly (1972) demonstrated that the faulting mechanism operated for this material and they showed that the results of their own study like those of Bayles *et al.* (1967) could be accounted for on this basis and not by Ostwald ripening. This conclusion was reinforced by a parallel study of the same eutectic grown carefully to avoid faulting. This fault-free material showed no detectable coarsening for the same conditions that had reduced the fibre density in the faulted material by 60%. Smartt, Tu and Courtney (1971) also suggested that fault migration was the origin of the coarsening process.

However, the important role of the interfacial orientation as originally suggested by Bayles *et al.* (1967) seems to gain some support from Nakagawa and Weatherly (1972). They found that in the occasional eutectic grain there was a rapid instability of the Rayleigh type (fig. 5.53) leading to very short rods (not true spheres), but that this behaviour was not typical, in that as previously discussed in most of the grains the rod phase maintained its high length to diameter ratio and coarsened solely by fault migration. Unfortunately, no orientation information was given for the unstable grains, so it is not known what was the significant orientation variable. However, on the basis of Jaffrey and Chadwick's experiment in which the $Al_3Ni$ had a constant fibre axis and did not show the Rayleigh instability but always showed a stable non-circular rod cross-section (that is, a cusped $\sigma$-plot), it is tempting to assume that the $\langle 010 \rangle$ growth of the fibres gives the stability irrespective of the orientation of the aluminium matrix (the rod–matrix interface contains the rod growth axis).

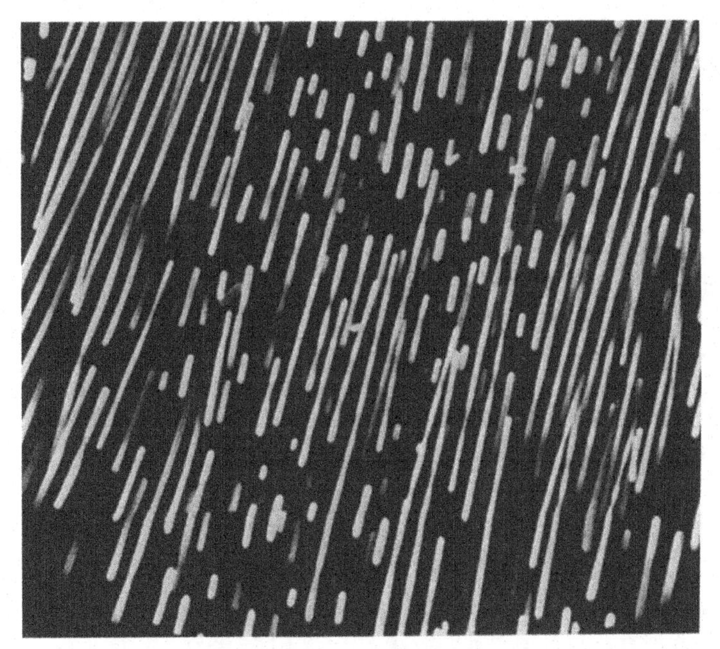

**Figure 5.53** Scanning electron micrograph of unidirectionally solidified $NiAl_3$–Al eutectic annealed at 998 K for 2 h showing the break-up of $NiAl_3$ rods by longitudinal perturbations. This behaviour occurred only in occasional grains – for the majority of the grains this type of instability was not seen. Magnification ×4900. (After Nakagawa and Weatherly 1972. Courtesy of *Acta Met.*)

Cline's suggestion that the Rayleigh instability is prevented by faceting seems to gain some measure of support from these observations. It is perhaps significant that the Rayleigh instability was found by Marich (1971) in the Fe–FeS system in which there is no orientation relationship between the phases and no evidence for faceting. That is, the $\sigma$-plot is there unlikely to have any cusps.

## 5.7.4 Observations of coarsening of lamellar eutectics

Graham and Kraft (1966), who investigated the lamellar Al–$Al_2Cu$ eutectic, found that in regions with a well-developed lamellar morphology the structure was very stable, but that all non-lamellar regions coarsened rapidly. Their microstructural observations (fig. 5.54) suggest that as expected, coarsening only occurred at lamellar faults. This was supported by their kinetic results which appeared to agree with their theory developed for coarsening by diffusion limited fault migration.

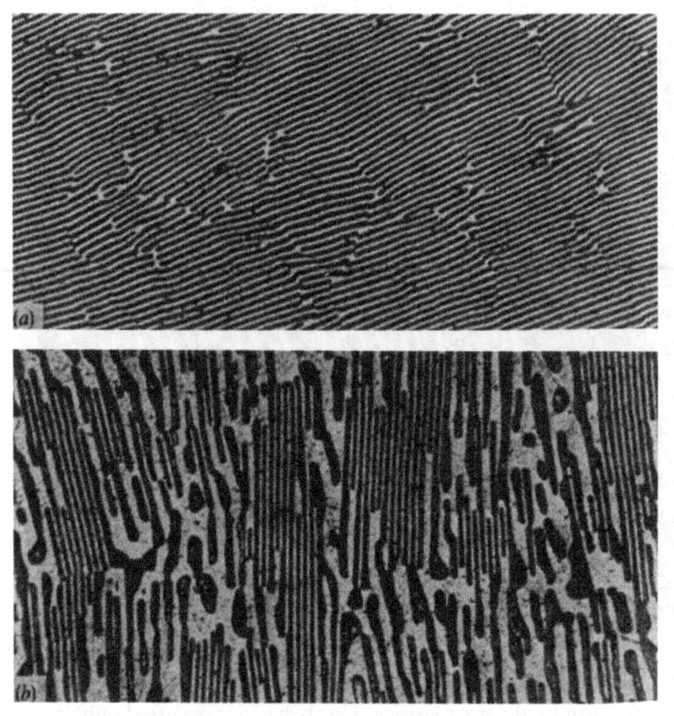

**Figure 5.54** Transverse sections of the lamellar eutectic of Al–Cu following unidirectional solidification and long anneals: (*a*) at 793 K for 24 h showing faults and well-aligned lamellae; (*b*) at 793 K for 1200 h showing unchanged regions of fault-free material surrounded by regions of coarsened and degraded lamellae. Magnification ×3200. (After Graham and Kraft 1966. Courtesy of AIME.)

## 5.7.5 Effect of stress and strain on the stability of eutectic microstructures

The creep strength of aligned eutectics was investigated by Salkind, Leverant and George (1967) for Al–$Al_3Ni$ and the expected resistance to coarsening was found in contradiction to an earlier report by Cratchley (1965). At medium temperatures (625 K) there was no spheroidisation even of the ends of the broken fibres close to the fracture surface, but at the higher temperature of 823 K there was some spheroidisation of the broken fibre ends. However, this instability followed failure and is not likely to have been a cause of it. Many other investigations listed by Soutiere and Kerr (1969) also show promising results for the stability of the structure during high temperature creep. Extensive *prior* plastic deformation, however, has been shown to be a potent cause of subsequent thermal coarsening. Salking *et al.* (1969) cold-rolled the Al–$Al_3Ni$ eutectic, and after deformations greater than 30–40% which induced considerable 'waviness' of the fibres, there was then a rapid spheroidisation of the structure and the micrographs suggest the Rayleigh instability.

Soutiere and Kerr (1969) compressed lamellar Cu–Zn eutectic samples normal to their growth direction, producing kinked eutectic deformation bands. Subsequent annealing caused rapid spheroidisation and coarsening at these deformation bands. Two possible explanations were offered. One was that the kinking, possibly followed by recrystallisation, destroyed the orientation relationship between the phases and thus raised the interfacial free energy (also this process might move the interface away from a 'cusp' orientation with its stabilising 'torque' term, $d\sigma/d\theta$). Alternatively, the kinking could have caused fracture of the lamellae, thereby promoting the migrating fault mechanism. However, since the system appeared initially to have many faults, the change of orientation is most likely to be the factor causing the major accleration of coarsening.

Similar results have been found also in the lamellar Al–$Al_2Cu$ eutectic by Butcher, Weatherley and Petit (1969) and in the lamellar pearlite structure in Fe–C eutectoid by Chojnowski and McTegart (1968). After hot tensile strain in the Al–Cu system, local spheroidisation of the structure was found associated with low angle polygonisation boundaries (fig. 5.55) which by formation of boundary grooves in the Al–$Al_2Cu$ interface (fig. 5.2(*b*)) caused the curvature which could lead to spheroidisation. In the steel structure the rate of spheroidisation in undeformed samples at 700 °C was increased by a factor of 100 by cold work followed by annealing at 700 °C, and by a factor of $10^4$ by hot torsion at 700 °C. Spheroidisation of the eutectoid material was clearly heterogeneous, fig. 5.56. In both the last two studies the suggestion was made that accelerated diffusion in the low angle boundaries was an important contribution to the observed spheroidisation, but no direct evidence for this was offered.

The overall effect of deformation appears to be that strain, particularly strain at high temperatures, causes the stable aligned structure to spheroidise. Which of the various possibilities is the most important is not clear, but the candidates include the loss of the low energy orientation relationship, the polygonisation induced interface grooving observed by Butcher *et al.* (1969), and the fracture of one of the phases.

In summary it appears that for most eutectoids and eutectics with a low energy, possibly faceted interface, the microstructure can only coarsen at morphological faults. Carefully grown material with few such faults would appear therefore to be very stable indeed. The resistance of the faceted interface to the Rayleigh instability is probably due, as suggested by Cline (1971), to the anisotropy of $\sigma$ which makes it reluctant to *rotate* away from the cusp orientation, but alternatively, as initially suggested by Aaronson (1962), such an interface may *migrate* only with difficulty due to the need for growth ledges.

There has been little quantitative testing of the qualitative ideas so far discussed. The expected form of the kinetics of spheroidisation appears to have been found in some cases (Graham and Kraft 1966; Nakagawa and Weatherly 1972) but unlike the situation for Ostwald ripening of precipitates, no exact

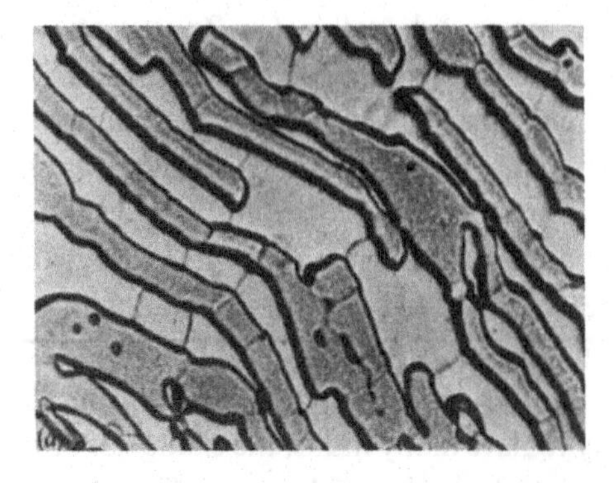

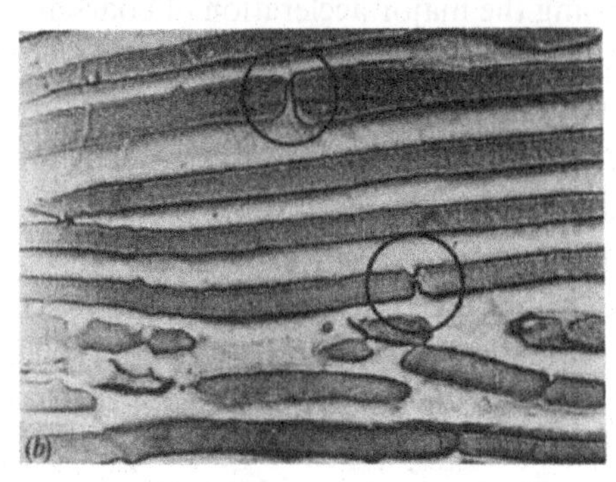

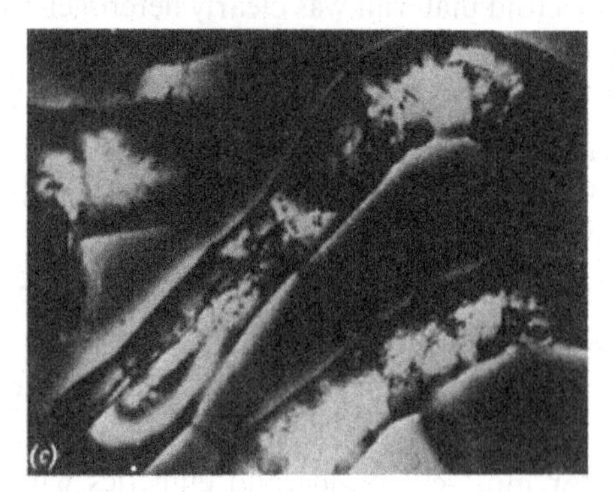

**Figure 5.55**
Unidirectionally solidified Al–$Al_2Cu$ eutectic following high temperature tensile deformation. (*a*) Deformed at 777 K showing polygonisation boundaries in both phases. Magnification ×1000. (*b*) Later stages of spheroidisation. Magnification ×1000. (*c*) Transmission micrograph of material deformed at 573 K showing subgrain boundaries in $Al_2Cu$ giving rise to grooving of the phase boundaries. Magnification ×20 000. (After Butcher *et al.* 1969. Courtesy of the Metals Society.)

experimental testing of the theories, for example, prediction of the interfacial energy, has yet been made, nor is there yet any hard experimental or theoretical support for the plausible suggestion that cusps in the $\sigma$-plot may stabilise the interface against perturbation.

## 5.7.6 Discontinuous coarsening of polycrystalline lamellar structures

An alternative method by which lamellar microstructures may coarsen was described and discussed by Livingston and Cahn (1974). This is shown in fig. 5.57 for the Co–Si eutectoid structure sectioned transverse to the original growth direc-

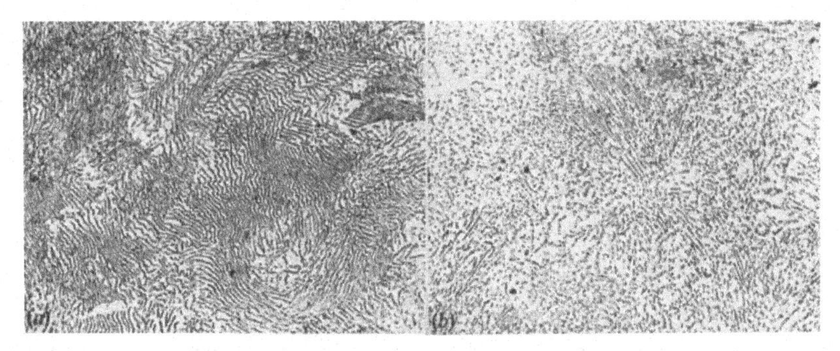

**Figure 5.56** Accelerated spheroidisation of eutectoid lamellae in Fe – 0.8 wt% C following cold deformation. (*a*) The distorted lamellae after 50% compression at room temperature. (*b*) The same material but after 30 min annealing at 973 K, showing the heterogeneous nature of spheroidisation of the deformed material with the unchanged regions those with the smallest distortions. Magnification ×1000. (After Chojnowski and McTegart 1968. Courtesy of the Metals Society.)

tion – the process of coarsening occurs only at migrating grain boundaries with the coarsening occurring in the growing grain and providing the driving force for the reaction. To this extent it is very similar to discontinuous precipitation, and has been named *discontinuous coarsening*. The motion of the grain boundary was *into* the grain whose lamellae were most nearly normal to the boundary plane as can be seen in fig. 5.57. Livingston and Cahn analysed the process on the basis that the driving force is the reduction of interfacial energy with the principal solute path being diffusion along the migrating grain boundary. Their results were in good agreement with the observations for the three eutectoid systems, Co–Si, Cu–In and Ni–In, that they examined. An interesting and important kinetic result was that the coarsening ratio, $r'$, of the interlamellar spacing after coarsening to that before coarsening was large (5–20), this ratio is much greater than that expected for growth at either the maximum possible velocity or growth at the maximum possible rate of entropy production. The two postulates are those frequently made for many types of phase transformation and are apparently invalidated by the results of discontinuous coarsening. From the point of view of microstructural stability, the existence of this discontinuous mode of coarsening of aligned lamellar structures implies that the full resistance to degradation of these types of structures can only be achieved by the use of 'single crystal' eutectics (that is, two interpenetrating single crystals) for high temperature use.

## 5.8 Microstructural change due to grain boundary energies

### 5.8.1 Normal grain growth

The microstructural feature that is present in almost all metals is that of their grain boundaries – the interfaces between crystals of the same composition

**Figure 5.57** Transverse section of unidirectionally grown polycrystalline eutectoid Co–Si showing discontinuous coarsening after annealing at 1273 K for 96 h. The migrating grain boundaries responsible for the instability are only revealed in the grey $Co_2Si$ phase. The lower grain has invaded the upper grain on the left and the upper grain has invaded the lower grain on the right – in both bases the invaded grain has its lamellae most nearly normal to the grain boundary. Note the large ratio ($r'$) of lamellae spacing after growth to that before. (After Livingston and Cahn 1974. Courtesy of *Acta Met.*)

differing only in orientation. These boundaries have energies (table 5.1) that are usually about one-third of the surface energy, but for both low angle boundaries and some coincidence site boundaries (twin boundaries) the energies will be much less (Byrne 1965; Shewmon 1966; Hondros 1969). For some boundaries, such as the twin boundary between two fcc crystals with an orientation relationship of 60° rotation about a common (111) axis, it is well known that there is a considerable anisotropy of boundary energy with the orientation of the boundary plane. That is, the 'incoherent' twin boundaries have energies much greater than the 'coherent' boundaries (Shewmon 1966). Also, Bishop, Hartt and Bruggeman (1971) have observed 'facetting' of an initially smooth grain boundary plane in zinc bicrystals – this phenomenon, as in the earlier discussion (§5.4.2), arises from anisotropy in the $\sigma_b$-plot, that is, with variation of the energy as the boundary plane varies but at a fixed misorientation between the crystals. Despite these known anisotropies in energy with both crystal misorientation and variation in orientation in the grain boundary plane, there has been no published theory that attempts to deal with the effect of variation of $\sigma_b$ on grain growth. We shall therefore follow the conventional approach (Beck, Kramar, Demer and Holzworth 1948; Byrne 1965; Higgins 1974) and discuss grain growth without considering this complication. That is, the assumption is that all high angle grain boundaries have equal energies. As has already been shown in §1.2, the total driving force for grain growth is much smaller (100 times) than that for grain boundary migration in the recrystallisation of cold-worked metal. So, grain growth will occur more slowly, or at higher temperatures, than recrystallisation and will be much more retarded by solutes and dispersed second phases. Atkinson (1988) has provided a very detailed review of the various models for grain growth.

In this section the process of so-called 'normal' grain growth will be discussed. In normal grain growth, as in the models for particle coarsening, the microstructure grows with a near steady state distribution of, here, the grain sizes. This is a further example of a 'self similar' microstructure – that is, a microstructure that when observed at any stage it is essentially unchanged – apart, of course, from an increase in the length scale. The microstructure, after different periods of growth, will appear unchanged – provided that the magnification of the viewing microscope is appropriately reduced to give the same apparent mean grain intercept. In the case of normal grain growth, experimental observations (Feltham 1957; Rhines and Patterson 1982), theoretical analyses (Kurtz and Carpay 1980) and computer simulation (Anderson, Grest and Srolovitz 1989a) all suggest that the grain size distribution is approximately log-normal. A log-normal distribution is one with a normal (Gaussian) distribution, when the grain frequency is plotted against the *natural logarithm* of the appropriate measured grain dimension. This dimension can be the linear grain intercept, the grain area or the grain volume. The linear intercept and the apparent grain area are usually measured on two-dimensional sections through the three-dimensional structure while the true grain volume requires some type of three-dimensional viewing. The three-dimensional analysis is readily accomplished in computer models but in actual experiments either serial sectioning (Rhines and Craig 1974) or grain disintegration using liquid metal wetting of the grain boundaries is required (Williams and Smith 1952; Rhines and Patterson 1982).

An alternative type of grain growth, described either as 'abnormal' grain growth or as 'secondary' recrystallisation, involves the growth of a few grains into a matrix of non-growing or very slowly growing 'matrix' grains. In this process the grain size distribution becomes bimodal – at least while some of the slowly growing matrix grains still survive. Abnormal grain growth is discussed in §5.8.4.

Grain growth is most simply analysed for a thin sheet material in which the sheet thickness is much less than the average grain diameter, that is, all the grain boundaries are normal to the sheet surface. If the triple point angles have the equilibrium angle of 120° then only six-sided grains have straight sides, all others will be curved (fig. 5.58). A grain with less than six sides will 'bulge' outwards, or more accurately the centre of curvature of its boundaries will lie on the small grain side of the boundary. A grain with more than six sides will bulge inwards with the centres of curvature of its boundaries outside the big grain. For a

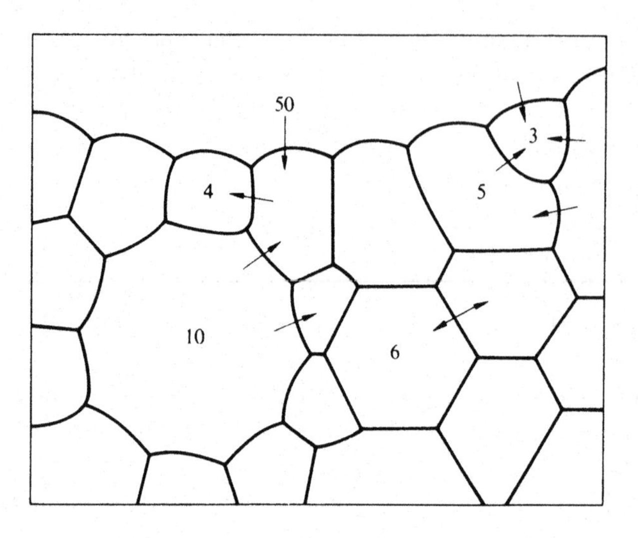


**Figure 5.58** Grain shape equilibrium in a two-dimensional sheet. Since the triple points have angles of 120°, grains with less than six sides have a positive radius of curvature and will shrink, those with six sides are stable, and those with more than six sides have a negative radius and will grow. The numbers give the number of grain sides and the arrows indicate the expected direction of boundary migration. (After Coble and Burke 1963. Courtesy of *Progr. Ceramic Sci.*)

particular distribution of grain size, the mean radius of curvature $\bar{R}_c$ of all boundaries will be $\beta\bar{D}$ where $\bar{D}$ is the mean grain diameter and $\beta$ is dependent on the dispersion in grain size (or rather the dispersion in number of neighbours). With a constant distribution in grain sizes and shapes $\beta$ should be constant. The symbol $R$ will be used for the radius of grains keeping the symbol $r$ for the radius of any second-phase particles present, see §5.8.4.

Across a curved boundary there is a difference in chemical potential per atom, $\Delta\mu$, given by

$$\Delta\mu=\sigma_b V_a/R_c=\sigma_b V_a/\beta\bar{D} \tag{5.10b}$$

The factor 2 has gone because in a sheet specimen the surfaces are cylindrical, not spherical, in shape. $V_a$ is the volume per atom, whereas in eq. (5.10a) $V_m$ was the molar volume. In three dimensions:

$$\Delta\mu=2\sigma_b V_a/R_c=\sigma_b V_a/\beta\bar{D} \tag{5.10c}$$

The migration velocity $v$ should be proportional to this driving force (eq. (4.52), that is, exhibiting a constant mobility, unless the driving force is in the transition region between the two limiting regions of boundary mobility as affected by solute (see fig. 4.42).

$$v=M\Delta\mu$$

Therefore

$$v = M\sigma_b V_a / \beta D \approx d\bar{D}/dt \qquad \text{(two-dimensional case)} \qquad (5.32)$$

or

$$v = 2M\sigma_b V_a / \beta D \approx d\bar{D}/dt \qquad \text{(three-dimensional case)} \qquad (5.32a)$$

Integration of eq. (5.32) from time=0, when $\langle D \rangle$ is $\langle D_0 \rangle$, to a time $t$, when $\langle D \rangle$ is $\langle D_t \rangle$, gives:

$$\bar{D}_t^{\,2} - \bar{D}_0^{\,2} = 2M\sigma_b V_a t/\beta \qquad \text{(two-dimensional case)} \qquad (5.33)$$

$$\bar{D}_t^{\,2} - \bar{D}_0^{\,2} = 4M\sigma_b V_a t/\beta \qquad \text{(three-dimensional case)} \qquad (5.33a)$$

Note that, as might have been expected, these kinetic equations have the same form (eq. (5.19)) as that of interface controlled coarsening of precipitates. The only rate limiting step in grain growth is atomic transport across the interface – here the grain boundary. The identification of grain growth with precipitate coarsening was extended by Hillert (1965) who used the idea to derive the steady state distribution of grain sizes during grain growth; this distribution was, however, not the log-normal form expected for grain growth.

Hillert also derived a similar expression to eq. (5.33) for two-dimensional grain growth using a topological 'defect' (consisting of two grains, one with seven sides and one with five sides – fig. 5.59(i) – in a background of the ideal six-sided grains of a sheet specimen). In the *absence* of such defects all the six-sided grains, whatever the variation in grain size, would have straight sides (fig. 5.58) and there would then be no grain growth since $\beta \rightarrow \infty$. A consequence of this model is that the rate of growth of the average grain size in a sheet specimen will depend on the concentration of such defects, which would presumably be affected by the details of the previous heat-treatment (for example, recrystallisation, solidification, solid state transformation). This will occur through modification of the $\beta$ factor of eq. (5.32). Cahn and Pedawar (1965) pointed out that 7:5 grain defect has all the topological properties of a dislocation (fig. 5.59(ii)) and they considered some interesting developments of this idea.

For a three-dimensional structure the situation may be different since, as suggested by Lord Kelvin (1887), the nearest approach to surface energy equilibrium would be achieved if each grain were a distorted cubo-octahedron (a tetrakaidecahedron) where the sides of the square faces have the same length as all the sides of a hexagonal face (fig. 5.60). The required stacking of such polyhedra is as a bcc array. It can be seen that the six square faces are all equivalent to '{100}' planes of a cubic lattice, the eight hexagonal faces to '{111}' planes,

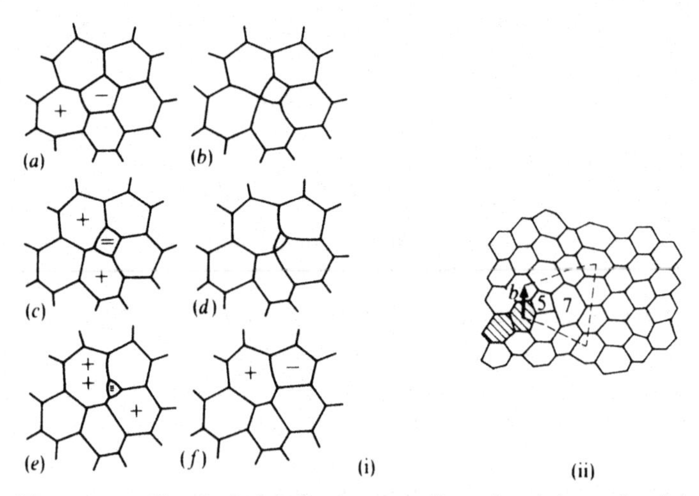


**Figure 5.59** Topological defect in a two-dimensional sheet of mainly six-sided grains: (i) The process of grain growth associated with the defect with the deviation from the ideal number of sides given by the number of + and − signs. (*a*)–(*f*): loss of one grain while the defect remains present. (Hillert 1965.) (ii) Dislocation-like nature of this defect. The extra 'half-row' of grains is shown shaded, the triangular 'burgers' circuit is dashed, and the closure failure is indicated by the arrow. Loss of one shaded grain would correspond to the 'climb' of the dislocation. (After Cahn and Pedawar 1965. Courtesy of *Acta Met.*)

and the edges are in '⟨110⟩' directions. Equilibrium is only achieved if the four grain edges, which meet at a grain corner, have the angle of 109.5° to each other. However, as is shown in fig. 5.61, the angles between planes meeting at a grain edge (normal to the figure) are 125°16′, 180° less the angle between adjacent {100} and {111} planes, and 109°, 180° less the angle between adjacent {111} planes. These angles can only be altered to the equilibrium triple point angles of 120° by the introduction of curvature into the boundaries of the grains. Equivalent distortions are needed at grain corners where the ⟨110⟩ grain edge directions would meet at 90° or 120° rather than the angle of 109° 50′ required for surface tension equilibrium.

The required distortions are that the grain edges are curved and the hexagonal faces have a double curvature yielding a zero net curvature in the perfectly stacked bcc array of fig. 5.60. This was found in the soap film model described by Smith (1964). However, the actual shapes of metal grains appear to deviate somewhat from this possible shape with the average grain having less than the 14 faces of the Kelvin tetrakaidecahedron. Williams and Smith (1952) found that the average number of faces per grain was 12.5 in the 100 aluminium grains that they studied. Randomly formed microstructures are also very unlikely to be stacked in the required bcc array.

It is not clear at present if for three-dimensional grain growth there is the same need for the Hillert type of grain configurational defect as in two-dimensional

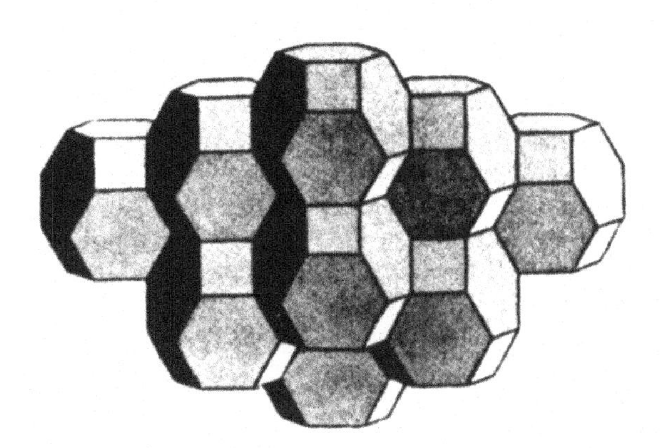

**Figure 5.60** Array of cubo-octohedrons showing hexagonal '{111}' faces and square '{100}' faces. (After Weyl 1957. Courtesy of Princeton University Press.)

growth, or whether because of the inevitable grain boundary curvature any dispersion of grain size will automatically lead to grain growth. The same question can be posed in other words as: is there a larger value of $\beta$ in three-dimensional growth? Morral and Ashby (1974) successfully extended Hillert's idea of topological defects with a dislocation-like character to a three-dimensional array of grains. They concluded that glide of such defects would produce grain boundary sliding and that climb of dislocations would produce grain growth. They pointed out that if the density of grain dislocations changed with time then the $D^2$ relationship of eq. (5.33) would be lost (this should be equivalent to a change of $\beta$ with time or a change in the grain size distribution).

It is reported (for example, Beck *et al.* (1948) and Burke and Turnbull (1952)) that grain growth appears to slow down and stop when, after a period of three-dimensional growth in a sheet material, the grain size becomes slightly larger than that of the sheet thickness, so the growth becomes two-dimensional. This appears to support the idea that there is a structural difference between two- and three-dimensional grain growth. Some reduction is to be expected since cylindrical curvatures will be less than spherical ones and also since thermally etched grooves (see fig. 5.2(b)) at the surface–grain boundary intersection will also provide a significant drag. In 'one-dimensional' wire specimens, grain growth is inevitably going to cease once the bamboo structure is achieved as is found in zero creep experiments on wires (Udin *et al.* 1949). This arises because when the boundaries lie normal to the wire surface, fig. 5.2(*a*), there will be no boundary curvature whatever variation in grain lengths there might be.

A completely different approach to the theory of grain growth was proposed by Rhines and Craig (1974) on the basis of grain topology. Instead of considering the apparent grain diameter, $\lambda$, measured from the number of boundary intercepts per unit length on a polished *surface*, they suggest that the mean grain volume $\bar{V}$, measured as the reciprocal of $N_v$, the number of grains per unit volume, should be used. $N_v$ can be obtained by 'serial sectioning' (Rhines 1967) to investigate the structure in depth below a polished surface. Other parameters

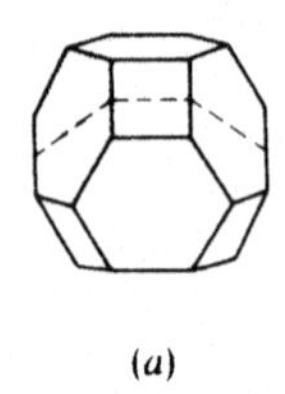

(a)

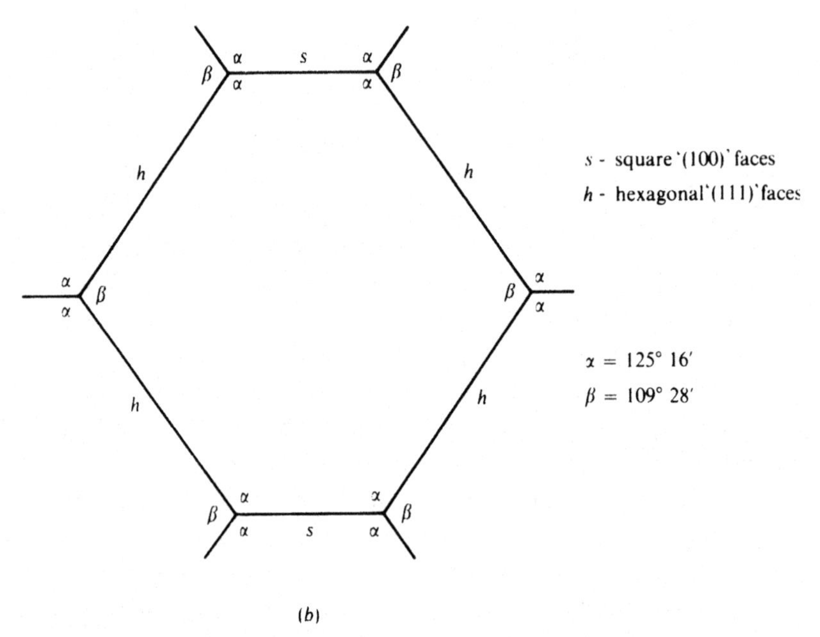


(b)

**Figure 5.61** (*a*) Section equivalent to a '{110}' plane through a cubo-octahedron normal to six of the grain edges. (*b*) Two angles $\alpha$ and $\beta$ between the square faces and the hexagonal faces. Since $\alpha \neq \beta \neq 120°$, surface tension equilibrium can only be achieved by the introduction of curvature into the faces even if grains are all of the same size.

that can be measured during serial sectioning are the surface area per unit volume $S_v$ ($S_v = 2/\lambda$) and the total curvature per unit volume $M_v$ (DeHoff and Rhines 1968). An apparent value of $M_v$ can be obtained from $T_A$ the number of tangencies produced between grain boundary and unit length of a test line swept through unit area of a two-dimensional section. It should be noted, however, that this method of measuring curvature is likely to be suspect since it only measures one of the two radii of curvature that make up the total curvature $(1/R_1 + 1/R_2)$ of a curved surface. Rhines and Craig also defined a dimensionless ratio $\Sigma$, which they called the 'structural gradient', as

$$\Sigma = M_v S_v / N_v = M_v S_v \langle V \rangle$$

In their measurements, the structural gradient, $\Sigma$, changed somewhat especially in the early stages of their observed grain growth in aluminium. Subsequent analysis by Kurtz and Carpay (1980) showed that if the grain size distribution were to remain constant, as required for true steady state growth, then the structural gradient will also stay constant. The changes in $\Sigma$ observed by Rhines and Craig indicate that during their period of observations of grain growth a true steady state grain size distribution had not been achieved.

The other concept that Rhines and Craig introduced was that of the 'sweep constant' $\Theta$, which they defined as the number of grains lost when grain boundaries sweep through a given volume of material. The idea here is that although grain growth occurs by the disappearance of the smallest grains with the fewest number of faces per grain (the smallest grains seen were grains with three faces at grain edges between three larger grains) the *distribution* of numbers of faces per grain stayed constant. This observation means that, for a small grain to disappear, not only must its boundaries migrate but there must be migration of the boundary of the surrounding grains to maintain the distribution of grain shapes. Doherty (1975) suggested that the sweep constant should be more accurately defined as $\Theta^*$, being the number of grains lost when grain boundaries sweep through *a volume equivalent to the mean volume of a grain*, $\bar{V}$. That is, the volume that must be swept by migrating boundaries to eliminate one grain should increase as $\bar{V}$ increases as the grains grow. For a 'self similar' structure, expected for normal growth, the volume swept by the moving grain boundaries for each lost grain would clearly be expected to scale with the mean grain volume. Hunderi (1979), using Hillert's grain growth model, showed that $\Theta^*$ was indeed constant ($\Theta^*=1.76$) during grain growth, while $\Theta$ fell. $\Theta^*$ was also shown to be constant by use of Hunderi's computer model for grain growth. In Hunderi's computer model, $\Theta^*$ was found to be 1.67.

On the basis of their incorrect definition of the sweep constant, Rhines and Craig (1974) suggested that the mean grain volume $\bar{V}$ should increase linearly with annealing time, that is,

$$d\bar{V}/dt=a\Theta M\Sigma\sigma_b \tag{5.34}$$

where $M$ is the grain boundary mobility as defined by eq. (4.49) and $\sigma_b$ is the boundary energy. This result was supported by their experimental observations. However, a linear increase of grain volume with time implied that the *cube* of grain size (measured either as the mean intercept length $\lambda$, or the true mean grain radius $R$) will increase linearly with time. This result is clearly incompatible with the other analysis of 'self similar' normal grain growth with a really steady state grain size and shape distribution – see, for example, Mullins and Vinals (1989). However, as pointed out by Doherty (1975), Rhines and Craig made an additional error in their definition of grain boundary mobility. Vandermeer (1992) reanalysed the original Rhines and Craig topological model using the correct definitions for the sweep constant, $\Theta^*$, and the boundary mobility and keeping the structural gradient $\Sigma$ constant as required by Kurtz and Carpay (1980) and showed that the grain size intercept length, $\lambda$, increased in the expected manner (5.35):

$$\lambda^2 - \lambda_0^2 = (8\Theta^* M \Sigma \sigma_b V_a^{2/3} / 3F^3)t \tag{5.35}$$

where $F$ is a dimensionless number defined as the ratio of the boundary area per unit volume to the cube root of the number of grains per unit volume, $F = S_v / N_v)^{1/3}$. For tetrakaidecahedral grains $F=2.66$. Vandermeer (1992) showed that this value of $F$ was close to that found in a range of experimental grain structures.

## 5.8.2 Computer modelling of grain growth

Interesting new insight into many aspects of grain growth has been given by various computer simulation techniques. Examples include Hunderi, Ryum and Westengen (1979), Novikov (1978), Weare and Kermode (1983), Fradkov, Shvindlerman and Udler (1985), and Frost and Thompson (1986). Many of these models have been discussed in seminars on computer modelling of microstructure, for example, Rollett and Anderson (1990). However, perhaps the most successful and influential model has been the Monte Carlo or Potts model introduced by Anderson, Srolovitz, Grest and Sahni (1984) and Srolovitz, Anderson, Sahni and Grest (1984a) for two-dimensional grain growth and extended to three-dimensional grain growth by Anderson, Grest and Srolovitz (1985, 1989a). In this Monte Carlo simulation, a two- or three-dimensional array of points is set up and each point is assigned a random number, $Q$, usually set as $1 \leq Q \leq 48$, see fig. 5.62. An energy rule is imposed that assigns an increase of energy, $J$, between any two adjacent or nearly adjacent points if they have different values of $Q$. As a result of this rule a simulated grain boundary is produced between points with different $Q$ values. Adjacent points with the same number have no increase of energy, that is, they behave as though they were part of the same grain. By making appropriate choices of both the geometric arrangement of points and the numbers of nearest neighbours considered, the simulations produce very realistic grain structures, see Anderson *et al.* (1984, 1989a). The grain microstructure can be readily displayed by the computer by the simple means of drawing a line between all adjacent points that have different $Q$ values, fig. 5.62. The appropriate Monte Carlo rules can give the correct geometry of grain boundaries meeting at the grain edge angle of 120° and as a result, the appropriate curvatures for producing grain growth are given to the grain boundaries. A typical section in one of the three-dimensional simulations is shown in fig. 5.63.

The conditions at the edges of the lattice are that the lattice is assumed to be surrounded by identical lattices. That is, the configuration at each point on the surface (for example 1, $y$ and $z$) is set to be that at the matching point (101, $y$ and $z$) in a $100 \times 100 \times 100$ cubic lattice. The effect of this can be seen in the sections through the simulated three-dimensional simulation shown in fig. 5.63. The edge of each grain on the left hand side of the figure matches that on the right hand side and the same feature can be seen at the top and bottom of each simulated microstructure.

In both the initial two-dimensional and three-dimensional grain growth simulations, the usual starting structure was with random $Q$ values assigned to each lattice point. That is, all the grains have the same initial size – either the area (in two dimensions) or the volume (in three dimensions) of one lattice point. This means that initially there will be *no spread in the grain size distribution*. The Monte Carlo model is run by randomly picking points, one at a time, and, again randomly, switching the $Q$ values. In the grain growth simulations, a switch is accepted if there is either no change of energy or if the energy decreases. The effect of this rule is that single point grains grow to two, then to three and on to higher numbers of points and in doing so not only does the mean grain size grow but a grain size distribution evolves. After an initial transient, the system tends to a stationary grain size distribution. Critically the Monte Carlo simulated distribution appears to match closely the experimentally observed log-normal form of the size distribution – see fig. 5.64 (Anderson *et al.* 1989a). Once this has been achieved, the grain growth kinetics fit eq. (5.36) with the observed value of the exponent being close to the expected value of $m=2$ ($n=0.5$), within the statistical scatter of the data (Grest, Anderson and Srolovitz 1988)

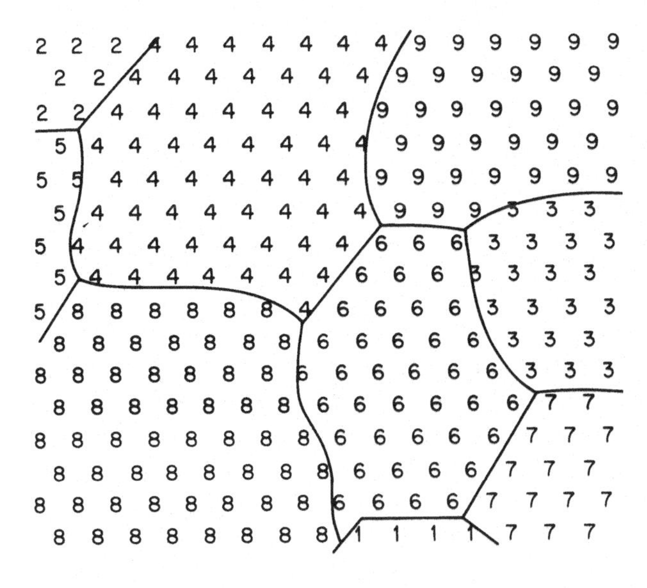


**Figure 5.62** The two-dimensional hexagonal lattice used for the Monte Carlo simulations. Regions in which all the lattice points have the same $Q$ value are assumed to be within one grain. Grain boundaries, of energy $J$, are found between adjacent lattice points with different $Q$ values. (Anderson *et al.* 1984. Courtesy of *Acta Met.*)

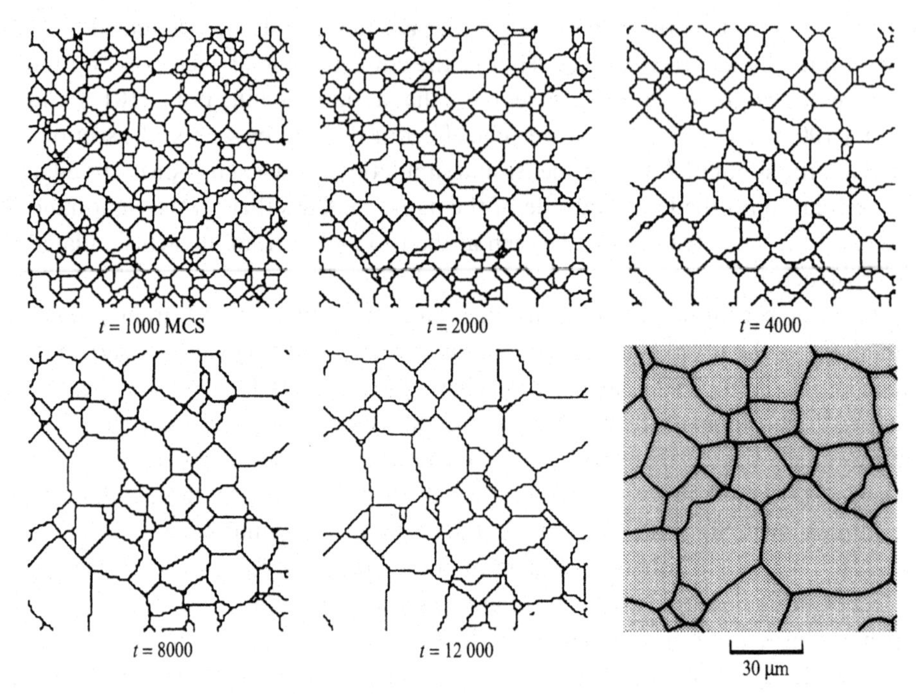


**Figure 5.63** Three-dimensional grain growth simulation on a 100×100×100 lattice seen on one cross-section at increasing times of grain growth, measured in MCS. The last frame is for pure iron deformed 50% and annealed at 700 °C. (Anderson *et al.* 1985). Courtesy of *Scripta Met.*)

$$\bar{R}_t^{\,m} - \bar{R}_0^{\,m} = k't \tag{5.36}$$

so for $\bar{R}_t \gg \bar{R}_0$

$$\bar{R}_t \approx (k't)^{1/m} = (k't)^n \approx (k't)^{0.5} \tag{5.36a}$$

The mean radius, $\bar{R}$, is taken from the mean area in two dimensions, as $\bar{R}=(\bar{A}/2\pi)^{1/2}$, or from the mean volume in three dimensions, as $\bar{R}=(3\bar{V}/4\pi)^{1/3}$. One Monte Carlo step (MCS), defined as occurring when the number of attempted switches is equal to the number of points in the lattice, is used on the timescale for the simulations. For both two- and three-dimensional Monte Carlo simulations the computer model appears to capture all the essential features of grain growth. This is well demonstrated by the match reported by Ling, Anderson, Grest and Glazier (1992) between the two-dimensional computer simulations and soap froth models of grain growth (Glazier, Gross and Stavens 1987, Weare and Glazier 1992). This is illustrated by figs. 5.65 and 5.66. In the study a two-dimensional soap froth of initially very uniform bubble size was allowed to evolve – going through an initial period when the bubble size dis-

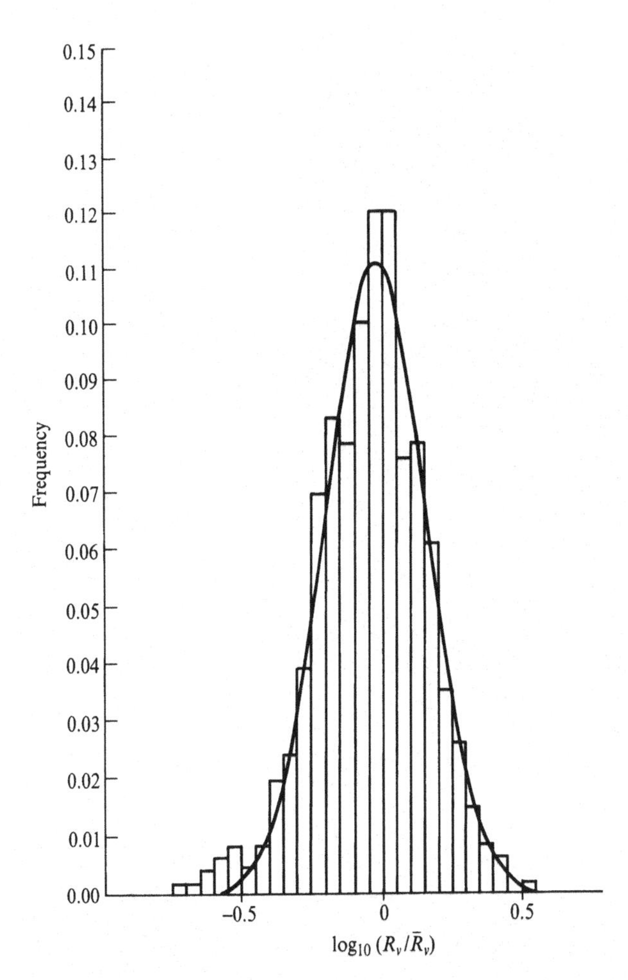


**Figure 5.64** Grain size distribution, the histogram, from the three-dimensional simulation showing the good fit to the log-normal function plotted as the solid line. (After Anderson *et al.* 1989a. Courtesy of *Phil. Mag.*)

tribution goes bimodal before it achieves a steady size distribution. The Monte Carlo (Potts) model was mapped with the initial soap froth distribution and in a second independent simulation also mapped onto the Monte Carlo model with the soap bubble array after 2044 min of evolution in the froth. The microstructures all evolved in very similar ways, fig. 5.65, and showed growth kinetics that, within the scatter of an individual run, fig. 5.66, displayed the required linear increase of mean grain *area* with time (eq. (5.36)) measured either in minutes (the soap froth) or in an adjusted number of MCS (the Potts model).

A significant limitation of the simulation at present, at least for the three-dimensional model, is the restriction of the number of points in the lattice. For most three-dimensional simulations run so far (Anderson *et al.* 1989a,b; Doherty *et al.* 1990; Li 1992), the lattice used was only $100\times100\times100$ points. This means that there is only a rather brief opportunity for the simulation to run after the simulation has achieved a nearly stationary grain size distribution and before the number of grains falls to too small a number to maintain a realistic

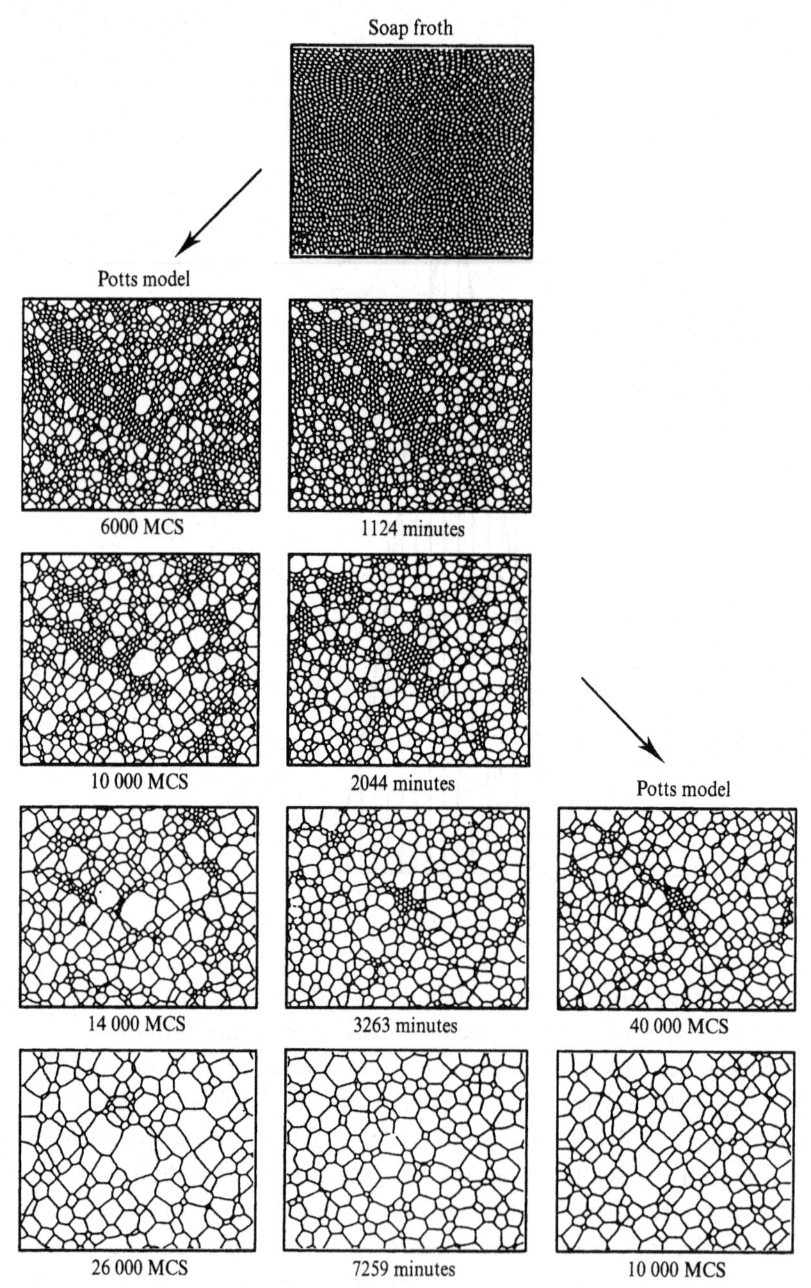


**Figure 5.65** Comparison of the two-dimensional grain growth simulations: a soap froth (centre) and the Monte Carlo method (Potts model). The Monte Carlo modes used, for the initial grain structures, the soap froth images at $t=0$ min (left) and $t=2044$ min (right). (After Ling *et al.* 1992. Courtesy of Trans Tech Publications.)

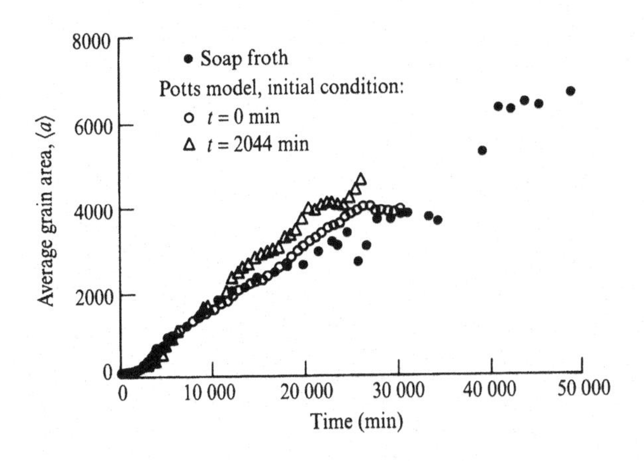


**Figure 5.66** The grain growth kinetics, as mean grain area against time (so $R^2$ vs $t$) for the microstructures shown in fig. 5.65 for both the soap froth and the two Monte Carlo simulations. (After Ling *et al.* 1992. Courtesy of Trans Tech Publications.)

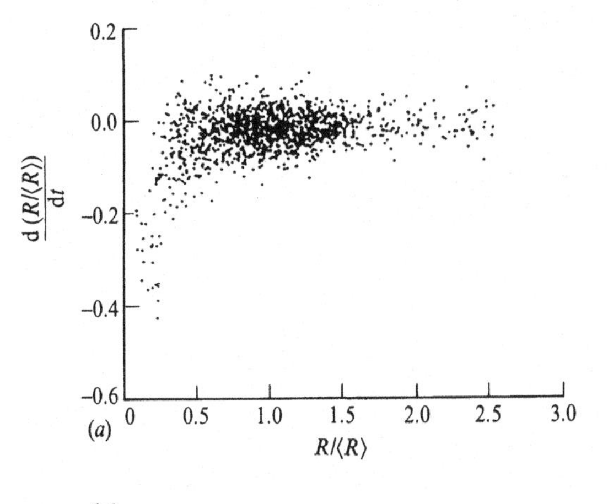


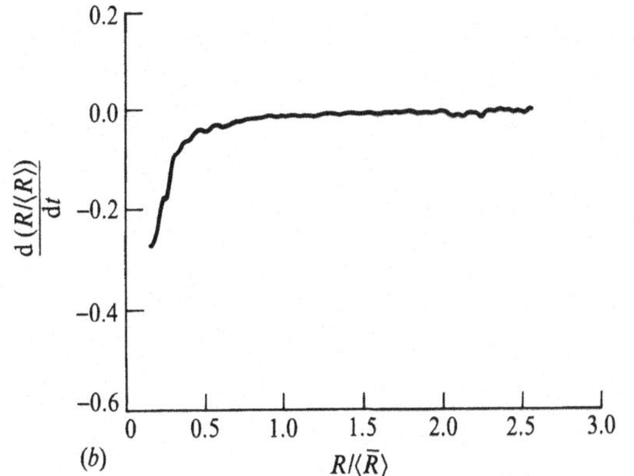


**Figure 5.67** The rate of growth of the normalized grain radius, $d(R/\langle\bar{R}\rangle)/dt$, as a function of the normalised grain radius, $R/\langle\bar{R}\rangle$. (*a*) The scatter in the individual results due to local effects. (*b*) The average result (from about 10 times the number of grains in (*a*)). It can be seen that large grains, those with $R>1.3\langle R\rangle$, grow at the same rate as the average grain size. (Srolovitz *et al.* 1984a. Courtesy of *Acta Met.*)

microstructure. An indication of this lattice size limitation is the jagged shape seen in the simulated grain boundaries in fig. 5.63. The jagged shape arose from the lattice size which is clearly not negligible compared to the grain size.

An interesting result shown in models of simulation of two-dimensional grain growth (Weare and Kermode 1983; Srolovitz *et al.* 1984a) is shown in fig. 5.67. This shows the scatter in normalised growth rates, $d(R/\bar{R})/dt$, as a function of the normalised radius, $R/\bar{R}$. Small grains shrink rapidly while large grains grow at a rate which is essentially that of the growth rate of mean grain size, fig. 5.67(*b*), $d(R/\bar{R})/dt=0$. However, there is a wide scatter of individual growth rates that appears to be dependent on the *local environment of each grain*. Some grains smaller than the average are seen to grow while some grains larger than the average are seen to shrink. This type of variability was not considered in the early models of grain growth (Hillert 1965) but later 'stochastic' models, for example those of Louat (1974), Pande (1987), attempted to model this phenomenon. The scatter revealed by fig. 5.67(*a*) can be compared to similar localised variability in individual coarsening rates of precipitates – see figs. 5.43 and 5.45. Palmer, Frodkov, Glicksman and Rajan (1994) found a similar scatter in directly observed grain growth studies in two dimensions.

### 5.8.3 Experimental results

Despite the availability of the theoretical prediction of eq. (5.33) most experimental studies of grain growth follow the original suggestion of Beck *et al.* (1948) that grain growth kinetics follow a relationship of the form of eq. (5.36a):

$$\bar{D}_t=(k't)^n$$

and could therefore be studied by means of plots of log $D$ versus log $t$ to determine the exponent, $n$. This can only be related to the theoretical expression if $\bar{D}_t \gg \bar{D}_0$ or if appropriate corrections are applied to the value of time $t$ to correct for the time to grow from $\bar{D}=0$ to $\bar{D}_t=\bar{D}_0$ and also for the time delay on heating up (Grey and Higgins 1973). Early results on impure metal (Burke 1949, Beck *et al.* 1948) found values of $n$ that were well below the expected value of 0.5. For zone-refined metals where the solute drag at the boundaries is less, values of $n$ nearer (but not exactly) 0.5 are found (for example, Bolling and Winegard (1958), Drolet and Galibois (1971)), fig. 5.68. Godwin (1966) showed that for very high purity cadmium doped with traces of lead, $n$ only became 0.5 at high temperatures close to the melting point of cadmium (for 20 ppm lead this was $T>0.95T_m$, while for only 1 ppm lead, $T>0.75T_m$). Values of $n$ less than 0.5 will be expected if the velocity of grain boundary migration ($v$) was not a linear function of the driving force, $\Delta\mu$, that is, if the mobility $M$ of eq. (5.23) was a function of $\Delta\mu$, and therefore also of $\bar{D}$. Such a variation of mobility could arise according to the solute drag theory (chapter 4), during the transition from the high velocity to the low velocity extreme (fig. 4.47) as suggested by Gordon and El-Bassyouni (1965). Such a variation of mobility was found in the studies of Rath and Hu (1969) who observed the migration of a boundary in a wedge-shaped bicrystal of zone-refined aluminium and reported that

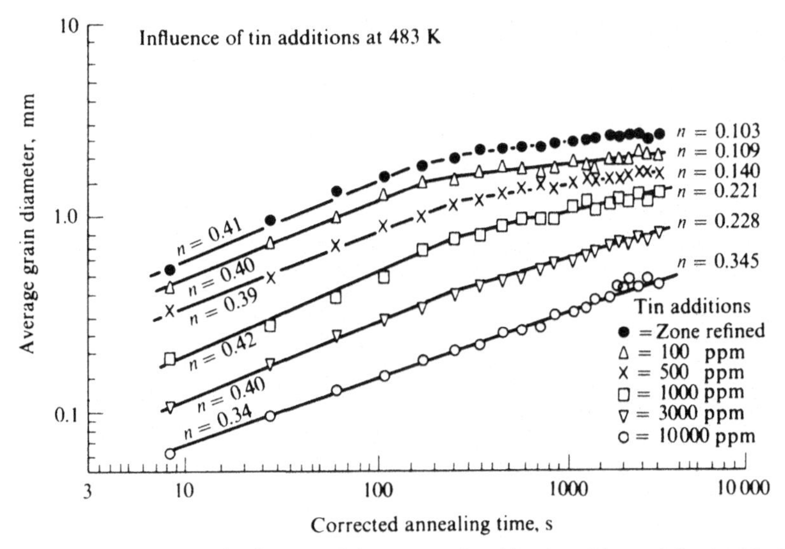


**Figure 5.68** Grain growth in zone-refined lead and in such lead with tin additions, showing the deviation of the exponent $n$ from the 'ideal' 0.5 and also that the deviation increases as the grain size becomes larger. (Data first given by Drolet and Galibois 1971. After Grey and Higgins 1973. Courtesy of *Acta Met.*)

$$v=k'(\Delta\mu)^p$$

(where $p$ is the empirically measured exponent) which, if $v=M\Delta\mu$, implies that the boundary mobility $M$

$$M=k''(\Delta\mu)^{p-1} \tag{5.37}$$

where $p=5.5$, not the expected value of $p=1$ for either the low velocity or high velocity extreme of the Cahn theory (1962). ($k'$ and $k''$ are proportionality constants.) A value of $p>1$ implies that the value of $n$ ($n=1/(p+1)$) will be less than 0.5 in eq. (5.36a). This can be seen by using the expression for mobility of eq. (5.37) in deriving eq. (5.24).

The fall of $n$ found for the later stages of growth (fig. 5.68) may be due to the fall in driving force as $\bar{D}$ becomes larger (eq. (5.22)) so passing from the high velocity extreme to the 'transition' region. Alternatively Grey and Higgins (1972, 1973) proposed that the empirical relationship between $v$ and $\Delta\mu$ is, in fact

$$v=M(\Delta\mu-\Delta\mu^*) \tag{5.38}$$

The $\Delta\mu^*$ here differs from the $P_i$, the impurity-dependent drag term of Cahn's theory in eq. (4.54), in that it is *independent* of the boundary velocity and is equal to the minimum driving force below which no boundary migration whatsoever can occur.

In the low velocity extreme of the Cahn (1962) theory (eq. (4.55)) where $\beta^2 v^2 \ll 1$, $P_i$ is linearly related to $v$. (NB $\Delta\mu$ is $\Delta F$ and $M$ is $\lambda^{-1}$.)

$$P_i = \alpha C v \text{ (eq. (4.55))}$$

$$v = M(\Delta\mu - P_i) = M(\Delta\mu - \alpha C v)$$

Therefore

$$v = M\Delta\mu/(1 + \alpha C M)$$

so that the solute drag only *reduces* the *mobility* and does not act as a *threshold limit* to migration.

Eq. (5.38) was obtained from bicrystal experiments using a different geometry from that of Rath and Hu (1969) but the empirical relationship of eq. (5.38) was also supported by grain growth experiments in zone-refined metals. (The reduction in net driving force as $\bar{D}$ becomes large results in the value of $n$ falling below 0.5.) The explanation of $n<0.5$ due to the $\Delta\mu^*$ term is an attractive proposal since during grain growth, with much lower driving forces than those present during recrystallisation, all the results would be expected to be in the *low velocity extreme* of Cahn's theory and not in the high velocity region/transition region as proposed by Gordon and El-Bassyouni (1965) to account for $n<0.5$. The small $\Delta\mu^*$ term is more likely to be a significant fraction of the total driving force during grain growth than during recrystallisation. The existence of a finite value of $\Delta\mu^*$ suggests that grain growth will cease at a finite grain size $\bar{D}_f$. Here we have

$$v = M[(2\sigma_b V_m/\beta\bar{D}_f) - \Delta\mu^*] = 0$$

Therefore

$$\bar{D}_f = 2\sigma_b V_m/\beta\Delta\mu^*$$

Although there have been a couple of reports of grain growth ceasing in high purity single phase materials (Andrade and Aboav 1966; Fiset, Braunovic and Galibois 1971), this does not seem to be a general phenomenon. An alternative solution to this problem was put forward by Louat and Imam (1990). They pointed out that if the moving grain boundary is treated as being flexible rather than rigid as in the Cahn model of solute drag, then under low driving forces of grain growth a non-linear relationship between driving force and velocity would indeed be expected. Consequently the grain growth exponent $n$ will be, in general, less than the value of 0.5 expected for a constant mobility. Only under the conditions of very high intrinsic mobility at temperatures close to the melting point of the metal and very low impurity contents will the experimental data be expected to give an exponent $n$ of 0.5. Louat and Inman (1990) suggested that the solute drag model is not correct in predicting a constant mobility in the low driving force regime. Their model suggests that only in exceptionally high purity materials would a constant mobility be expected in grain growth.

In an important experimental paper, Rhines and Patterson (1982) investigated experimentally the influence of initial grain size distribution on the kinetics of grain growth. They found that a wide range of different grain size distributions could be achieved by primary recrystallisation after different amounts of deformation. Notably they found that recrystallisation after a heavy cold reduction that produced a fine recrystallised grain size also had a *narrow* spread in grain size while recrystallisation after a lighter deformation led not only to a coarser grain size but also to a *relatively wider* size distribution. Surprisingly these differences in relative grain size distributions were found to be maintained even after extensive periods of grain growth. Less surprising was that these differences in the size distributions were shown to have a direct effect on the rate of grain growth: microstructures with wider grain size distributions grew faster than those with narrower distributions – even after growing to the same mean grain size, see fig. 5.69. It is tempting to conclude that this effect could arise from a reduction in the $\beta$ parameter of eqs. (5.10b) and (5.33). This may be in error since the approximation leading to eq. (5.33) also included the assumption that

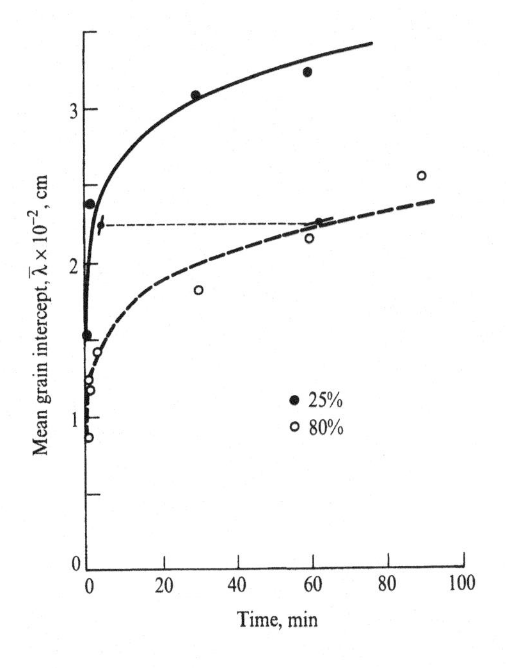


**Figure 5.69** Growth of the mean grain size in recrystallised aluminium following two different cold-rolling reductions that produced reduced initial grain sizes and narrower grain size distributions at the higher reduction. Tangents at the same grain size indicate the slower growth rate of the sample with the narrower size distribution. (After Rhines and Patterson 1982. Courtesy of *Metal. Trans.*)

$d\bar{D}/dt = \bar{v}$ and this assumption is also very likely to be modified by changes in the grain size distribution. Irrespective of its analytical description it is interesting to find in grain growth (Rhines and Patterson 1982) as in precipitate coarsening (Fang and Patterson 1993) that during experimentally accessible times of coarsening the observed growth kinetics are accelerated by wider spreads in the initial grain size distribution. In particle coarsening, however, there is clearly a tendency of different particle size distributions to converge during coarsening. The results of Rhines and Patterson suggest that the initial grain size distributions in grain growth are rather stable.

Glicksman *et al.* (1992) directly observed two-dimensional grain growth in high purity succinonitrile, a transparent 'plastic' organic material often used in metal simulations. They observed that after a short transient to grow a uniform two-dimensional grain structure the material underwent grain growth in ways that matched the two-dimensional Monte Carlo simulations. Firstly they observed variations of individual grain growth rates at identical values of the normalised grain size, $A/\bar{A}$ or $R/\bar{R}$. This experimental behaviour was very similar to the simulations shown in fig. 5.67(*a*). They also observed that though there were some occasional changes in the number of grain neighbours, for example, a six sided grain became a five-sided grain by the vanishing of one of its neighbours, most small grains vanished within their topological class. That is, most four-sided grains shrank and vanished without becoming, even for a brief period, three-sided grains. Further analysis of this study has been given by Palmer *et al.* (1994).

## 5.8.4 Influence of a dispersed second phase on grain growth

It was originally pointed out by Zener (1948) that grain boundaries would be held back by second-phase particles (fig. 5.70). As discussed in §4.3.4, fig. 4.50 shows this pinning process schematically, and the bulging boundary at an angle $\theta'$ is held back by a force ($B$) due to the grain boundary tension $\sigma_b$, acting along a circle of radius $r' = r\cos\theta'$. After resolving the components of this force in the vertical direction, the direction of motion, Zener obtained

$$B = 2\pi\sigma_b r\cos\theta' \sin\theta' = \pi\sigma_b r\sin 2\theta' \tag{5.39}$$

The maximum value of $B$ is therefore at $\theta' = \pi/4$:

$$B = \pi\sigma_b r \tag{5.39a}$$

Ashby (1980) obtained eq. (5.39a) by an alternative energy argument. He noted that when a grain boundary lies symmetrically on the diameter of a spherical

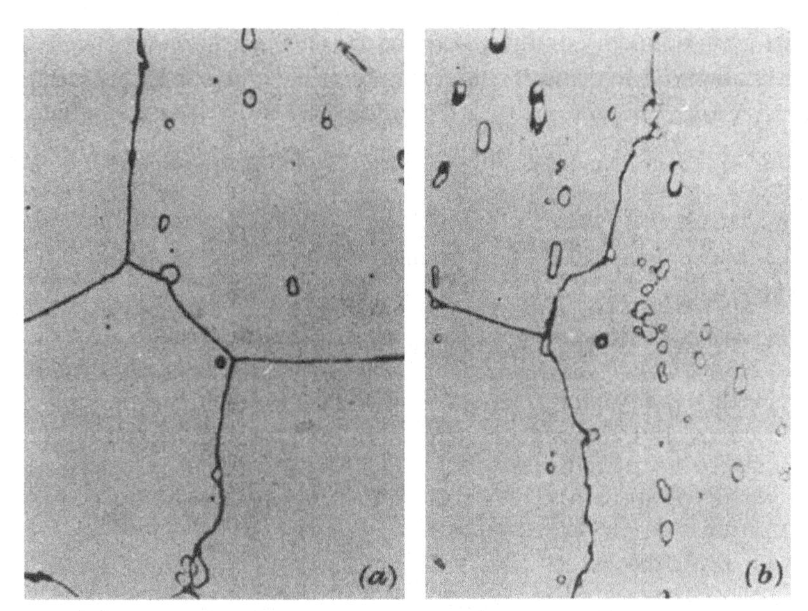


**Figure 5.70** Optical micrographs of migrating grain boundaries pinned by second-phase precipitates. (After Gladman 1966. Courtesy of the Royal Society.)

particle, an area $\pi r^2$ of boundary disappears – so lowering the energy by $\pi r^2 \sigma_b$. A mean force $B$ must be applied by the boundary as it moves over a distance of the radius $r$ to overcome this energy loss. That means that $B \approx \pi r^2 \sigma_b / r$. This simplified analysis is of great use in dealing with more complex problems – for example, when the precipitate has relaxed from a spherical shape to that of a lens of two spherical caps on a grain boundary at an equilibrium contact angle, $\theta$, fig. 5.25, where $\theta = \cos^{-1}(\sigma_b / 2\sigma_{\alpha\beta})$.

Each cap has a radius of $r'$, a volume of $2\pi r'^3(1-\cos^2\theta)(2-\cos\theta)/3$, a height $h = r'(1-\cos\theta)$, and an area of $\pi(r' \sin\theta)^2$ (see, for example, Christian (1975)). The drag force for the relaxed precipitate shape, $B_R$, is then given by

$$B_R = \pi\sigma_b (r' \sin\theta)^2 / r'(1-\cos\theta) = \pi\sigma_b r'(1+\cos\theta) \qquad (5.40)$$

To compare this value of the drag force to that for a spherical precipitate of equal volume and radius $r$ (eq. (5.39a)) we need to relate $r'$ to $r$ using the equal volume condition:

$$2\pi r'^3(1-\cos^2\theta)(2-\cos\theta)/3 = 4\pi r^3/3$$

This yields for $B_R$ as a function of the spherical radius $r$

$$B_R = \pi\sigma_b r(1+\cos\theta)[2/(1-\cos^2\theta)(2-\cos\theta)]^{1/3} \qquad (5.40a)$$

If the grain boundary and the phase boundary have the same energy, then the contact angle $\theta=\pi/3$, and $B$ is increased from $\pi\sigma_b r$ for a spherical precipitate to $B_R=1.82\pi\sigma_b r$, an increase of 80% in the drag force. The magnitude of this increase rises further as the contact angle falls, so that as $\cos\theta\rightarrow 1$, the drag force $B_R\rightarrow\infty$.

We can also easily use the Ashby model to consider the drag force needed to pull a three-grain edge or a four-grain corner from a spherical particle of radius $r$. The results are:

For a *spherical* particle on a three-grain edge:

$$B=(3/2)\pi\sigma_b r \qquad (5.41)$$

and for a *spherical* particle on a four-grain corner:

$$B=6(109^\circ\ 28'/360^\circ)\pi\sigma_b r=1.82\ \pi\sigma_b r \qquad (5.42)$$

If the spherical precipitates relax to the expected equilibrium shapes then the drag forces exerted by the particles on the grain edges and grain corners will increase still further. The values of $B$ can, for the equilibrium shaped particles, be calculated using the geometric values of precipitate volumes and surface areas given by Christian (1975).

In all the analyses above, it was assumed that there was a single value of the interfacial energy between the $\alpha$ matrix and the $\beta$ precipitate, that is, $\sigma_{\alpha\beta}$ was assumed to be constant irrespective of the relative orientation of the two solid phases. Under these circumstances the grain boundary falls normally on a spherical precipitate, cutting it at the diameter (fig. 4.50), or symmetrically for the equilibrium lens shaped precipitate (fig. 5.25). It is usually the case that precipitates, if they have formed by nucleation within one grain, will have an orientation relationship with that grain giving a low energy interface. In the extensively studied Ni–$Ni_3Al$ ($\gamma/\gamma'$) the coherent 'cube–cube' orientation gives a very low energy $\sigma_{coh}$, see table 5.1. If a high angle grain boundary passes such a precipitate, the orientation relationship will be lost giving rise to the much larger value of the incoherent energy $\sigma_{incoh}$ that is likely to correspond to the matrix grain boundary energy, $\sigma_b$ (Doherty 1982)

$$\sigma_{incoh}\approx\sigma_b\gg\sigma_{coh}$$

Under these circumstances the value of $B$ doubles as shown either by the detailed force argument (Ashby, Harper and Lewis 1969) or by the energy argument

$$B=4\pi r^2(\sigma_{incoh}-\sigma_{coh})/2r\approx 2\pi r\sigma_b \qquad (5.43)$$

By use of soap films or two liquid interfaces to simulate either the fully incoherent situation or the transition from coherent to incoherent Ashby *et al.* (1969) were able to confirm experimentally the validity of eqs. (5.39a) and (5.43). If the analysis is extended for this coherent to incoherent transition with the precipitate allowed to relax to its equilibrium shape, a sphere to a hemisphere (Doherty 1982), this leads to a rise in the radius to $r'=(2)^{1/3}r$ and gives a further increase of 25% to the drag force of eq. (5.43).

If the volume fraction of randomly dispersed precipitates is $f$ and the particles all have the same radius $r$, and also *if the interaction between particles and grain boundaries is assumed to be random*, as in Zener's classic analysis, the number of precipitates intersected by unit area of grain boundary, $N_A$, as shown by §4.3.4, is

$$N_A = 3f/2\pi r^2 \tag{5.44}$$

The Zener drag pressure, the force per unit area $Z$ is then:

$$Z = N_A B \tag{5.45}$$

For the case of spherical particles, $B=\pi\sigma_b r$, and we obtain the classical Zener result

$$Z = 3f\sigma_b/2r \tag{5.46}$$

For the case of a boundary being driven past particles by an imposed growth pressure, Ashby *et al.* (1969) were able to confirm experimentally the validity of all aspects of the Zener analysis, including the random number density (eq. (5.44)) and consequently the Zener drag analysis (eq. (5.46)). Rollett *et al.* (1992) simulated, using the Monte Carlo method, the recrystallisation of a two-dimensional grain structure containing particles. They found that for the case of a boundary moving under a volume free energy driving force (a prescribed stored energy on each 'deformed' lattice site) the fraction of the grain boundary sites occupied by particles was indeed a random value which is just the volume fraction of particles $f$, see figs. 5.71 and 5.72. However, as can also be seen in fig. 5.71, the intersection density of particles on the grain boundaries was only *random for the recrystallisation front*. This was *not true generally for all grain boundaries*. In the partially recrystallised state there are three types of grain boundaries:

(i) Boundaries between recrystallised and unrecrystallised grains – the volume pressurised recrystallisation boundaries which show a random fraction of particles. The frequency of these recrystallisation boundaries

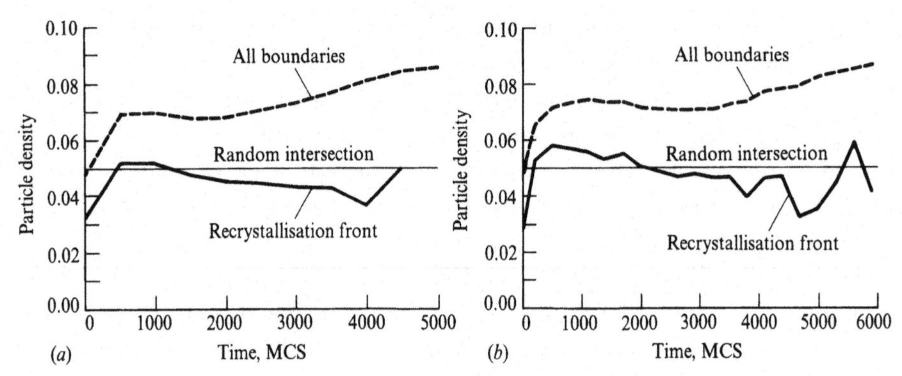


**Figure 5.71** Plot of the density of particles on grain boundary sites during a simulation of recrystallisation in a two-dimensional microstructure, containing 5% particles, for two different densities of recrystallisation nuclei (*a*) 100 and (*b*) 1000. The solid lines represent the densities on the recrystallisation front (boundaries between growing new grains and deformed grains) which have a *random density* (5%) and the dashed lines are for all boundaries which show, once the microstructure has started to evolve, *a higher than random density* of particles. (After Rollett *et al.* 1992. Courtesy of *Acta Metall. et Mater.*)

initially rises and then falls as recrystallisation begins and then goes to competition, as the number of MCS (the time) increases.

(ii) Boundaries between unrecrystallised grains. The frequency of these boundaries falls with recrystallisation.

(iii) Boundaries between recrystallised grains. The frequency of these boundaries increases with recrystallisation.

From the behaviour of the general boundaries in fig. 5.71, it is clear that *the non-pressurised boundaries*, (ii) and (iii), have *a higher than random density of particles* at least after a short period of growth. The particles were initially added at random to the 'grain structure'. These grain boundaries will be migrating, or attempting to migrate under their own curvature effect – that is, they will be undergoing grain growth. From the Ashby *et al.* (1969) experiments and the Rollett *et al.* (1989) simulations it appears that the Zener analysis leading to eq. (5.46) is valid for pressurised boundaries. From the Rollett *et al.* simulations (and from other evidence discussed below) *the Zener assumption of random particle boundary interaction appears invalid for grain growth.* This result has very important consequences as is discussed below.

### Particle limited grain growth

Zener (1948) in his original analysis of grain growth in the presence of particles assumed that eq. (5.46) was valid for grain growth and he used it to determine the limiting grain size at which normal grain growth should cease.

The free energy difference per atom across a curved grain boundary of radius of curvature $R_c$ ($R_c = \beta\bar{D}$) is given by eq. (5.10c). Migration of an atom of volume,

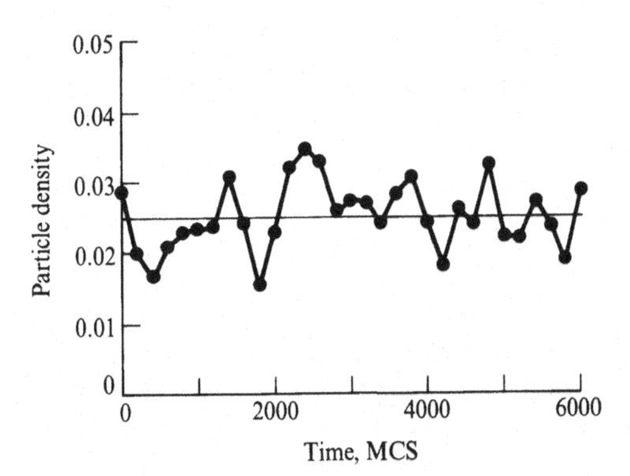


**Figure 5.72** Plot of the density of particles on the recrystallisation front in a simulation of recrystallisation of a deformed single crystal containing a fraction of particle sites of 0.025. In this simulation the stored energy was twice that in the simulations shown in fig. 5.71. The recrystallisation front is in contact with *a random density* of particles. (After Rollett *et al.* 1992. Courtesy of *Acta Metall. et Mater.*)

$V_a$, against the Zener pressure $Z$ causes an opposing *increase* of free energy, $ZV_a$, so that grain boundary migration should cease when:

$$2\sigma_b V_a / R_c = 2\sigma_b V_a / \beta \bar{D} = 3f\sigma_b V_a / 2r \tag{5.47}$$

Rearrangement of eq. (5.47) yields the frequently quoted value, with $\beta=1$, for the Zener limited grain size in normal grain growth:

$$\bar{D} = \bar{D}_Z = 4r/3f \tag{5.48}$$

Various modifications to this analysis have been offered in the metallurgical literature, for example, by Gladman (1966), Hazzledine, Hirsch and Louat (1980), Nes, Ryum and Hunderi (1985) and Hillert (1988). These modifications, in general, make little change to the general form of the much quoted eq. (5.48). In 1984, however, Srolovitz *et al.* used the two-dimensional Monte Carlo model to determine by simulation how grain growth would be modified by inserting inert particles into the lattice of fig. 5.62. The particles were introduced by assigning to a fraction of the lattice sites, chosen at random, a particular $Q$ value, usually $Q=49$, that *was not allowed to switch*. The same technique was subsequently used by Rollett *et al.* (1992) in the simulation of recrystallisation in a particle containing matrix, as discussed above. The introduction of inert particles in this way produces two effects: firstly, the particle–matrix energy is set at the same value as the grain boundary energy, and secondly the particles have a size equal to the lattice size that in the Srolovitz simulation was also the starting grain size. This latter result means that the fraction, $\phi_p$, of particles that lie on grain boundaries at the start of the simulation will be initially 1. The early stages of grain growth in the presence of particles were found to occur at a rate almost identical to that in their absence. However, after some time, grain growth was found to stop, for the two-dimensional simulation, at a limiting mean area, $\bar{A}_{2D}$, given by:

$$\bar{A}_{2D}=2.9a/f \text{ so } \bar{R}_{2D}=(2.9r^2/f)^{1/2}=1.7r/f^{1/2} \qquad (5.49)$$

Since the volume fraction of particles $f \ll 1$, the two-dimensional, Monte Carlo simulation result (eq. (5.49)) indicates *a much smaller grain size* than the Zener result (eq. (5.48)). For example, at $f=0.01$ (1%) the simulation result is for a limiting grain size about 10 times smaller than the Zener result. The Zener model for two-dimensional grain growth is readily shown (see Srolovitz *et al.* (1984b) or Doherty, Srolovitz, Rollett and Anderson (1987)) to predict a drag force per particle, $B$, of $B=2\sigma_b$, a random value of the number of particles on unit length of the two-dimensional boundary, $N_L$, of $2f/\pi r$, and so the Zener limiting grain size in two-dimensions $\bar{D}_{Z(2D)}$, is

$$\bar{D}_{Z(2D)}=\pi r/4f \qquad (5.48a)$$

Doherty *et al.* (1987) pointed out, using data from Srolovitz *et al.* (1948b), that the immediate physical origin for the discrepancy between eqs. (5.48a) and (5.49) was *the higher than random density of particles on the grain boundaries* seen in the Monte Carlo simulations of Srolovitz *et al.* (1984b). The fraction of particles on the grain boundaries if there was a random correlation of particles and grain boundaries, in two dimensions, $\phi_p(R)$ was $2r/\bar{R}$. At the limiting grain radius found in the simulations (eq. (5.49))

$$\phi_p(R)=1.4f^{1/2} \qquad (5.50)$$

The observed values of $\phi_p$ were much larger (by 3–10 times) than the random values given by eq. (5.50) with 80–90% of the particles in contact with the grain boundaries when grain growth stopped. The observed fact that $\phi_p > \phi_p(R)$ shows *how* the Zener model fails to predict the actual limiting grain size; however, it does not explain *why* the Zener model failed. The important insight to this question was provided by Srolovitz *et al.* (1948b) and Hillert (1988) for two-dimensional grain growth and is illustrated by fig. 5.73. The important idea is that particles will have a more fundamental effect in opposing grain coarsening than merely providing a drag force. The particles actually *remove the grain boundary curvature* that drives grain growth. This was also clearly seen to apply in the two-dimensional simulations where the pinned grain structures *show no curvature*, fig. 5.74. In two dimensions, three particles per grain will remove all curvature, fig. 5.73. So in the Srolovitz *et al.* (1984b) simulation, in which grains grew from a size much smaller than the interparticle spacing, $\Delta_2$, which in two dimensions is $\approx r/f^{1/2}$, each growing grain needed only to contact about three particles for grain

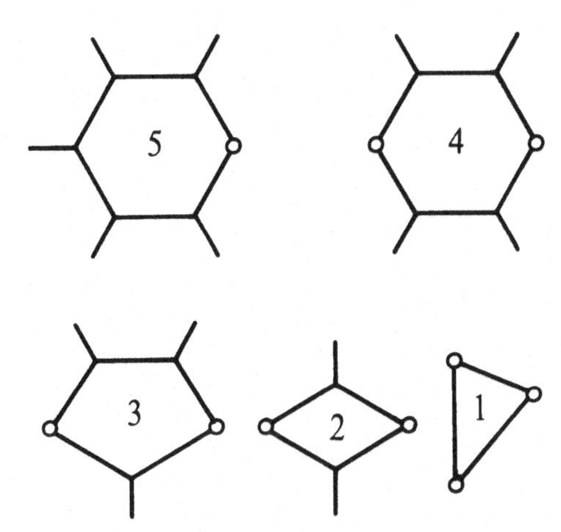


**Figure 5.73** Schematic two-dimensional grain structures showing how particles can eliminate the curvature of grains having less than six neighbours. (Hillert 1988. Courtesy of *Acta Met.*)

growth to cease. This idea immediately yielded the result, given by Srolovitz *et al.* (1984b), that $\bar{A}=3\,a/f$. This result comes from the fact that the average number of particles per grain (required to be three) is the product of the mean grain area $\bar{A}$ and the number of particles per unit area, $N_A$, where $N_A=f/a$, with $a$ the area per lattice point. This predicted pinned grain area, within the statistical scatter, agrees with the experimental result obtained using eq. (5.49), $\bar{A}=(2.9\pm0.2)\,a/f$.

Although the two-dimensional simulations provided a very clear picture of the fundamental influence of particles on grain growth it was not immediately obvious that the same situation would apply in the more usual case of a three-dimensional grain structure. The idea of non-random particle grain boundary correlation put forward by Doherty *et al.* (1987) for the two-dimensional case was immediately challenged by Hillert (1988) as being not generally appropriate for the more important three-dimensional case.

Anderson *et al.* (1989b) responded to this question by running three-dimensional simulations of particle limited grain growth. The three-dimensional simulations were exactly those of the single phase simulations (Anderson *et al.* 1989a), but with a range of volume fractions of single point particles. The results of this set of simulations are shown in figs. 5.75–5.78. Fig. 5.75 shows the steady increase of mean grain volume $\bar{V}$ as a function of time, in MCS, for a range of volume fractions, including the single phase microstructure previously simulated. Initially the particle containing structures had the same rate of grain growth but after a period of growth that was shorter for the highest density of particles, the microstructures became stabilised at a limiting grain size. The limiting mean grain volume $\bar{V}_p$ is plotted against the volume fraction of randomly placed particles $f$ in fig. 5.76; this is log $\bar{V}_p$–log $f$ plot. The line in this figure is given by:

$$\bar{V}_p=(91\pm17)v/f^{0.9\pm0.05} \tag{5.51a}$$

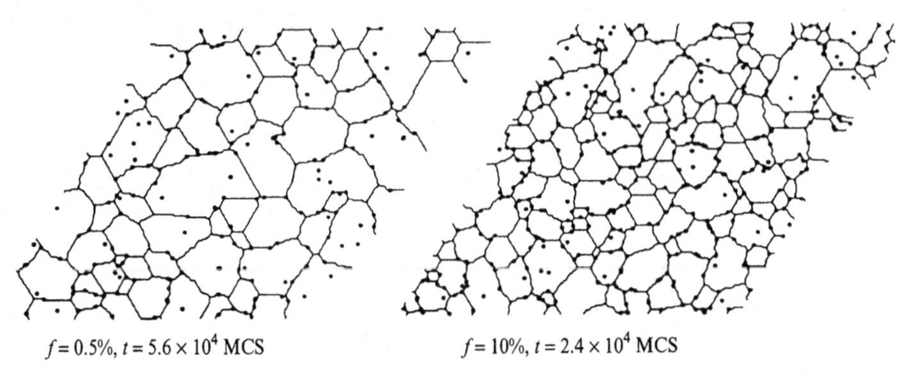


**Figure 5.74** Pinned two-dimensional grain structure at two values of the particle fraction, $f$=0.5% and 1%. (After Srolovitz *et al.* 1984b. Courtesy of *Acta Met.*)

or

$$\bar{R}_p=(4.5\pm0.8)r/f^{0.31\pm0.02} \tag{5.51b}$$

Examples of the pinned microstructure from the three-dimensional simulations are given in fig. 5.77. These are clearly different from the two-dimensional pinned structures of fig. 5.74. In three dimensions there are many more particles inside the grains. So, it appears that in three dimensions it is easier for a grain boundary to pass a particle than in two dimensions. In three dimensions there is also some apparent (two-dimensional) retained curvature on some of the boundaries. Since grain growth has actually stopped, this cessation of boundary migration requires that the total curvature, $\kappa$, must be zero. The curvature, $\kappa$, is determined by any two perpendicular radii of curvature, $R_1$ and $R_2$

$$\kappa=(1/R_1+1/R_2)=0 \tag{5.52a}$$

If the boundaries are stationary then there can be no net curvature and so for all points on the grain boundaries, $R_1=-R_2$. That is, any curvature seen in the plane of the section should be balanced by an equal and opposite radius of curvature normal to the section. Li (1992) in a detailed analysis of the microstructures from the simulation has shown that this assumption does appear valid. However, this conclusion was difficult to confirm fully due to the limited size of the lattices used in the three-dimensional simulation. The small size of the lattice means that the pinned grains, even at the smallest values of $f$, have such a small number of lattice sites on each grain face that the curvatures in fig. 5.77 are too 'noisy' to be measured reliably. Li, however, demonstrated the central feature of the stagnation of grain growth in a set of three-dimensional simulations using a small central grain surrounded by different numbers of other grains. Experiments with: (i) four neighbouring grains giving a central tetrahedron; (ii) six neighbouring grains

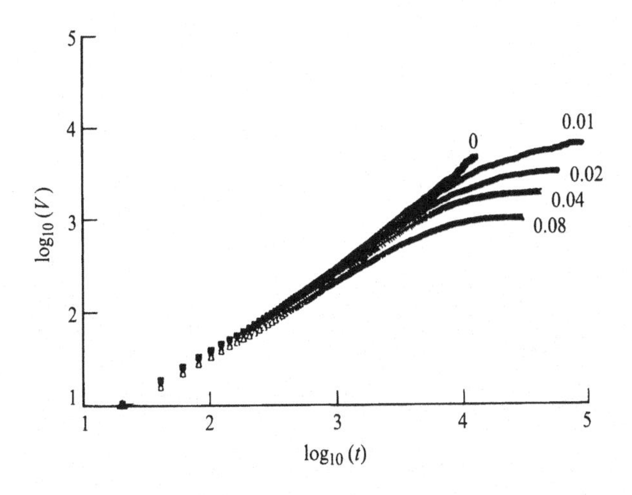


**Figure 5.75** The growth of the mean grain volume, $\bar{V}$, with time in MCS for three-dimensional grain growth simulations at five values of the particle fraction, 0, 0.01, 0.02, 0.04 and 0.08 showing the pinning of grain growth by particles. (After Anderson *et al.* 1989b. Courtesy of *Scripta Metall.*)

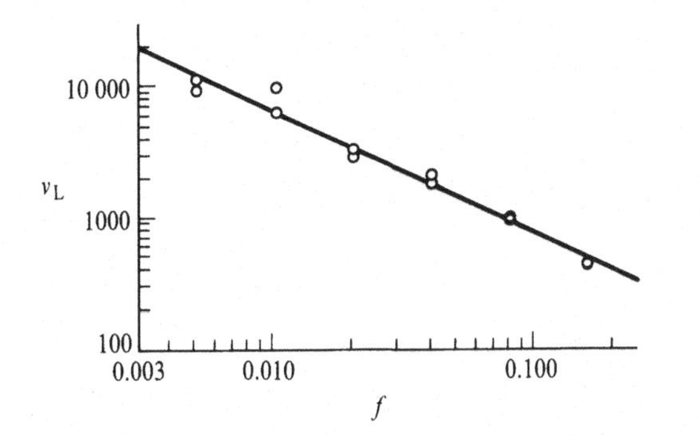


**Figure 5.76** Mean limiting grain volume, $\bar{V}$, as a function of particle fraction, *f*, for three-dimensional grain growth simulation. (After Anderson *et al.* 1989b. Courtesy of *Scripta Metall.*)

giving a hexahedron; and (iii) eight giving an octahedron were successively simulated. Since all these grains have fewer faces than the fourteen faces of the Kelvin tetrakaidecahedron they were all found to shrink and vanish as expected in the absence of any particles. The tetrahedral and hexahedral grains still shrank and vanished even when particles had been placed on all the grain corner sites. There are four corner sites on the regular tetrahedron, and eight on the hexahedron. The octahedral grain, however, was found to resist shrinking when particles were placed on all twelve of its corner sites. This resistance to shrinkage was maintained even with particles on only eight of its twelve corner sites. Fig. 5.78 shows the computer image of structure of the pinned octahedral grain. The combination of the interfacial tensions of the eighteen grain boundary edges from the neighbouring eight grains in combination with the particles at eight of the twelve corners of the octahedral grain was found to have removed *all curvature from the octahedral grain surfaces*. These boundaries, shown in fig. 5.78, are indeed flat and in any case since there is no further grain boundary movement this again indicates that all of the curvature that would cause the boundaries to migrate must have been eliminated.

Further insights, given by the Monte Carlo simulations into the origin of the

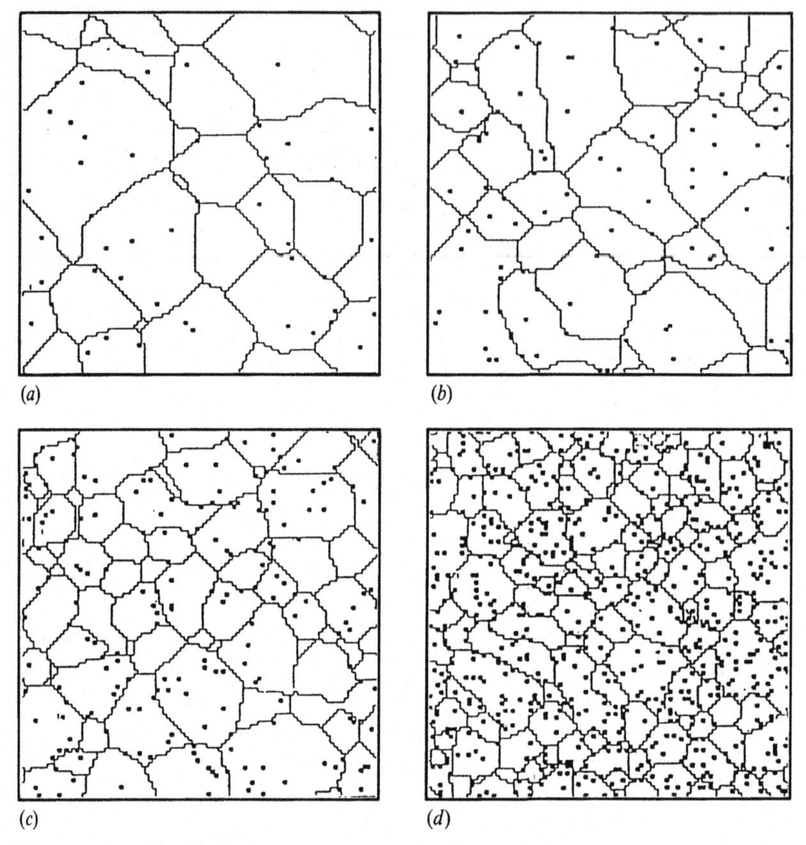


**Figure 5.77** The pinned grain structures for the three-dimensional grain growth simulations at four values of the particle fractions, $f$: (*a*) 0.005, (*b*) 0.01, (*c*) 0.02 and (*d*) 0.08. (After Anderson *et al.* 1989b. Courtesy of *Scripta Metall.*)

fine limiting grain size, were given by the changes in the grain size distribution produced by the presence of particles (Doherty *et al.* 1989; Li 1992). Fig. 5.79 shows the grain size distribution at comparable mean grain size for a single phase structure and for a pinned microstructure at $f$=0.02. Similar plots were found for other values of $f$ by Li (1992). The single-phase distribution shows the typical form analysed by Anderson *et al.* (1989a) with the smallest grains having radii as small as $0.1\bar{R}$. The pinned microstructure, however, showed no grains smaller than $0.4\bar{R}$. Both distributions show the statistical scatter arising from the limited number of grains in a single simulation. In the single-phase structure there are a significant number of small grains which are *all in the process of shrinking* so decreasing the number of grains. The continued loss of the smallest grains is the critical step by which the mean grain size grows. As these smallest grains vanish, they are replaced by shrinkage of the slightly larger grains, those with $R \approx (0.5\text{–}0.7)\bar{R}$. In the pinned structure, however, though all the grains smaller than $0.4\bar{R}$ have vanished *they were not replaced by the shinkage of the smallest*

**Figure 5.78** Computer image of pinned octahedral grain with particles on eight of the octahedral grain grain's twelve corners. Particles appear as spheres while lattice sites occupied by the pinned grains appear as faceted cubes. (After Li 1992.)

*surviving grains*, from a size just larger than $0.4\bar{R}$ to a size smaller than $0.4\bar{R}$. It is the stabilisation of the smallest surviving grains, those with $R \approx 0.5\bar{R}$, that causes normal grain growth to cease. Srolovitz *et al.* (1984b) in their simulations of two-dimensional, particle limited, grain growth also showed similar differences (figure 12 of their paper) in the particle size distribution between that the pinned grain structures and that for the growing single-phase grain structures. In the pinned two-dimensional structures there were no grains smaller than about $0.4\bar{R}$ while in the two-dimensional, single-phase microstructures, grains as small as $0.1\bar{R}$ were found. The significance of this difference in the two-dimensional size distributions at the smallest grain sizes was, however, not discussed by Srolovitz *et al.* (1984b).

Table 5.6 gives additional details of the pinned microstructures at different values of particle fraction, $f$, compared to an unpinned single-phase grain structure. The smallest 5 and 10% of the grains are found, as seen in fig. 5.79, to occur

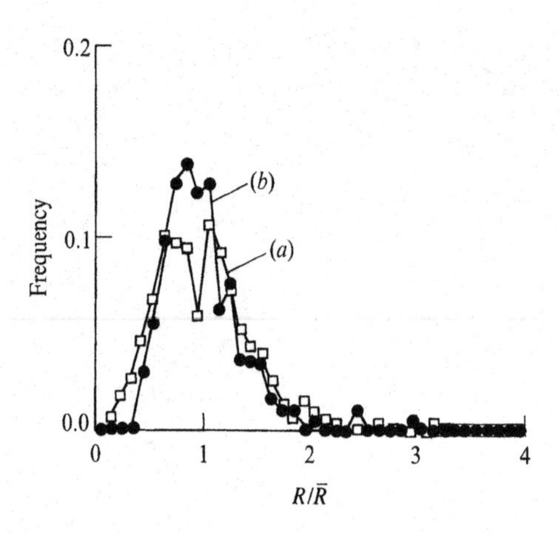


**Figure 5.79** The grain size distributions, at comparable mean grain sizes, for: (*a*) a single-phase grain structure; and (*b*) for a pinned particle containing microstructure at $f$=0.02, MCS=20 000. Notice the absence of grains in the two-phase structure for values of $R$ less than $0.4\bar{R}$. (After Li 1992.)

at larger values of the mean grain size for the pinned grain structure than in the still evolving the single-phase grain structure. These smallest grains in the single-phase structure have only seven neighbours ($N_N \approx 7$). This is fewer than the eight to nine neighbours found for the smallest grains in the pinned structure ($N_N \approx 8$–9). The pinned grain shown in fig. 5.78 has an 'octahedral' shape with the imposed eight neighbours. The smallest grains in the single-phase structure are only partially inhibited from shrinking by the boundary tension of the limited number of grain edges and this inhibition is insufficient to stop the small grains from shrinking. The smallest surviving grains in the pinned structure are prevented from shrinking by a *combination* of the larger number of impinging grain boundaries and by the particles on the two-grain boundaries of the small grain, $n_2$, by particles on the three-grain edges, $n_3$, and by those on the four-grain corners, $n_4$. It may be noted that the *smaller* pinned grains, at higher values of $f$, need *higher numbers of pining particles*, $n_2$–$n_4$, to remove the curvature and prevent shrinkage than do the larger grains at the smallest values of $f$.

As in the pinned octahedral grain in fig. 5.78, the smallest grains pinned, whose shrinkage has ended normal grain growth, are approximately octahedral in shape, $N_N \approx 8$. That is, pinning in the general microstructure is achieved by a combination of particles on the grain faces, edges and corners and the surrounding grain edges. Both particles and grain edges combine to remove the grain boundary curvature.

### Comparison of the computer simulation, the experimental observations and the Zener model

Fig. 5.80, which shows a large amount of relevant experimental data, is taken from a detailed review by Olgaard and Evans (1986) but it has been slightly modified by Li (1992). In the figure, experimental points have been plotted from a wide range of experimental metallurgical and sintered ceramic studies. In two of the metallurgical studies, those of Tweed, Hansen and Ralph (1983) and Hsu

**Table 5.6** *Topological analysis of the three-dimensional simulated microstructures. For $f$=0.00, 0.005, 0.01 and 0.04 this is for the pinned grain structure. (Li 1992.)*

| | $f$=0.00 $\bar{V}$=4.525 | $f$=0.005 $\bar{V}$=11.111 | $f$=0.01 $\bar{V}$=6.536 | $f$=0.04 $\bar{V}$=1.938 |
|---|---|---|---|---|
| Smallest | $V/\bar{V}$=0.03 | $V/\bar{V}$=0.08 | $V/\bar{V}$=0.06 | $V/\bar{V}$=0.10 |
| 5% of grains | $N_N$=7.2 | $N_N$=8.1 | $N_N$=8.7 | $N_N$=9.1 |
| | | $n_1$=0.9 | $n_1$=0.6 | $n_1$=0.65 |
| | | $n_2$=6.4 | $n_2$=3.9 | $n_2$=6.3 |
| | | $n_3$=1.2 | $n_3$=5.9 | $n_3$=12 |
| | | $n_4$=1.0 | $n_4$=2.0 | $n_4$=9.5 |
| Smallest | $V/\bar{V}$=0.03 | $V/\bar{V}$=0.11 | $V/\bar{V}$=0.08 | $V/\bar{V}$=0.12 |
| 10% of | $N_N$=7.6 | $N_N$=8.5 | $N_N$=9.1 | $N_N$=9.2 |
| grains | | $n_1$=1.4 | $n_1$=1.6 | $n_1$=1.1 |
| | | $n_2$=7.4 | $n_2$=6.8 | $n_2$=9.2 |
| | | $n_3$=4.7 | $n_3$=6.5 | $n_3$=13.0 |
| | | $n_4$=1.1 | $n_4$=1.9 | $n_4$=9.2 |
| 45–55% of | $V/\bar{V}$=1.0 | $V/\bar{V}$=0.65 | $V/\bar{V}$=0.50 | $V/\bar{V}$=0.51 |
| grain size | $N_N$=16.6 | $N_N$=13.6 | $N_N$=12.9 | $N_N$=12.6 |
| distribution | | $n_1$=26.0 | $n_1$=20.5 | $n_1$=27.1 |
| | | $n_2$=33.7 | $n_2$=31.0 | $n_2$=56.6 |
| | | $n_3$=12.2 | $n_3$=13.9 | $n_3$=31.2 |
| | | $n_4$=2.0 | $n_4$=3.8 | $n_4$=12.5 |
| grain with | $V/\bar{V}$=1.0 | $N_N$=16.0 | $N_N$=16.0 | $N_N$=16 |
| $V/\langle V\rangle$=1 | $N_N$=16.6 | $n_1$=32.6 | $n_1$=34.4 | $n_1$=36.0 |
| | | $n_2$=36.0 | $n_2$=13.9 | $n_2$=60.0 |
| | | $n_3$=13.0 | $n_3$=18.9 | $n_3$=43.0 |
| | | $n_4$=2.0 | $n_4$=4.1 | $n_4$=16.0 |
| Largest 5% | $V/\bar{V}$=4–13 | $V/\bar{V}$=3–7 | $V/\bar{V}$=3–15 | $V/\bar{V}$=3–30 |
| of grain size | $N_N$=34.8 | $N_N$=29.4 | $N_N$=34.0 | $N_N$=32.5 |
| distribution | | $n_1$=188 | $n_1$=170 | $n_1$=203 |
| | | $n_2$=118 | $n_2$=118 | $n_2$=208 |
| | | $n_3$=168 | $n_3$=37.8 | $n_3$=115 |
| | | $n_4$=4.8 | $n_4$=6.7 | $n_4$=34.5 |

$N_N$=number of neighbour grains/grain; $n_1$=number of particles within a grain; $n_2$=number of particles on a two-grain boundary etc.

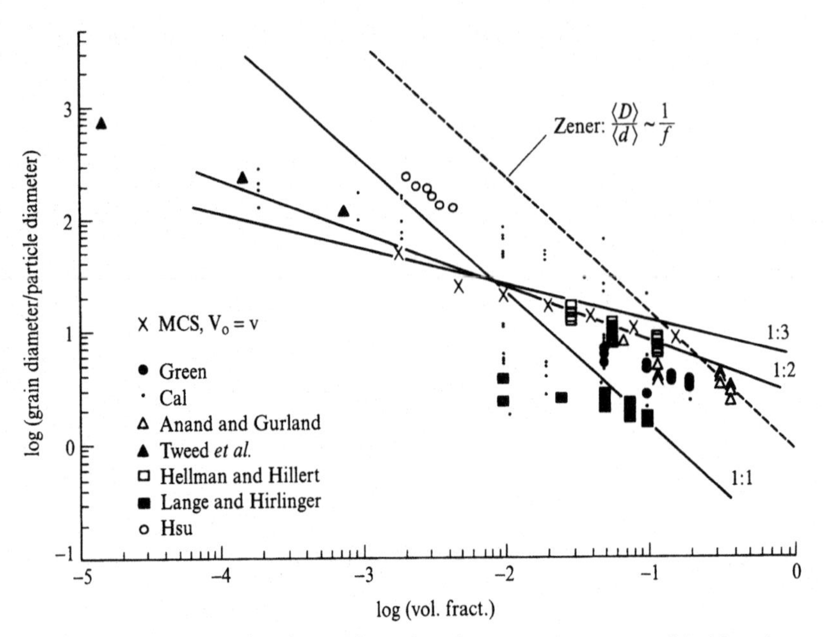


**Figure 5.80** A plot, from Olgaard and Evans (1986) as modified by Li (1992), of the ratio of the mean grain to the mean particle diameter $\langle D\rangle/\langle d\rangle$ in pinned structures as a function of the volume fraction of particles. The experimental values are those of the cited authors, with 'cal' the data for calcite from Olgaard and Evans. The three solid lines are possible dependences of $\langle D\rangle/\langle d\rangle$ on $1/f^n$, with $n=1$, 2 or 3. The Zener prediction (5.48) is the dashed line and the heavy crosses are the three-dimensional computer simulation results of Anderson *et al.* (1989). (After Olgaard and Evans 1986 and Li 1992. Courtesy of *J. Am. Ceram. Soc.*)

(1984), grain structures were produced by deformation and recrystallisation of an existing two-phase structure so that an initially random correlation of particles and grain boundaries is likely, see Ashby *et al.* (1969). In the study by Anand and Gurland (1975), cementite particles were precipitated on the fine ferrite grain structure in tempered martensite and were expected to be preferentially nucleated on the ferrite grain boundaries. In the studies of sintered ceramic materials, fine powders of the components were mixed together and sintered to full density by hot pressing. In these experiments it is likely that the fine minority particles will be located at the boundaries, edges and corners of the grain structure of the majority component. That is, an initially higher than random correlation of particles and grain boundaries will be found in both the tempered martensite and the sintered ceramic microstructures. In all the experimental studies, the starting grain sizes were larger than the sizes of second-phase particles. Also plotted in fig. 5.80 are three possible dependences of log $(\bar{D}/\bar{d})$ on log $f$; $\bar{D}/\bar{d}$ is the ratio of mean grain pinned diameters to mean particle diameters. The line marked 1:1 is the expected slope of a $1/f$ dependency, 1:2, the expected slope for $1/f^2$ and 1:3 that for $1/f^3$. These features were presented on the original

Olgaard and Evans (1986) plot. Additional data from Li (1992) are the results from his three-dimensional simulations represented as heavy crosses and also the actual Zener plot of eq. (5.48).

Various features can be seen in fig. 5.80. These include that:

(i) Almost all experimental points fall *below the Zener predicted values* and at the lowest values of $f$ the Zener prediction *is very seriously in error*.

(ii) For any individual set of data, the fit to either the $1/f^2$ or $1/f^3$ plots is clearly superior to that to the $1/f$ plot.

(iii) There is a very large spread of the experimental data both between data sets and also within a data set. This large spread indicates that the limiting grain size must depend not only on the mean particle size and particle fraction but also on some other feature, or features, in the analyses or simulations.

Doherty *et al.* (1989), Li (1992) and Li, Anderson and Doherty (1994) have considered two of the possible features in particle limited grain growth not previously modelled. These are: (*A*) the original grain size at the start of the grain growth; and (*B*) the randomness of the particle–grain boundary interaction at the start of a grain growth experiment or simulation.

In all the experiments in fig. 5.80, but not the simulations of Li, the starting grain structure size was much larger than the size of the dispersed second-phase particles – both with recrystallisation experiments in deformed structures (Tweed *et al.* 1983; Hsu 1984) and with the ceramic sintering studies.

During recrystallisation of deformed two-phase metallic systems the grain boundaries can usually readily pass the pinning particles since the stored energy per unit volume, $E_s$ J m$^{-3}$ (a pressure in N m$^{-2}$) is almost always larger than the Zener drag (eq. (5.46)). So as discussed above, this results in the particle–grain boundary interactions being close to random. Consequently the usual metallic microstructure of grains produced by recrystallisation in the presence of a dispersed phase will, at the start of grain growth, have the grain boundaries and the particles almost randomly correlated – the result found by Ashby *et al.* (1969). In sintered microstructures in which insoluble single-crystal powders are mixed and hot compressed to full density, all the dispersed phase particles of the minority component, which has a much smaller particle size, will be located at matrix grain boundaries, grain edges and grain corners. That is, the initial fraction of particles at grain boundaries should be larger than the random value. The same type of high correlation of particles with grain boundaries will be found in a fine grain single-phase structure in which second-phase precipitates form by heterogeneous nucleation, §2.2.3. An example is the study of tempered martensite by Anand and Gurland (1975), in which the authors reported that all carbide particles were at the $\alpha$ ferrite grain boundaries.

Doherty *et al.* (1989) studied, using two-dimensional Monte Carlo simula-

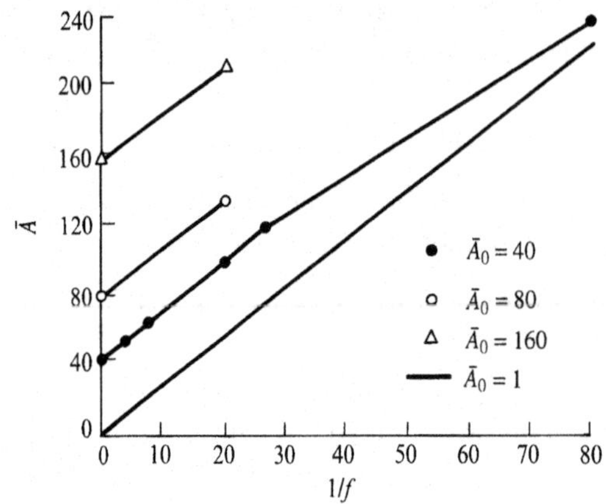


**Figure 5.81** Simulated mean pinned grain area, $\bar{A}$, as a function of the particle fraction $f$ in two-dimensional simulations for different initial mean grain areas $\bar{A}_0$: $\bar{A}_0=160a$, $80a$ and $40a$. Here $a$ is the area of one lattice point (the area of one particle). The values for $\bar{A}_0=a$ are those of Srolovitz *et al.* (1984b). (Doherty *et al.* 1989. Courtesy of Riso National Laboratory.)

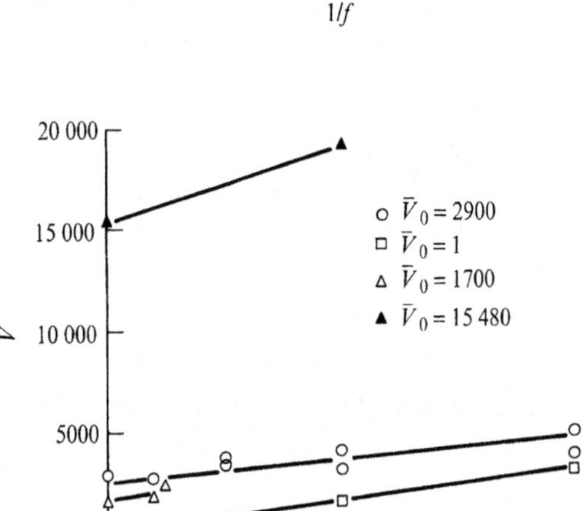


**Figure 5.82** Simulated mean pinned grain volumes, $\bar{V}$, as a function of particle fraction $f$ in three-dimensional simulations for different initial mean grain volumes $\bar{V}_0$: $\bar{V}_0=15\,480v$, $2900v$ and $1700v$. Here $v$ is the volume of one lattice point (the volume of one particle). The values for $\bar{V}_0=v$ are those of Anderson *et al.* (1989b).

tions, the first of these influences – that of having a starting grain size larger than the particle size. Li (1992) carried out a similar study in his three-dimensional simulations. Figs. 5.81 and 5.82 show the important results of these studies. In the two-dimensional simulations (fig. 5.81) and three-dimensional simulations (fig. 5.82) the grains were grown to different starting areas or volumes by the usual grain growth simulations. The simulations were stopped, a predetermined fraction of particles was inserted randomly into the microstructure, the MCS clock was reset to zero and grain growth allowed to continue. The mean grain size initially increased but pinned out at the values shown in figs. 5.81 and 5.82. It is clear from these results that the pinned grain size is a function of the starting grain size as well as the particle fraction $f$.

If in two dimensions the initial grain area $\bar{A}_0 \gg a$, the area of an individual lattice point, or in three dimensions the initial grain volume $\bar{V}_0 \gg v$, the volume of an individual lattice point, which is the particle size, the limited results from the two sets of simulations appear to be:

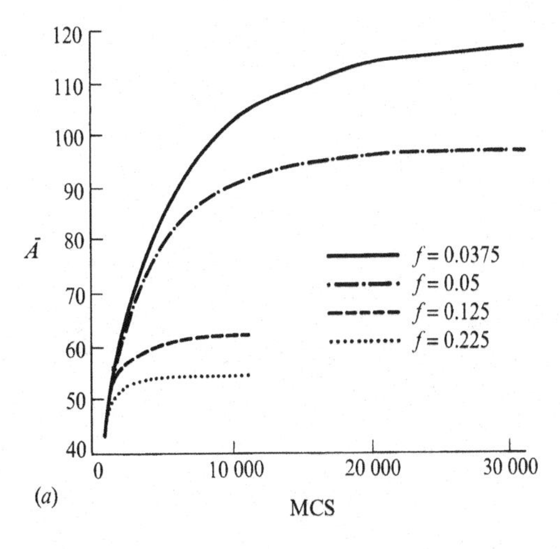


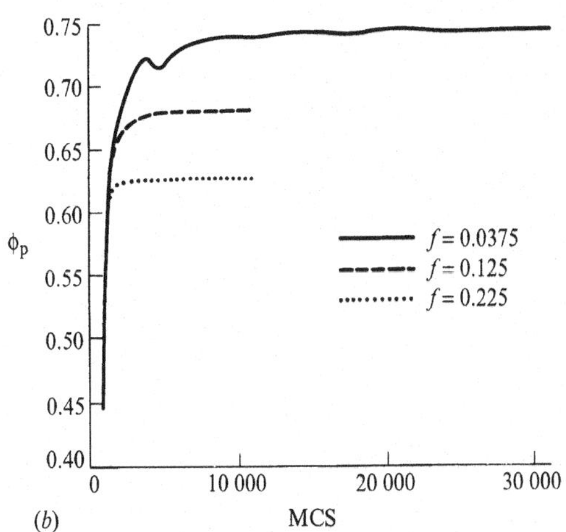


**Figure 5.83** Two-dimensional grain growth simulations from an initial grain area, $\bar{A}_0=42a$, at different volume fractions, $f$=0.0375, 0.05, 0.125 and 0.225, as a function of the time of growth in MCS. (*a*) The mean grain area $\bar{A}$ and (*b*) the fraction of particles, $\phi_p$, on the grain boundaries. (Doherty *et al.* 1989. Courtesy of Riso National Laboratory.)

$$\bar{A}_{2D} \approx \bar{A}_0 + a/f \tag{5.49a}$$

$$\bar{V}_{3D} \approx \bar{V}_0 + 100v/f \tag{5.51c}$$

If these results are added to the Olgaard and Evans plot of fig. 5.78, they would be found to be *above* the trend line for the MCS simulations with $\bar{V}_0=v$, (Li 1992).

A further result from the two-dimensional simulations of Doherty *et al.* (1989) is shown in fig. 5.83. This figure plots the change of mean grain area $\bar{A}$ and the fraction $\phi_p$ of particles on the grain boundaries during the simulated grain growth from a finite grain size, $\bar{A}_0=42a$. It can be seen that the fraction of particles of the boundaries, $\phi_p$, rose rapidly from the initial value, which was the *random value* for the starting grain size. This rise in $\phi_p$, shows that during grain

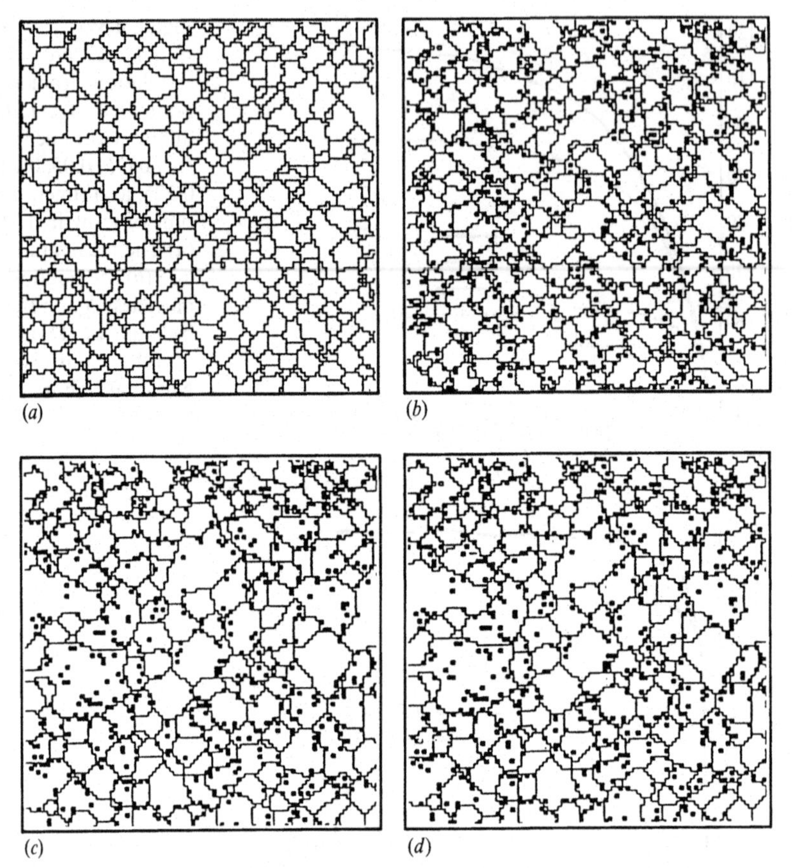


**Figure 5.84** Simulated three-dimensional grain growth in a 'sintered' structure with an imposed initial full correlation of particles on grain boundaries: (*a*) the starting single-phase grain structure, with $\bar{V}_0=150v$, before addition of particles onto the grain boundaries, (*b*) after 10 140 MCS of growth, (*c*) after 16 140 MCS of growth and (*d*) the pinned structure (after 20 140 MCS). (After Li 1992.)

growth boundaries migrate up to particles and are then held so the Zener assumption of random correlation again fails in a grain growth situation. Interestingly, the $\phi_p$ values saturated more quickly than did the mean grain sizes in fig. 5.83. The origin of this effect appears to be that only a small amount of grain boundary migration is needed to raise $\phi_p$ rapidly above the random value but grain growth continued in the simulation by loss of many of the small grains that did not contain the three particles needed to stop their disappearance in two dimensions.

A different effect was found in the limited number of three-dimensional simulations carried out for 'sintered' microstructures by Li (1992). Fig. 5.84 shows one of the runs in which for a finite $\bar{V}_0=150v$, a fraction $f=0.08$ was placed in a highly non-random manner, entirely at grain corners or at grain edges, so that initially $\phi_p=1$. During the subsequent simulated grain growth many small grains vanished giving a considerable increase in the mean grain volume and the fraction of particles on grain boundaries fell to $\phi_p=0.85$. The pinned grain size produced, $\bar{V}_p=380v$, was, however, significantly smaller than the value found in the initial grain growth simulations with $\bar{V}_0=v$ where at the same volume fraction, $f=0.08$, the pinned value of $\bar{V}_p$ was $930v$. So, if the results for these 'sintered' or non-random simulations are added to the Olgaard and Evans plot of fig. 5.78, they would be found to be *below* the trend line for the MCS simulations with $\bar{V}_0=v$, (Li 1992), even though there was a large initial grain volume.

The limited simulations in both two and three dimensions (*A*) subject to a large initial grain size or (*B*) with non-randomly placed particles strongly suggest that the first of these factors (*A*) gives a larger limiting grain size and that the second factor (*B*) leads to a smaller limiting grain size than found for standard runs. It would then appear that much of the large experimental scatter in the Olgaard and Evans plot of limiting grain-particle size as a function of particle fraction could result from the two effects, (*A*) and (*B*), given above. If the initial grain size is larger than the particle size, but there is a random correlation of particles and grain boundaries then the pinned grain size will be *larger* than the Monte Carlo simulation result of $\bar{R}_p=4.5r/f^{1/3}$. However, if despite a large starting grain size there is a strong initial correlation of particles at grain boundaries, such as is found in most sintering experiments on tempered martensite, then the limiting grain sizes may be even smaller than the initial prediction of the simulations.

The current set of experimental results suggest that the limiting grain size produced by normal grain growth is indeed a function of particle size and volume fraction (as $r/f^{1/3}$) but this simple result will be modified by (*A*) a large starting grain size, so increasing the expected limiting grain size, and (*B*) any initial correlation of particles and grain boundaries which will reduce the final pinned grain size. In these circumstances the large scatter shown by Olgaard and Evans (1986) in fig. 5.78 is not surprising.

### Coupled grain and particle coarsening

Hillert (1965) and Gladman (1966) discussed how Ostwald ripening of precipitates should allow a steady increase of the limiting grain size pinned by the particles. If the mean precipitate radius $r$ increases either at the lattice diffusion controlled rate ($r \approx kt^{1/3}$) or at the rate controlled by grain boundary diffusion ($r \approx kt^{1/4}$) then the observed grain size should coarsen with the same kinetics. The grain size appears, see Olgaard and Evans (1986), to be directly proportional to the particle radius at a fixed volume fraction.

Higgins, Wiryolukito and Nash (1992) and Yang, Higgins and Nash (1992) observed significantly faster coarsening of coupled two-phase structures in $\alpha/\beta$

brass (Cu–Zn) and in Ni–Ag alloys than could occur by solute transport between the dispersed phase particles even taking into account the faster solute diffusion along the matrix grain boundaries. They developed an interesting model in which particle coarsening occurred by means of particle coalescence. The particles situated at grain boundary edges and corners were pulled by the grain boundary tensions into contact with each other, fig. 5.85. In this model particle motion occurs by diffusion within, or along, a particle by differences in the curvature of the interfaces on different faces of a particle. The differences in particle curvature arose from the removal of grain boundary curvature by the particles – fig. 5.86. That is, the curvatures in the grain boundaries are transferred to the particle faces while maintaining equilibrium at the particle–grain boundary edges.

The rate of particle coarsening by this process, with solute transport around the particle–matrix interface, involving an interface diffusion coefficient, $D_{\alpha\beta}$, is given by Higgins *et al.* (1992), as

$$\bar{r}_t^4 - \bar{r}_0^4 = 16 w D_{\alpha\beta} \sigma N_\alpha(\mathrm{pb}) V_\mathrm{m} t / RT \tag{5.52b}$$

As in eq. (5.22) $w$ is the width of the phase boundary in which accelerated diffusion occurs and $N_\alpha(\mathrm{pb})$ is the solute content of the phase boundary. This particle drag model for coupled grain growth and particle coarsening was shown by Higgins *et al.* (1992) to fit the experimental results for the systems studied much better than did the simple particle coarsening model (eq. (5.22)).

## 5.8.5 Abnormal grain growth

Fig. 5.58 shows that a grain which is much larger than average will have many more than six neighbouring grains and should then be able to grow more rapidly at a rate determined, not by its own size, $D_\mathrm{A}$, but by the size of the matrix grains $\bar{D}_\mathrm{m}$. In this case eq. (5.32a) becomes

$$\mathrm{d}D_\mathrm{A}/\mathrm{d}t \approx 2M\sigma_\mathrm{b} V_\mathrm{a} / \beta \bar{D}_\mathrm{m} \tag{5.53}$$

If $\bar{D}_\mathrm{m}$ does not increase, for example, because of particle pinning, this rate of growth will be constant. This type of so-called 'abnormal' grain growth requires first the selection or 'nucleation' of the occasional existing grain which is to become a successful growing grain and then its subsequent growth into a matrix of grains which is held at a constant grain size $\bar{D}_\mathrm{m}$. Abnormal grain growth has transformation kinetics and resulting grain morphologies similar to those seen in primary recrystallisation of a cold-worked matrix. This similarity has resulted in the alternative name of 'secondary' recrystallisation for the process. This similarity has led to some confusion. Petrovic and Ebert (1972a,b) showed that the morphological changes in deformed thoria-dispersed nickel that were initially thought to be primary recrystallisation were, in fact, those of abnormal grain growth. The initial recrystallisation of the deformed material resulted in grains that were of the order of 1 $\mu$m in size but whose subsequent normal growth was inhibited by the dispersed thoria particles until abnormal grain growth took place. Abnormal grain growth can be a very serious drawback to the use of second-phase precipitates to control and limit grain size, since the onset of this process of abnormal grain growth leads to the complete destruction of the fine grain size and its replacement by a structure with a grain size that can be very much larger. In fact, if only one 'nucleus' grain develops the material will become a single crystal, and not even the sheet thickness or wire diameter will

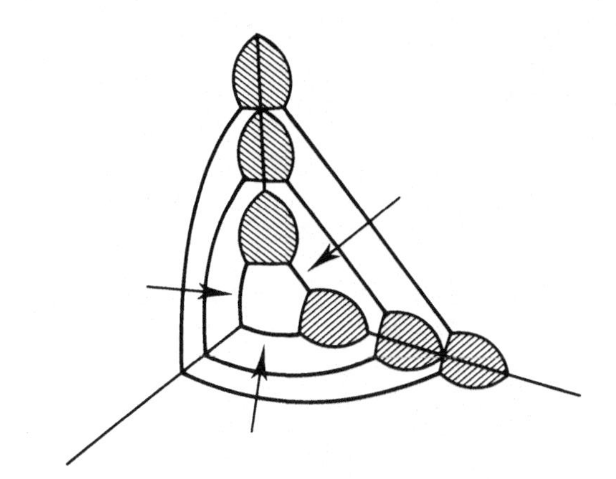

**Figure 5.85** Schematic of the motion of two particles at grain corners moving into coalescence as the small grain on which they sit shrinks and finally vanishes. (Higgins *et al.* 1992. Courtesy of Trans Tech Publications.)

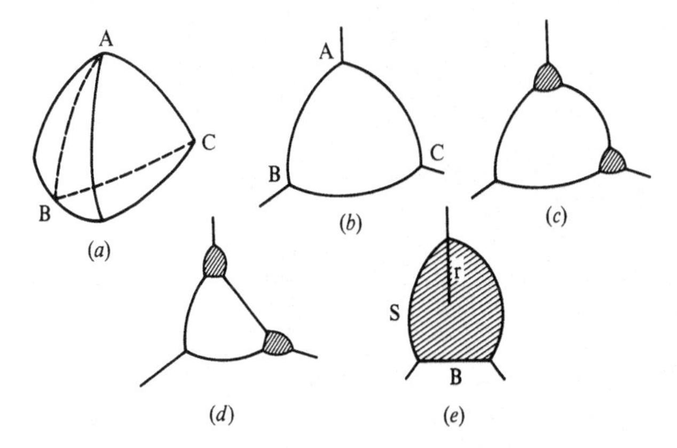


**Figure 5.86** (*a*) Three-dimensional view of a shrinking tetrahedral grain, (*b*) section through the grain, (*c*) as (*b*) but with particles at the grain corners, (*d*) the situation with *mobile* grain boundaries when the boundary curvature has been removed by the introduction of *differential curvatures in the particle–grain interfaces* and (*e*) an enlarged view of one of the particles in (*d*). Solute transport driven by the differential curvature will move the particles towards each other, as sketched in fig. 5.85. (Higgins *et al.* 1992. Courtesy of Trans Tech Publications.)

limit the grain size. A clear example of this process is shown in fig. 5.87 for iron containing a dispersion of $Fe_3C$ precipitates. The growing 'abnormal' grains are seen to have developed near the sheet surface and to be growing inwards.

Beck, Holzworth and Sperry (1949) provided an early review of abnormal grain growth and reported their own experiments using high purity Al–Mn alloys. They showed that the required inhibition of normal grain growth was most often provided by a low volume fraction of second-phase particles but that a strong 'cube' {100} ⟨001⟩ texture that left many of the grain boundaries as low angle–low mobility boundaries was also responsible in a few cases in the absence of second-phase particles. They found that annealing under conditions *close to, but not above*, the solvus temperature usually led to abnormal grain growth par-

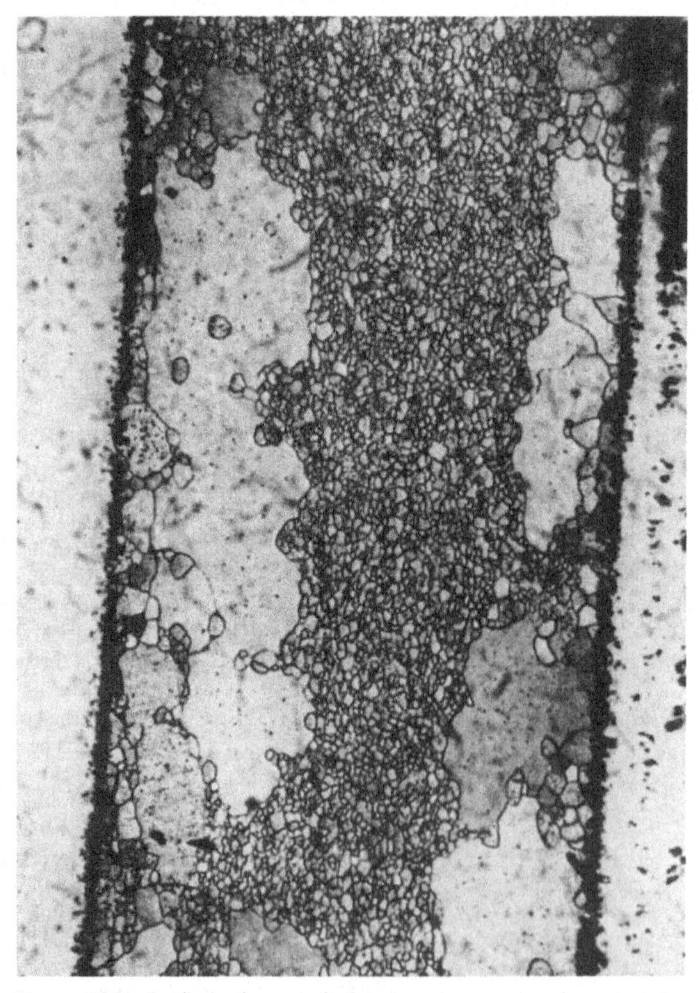

**Figure 5.87** Optical micrograph showing abnormal grain growth in a fine grain steel containing 0.4 wt% carbon. The matrix grains are prevented from growing by a fine dispersion of carbide particles that are not revealed. Magnification ×135. (After Gawne and Higgins 1971. Courtesy of the Metals Society.)

ticularly for an initially fine grain structure. In addition they reviewed earlier studies in which a gas-phase reaction preferentially dissolved pinning particles, for example, oxidation to remove carbides or hydrogen treatment to remove oxides led to surface nucleated abnormal grain growth as the volume fraction of particles at the surface fell towards zero.

Hillert (1965) and Gladman (1966) subsequently discussed the problem of how abnormal grain growth might be initiated by the Ostwald ripening of the precipitates. Petrovic and Ebert (1972a,b) have shown for nickel alloys containing fine dispersions of thoria (TD nickel) that abnormal grain growth does not always immediately result from inhibition of normal grain growth but did not suggest what might prevent abnormal grain growth from starting. Bjorklund and Hillert (1974), using the analysis developed earlier by Hillert (1965), have suggested one possible answer in terms of the microstructure produced by primary recrystallisation. Since the driving force of stored energy is much larger than that of grain growth, primary recrystallisation can occur in the presence of precipitates to give a grain size which may be larger than the limiting grain size of eqs. (5.31). Under these circumstances not only will normal grain growth be impossible but the abnormal growth of even the largest grain in the distribution may also be prevented. Their analysis suggests that if the mean grain size produced by primary recrystallisation is about three times the limiting value given by eq. (5.31), then abnormal grain growth will also be prevented. However, they suggested that if the grain size is between the limiting grain size and three times the limiting grain size, then abnormal grain growth can occur. In a situation where both normal and abnormal grain growth is prevented but precipitate coarsening can occur, this coarsening will reduce the precipitate drag and increase the limiting grain size. Continued annealing at high temperature will then result in abnormal grain growth. Some experimental evidence for this was found in a study of structural change in Cu–Cd alloys. A similar effect is obviously possible – in addition to coarsening a reduction of the volume fraction can occur and lead to abnormal grain growth.

Beck and Sperry (1949) pointed out that in a material with a very sharp annealing texture normal grain growth should also be inhibited. This can arise since a sharp texture means that the boundaries between adjacent grains will often be of a low average misorientation and this results in a reduction in both the grain boundary energy (the driving force) for grain growth and also in the grain boundary mobility. It is therefore reasonable that, as in some of the work they reviewed, the texture inhibition can produce the stabilised matrix grain structure which allows a 'rogue' grain of a different orientation to grow by abnormal grain growth. 'Nucleation' in such a process would be very similar indeed to nucleation in primary recrystallisation. It should be noted that only if the texture has a single variant of a single texture component, for example, the $\{100\}\langle 001\rangle$ 'cube' texture in fcc or bcc materials, will the necessary condition of general low angle boundaries be achieved. Between different variants of less symmetrical texture component, such as the two {112}⟨111⟩ 'copper' components in fcc metal, there are high angle and thus mobile grain boundaries.

One of the most important and frequently studied materials is Fe–3 wt% Si. Secondary recrystallisation of this material in sheet form leads to the desired texture for the soft magnetic properties of iron for transformer cores. May and Turnbull (1958) investigated both the texture changes and the influence of manganese sulphide inclusions and showed quite clearly the vital role of these inclusions for the occurrence of abnormal grain growth – their results are given in fig. 5.88, which illustrates many of the features discussed in this section. A full review of abnormal grain growth with particular emphasis to Fe–3 wt% Si was given in a major review article by Dunn and Walter (1966).

Another piece of work that confirms the importance of the dispersed phase in allowing abnormal grain growth to develop is that of Calvet and Renon (1960). They studied recrystallisation and grain growth in Al–Cu and other alloys, and

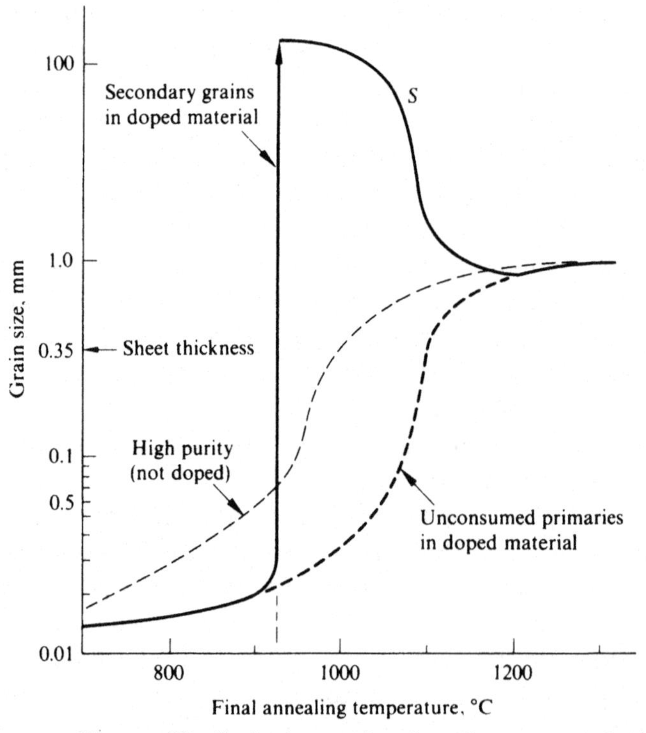


**Figure 5.88** Grain size as a function of temperature for 1 hour anneals of pure and MnS-doped Fe–Si following 50% reduction by cold-rolling to give a sheet of 0.35 mm thickness. Above a critical temperature a few grains grow by secondary recrystallisation to a size much larger than the specimen thickness. The fall in the curve at $S$ is due to the large number of secondary grains that grow at higher temperatures. Above 1100 °C all grains are limited to a size controlled by the specimen thickness. (After May and Turnbull 1958. Courtesy of AIME.)

found, as shown by Beck *et al.* (1949), that where the alloy was annealed at a temperature close to the solvus, abnormal grain growth always resulted; while annealing in the single-phase field gave normal grain growth. The correlation of abnormal grain growth with annealing at temperatures close to the solvus temperature strongly suggests that conditions in which $f \rightarrow 0$ are very favourable for promoting abnormal grain growth. This has been confirmed by an unanticipated occasion of abnormal grain growth. Liu (1987) attempted to produce, for a recrystallisation study, a range of two-phase Al–$Mg_2Si$ alloys with as wide a range of values of $f$ and $r$ as possible. In the course of some preliminary work, an alloy containing the maximum possible amount of $Mg_2Si$, but with a trace of iron as $Al_{12}Fe_3Si$, was, after casting, homogenised at the Al–$Mg_2Si$ eutectic temperature, which for the alloy used was also the solvus temperature for $Mg_2Si$. This anneal produced abnormal grain growth with some grains larger than 10 mm. The equiaxed as-cast grain structure was texture-free and had a typical DC cast grain size for aluminium alloys of about 150 $\mu$m. Dillamore (1978) in a review of abnormal grain growth in Fe–3 wt% Si pointed out that the critical anneals are usually carried out in an atmosphere of hydrogen that will act to remove sulphur from the surface of the sheet, so tending to reduce the fraction of MnS particles at the surface towards zero. In May and Turnbull's classic study of abnormal grain growth (1958) in Fe–Si alloys they showed that comparing two alloys with different amounts of MnS, the alloy with the larger amount of MnS had to be annealed at a higher temperature to initiate abnormal grain growth. Their anneals were carried out in dry hydrogen. All these studies lead to the conclusion that a necessary condition for abnormal grain growth is annealing in conditions where $f$ is finite but very small. Only for materials with very sharp cube textures that may be inhibited by low mobility of the boundaries has abnormal grain growth been demonstrated in the absence of second-phase particles.

A difficulty has, however, always existed in the understanding of abnormal grain growth in the presence of particles. This difficulty is that the derivation of the Zener relationship for a grain size stable against normal grain growth (eq. (5.48)) appears to apply to both normal and abnormal grain growth. This had been pointed out by, for example, Doherty (1984). In both cases, stability aganst grain growth is found when the Zener drag ($Z=3f\sigma_b/2r$) just balances the pressure from the curvature provided by the size of the average grain, $P=2\sigma_b/\beta\bar{D}$. However, the demonstration, by Monte Carlo simulations (Anderson *et al.* 1989b), of the failure of the Zener equation to predict the limiting grain size in normal grain growth has provided a possible solution to this difficulty (Doherty *et al.* 1990; Li 1992). This idea is discussed below.

In order to test the validity of the Monte Carlo predicted limits of normal grain growth, eq. (5.51b), against the Zener limit, eq. (5.48), a series of grain growth experiments was carried out by Doherty *et al.* (1989). An Al–Fe–Si alloy was used; this contained a small fraction of eutectic $Al_3Fe$ particles ('constituents' in the language of aluminium casting technology) with a very low value of $f$(0.002). This small value of $f$ was chosen to produce a large difference in the relevant parameters of the two equations – $r/f$ and $r/f^{1/3}$. These low solubility large particles coarsened very slowly even at temperatures close to the melting point of aluminium. The Zener predicted grain size (eq. (5.48)) was over 2 mm while that predicted by the Monte Carlo simulation was only 60 $\mu$m (eq. (5.51a)) if the initial grains were the size of the particles (2 $\mu$m) or somewhat larger if the starting grain size was larger (eq. (5.51b)). Table 5.7 gives a summary of the results obtained in this test. Various important features can be seen in table 5.7. Firstly, in all cases, there is a small amount of general matrix grain growth from the initial grain size, just as in the computer simulation from a finite grain size (figs. 5.79 and 5.80). Beck *et al* (1949) reported a similar result. Secondly, this matrix grain growth stopped at grain sizes much smaller than the Zener limit of about 2 mm. Finally, and most significantly, it was found that with a small initial grain size (<150 $\mu$m) abnormal grain growth always occurred. Subsequent studies confirmed the absence of texture in the recrystallised structures both before and after abnormal grain growth. The grains grew in many cases to sizes of the order of $10^4$ $\mu$m or more, very much larger than the Zener limit. Subsequent studies by Doherty *et al.* (1990) demonstrated that, even in samples with a recrystallised grain size as large as 250 $\mu$m, abnormal grain growth could be stimulated by applying a hardness indentation. During annealing, localised recrystallisation took place in the vicinity of the hardness indentations. This *local* recrystallisation produced a heterogeneity of grain sizes in the region of the indentation and this appeared to promote the 'nucleation' of one or two abnormal grains. This brief set of experiments indicates that a combination of a low

**Table 5.7** *Data on grain sizes (in $\mu$m) in Al – 0.15 wt% Fe – 0.07 wt% Si. Samples deformed various amounts, recrystallized and the subjected to extended anneal at 620 °C. (Doherty et al. 1989.)*

| Reduction (%) | Rex. temp (%) | As-rexed* | 1 h | 4 h | 8 h | 20 h | 112 h |
|---|---|---|---|---|---|---|---|
| 20 | 500 | 280 | 254 | 274 | | 270 | 272 |
| 30 | 500 | 176 | 186 | | | 208 | |
| 34 | 450 | 200 | 208 | | | 220 | |
| 40 | 450 | 132 | 208 | 234 | A | A | |
| 60 | 400 | 85 | 200 | A215 | A192 | A208 | |

A – indicates that abnormal grain growth has occurred in part or most of the sample.
* – Grain size at end of recrystallisation before annealing at 620 °C.

value of $f$ and a fine recrystallised grain size significantly smaller than the Zener limit appear to give rise to abnormal grain growth – particularly if a local grain size heterogeneity is also present.

Li (1992) attempted to study the onset of abnormal grain growth in the three-dimensional Monte Carlo simulations, as initially described by Doherty *et al.* (1990). The starting point for these simulations was the pinned grain structures from the initial simulations described by Anderson *et al.* (1989b), see figs. 5.75–5.77. The first abnormal grain growth simulations were carried out by placing a large spherical grain into the centre of the pinned grain structure. This set of simulations was, however, not successful. The added grains, if only a little larger than the matrix grains, only grew a little and then stopped. When the added grain was made larger, it soon reached the surfaces of the cubic array of lattice points and, due to the boundary conditions used, the grain then 'joined up' on itself and, as a consequence, the net grain curvature 'inverted' as it grew in an abnormal manner into the corners of the cubic lattice. This process was clearly an artefact of the lattice size used in the simulation.

In order to overcome the difficulties of simulating the 'nucleation' of abnormal grain growth, a set of simulations of the post nucleation growth was attempted. Onto the pinned grain structures, such as those of fig. 5.77, a layer, two lattice planes thick, of a single grain was placed on the face of the cube of lattice points. This insertion simulated the artificial nucleation of a giant grain whose initial radius of curvature was infinite. The model requires symmetry on opposite surfaces so the artificially nucleated giant grain was placed on two opposite faces of the grain structure. Figs. 5.89–5.91 show what occurred in three of the simulations run at different values of the particle fraction $f$. At the lowest value of $f$, $f=0.005$, seen in fig. 5.89, the interface rapidly corrugated on a grain scale as the large grain migrated locally into the grain boundaries of the matrix grains. This produced the required local radii of curvature, on the scale of the pinned matrix grain size, at the boundaries of the giant grain. Under this local driving force the giant grain grew forward, at a nearly constant velocity, and entirely consumed the structure. That is, continued *growth* of an abnormal grain was shown, by the simulation, to occur under the conditions of a low value of $f$, with a matrix grain size pinned at the size given by eq. (5.51a) *provided a giant grain was nucleated.* A similar behaviour was found for $f=0.01$ (Li 1992). At a larger value of the density of particles, $f=0.08$ shown in fig. 5.90, abnormal grain growth started, consumed 2–3 layers of the matrix grains but then stopped. It can be seen at $GG'$, in fig. 5.90, that some *matrix grain growth* has been stimulated by the approach of the giant grain. The same type of behaviour was also found at $f=0.04$. At the highest value of the fraction of particles, $f=0.16$, seen in fig. 5.91, the giant grain face merely adjusted locally to the neighbouring grains but did not consume more than about a half layer of the adjacent matrix grains. So, the simulations showed that:

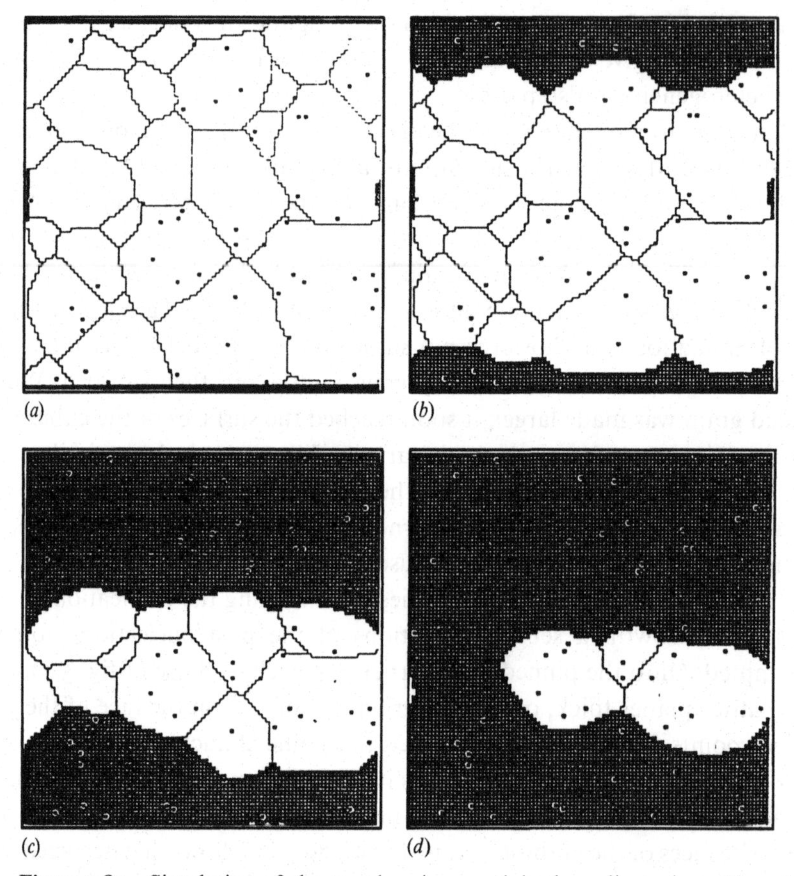


**Figure 5.89** Simulation of abnormal grain growth in three dimensions. Pinned grain structure at $f=0.005$ with addition of two grains with infinite radii at the top and bottom surfaces: (*a*) $t=0$ MCS; (*b*) $t=6000$ MCS; (*c*) $t=37\,000$ MCS; and (*d*) $t=43\,400$ MCS. (After Li 1992.)

(i) At low values of $f$ (0.005 and 0.01) artificially nucleated large grains could grow readily into the pinned structures giving complete abnormal grain growth.

(ii) At intermediate values of $f$, abnormal grain growth from an artificially nucleated grain could start but that it was stopped, apparently inhibited by normal grain growth, just ahead of the growth front.

(iii) At the highest value of $f$, abnormal grain growth even from the artificial giant grain did not even start.

The reduced tendency at higher values of $f$ for abnormal grain growth into the pinned structures occurred *despite the fall in matrix grain size with larger f*, $\langle R\rangle=4.5r/f^{1/3}$.

Detailed analyses of these abnormal grain growth simulations by Li (1992) have provided some important insights into this phenomenon, table 5.8. Firstly,

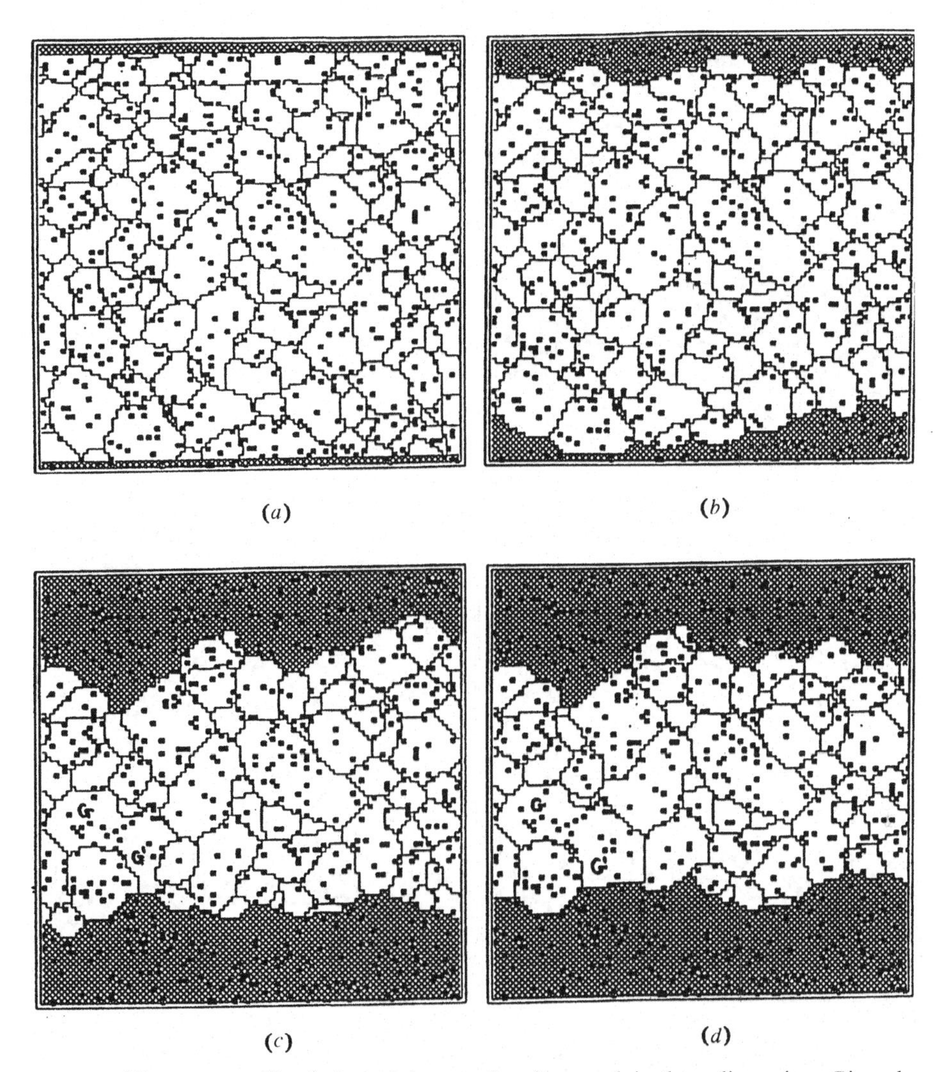


**Figure 5.90** Simulation of abnormal grain growth in three dimensions. Pinned grain structure at $f=0.08$ with addition of two grains with infinite radii at the top and bottom surfaces: (*a*) $t=0$ MCS; (*b*) $t=4700$ MCS; (*c*) $t=198\,000$ MCS; and (*d*) $t=410\,400$ MCS. Note that while abnormal grain growth occurred for a short distince it finally halted. Some matrix grain growth at *GG'* can be seen, apparently triggered by the arrival of the giant grain. (After Li 1992.)

it can be seen that before the unconstrained abnormal grain growth at $f=0.005$, the density of particles on the abnormal grain faces, $\phi_f$, on its grain edges, $\phi_e$, and on its corners, $\phi_c$, had the expected random values, $\phi_f \approx \phi_e \approx \phi_c \approx f$. These densities are defined as the fraction of boundary lattice sites occupied by particles. At $f=0.005$, there were so few particles on the limited number of corners that the value of $\delta_c$ (though$\approx f$) was statistically unreliable. Secondly on the abnormal grain, *the fraction of particles was found to remain close to the random*

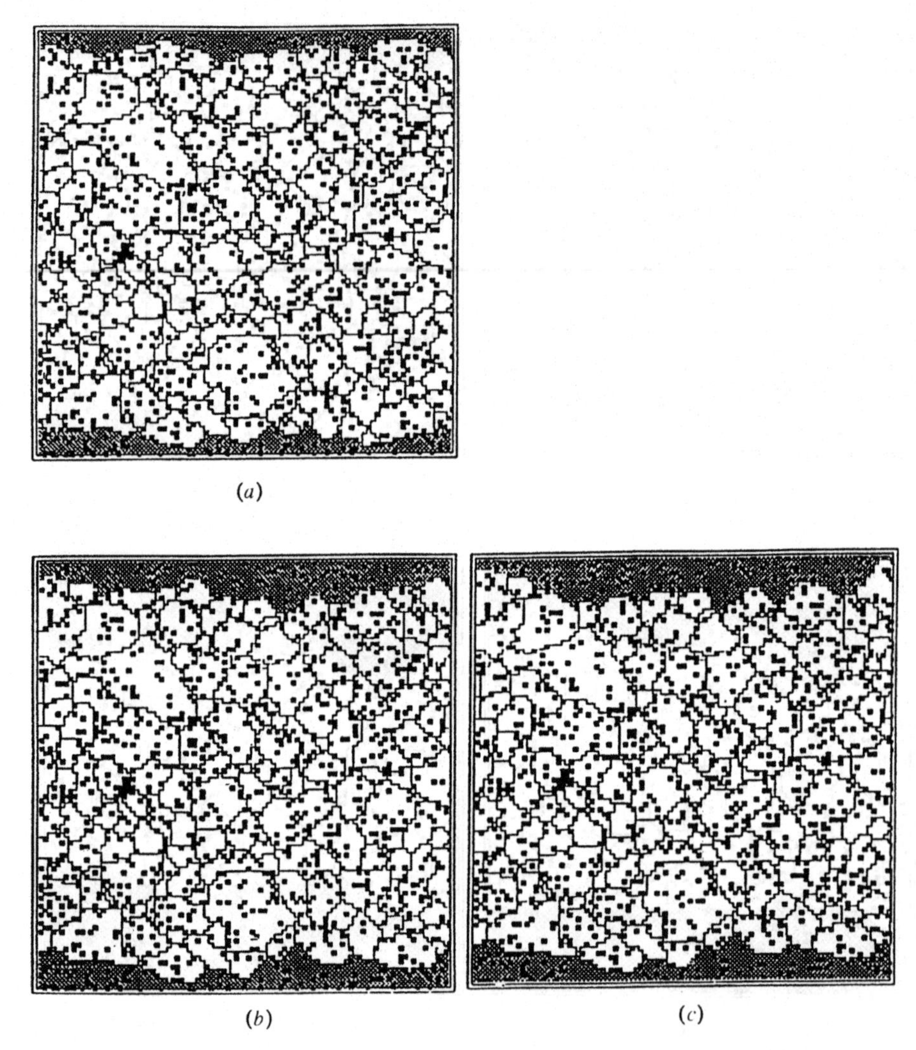


**Figure 5.91** Simulation of inhibited normal grain growth in three dimensions. Pinned grain structure at $f$=0.16 with addition of two grains with infinite radii at the top and bottom surfaces: (*a*) $t$=1840 MCS; (*b*) $t$=7040 MCS; (*c*) $t$=10 380 MCS. In this case after an initial relaxation of the abnormal grain interface no further boundary migration occurred.

*value during this simulated abnormal grain growth.* That is, during abnormal grain growth the migrating interface samples a random density of particles. This the same result as was found in both the experimental (Ashby *et al.* 1969) and the simulated (Rollett, Srolovitz, Anderson and Doherty 1992) studies of recrystallisation. That is, the 'pressurised' migrating grain boundary passing through the particle dispersion appears to sample a random density of particles. This random sampling means that the Zener assumption (eq. (5.44)) of random interaction is valid not only in the case of recrystallisation (volume driving force)

**Table 5.8** *Microstructural details of the boundaries of the giant grains in simulated abnormal grain growth. $f$ is the volume fraction of randomly dispersed particles, MCS is the time of simulation, $V_a$ the volume of the two abnormal grains, NB is the number of matrix grains next to the giant grains, F is the number of face sites, E the number of edge sites and C is the number of corner sites. $\phi_f$, $\phi_e$ and $\phi_c$ are the densities of particles on the faces, edges and corners of the giant grains. (Li 1992.)*

| $f$ | MCS ×1000 | $V_a$ ×1000 | $NB$ | $F$ | $E$ | $C$ | $\phi_f$ | $\phi_e$ | $\phi_c^a$ |
|---|---|---|---|---|---|---|---|---|---|
| 0.5% | 0 | 40.0 | 53 | 19859 | 4162 | 412 | 0.0056 | 0.0054 | 0.0066 |
| | 0.2 | 44.5 | 57 | 23683 | 4777 | 448 | 0.0051 | 0.0053 | 0.0072 |
| | 0.44 | 49.3 | 52 | 25662 | 4650 | 462 | 0.0054 | 0.0058 | 0.0059 |
| | 0.94 | 60.2 | 55 | 28257 | 4525 | 427 | 0.0056 | 0.0062 | 0.0052 |
| | 1.94 | 83.1 | 59 | 30499 | 4534 | 406 | 0.0059 | 0.0056 | 0.0073 |
| | 5.8 | 163.4 | 57 | 31086 | 4440 | 430 | 0.0064 | 0.0066 | 0.0064 |
| 8% | 0 | 36.8 | 469 | 18308 | 7443 | 1494 | 0.080 | 0.071 | 0.084 |
| | 0.5 | 67.0 | — | — | — | — | 0.091 | 0.093 | 0.096 |
| | 1.0 | 83.6 | 214 | 29417 | 7434 | 1227 | 0.093 | 0.095 | 0.100 |
| | 9.6 | 229.2 | 183 | 34163 | 6602 | 924 | 0.093 | 0.103 | 0.128 |
| | 198.4 | 335.2 | 157 | 35416 | 6305 | 800 | 0.093 | 0.104 | 0.127 |
| | 414.0 | 388.2* | 139 | 36452 | 5775 | 640 | 0.093 | 0.103 | 0.119 |
| 16% | 0 | 33.52 | 653 | 16626 | 8037 | 1881 | 0.161 | 0.143 | 0.169 |
| | 1.8 | 70.93 | 261 | 26391 | 6768 | 1151 | 0.181 | 0.188 | 0.209 |
| | 3.6 | 86.03 | 235 | 27574 | 6404 | 1013 | 0.182 | 0.194 | 0.218 |
| | 5.3 | 95.09 | 225 | 27956 | 6236 | 964 | 0.184 | 0.201 | 0.221 |
| | 70.4 | 101.02 | 230 | 28395 | 6169 | 941 | 0.185 | 0.203 | 0.229 |
| | 87.0 | 104.5 | 230 | 28599 | 6113 | 904 | 0.185 | 0.204 | 0.234 |
| | 103.8 | 107.9* | 211 | 28903 | 6067 | 883 | 0.184 | 0.200 | 0.229 |

*Giant grain pinned; [a]Large statistical scatter for 0.5% – so the differences from $f$ are not significant.

but also for abnormal grain growth. The validity of the Zener assumption in abnormal grain growth is in marked contrast to its failure for normal grain growth. The difference appears to be that for the abnormal grain growth case the size of the giant growing grain *is very much larger* than the interparticle spacing, $\Delta_3$ ($\Delta_3 \approx r/f^{1/3}$). Normal grain growth ceases after the grains have grown by this interparticle spacing. It may be noted that during recrystallisation of deformed high stacking fault energy metals such as aluminium, recovery reduces the distributed dislocations to a set of subgrain boundaries. That is, the volume stored energy of deformation has been replaced by an area of subgrain boundaries, with a certain area per unit volume, $S_v$, and the subboundaries have an energy per unit area, $\sigma_s$. So, during recrystallisation of such a microstructure in the presence of particles, (see, for example, Doherty and Martin (1962/3) or Hutchinson and Duggan (1978)) the microstructure at the recrystallisation front is very similar to that simulated in fig. 5.89 for abnormal grain growth. Most of the subgrain boundaries are pinned by particles, while the recrystallisation occurs, driven by the local curvatures of the recrystallisation front arising from the fine subgrain size. This was, indeed, the situation found experimentally by Ashby *et al.* (1969) in which a random density of particles on the recrystallisation front was seen.

If abnormal grain growth occurs, as it appears to, with a random correlation of particles with the growth front, then we can use the Zener value of particle drag (eq. (5.46)) for this situation. The condition that the driving pressure for abnormal grain growth comes from the matrix grain size (eq. (5.53)) then leads immediately to eq. (5.48), that is, to $\bar{D}_Z = 4r/3f$ (for $\beta = 1$). *This expression is now the limit of the matrix grain size that could support abnormal grain growth.* It is, however, *not the limit for normal grain growth* (Doherty *et al.* 1990). The idea that eq. (5.48) gives the limit to abnormal growth was tested by Li (1992) in the following way. In the three-dimensional simulations the grain radius at which normal grain growth ceased is given by eq. (5.51a)

$$\bar{R}_p = 4.5r/f^{1/3}$$

If this is the matrix grain size into which abnormal grain growth cannot occur then by combining eqs. (5.48) and (5.51) we can obtain the critical value of $f$, $f^*$, that should just prevent the growth of an artificially nucleated grain into these microstructures.

$$4r/3f^* \approx \bar{R}_p \approx 4.5r/f^{*1/3} \qquad (5.54)$$

so that

$$f^* \approx (4/13.5)^{3/2} = 0.16 \qquad (5.54a)$$

This prediction was made before the simulation shown in fig. 5.91 (Li 1992) was carried out. The simulation matched the prediction to the extent that at $f$=0.16, abnormal grain growth did not even start, although it was seen to start at smaller values of $f$.

The initiation of abnormal grain growth followed by its inhibition at the intermediate values of $f$, fig. 5.90, appears to arise because the movement of the boundary of the large 'abnormal' grain appears to trigger local grain growth. This was seen, for example, at $GG'$ in fig. 5.90. It can also be seen in table 5.8, where for all values of $f$ there was an immediate rise in the number of 'face' sites on the abnormal grain from MCS=0. This occurs as the artificial flat interface adjusts to the locally curved interface after a short period of growth. For the freely migrating interface at $f$=0.005, there was also an increase in the number of edge and corner sites. These edge and corner sites are points at which the interface is being pulled forward by matrix grain boundaries ahead of the growth front, figs. 5.89–5.91. For both $f$=0.08 and $f$=0.16, although the number of face sites increased *the number of edge and corner sites fell.* This fall in the edge and corner sites is a direct result of the preferential loss of small grains in the vicinity of the growing abnormal grain. This loss of small grains occurs by two different mechanisms. Firstly there is the preferential consumption of the smallest grains by the abnormal grain. This is likely to be seen only when the abnormal grain is becoming almost stationary. The other mechanism is that seen at $GG'$, the consumption of small matrix grains *by other matrix grains* apparently due to local unpinning of the previously stationary matrix grain boundaries triggered by the arrival of the abnormal grain front. In both microstructures, $f$=0.08 and 0.16, analysed in table 5.8, the driving pressures, measured by the number of grain edge and corner sites, fell by the loss of some of the smaller matrix grains. As this occurred and the rate of boundary movement fell, there was a corresponding small increase in the density of particles pinning the boundary of the giant grain. That is, $\phi_f$ and particularly $\phi_e$ and $\phi_c$ all rose as the boundary slowed. Finally this increase of particle density combined with the fall in matrix grain boundaries appeared to remove the residual boundary curvature needed for continued growth.

It should be noted that these insights into abnormal grain growth were all derived from a rather limited set of simulations using a rather small three-dimensional lattice. It would be good for them to be repeated and extended. However, the current tentative ideas seem to be capable of explaining, at least qualitatively, a large amount of current experimental data for which there appears to be no other obvious explanation.

The current picture generated from the computer models shows that:

(i) Normal grain growth can be stopped at a grain size much smaller than the Zener limit of $\bar{D}\approx 4r/3f$.

(ii) The increment of grain growth upto a limiting structure scales with the particle dispersion as $r/f^{1/3}$ in three dimensions, and as $r/f^{1/2}$ in two dimensions, the length scales of the mean interparticle spacings.

(iii) The limiting grain size increases with initial grain size.

(iv) The limiting grain size falls if the particles are initially highly correlated with grain boundaries. That is, when the density of particles on the grain boundaries is higher than random there is a smaller increment of grain growth than when the initial fraction of the grain boundary occupied by particles is the random value – the volume fraction of particles.

(v) The Zener limiting grain size appears to be that at which a microstructure is fully stable against abnormal grain growth.

(vi) The difference between the Zener size and the size limited by the interparticle spacing increases as the volume fraction of particles falls, so that at low volume fractions the system is increasingly unstable against abnormal grain growth. Abnormal grain growth requires that the grain size is small, only slightly larger than the particle spacing.

These insights appear to be fully compatible with the experimental observations. The experimental observations appear to show the highest susceptibility to abnormal grain growth at low values of the particle volume fraction and at the smallest initial grain sizes. Abnormal grain growth can be readily stimulated by local grain size inhomogeneities produced, for example, by recrystallisation from a hardness indentation.

Most of the known examples of abnormal grain growth occur for systems where a precipitate is present, but there are examples of abnormal grain growth reported in high purity metals. Simpson, Aust and Winegard (1971) did report abnormal grain growth following recrystallisation in cadmium and in lead, both pure to 1 ppm. However, no texture information was offered so it is not certain that texture inhibition was the cause, but clearly no second-phase particles were present since normal grain growth of the large grains produced by the abnormal growth stage occurred with the full grain growth exponent of $n=0.5$. The computer simulations of Srolovitz, Grest and Anderson (1985) and Rollett, Srolovitz and Anderson (1989) suggest that abnormal grain growth can only take place in single-phase microstructures under rather unusual conditions, such as a very strong single component texture. The computer results fit with most models of grain growth (Hillert 1965; Thompson, Frost and Spaepen 1987). In a microstructure where all the grain boundaries have similar mobility and energy then, even if a large grain whose radius $R_1$ is much larger than the mean size $\bar{R}$ is present, the rate of growth of the large grain, $dR_1/dt$, is actually smaller than the rate of growth of the mean grain size $d\bar{R}/dt$. Srolovitz *et al.* (1985) showed that under these circumstances a grain much larger than the mean grain size though it grows rapidly is still caught up by the growth of the mean grain size. This apparently surprising result occurs because the increase of mean grain size is actually determined by the rate of disappearance of the smallest grains in the system – this causes the number of grains per unit volume, $N_v$, to fall and the mean grain volume, $\bar{V}=1/N_v$, to rise. This disappearance of the small grains takes place in a time step, $t'$, determined by large shrinkage rate of small grains, fig. 5.67, and the small distance that their boundaries have to move (half the grain size) for the small grains to vanish.

$$t' \approx R/(dR/dt)$$

So, as shown analytically by Hillert (1965) and Thompson *et al.* (1987) and by the computer simulations of Srolovitz *et al.* (1985) and Rollett *et al.* (1989), during this time the growth of the largest grain is not fast enough to exceed the rate of growth of the mean grain size.

Rollett *et al.* (1989) showed, however, that abnormal grain growth could be produced in the two-dimensional computer simulations either by allowing a minority of grains to have low energy or, more realistically, by allowing a minority of grains to have higher mobility boundaries. It was thus confirmed by the simulation, as has been reported previously by Beck *et al.* (1949), that abnormal grain growth can occur in a material if it has recrystallised to a sharp texture in which many of the grain boundaries are low angle and thus low mobility. The microstructural and texture effects of this phenemonon have been explored in a statistical model by Abbruzzese and Lucke (1986) and Eichelkraut, Abbruzzese and Lucke (1988). They showed not only how texture can be changed during grain growth but how during such a texture change, abnormal grain growth of a small number of grains with the minority texture can consume a slowly growing matrix grain structure with a different texture. The statistical model was based on the differential mobility of grain boundaries between different texture components. The detailed computer model of Rollett *et al.* (1989) appears also to be applicable to the problem of nucleation of recrystallisation where the majority of boundaries are low angle subgrain boundaries and growth occurs from individual more highly misoriented subgrains, as discussed by Doherty (1978). Humphreys (1992) has specifically simulated this phenomenon and shown how misoriented subgrains with higher mobilities can quickly evolve into recrystallisation 'nuclei'. Geometrically, both the 'nucleation' and growth stages of primary recrystallisation in a well-recovered subgrain structure appear to be very similar to abnormal grain growth. In both phenomena a very small minority of grains (or subgrains) consume a large number of slowly growing neighbouring grains (subgrains). The only significant difference is that the preferential or 'abnormal' growth of a few subgrains with high mobility appears to be the *normal* means of recrystallisation nucleation since in most deformed subgrain structures most of the boundaries are low mobility. In grain growth the process

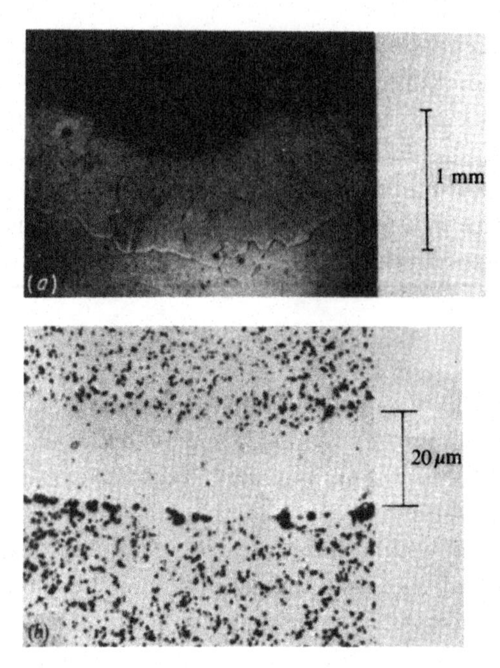


**Figure 5.92** Section through copper specimen initially a single crystal with a dispersion of silica particles, indented and then annealed at 1273 K for 2 h: (*a*) showing the indentation and recrystallised grains that have grown away from the indentation into progressively less deformed material; (*b*) an enlargement of the region seen in (*a*) showing the particle-free region swept by the boundaries furthest from the indentation. The boundary was shown by subsequent etching to lie along the dense line of particles at the bottom of the micrograph. (After Ashby and Centamore 1968. Courtesy of *Acta Met.*)

is abnormal since, in general, in grain structures most boundaries are very mobile so inhibition must be supplied by second-phase particles. Only in strongly textured materials will most of the boundaries be low mobility.

## 5.8.6 Dragging of particles by moving grain boundaries

With a high driving force, it is easy for a mobile grain boundary to migrate past second-phase inclusions. However, when the driving force is equal to or less than the drag due to the precipitates, boundary motion will cease – *unless the precipitates can be dragged along by the moving boundaries.* A clear early experimental demonstration of this effect was given by Ashby and Centamore (1968) using copper alloys internally oxidised to give dispersions of $B_2O_3$, $GeO_2$, $SiO_2$ or $Al_2O_3$. The samples were locally deformed by a hardness indentor and annealed at various temperatures. Nucleation of new grains took place in the heavily deformed regions close to the indentation and growth followed into the progressively less deformed metal away from the indentation (fig. 5.92). It is clear from these micrographs that over most of the distance that the boundaries had migrated, the particle dispersion was unaffected – for example, in fig. 5.92(*b*) – but that in the last 20 $\mu$m of boundary movement all the particles had been dragged forward by the migrating boundary. This seems reasonable: close to the

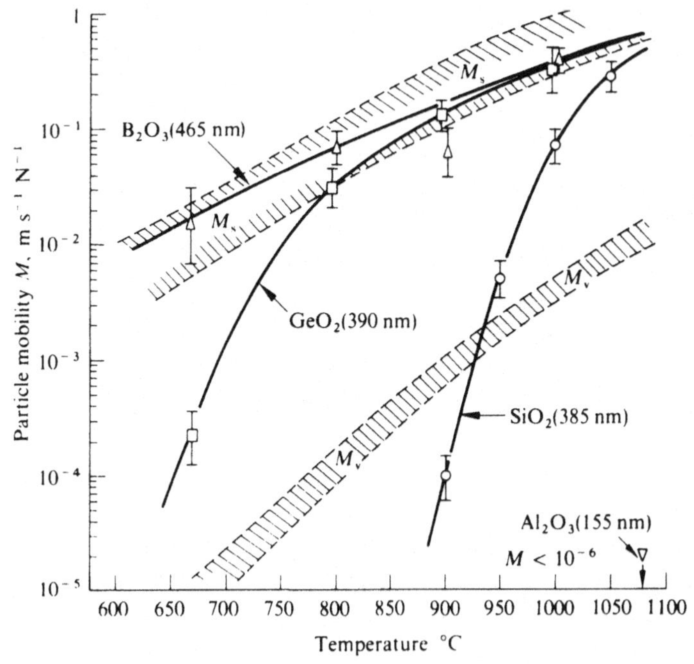


**Figure 5.93** The observed mobility of different particles plotted as full lines labelled by the type of particle and the average diameter. The expected mobility for volume diffusion control is shown as a shaded band labelled $M_v$ and that for surface diffusion control is shown labelled $M_s$ (After Ashby and Centamore 1968. Courtesy of *Acta Met.*)

indentation, the stored energy of the deformation was sufficiently high to be larger than the drag of the precipitates so the boundary could move easily past the particles. This same result was found by Rollett *et al.* (1992) in their simulations. However, at a critical distance from the indentation where the driving force just balanced the drag force, further boundary migration requires that the particles move. An analysis of the kinetics of the particle motion is summarised in fig. 5.93. Here the theoretically expected mobility of inclusions in copper is given for both surface diffusion control ($M_s$) and volume diffusion control ($M_v$) of the movement of copper atoms from in front of the particles to behind them as they move (see §6.2). It can be seen that the expected mobility due to surface diffusion is often higher than that found for the particle sizes and temperatures involved; moreover, only $B_2O_3$ has the expected mobility over the entire temperature range, $GeO_2$ only at the higher temperatures, $SiO_2$ only approaches the expected mobility at the highest temperature, while $Al_2O_3$ had an immeasurably small mobility at all temperatures (that is, no movement of $Al_2O_3$ could be detected). This discrepancy appeared to be due to the fluidity of the oxide – $B_2O_3$ was a liquid, $GeO_2$ becomes viscous at lower temperatures, $SiO_2$ becomes glassy and $Al_2O_3$ was crystalline at all temperatures. This oxide fluidity appears necessary for the copper atoms to leave the front of the inclusion and join the back of the inclusion, and when this becomes difficult the process is controlled not by the rate of copper diffusion but by plastic accommodation at the interface, an important element of *interface control.* Ashby and Centamore point out that the immobility they found for $Al_2O_3$ inclusions in internally oxidised copper differed from the mobility previously reported for $Al_2O_3$ after mechanical introduction into nickel and silver. They suggested that after internal oxidation a good interface bond is present between the metal and $Al_2O_3$, while in the mechanically mixed alloys the oxide may be sitting loosely in a hole and the mobility is merely that of the hole.

Higgins *et al.* (1992) and Yang *et al.* (1992) have, as discussed above, modelled the coupled coarsening of particles and grains by the process of particle drag by the boundaries, as previously shown in figs. 5.85 and 5.86. The grain boundaries and grain edges in the absence of particles would meet at the equilibrium angles of 120° or 109.5°. However, the presence of a particle at the edges or at the corners that holds back the movement of the grain edges will cause the grain boundary to migrate to a position of zero curvature, fig. 5.85. The zero curvature grain boundaries will then impose an unbalanced drag force on the particle. As is also shown by fig. 5.85, the removal of grain boundary curvature will impose a *difference of curvature on different parts of the particle–matrix interfaces*. Higgins *et al.* (1992) analysed the resulting particle motion due to the solute fluxes within or around the particle driven by these differences in curvature of an individual particle. With the fastest path for solute transport being diffusion along the interface, their analysis leads to particle coarsening at a rate predicted by eq. (5.52b). The reason for the accelerated coarsening by this model is that the required solute diffusion is over distances of the particle diameter not the particle spacings as would happen in conventional coarsening driven by curvature differences between particles of different sizes.

Annavarapu, Liu and Doherty (1992) studied grain growth in fine-grain-size, two-phase, solid plus liquid systems. The liquid phase behaves in a very similar way to the solid particles discussed above; that is, the liquid is dragged by the grain boundaries and as a result the rate limiting step will be the liquid phase mobility, controlled in this case by rapid solute transport within the liquid, so that length$^3$ versus time kinetics are expected for both the grains and the liquid films. The major difference between solid and liquid particles, at least in metallic systems, is that the liquid tends to wet the grain boundaries, $\theta \rightarrow 0°$, fig. 5.25 (since $\sigma_b \approx 2\sigma_{sl}$). As a result the liquid is present as very thin films along the two-dimensional grain boundaries and as star-shaped regions at triple points and grain corners, fig. 5.94. This micrograph also shows an interesting feature that is not yet fully understood:

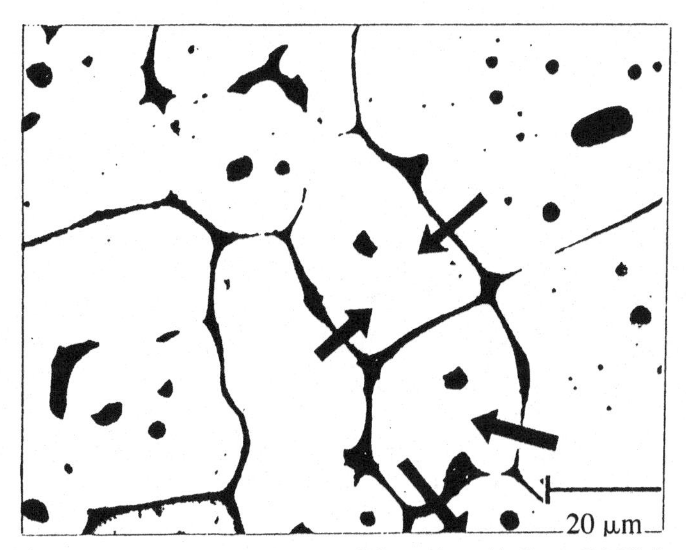


**Figure 5.94** Grain growth in partially molten Al–Cu at 600 °C. In most cases the molten liquid film migrates with the grain boundary but in a few cases liquid droplets are being left behind. (Annavarapu *et al.* 1992. Courtesy of Trans Tech Publications.)

this is that thicker regions of the liquid films can be apparently left behind by the migrating grain boundaries, even though the liquid wets the boundary. Annavarapu *et al.* suggested that this may arise when a large liquid droplet within the shrinking grain is captured by a rapidly moving boundary. The boundary appears to be moving past the droplet faster than the liquid in the droplet can spread itself along the boundary. Since the rate of boundary migration in the Annavarapu model is proportional to boundary thickness the thicker region of the boundary will be dropped behind by the more rapidly moving thin parts of the boundary. In coarser grain structures where the rate of boundary migration will be much slower, this bypassing process would not be expected to occur.

This phenomenon of particle drag can be both of considerable importance and a considerable nuisance, since its occurrence will lead to regions of the metal being depleted of the dispersion, and thus becoming likely sites for localised plastic deformation. In addition, continued drag will completely cover the boundary with a film of the second phase, which represents a potential hazard with respect to both corrosion and brittleness. The likelihood of a recrystallisation process having just the right energy to drag a boundary may be remote, except perhaps in lightly deformed polycrystalline specimens (Bjorklund and Hillert 1974). The possibility of particle drag is, however, always going to be present during grain growth. A general review of the topic of the movement of second-phase particles in solids was given by Gyeguzin and Krivoglaz (1971) and a discussion of the kinetics has been given by Ashby (1980).

## 5.9 Dendrite arm coarsening

Another process that appears to be very similar to the processes discussed in this chapter is the coarsening of dendrite arms. Kattamis and Flemings (1966) and Kattamis, Holmberg and Flemings (1967) showed that the spacing between adjacent arms of a solidification dendrite coarsened both during solidification and during isothermal holding. Kahlweit (1968) observed dendrite arm coarsening directly in the transparent system of ammonium chloride and water. He found that the fundamental process was the dissolution, at a diffusion controlled rate, of the material at the tips of the smallest diameter dendrite arms, the solubility being increased by the curvature as described by eq. (5.12). This phenomenon of dendrite arm coarsening is of great importance for the microstructure of cast metals (§2.12) and measurements of the arm spacing are widely used both to monitor the time spent freezing as-cast samples and in rapid solidification processing (Kurz and Fisher 1989). With rapid solidification the time spent coarsening is very small so that the dendrite structure is made much finer. This has two major advantages. The first is that very little post solidification homogenisation time is needed since the diffusion distances are very small (§2.1.3) and secondly the sizes of any insoluble second-phase eutectic-phase particles that solidify between the dendrite arms are also reduced. By rapid solidification insoluble impurity particles can be sufficiently refined that their spacing, approximately that of the dendrite arm spacing, can be sufficiently small to provide strengthening, see eq. (5.8). Equivalent coarsening, but here of spherical grains in a liquid matrix occurs during liquid-phase sintering (Hardy and Voorhees 1988) and during the post deposit solidification in the spray casting process, see Annavarapu *et al.* (1992) and Annavarapu and Doherty (1995).

# Chapter 6

# Other causes of microstructural instability

## 6.1 Introduction

The earlier chapters of this book have dealt with the major causes of instability in the microstructure of metals and alloys, but there remains a series of other influences which can also modify the structure. Plastic deformation and irradiation can totally alter the defect structure in metal crystals, and corrosion can completely destroy not only the metallic microstructure but the metal itself. These subjects are, however, too extensive in scope and too important to materials science to be dealt with in a brief chapter. In addition they do not fall within a reasonable definition of the *stability of microstructure*. However, there have been some interesting and important investigations of the stability of dispersed second-phase precipitates under conditions of plastic deformation, and to a smaller extent, of their stability under irradiation. These instabilities will be discussed here. Other types of external influences include: temperature gradients; gravitational, electrical and magnetic fields; and, finally, annealing under imposed elastic stresses. All of these influences have been found to cause changes in precipitate morphology, and will form the subject of this chapter.

Many of the most informative experiments in this area have been carried out on transparent non-metallic materials such as ice and potassium chloride. The results of these investigations appear to be of direct application to metals and so, for the purposes of this chapter, any material subjected to a relevant and interesting experiment will be considered metallic.

## 6.2 Migration of second-phase inclusions in a temperature gradient

Arctic explorers have known for a long time that salt polar ice becomes steadily purer and more drinkable at its cold upper surface. Whitman (1926) suggested that this might result from the migration of the brine inclusions in the temperature gradient between the cold air and the relatively warm sea beneath the ice. Such migration of liquid drops in a solid material was discussed in more general terms by Pfann (1955) as *temperature gradient zone melting*. The mechanism is shown in fig. 6.1, where the liquid in contact with the upper solid at the higher temperature $T_1$ has the lower solute content $C_l(1)$ while at the lower interface there is a higher solute content $C_l(2)$. This difference results in a concentration gradient in the liquid $dC/dz$ given by:

$$dC_l(1)/dz=[dC_l(1)/dT][dT/dz]=G/m \tag{6.1}$$

$dC_l(1)/dT$ is the reciprocal of the slope of the liquidus ($m$), and $dT/dz$ is the temperature gradient ($G$).

There is then an upward flux of solute, $J$, given by:

$$J=-DG/m \tag{6.2}$$

This flux causes the upper surface to melt and the lower to freeze, resulting in the upward movement of the liquid inclusion towards the hotter end of the sample.

If the rate controlling step is diffusion then at the melting surface an amount $C_l(1-k_0)$ of solute must be provided (where $k_0$ is the partition coefficient, the ratio of solute in the solid to that in the liquid). The migration velocity, $v$, will thus be:

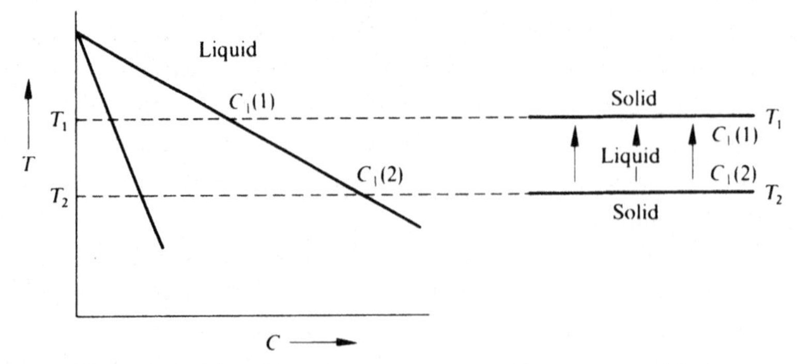


**Figure 6.1** The effect of a temperature gradient upon equilibrium composition.

$$v = -D_l G / C_l (1 - k_0) m \tag{6.3}$$

which is independent of the droplet size or shape. $D_l$ is the diffusion coefficient in the liquid.

Tiller (1963) and Shewmon (1964) both discussed the general case of the velocity of spherical gas or liquid inclusions in a solid subjected to a thermal gradient, under conditions of volume diffusion control, giving results equivalent to eq. (6.3), and also for conditions of surface diffusion control and for control by the reactions at the interface (interface control). For surface diffusion, expected for small gas bubbles in metals (with a radius $r$ of 1 $\mu$m), the velocity should be:

$$v = 2 w D_s Q_s^* G / k T^2 r \tag{6.4}$$

where $w$ is the thickness of the high diffusivity surface region, $D_s$ the surface diffusion coefficient, $Q_s^*$ the heat of transport per atom (see §6.2.3), and $r$ the droplet radius.

For any condition of *interface* control, where there is a power law relationship between interface velocity, $v_i$, and driving force, $F$, with an exponent $n$ (that is, $v_i - aF^n$), then the inclusion velocity should be

$$v = a' G r^n \tag{6.5}$$

where $a'$ is a proportionality constant. The physical reasons for these results are readily explained: for small bubbles moving under surface diffusion control, the ratio of surface to volume is $1/r$, so as the radius falls the parameter, $1/r$, will determine the number of surface diffusion paths and thus the velocity of the bubble. For interface control as the bubble size increases, there will be a larger driving force on the two interfaces, so driving the bubble faster. The various possibilities are shown in fig. 6.2. (Nichols (1972) developed a generalised analysis for the diffusion limited movement for any defect under any driving force.)

A bubble containing the vapour of the surrounding solid in a temperature gradient will be expected to move towards the hot end of the solid sample. This occurs since the vapour pressure rises as the temperature rises so producing a concentration gradient in the bubble moving the atoms from the hot surface to the cold. Similarly a solid inclusion, whose solubility in the solid matrix increases with temperature, should in a temperature gradient also migrate up the gradient under the influence of the solute concentration gradient produced by the temperature variation.

Before leaving this general discussion it should be pointed out that the derivation of eq. (6.3) only described the *mechanism* of migration in a temperature

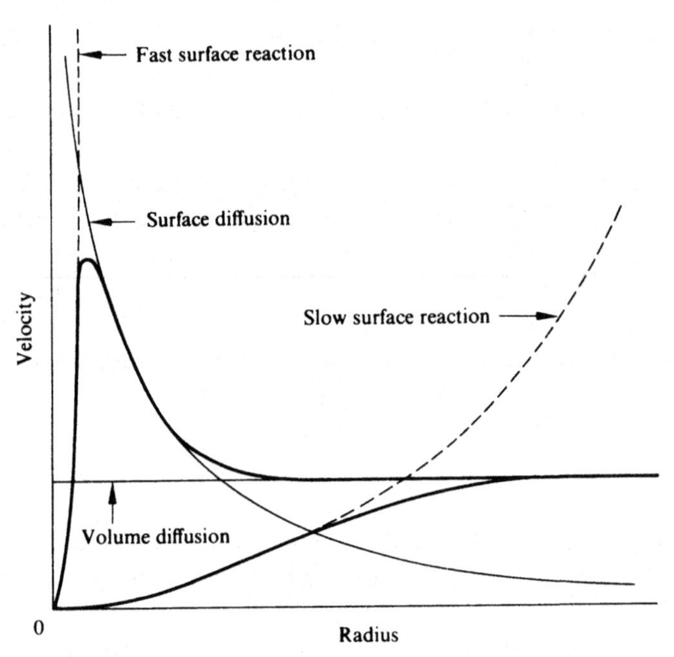


**Figure 6.2** The variation of velocity with radius for a given $D_s$ and $D$ combined with either a slow surface reaction or a fast surface reaction. (After Shewmon 1964. Courtesy of AIME.)

gradient, and not the thermodynamic driving force. The ultimate driving force comes from the increase of entropy of *heat flowing down the temperature gradient*. The heat flow is achieved by melting or evaporation at the high temperature side of the bubble and freezing or condensing at the cold side, thereby transferring the latent heat from the upper hot interface to the lower cold interface. That is, the essential thermodynamic process *is the transport of the heat down the temperature gradient*. The mechanism and the kinetics are controlled by the diffusive flux of atoms. It might be noted that the reciprocal of slope of the solidus in fig. 6.1 ($dC_s/dT<0$), transports solute up to the lower interface and away from the upper interface and so should try to move the droplet down the gradient but the opposing flux in the liquid is larger due to both a larger diffusion coefficient $D_l \gg D_s$ and a larger change of solute content with temperature, $dC_l/dT > dC_s/dT$, fig. 6.1. A similar analysis shows that a solid inclusion in a liquid will also move *up* the gradient. In a eutectic microstructure *both* solid phases move up the gradient (McLean 1975). A grain boundary in a single-phase solid might also be expected to move up the gradient since the concentration of vacancies rises with temperature so there should be a flux of vacancies down the gradient. This flux of vacancies will remove atoms on the high temperature side of any grain boundary and add atoms on the low temperature side so moving the grain boundary in the direction of increasing temperature. See, however, the discussion of the Soret effect in §6.2.7.

## 6.2.1 Migration of liquid inclusions

There has been only a limited amount of work published on the migration of liquid inclusions in metals. Pfann (1966) gave an early review of the field and also pointed out the possibilities of using the process for the manufacture of semiconducting devices, for controlled crystal growth and for improved joining techniques where the 'solder' might be subsequently removed from the joint. He also discussed the injection of material through a container wall into a heated interior. During growth of constitutionally supercooled crystals, small liquid inclusions are often found trapped in the solid behind the cell grooves. Such inclusions migrate in the temperature gradient behind the main solidification front, leaving damaged crystal and a 'solute trail' behind them (Bardsley, Bolton and Hurle 1962).

Wernick (1956, 1957) investigated the migration of aluminium rich liquid inclusions in germanium and, from the speeds of migration of droplets of different sizes, suggested that some measure of interface control was operating. That is, the larger inclusions migrated faster. This observation was confirmed by Tiller (1963) who also discussed the instability expected in the droplets if the interfacial kinetics were anisotropic. In an investigation of the migration of liquid droplets in the completely metallic system of Al–Pb, McLean and Loveday (1974) observed the movement directly by transmission electron microscopy. They found that inclusions larger than 0.8 $\mu$m in diameter moved at the rate expected for volume diffusion control, but smaller droplets moved more slowly, and ones smaller than 0.1 $\mu$m *did not move at all*. This result was treated as suggesting that at small sizes, the melting or freezing reactions at the solid–liquid interfaces are inhibiting the migration. The immobility of the very small droplets in the low temperature gradient of this experiment indicated that undercoolings of the order of $10^{-7}$ K were required to drive metallic solidification. Doherty and Strutt (1976), however, challenged this analysis. They pointed out that there was a large body of evidence from metallic solidification (see for example, Jackson, Uhlmann and Hunt (1967), Flemings (1974) and Kurz and Fisher (1989)) that metallic solid–liquid interfaces being diffuse at the atomic scale should always be able to migrate easily. Doherty and Strutt (1976) proposed an alternative mechanism whereby small inclusions could be stopped from migrating in a crystal – this is the drag exerted by any dislocation captured by the liquid droplet, fig. 6.3. Such a drag was shown to be capable of providing sufficient drag to arrest completely the motion of the small droplets in the temperature gradients studied by McLean and Loveday (1974).

Many more experiments have been carried out on transparent materials where direct observation by optical microscopy is possible, notably on ice crystals with brine inclusions and also ionic salt crystals with aqueous inclusions. Many of the references to this work are given by Tiller (1969) and Anthony and Cline (1971).

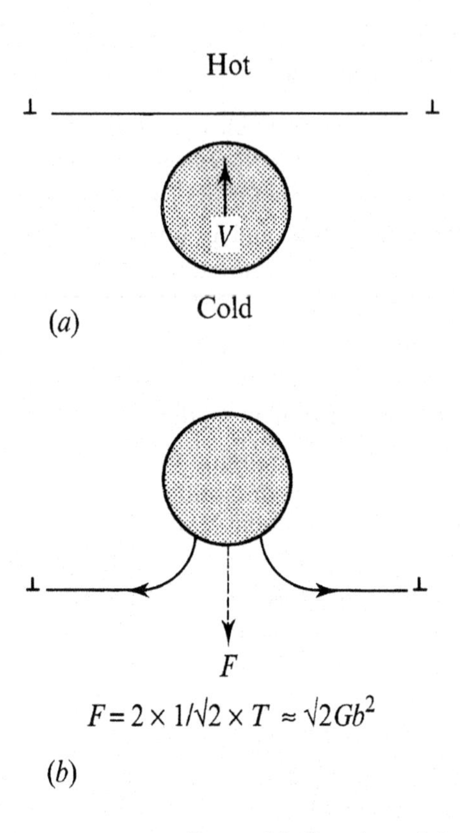


**Figure 6.3** Liquid inclusion migrating up a temperature gradient: (*a*) just before meeting and (*b*) being held back by a dislocation.

In these systems, crystallographic faceting of the solid–liquid interface is found. This is expected since the entropy of crystallisation is large ($\Delta S/R>4$) so the interface is unlikely to be diffuse as predicted in the simple model of Jackson (1958). The tendency for the solid–liquid interface to be atomically sharp then requires some source of growth or dissolution ledges, see §1.3, and this will provide some small but significant interfacial barrier to migration as demonstrated for example by Jackson *et al.* (1967) and reviewed by Woodruff (1973).

The speeds of migration of brine droplets in ice from several investigations were analysed by Tiller who showed that, although volume diffusion in the brine was the most important factor, some of the results suggested a measure of interface control. The deduced form of the interface kinetics was linear, as previously reported from direct solidification experiments by Hillig (1958); that is,

$$v_i = \mu_i \Delta T \tag{6.5a}$$

with $\mu_i$ about $10^{-5}$ m s$^{-1}$ K$^{1}$.

Anthony and Cline (Anthony and Cline 1971; Cline and Anthony 1971a,b, 1972) investigated the behaviour of aqueous inclusions in potassium chloride at room temperature ($0.3T_m$ of KCl). The inclusions were introduced by drilling a hole into a potassium chloride crystal, adding water and sealing with glue.

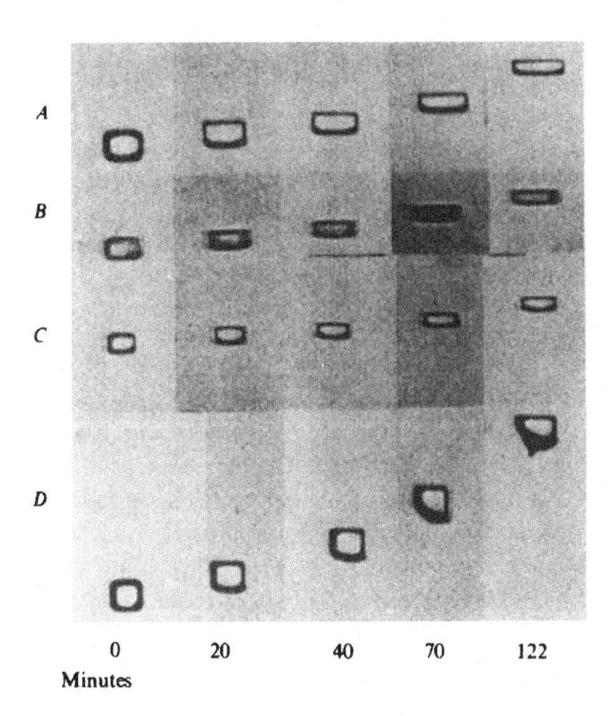


**Figure 6.4** The shape of brine droplets *A*, *B*, *C* and *D* after 0.20, 40, 70 and 122 minutes of thermomigration, respectively. (After Anthony and Cline 1971. Courtesy of *J. Appl. Phys.*)

Under the influence of a temperature gradient this large inclusion became unstable and broke up into numerous small droplets (see §6.2.3). All the droplets were faceted, being bounded by the low energy {100} planes.

Following migration, the inclusions did not have the expected equilibrium shape which should have cubic symmetry (§5.4) but were flattened in the direction of the temperature gradient which was imposed along ⟨100⟩, as is shown in fig. 6.4. After the crystals were removed from the temperature gradient, the inclusions were found to relax back towards the equilibrium cubic shape. However, many of the inclusions, those larger than about 20 $\mu$m, did not fully relax and they remained somewhat flattened. When the temperature gradients were reimposed, two changes occurred. Firstly, the droplets migrated up the temperature gradient, at a steadily increasing velocity up to a limiting value which varied from droplet to droplet. Secondly, during migration, the droplet shapes became reflattened. The terminal velocity of the droplets showed a large scatter in the velocity–size relationships, fig. 6.5, but the overall trend was for droplets smaller than about 15 $\mu$m to be immobile with a steady increase of droplet velocity as the liquid inclusions became larger.

Anthony and Cline have shown that all their results are in accord with the idea that there is some measure of interface control in the droplet kinetics. They suggest that the *dissolution* of potassium chloride is the problem since the sides of the

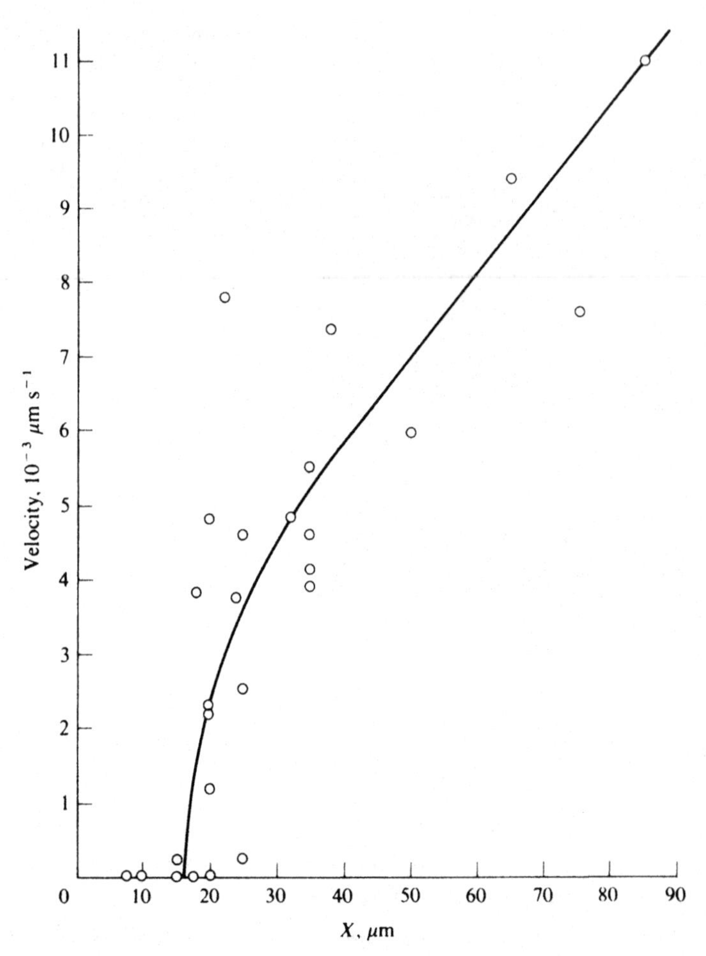


**Figure 6.5** The steady state velocities of brine droplets in potassium chloride as a function of droplet size. $X$, the steady-state width of the rectangluar droplet in the direction normal to the temperature gradient. The dimension $X$ is shown in figs. 6.8 and 6.10. (After Anthony and Kline 1971. Courtesy of *J. Appl. Phys.*)

droplet provide a source of 'ledges' on which crystallisation can occur (fig. 6.6). Dissolution, on the other hand, requires 'nucleation' of the steps at which ions may leave the crystal. Support for this idea is provided by the occasional inclusion such as $D$ in fig. 6.4, which stayed nearly cubic during migration, developed a trailing cusp and *moved at the higher velocity* expected for diffusion control. Such a bubble may be attached to a low angle boundary whose dislocations would aid ionic dissolution. Full analyses of three phenomena (the velocity of the slower moving average inclusion, and the shape changes when the temperature gradient was first applied and when it was removed) all suggested partial diffusion control, but with a part of the driving force required for dissolution. The free energy required for dissolution was given by the empirical equation:

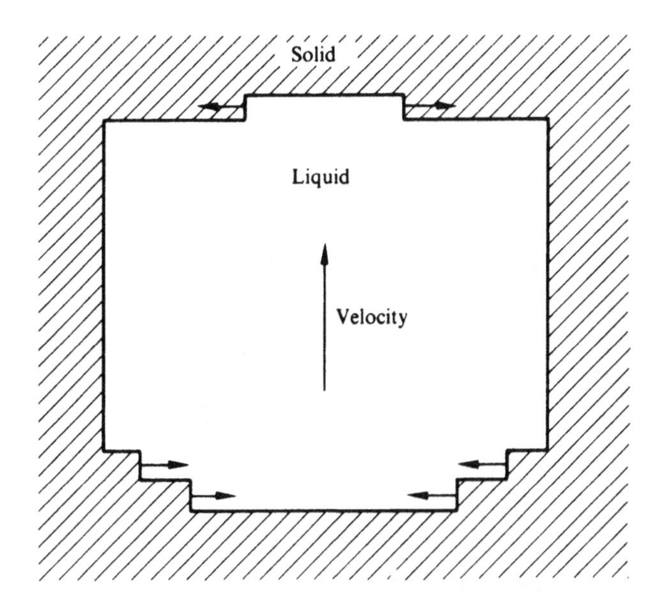


**Figure 6.6** Schematic diagram of the layer growth mechanism applied to droplet motion. (After Anthony and Cline 1970. Courtesy of *Phil. Mag.*)

$$\Delta G = \Delta G_0 + \alpha v_i$$

$$v_i = (\Delta G - \Delta G_0)/\alpha \tag{6.5b}$$

The critical value of $\Delta G_0$ ($2.4\times10^{-2}$ J mole$^{-1}$) is equivalent to an undersaturation ($\Delta C/C$) of the solution in contact with the dissolving face of $3\times10^{-5}$, or to a superheating of the face by $3\times10^{-3}$ K. It is this undersaturation during migration that causes the side faces of the inclusion to dissolve, giving the observed flattening (fig. 6.4). The existence of a critical value, $\Delta G_0$, of the driving force required to move the interface at all is a surprising result not normally expected in crystal growth or dissolution (Tiller 1963), but is required to account for the immobility of small droplets and the inability of the larger droplets to relax back fully to the equilibrium shape after the temperature gradients are removed. The explanation of an inhibition by dragged dislocation might explain the inhibition of migration of droplets here, as in McLean and Loveday (1974). In addition, dislocations captured during migration would, as shown in fig. 6.7, tend to pull the droplets back when the temperature gradient was removed and would also tend to fall on the droplet on its short sides so providing a drag against the droplet returning to its expected cubic shape.

Jones and Chadwick (1971) observed some element of interface control during thermal migration of liquid inclusions in salol, a faceted compound known to require screw dislocation ledge sources. They did not, however, find a critical minimum driving force. Camphene, however, did not show any indication of interfacial inhibition of migration. The experimental results of thermal

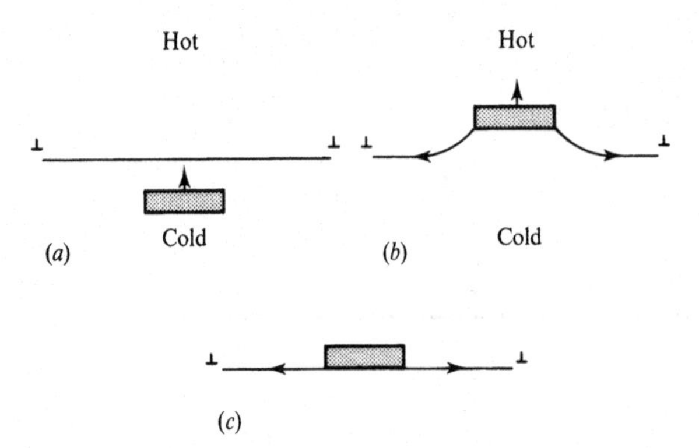


**Figure 6.7** Liquid droplet migrating in potassium chloride: (*a*) just before meeting a dislocation; (*b*) being held back by the dislocation; and (*c*) after removal of the temperature gradient when the droplet will be pulled back and hindered from achieving the expected cubic shape by the line tension of the dislocation.

migration in these organic compounds fit in with current ideas on solidification kinetics (Jackson *et al.* 1967; Woodruff 1973). Camphene, a material with a low entropy of melting, appeared to have only a small interfacial barrier to migration especially at temperatures close to the melting point of the pure material. Salol, on the other hand, a material with a high entropy of melting, showed the migration kinetics expected for interfacial control determined by *dissolution* at emerging screw dislocation sites. The relation between the driving force and velocity of the *dissolving* interface was the expected relationship for screw dislocation mechanism:

$$v_i = \mu_2 \Delta G^2 \tag{6.5c}$$

The role of dislocations in controlling migration was supported by the observations of the faceted form of the liquid droplets on their dissolving surface and more strikingly by the fact that a proportion of the droplets did not move at all, even though they were as large as others that did move. This random immobility is what would be expected for the occasional droplet that was *not* on a dislocation and thus had no source of ledges required for movement of a faceted solid–liquid interface.

## 6.2.2 Interaction of migrating inclusions with grain boundaries

In the same way as stationary inclusions can inhibit the movement of migrating grain boundaries, so immobile grain boundaries could be expected to have the same effect on inclusions moving, for example, in a temperature gradient. This has been elegantly studied by Cline and Anthony (1971b) using their potassium chloride/brine system. The brine droplets were made to migrate into a horizontal 15° twist boundary by a vertical temperature gradient, and provided the

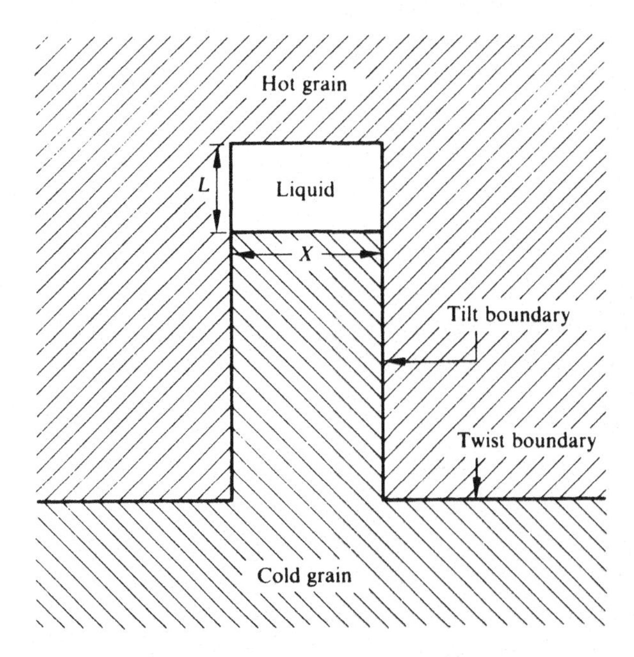


**Figure 6.8** A twist boundary of a potassium chloride bicrystal after penetration by a migrating brine platelet. The grain boundary formed behind the migrating platelet is a square, columnar, pure-tilt boundary. (After Cline and Anthony 1971b. Courtesy of *Acta Met.*)

gradient was larger than about 0.3 K mm$^{-1}$ the droplet could penetrate the upper grain; however, it pulled the grain boundary with it, see fig. 6.8. Since the grain boundaries in potassium chloride are immobile at room temperature ($0.3T_m$) a column of 15° tilt boundary developed behind the inclusion. The migrating droplet was held back by this boundary and if the temperature gradient was reduced to less than the critical amount then the droplet was pulled back to the plane of the original boundary. Measurement of this critical gradient for a droplet of known size gave a value of the grain boundary tension that was just balancing the driving force of the temperature gradient. The boundary energy was 0.03 J m$^{-2}$.

Evidence that supported the ideas of the dissolution difficulty of the crystal was obtained by the observation that it was always the dissolving face that remained faceted while the crystallising face was usually curved – this occurred whether the inclusion was moving up or down the temperature gradient.

## 6.2.3 Shape stability of migrating liquid inclusions

Several observations have been made of large migrating inclusions breaking up into many smaller droplets. Anthony and Cline (1973) discussed the phenomenon and showed that in their potassium chloride crystals this instability occurred for all droplets which were moving with a greater than some critical ratio of droplet velocity to temperature gradient. The origin of this instability was suggested to be due to the local distortion of the temperature gradient by the presence of an inclusion with a lower thermal conductivity than the crystal matrix. Because of this distortion the centre of the droplet tries to migrate faster than the edge since the temperature gradient in the liquid is largest at the centre. If the velocity of the centre exceeds the maximum possible velocity of the droplet, that which would occur in the absence of the interface barrier, then a trail of liquid is left at the edge of the droplet which breaks up into many small droplets. This model was able to predict successfully the conditions of instability. Anthony and Cline pointed out that the unstable droplet shapes observed in potassium chloride were very similar to shapes previously reported for gas-filled voids in nuclear fuel rods and in liquid droplets found in silicon and germanium.

### 6.2.4 Migration of gas bubbles

The movement of gas bubbles in solids has received a great deal of attention since Barnes and Mazey (1963) first reported their observations of the migration of small, helium filled voids in copper in the temperature gradient of a hot stage electron microscope (for example, Nichols (1969), Anthony and Sigsbee (1971) and Sens (1972)). This subject has considerable technical importance for nuclear fuels, since some of the fission products of uranium are gases (krypton, xenon) and these are extremely insoluble in both metallic and ceramic fuels. This insolubility results in both the precipitation of finely dispersed gas bubbles and their resistance to growth by diffusion-controlled Ostwald ripening (§5.5). Growth of the bubbles occurs instead by collision between migrating bubbles and their subsequent coalescence. This growth leads to rapid swelling of the fuel element. The origin of the swelling is the increasing volume of the gas as the capillarity pressure ($2\sigma/r$) of small bubbles is relaxed by growth. For a surface energy, $\sigma$, of 1 J m$^{-2}$ this pressure will be $10^3$ atm in a bubble with a radius of 10 nm. Coalescence of two bubbles increases the radius, lowers the pressure and thus leads to a growth in the volume of the gas filled bubbles. If the gas is assumed to be ideal, that is,

$$PV=nRT \qquad (n \text{ is the number of g-moles of gas})$$

Therefore

$$(2\sigma/r)(4\pi r^3/3)=nRT$$

and

$$r^2=3nRT/8\pi\sigma \tag{6.6}$$

Coalescence of two bubbles of radii $r_1$ and $r_2$, containing $n_1$ and $n_2$ moles of gas, will produce a bubble whose radius $r_3$ will be given by:

$$r_3^2 = r_1^2 + r_2^2 \tag{6.7}$$

If $r_1 = r_2$, then $r_3/r_1 = \sqrt{2}$, and coalescence will cause a volume increase $\Delta V$ of the solid given by

$$\Delta V = 4\pi(r_3^3 - 2r_1^3)/3 = 8\pi r_1^3(\sqrt{2} - 1)/3 = 3.5r_1^3 \tag{6.8}$$

More meaningful is $\Delta V/V$:

$$\Delta V/V = 3.5r_1^3/[4\pi(2r_1^3)/3] = 0.41 \tag{6.8a}$$

So each such coalesence leads to a local 41% increase in volume that results in local swelling of the solid.

To prevent the bubble migration in uranium fuel rods that would lead to this type of swelling, the uranium is commonly alloyed with iron and aluminium to give a fine precipitate that anchors the gas bubbles. This problem and its solution were discussed by, for example, Bellamy (1962) and Barnes (1964).

In the original observation of the movement of helium filled bubbles, Barnes and Mazey reported that the migration velocity was fastest for smaller bubbles, indicating surface diffusion control (Shewmon 1964). It was also reported that coalescing bubbles obeyed the relationship given by eq. (6.7), and furthermore that adjacent bubbles of different sizes that were not actually in contact showed no sign of any Ostwald ripening by solid state diffusion of the gas atoms as solute in the solid.

The conclusion that surface diffusion is totally rate controlling for bubble migration in metals such as copper and gold was revised by Willertz and Shewmon (1970). They studied bubble migration in thin foils of these two metals, in the absence of a temperature gradient, by the process of random 'Brownian' motion of bubbles. The bubbles should have a bubble diffusion coefficient $D_b$ related to the surface diffusion coefficient $D_s$ by eq. (6.9), due to Gruber (1967),

$$D_b = 0.3(a_0/r)^4 D_s \tag{6.9}$$

where $a_0$ is the atomic diameter and $r$ the bubble radius.

Willertz and Shewmon found that the bubbles behaved as expected, except that for copper the value obtained for $D_s$ was 10–100 times smaller than the normal value, and for gold the value was too small by a factor of $10^4$–$10^5$. These discrepancies appeared to be due to some small measure of interfacial reaction control which was interpreted as the difficulty of nucleating steps on the (111) facets of the bubbles. The larger inhibition of migration of the bubbles in gold was explained by the larger anisotropy in surface energy of gold than of copper.

As for liquid inclusions, an effective way of studying migration of larger gas bubbles is to use transparent materials and optical microscopy. Anthony and Sigsbee (1971) observed the movement of spherical gas bubbles in polycrystalline camphor in a temperature gradient and found that the migration rate was controlled by vapour phase diffusion for larger bubbles, but that bubbles smaller in diameter than 7 $\mu$m had some measure of interface control in that they moved more slowly than larger inclusions. Other interesting observations were made in this study: the bubbles became attached to transverse grain boundaries and migrated, dragging the boundaries with them. This had two consequences: firstly, the spherical bubbles on such transverse boundaries were seen to coalesce to a lenticular shape, and secondly, such enforced unidirectional grain boundary migration would, if long continued, give a columnar microstructure. Both of these microstructural features are well known in ceramic $UO_2$ nuclear fuel rods (Sens 1972). The process of migration of gas bubbles in such fuel rods occurs rapidly by virtue of the high radial temperature gradient and both the initial 'sintering' pores and fission products migrate to the centre of the rod, producing a pore-free region where the temperature had been high enough, a large central cavity into which the pores have gone and radial columnar grain structure (fig. 6.9).

## 6.2.5 Migration of two-phase (liquid and vapour) inclusions

The usual driving force for migration in a temperature gradient is the transfer of heat down the temperature gradient, so causing the liquid or gas bubble to migrate up the gradient. An additional factor will be any change in the interfacial free energy, $\sigma$, with temperature and it is this factor that apparently causes two-phase bubbles, containing both liquid and vapour, to migrate *down* the temperature gradient in potassium chloride (Anthony and Cline 1972). The phenomenon is shown in fig. 6.10, where in addition to the normal diffusive flux of potassium chloride ($J_D^{KCl}$) from the hot to the cold surfaces of the inclusion, there is a larger flux in the opposite direction by liquid flow ($J_L^{KCl}$) caused by the *fall* in the liquid vapour interfacial energy as the temperature falls. As discussed in §5.4.1, *negative* adsorption of charged ions at a water interface causes the value of $\sigma$ to rise with increases of salt concentration, so that as this concentration falls with the declining solubility of potassium chloride at lower temperatures, the value of $\sigma$ falls back towards that expected for pure water. The steadily falling value of surface energy and consequently of surface *tension* means that along the water–vapour interface there will be an out-of-balance

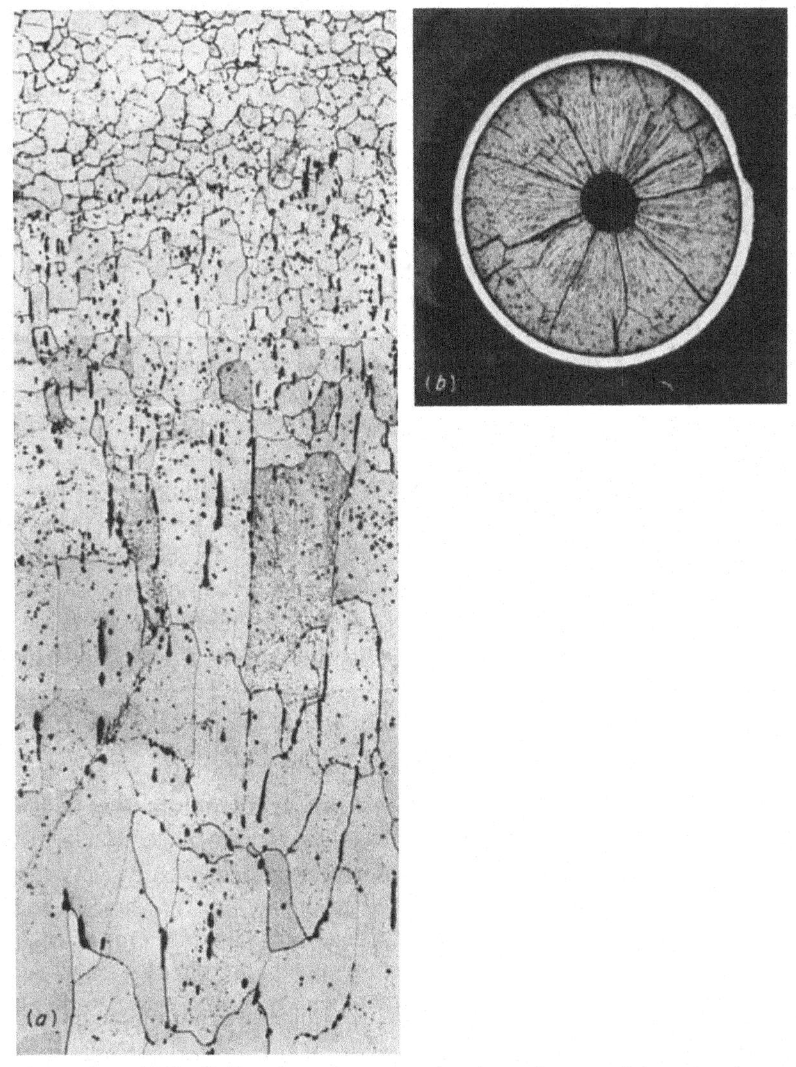


**Figure 6.9** (*a*) Radially oriented pores at the circumference of the columnar grain region, pointing towards the centre of the fuel (bottom). Magnification ×180. (*b*) Section through an irradiated fuel pin, showing columnar grains and collection of porosity as a central cavity. Magnification ×10. (After Sens 1972. Courtesy of *J. Nucl. Mat.*)

surface shear that must result in viscous flow of saturated brine from the cold to the hot side of the inclusion. This flow is the essential step in the bubble movement, since it carries the potassium chloride that will be deposited at the *hot* side of inclusion as the inclusion moves to the cold end of the crystal. The cycle is completed by the crystallisation of potassium chloride at the hot side of the inclusion as the water evaporates, and is transported by vapour diffusion to the cold side of the inclusion. Here it condenses, dissolving more potassium

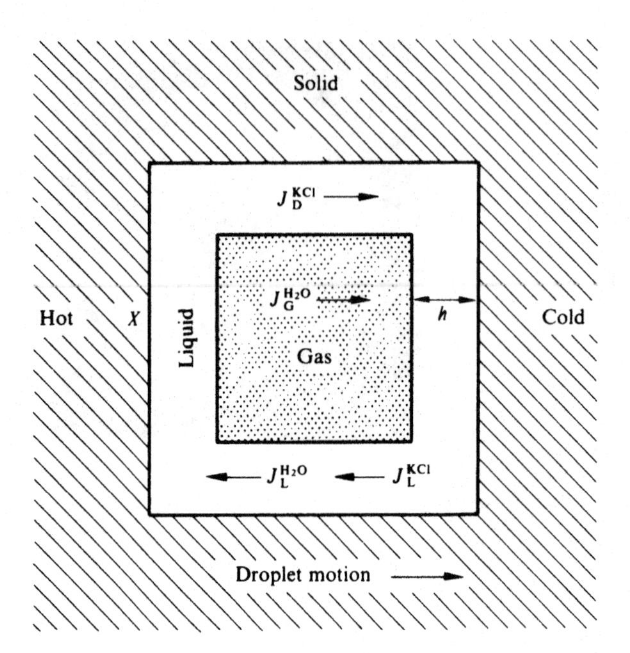


**Figure 6.10** A schematic diagram of a biphase vapour–liquid droplet migrating down a temperature gradient in a solid. (After Anthony and Cline 1972. Courtesy of *Acta Met.*)

chloride, and is then available for further transport of potassium chloride by the surface tension effect.

An analysis based on this model successfully predicted the observed migration velocities of the two-phase bubbles. Under conditions where the cross-sectional area of the bubble was mainly occupied by the vapour phase it was shown that the slowest, rate controlling step was the vapour-phase transport of water vapour, and the expected migration velocity was then given by:

$$v=(3/2)[A_g/(A_g+A_l)](C_l^{KCl}D_g^{KCl}/C_l^{H_2O}C_s^{KCl})(dC_g^{H_2O}/dT)dT/dx \qquad (6.10)$$

where $A_l$ and $A_g$ are the cross-sectional area of the inclusion occupied by liquid and gas phases, $C_l^{KCl}$ is the concentration of potassium chloride in the liquid phase, with the same notation used for the other parameters in the equation. This relation, which assumed the presence of a foreign gas at atmospheric pressure in the inclusion, gave values of the observed velocity close to those actually found. In the absence of any foreign gas, the vapour phase diffusion of water would have been about ten times faster.

Considerable practical interest attaches to these ideas, since there is the possibility that geological salt deposits might be used for the long term disposal of self heating radioactive wastes. Natural salt deposits contain aqueous inclusions which will migrate up the temperature gradient to any burial crypts containing the hot wastes. On reaching the crypt some of the water may evaporate resealing the by now two-phase inclusions, which will migrate away from the crypt *down* the temperature gradient, possibly carrying radioactive waste. The temperature gradients due to the radioactive self heating will only last a few years, during which the inclusions will have moved only a short distance (either 0.01 m or 0.6 m depending on the presence or absence of air in the inclusion). Subsequently, the inclusion might migrate large distances in the many years that the waste will remain radioactive, under the influence of the weak natural thermal and gravitational fields of the earth. However, danger of escape of activity out of the salt will be obviated by the grain boundaries in the polycrystalline salt. These boundaries would be expected, §6.2.2, to prevent any long range migration of the contaminated brine droplets in the weak natural thermal and gravitational gradients in the deposits.

### 6.2.6 Migration of solid phases

A few examples have been reported of the migration of solid–solid interfaces in metallic alloys (Jones 1974a; Jones and May 1975; McLean 1975). These investigations reported accelerated coarsening of fibrous eutectic microstructures annealed in a temperature gradient normal to the fibre axis. Jones (1974b) has proposed that this may be due to contact and subsequent coalescence between fibres moving with different velocities, but a more convincing mechanism has been proposed by McLean (1975). He pointed out that solid state diffusion has a high activation energy so the diffusion coefficient ($D$) will increase with distance causing all interfaces to move away from each other up the temperature gradient. This model accounts naturally for fibre coarsening without having to postulate either interface kinetics or surface diffusion to give different speeds of different size fibres. Some measure of quantitative agreement with his model was reported by McLean in the cobalt–chromium carbide eutectic system.

### 6.2.7 The Soret effect: thermal migration of atoms in a temperature gradient

In a temperature gradient movement of *atoms* is possible and occurs since there is, on any atom of an element A, a driving force, $f_A$, whose magnitude is given, for example, by Denbigh (1951) as

$$f_A=(Q^*/T)(dT/dx) \tag{6.11}$$

where $Q^*$ is the heat of transport, a quantity difficult to estimate in either magnitude or sign but which physically is the excess energy of an atom making a diffusive jump. As previously discussed, for a gas or liquid inclusion $Q^*$ is the latent heat of vaporisation or melting per atom, but no general rule for determining $Q^*$ seems available for atomic migration (Ho 1966; Nichols 1972).

There is, however, a simple model for the value of $Q^*$ for a metal diffusing by a vacancy mechanism which suggests that

$$Q^* = \beta E_m - E_f \tag{6.12}$$

where $E_m$ is the energy of vacancy motion, $\beta$ is a factor somewhat less than 1, and $E_f$ is the formation energy of a vacancy. This model is based on the idea that motion of an atom will carry down the temperature gradient a fraction of its activation energy for motion (the remaining fraction being carried by the surrounding atoms), but the motion requires the motion up the temperature gradient of the vacancy, which will carry its energy of formation. Ho pointed out that this model appeared to work reasonably well for metals with a simple electronic structure (gold, silver, copper, aluminium) and predicts a very small thermal migration since $E_m \approx E_f$, a conclusion supported by Swalin and Yin (1967). However, the large values of thermal migration found for some transition metals, such as platinum, iron, titanium and cobalt, was not compatible with the simple theory. Ho also pointed out that the rates of migration of metals in an electric field (§6.4) showed the opposite effect – simple metals moved rapidly while the transition metals were somewhat immobile. Vogel and Rieck (1971) have obtained measurements of thermal migration of interstitial solutes in some transition metals, but drew no general conclusions from their work.

It should be noted that the Soret effect in a solid or liquid solution may move different species at different rates, and even in different directions ($Q^*$ may be positive or negative), and this can complicate the motion of inclusions discussed earlier in this section. In their work on movement of brine inclusions in potassium chloride, Anthony and Cline were able to show that the Soret effect was negligible, but this may not always be the case.

## 6.3 Migration in a gravitational (acceleration) field

All the ideas developed in §6.2 apply equally well to the migration of second-phase inclusions and atoms in a gravitational field (solid state sedimentation) in which 'denser' inclusions and atoms sink and 'lighter' ones float upwards. The natural gravitational field of the earth is usually much too small to have any significant effects. However, much larger accelerational fields can be produced in modern centrifuges. Anthony and Cline (1970) studied the motion of brine droplets in potassium chloride in such a centrifuge at 53 000 times the earth's gravity, and found velocities of the order of 0.3 $\mu$m h$^{-1}$. As in their investigations of these droplets in a temperature gradient (§6.2.1), small droplets did not move while larger ones moved with velocities that increased with droplet size. Analysis of the kinetics was very similar, migration being controlled largely by volume diffusion but with a significant contribution from the interface reaction at the dissolving crystal interface.

The interface kinetics required a larger driving force for movement than was required for the same interfaces moving in a temperature gradient. (NB Anthony and Cline originally omitted a factor of 300 in their calculated interface kinetics.) Cline and Anthony (1971a) suggested that the difference between the two experiments might be due to a larger dislocation density in the crystals that had been centrifuged. However, since the difference between the two experiments is now found to be that the interfaces are *less* mobile in the acceleration field, this hypothesis would require that the dislocations acted as a drag on the migrating droplets (Doherty and Strutt 1976) and not as sites at which dissolution would be accelerated. Clearly, experiments involving actual determination of the dislocation configuration around moving bubbles, perhaps by transmission electron microscopy, are needed here.

Anthony (1970) also investigated the sedimentation of heavy gold atoms dissolved in indium at 40 °C in the accelerational field of a centrifuge. The system was chosen since gold was known to diffuse rapidly in indium by an interstitial mechanism (Anthony and Turnbull 1966). Following acceleration at $10^5$ times gravity for three days, a non-uniform distribution of gold was found in a previously homogeneous sample. The change of the atomic fraction of gold, $n_{Au}$, with the radial distance $r$ in the centrifuge was predicted and found to obey the relationship:

$$\mathrm{d}\ln n_{Au}/\mathrm{d}r=[M_{Au}-M_{In}(\bar{V}_{Au}/\bar{V}_{In})](\omega^2 r/kT) \qquad (6.13)$$

where $M_{Au}$ and $M_{In}$ are the atomic masses of the two elements, $\bar{V}_{Au}$ is the partial molar volume of gold, $\bar{V}_{In}$ the molar volume of indium, and $\omega$ the angular velocity of the centrifuge. The value of the partial molar volume of gold in indium was small and found to lie in the range $0.2\ \bar{V}_{In}>\bar{V}_{Au}>-1.7\bar{V}_{In}$ indicating, as expected, the interstitial nature of gold in dilute solution in indium.

## 6.4 Migration in an electric field – electromigration

Since inclusions are uncharged, no significant motion of second phases would be expected, but metallic ions in a metal are expected to move in an electric field since they will be subject to a force $f_A$ which, as discussed, for example, by Ho and Huntingdon (1966), is given by:

$$f_A=eEZ^*=eEZ(1-W) \qquad (6.14)$$

The charge on the electron is $e$, the electric field is $E$, $Z$ is the ionic charge, $Z^*$ is the apparent charge, and $W$ is the 'electron wind' caused by momentum transfer from scattering by the ion of the migrating electrons. For most metals the electron wind appears to be more important than the direct effect of the electric field on the ionic charge, since migration is normally found to occur towards the *anode* (the same direction as electron flow). This migration can only occur at a temperature where there is sufficient thermal activation to allow atomic diffusion. Using the Nernst–Einstein equation (mobility $M=D/kT$) there will be a flux of ions $J_A$ due to the force $f_A$ given by

$$J_A = C_A D_A f_A / kT = C_A D_A eEZ^* / kT \tag{6.15}$$

where $C_A$ is the concentration of A atoms per unit volume and $D_A$ their diffusion coefficient.

In their measurements of electromigration in silver, Ho and Huntingdon applied the needed electric field and also heated the sample by passing a large direct current down a thin sample. Two major changes occurred in the silver with this treatment; firstly, electromigration of silver towards the anode was found, indicating a value of $Z^*$ of $+25$ electron charges, a very large wind effect, and secondly, as the temperature and diffusion coefficient was highest at the centre of the sample, there was non-uniform migration. This meant that while silver atoms were deposited close to the anode a vacancy supersaturation led to *void precipitation* near the cathode. Similar microstructural changes have been found in aluminium foils by Bleach and Meieran (1969), and they pointed out that such void formation in aluminium connections is likely to be the cause of failure of electronic circuits made with thin metal connectors when subject to high current operation (Bleach and Meieran 1967; Ghate 1967). Under these conditions voids forming near the cathode in a temperature gradient may migrate and coalesce, breaking the connection.

Ho (1966) also investigated electromigration in cobalt which, like iron, is a 'p-type' electrical conductor – that is, most of the electrical conductivity is due to the motion of positive holes. Under such conditions a 'hole wind' would be expected to cause electromigration to the cathode; this was indeed found, though with only a small value of $Z^*$ ($-1.6$ electron charges).

### 6.4.1 Migration of grain boundaries under the influence of an electric field

Lormand, Rouais and Eyraud (1965, 1974) have observed that, in a large number of metals undergoing grain growth under the influence of an electric field and thus an electric current, the direction of motion of these boundaries was biassed so that preferential migration towards the cathode occurred. This effect was also found by Haessner, Hofmann and Seekel (1974) during the primary recrystallisation of gold single crystals under high direct electric current flow. Haessner *et al.*

analysed the results as indicating a drift of atoms in the normal way towards the anode, thereby displacing the grain boundary in the opposite sense. However, the observed effect was much larger (by $10^{-2}$) than that predicted, possibly due to higher values of the effective charge and resistivity at the grain boundary region.

## 6.5 Effect of magnetic fields on metallic microstructure

The energy of magnetic phases is altered by the presence of a magnetic field, and so it is not unexpected that the microstructure (and hence properties) of alloys that are or may become magnetic will be changed by application of a magnetic field under conditions in which atomic movement is possible. This subject has been discussed, for example, by Cullity (1972) and Luborsky, Livingston and Chin (1983) whose reviews should be consulted for a fuller account of the field.

Since there is no change of energy of a magnetic atom or inclusion with position in a uniform magnetic field, no migration is to be expected, in contrast to that found in thermal, accelerational or electric fields. (Migration in a non-uniform field would be possible.) However, there can be a reduction of energy as the orientation of a magnetic particle changes with respect to a magnetic field, as occurs in a compass needle, and this has been found to cause microstructural change (§6.5.1). In addition, phase transformations involving a magnetic phase will be modified by the presence of an external magnetic field (§6.5.2).

### 6.5.1 Effect of magnetic fields on the orientation of microstructural features

There are two distinct, though related, phenomena that should be considered here: changes in stable and in unstable solid solutions. In the first type of solid solution all that is changed is the arrangement of the different atoms, and such a possible change in 'directional order' is shown in fig. 6.11. In this situation, unlike that of long range order, there need be no change from the random situation in terms of the fraction of like and unlike nearest neighbour pairs; what alters is the proportion of like pairs that are aligned in the vertical direction which was the magnetic field direction. This is the hypothesis proposed by Chikazumi (1950) and then discussed by, for example, Graham (1959). The evidence that directional order has been produced by the usual treatment of allowing the magnetic sample to cool from above the Curie temperature in an external magnetic field is shown by changes in the magnetic properties. The main change is the development of a uniaxial magnetic anisotropy in the original field direction, such that it is easier to magnetise the sample in that direction than in any other (that is, the initial permeability is increased in the original field direction), and magnetostriction is reduced. Amongst the evidence that supports the directional order model of these effects is

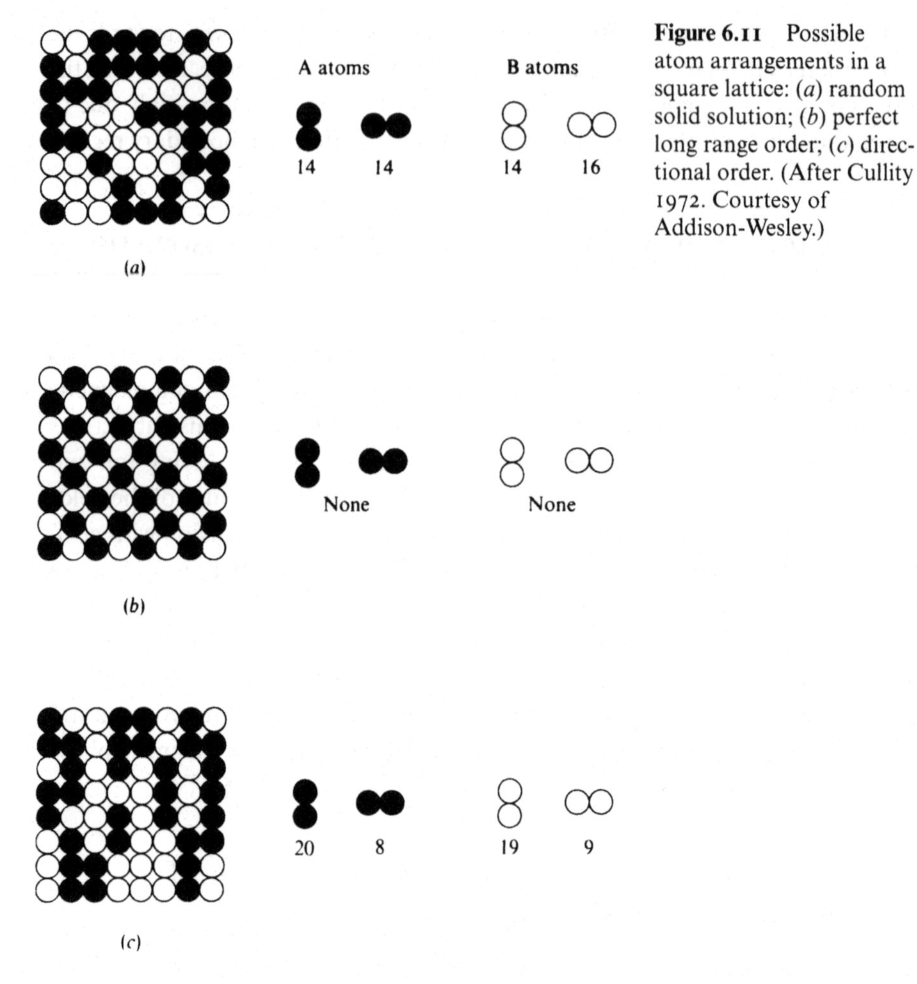


**Figure 6.11** Possible atom arrangements in a square lattice: (*a*) random solid solution; (*b*) perfect long range order; (*c*) directional order. (After Cullity 1972. Courtesy of Addison-Wesley.)

the observation that the effect is absent in pure metals and in alloys that are fully ordered. Furthermore, if the magnetic annealing is carried out isothermally, rather than by continuous cooling to below the temperature at which diffusion ceases, then the magnetic anisotropy is less and falls as the annealing temperature rises, owing to the randomising effects of thermal motion. Amongst the alloys that show this effect are many of the most useful 'soft' magnetic materials, Fe–Si, Fe–Ni and the ternary Fe–Ni–Co. As Graham (1959) pointed out, if these alloys are to be used as soft magnets for uniaxial magnetisation, then the increased permeability due to this directional order could be useful. Similar directional effects have been found for interstial solid solutions of carbon in iron by de Vries, Van Geest, Gersdorf and Rathenau (1959). A magnetic field applied along [100] caused a slow contraction of the crystal in that direction as the carbon atoms too large for the interstial holes preferred to occupy those interstial sites between iron atoms parallel to [010] and [001].

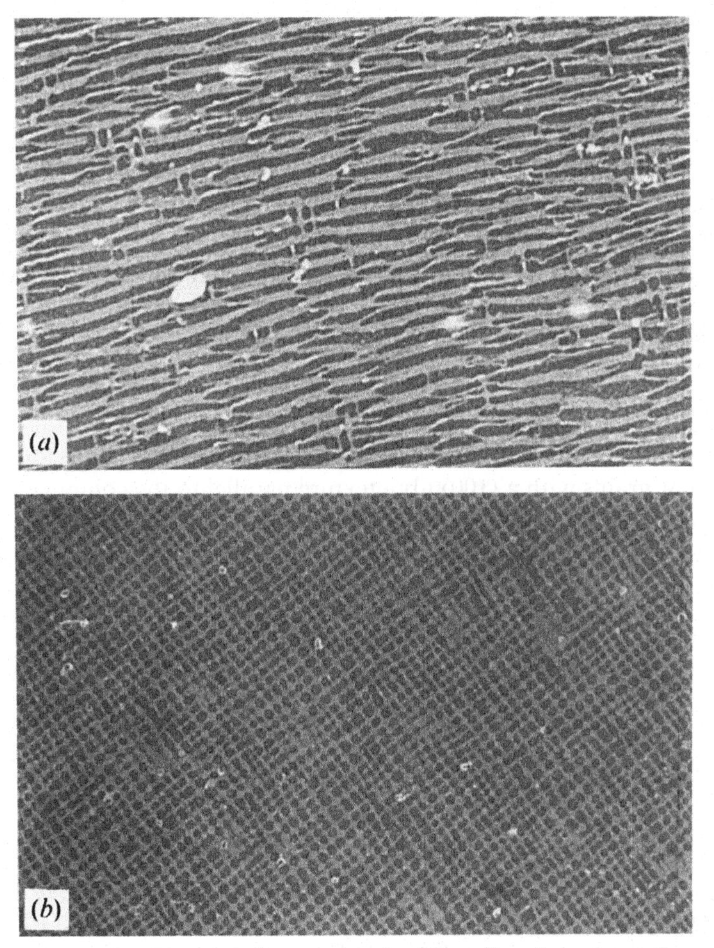


**Figure 6.12** Electron micrographs of oxide replicas from the surface of field-annealed Alnico 8: (*a*) section parallel to the annealing field; (*b*) section at right angles. Magnification ×50 000. (After de Vos 1969. Courtesy of Academic Press.)

## Spinodal decomposition of Alnico

Similar orienting effects of heat-treatment in a magnetic field are known to occur in the 'Alnico' series of permanent magnet alloys, but in these alloys the feature whose orientation is altered is second phase. De Vos (1969) reviewed the subject and showed that the normal heat-treatment produced, from the initially homogencous (bcc) $\alpha$ solid solution by slow cooling in a magnetic field, rods of strongly magnetic $\alpha'$ aligned in the field direction. For the optimum effect, the field direction should be parallel to $\langle 100 \rangle$. The aligned structure is shown in fig. 6.12. The matrix remains an $\alpha$ solid solution though depleted in both iron and cobalt, which concentrate in $\alpha'$, but enriched in the other constituents, nickel and aluminium, and is weakly magnetic. The microstructure of the material and the X-ray 'side bands' are typical of spinodal decomposition, and it is generally accepted that such decomposition has occurred during cooling. Cahn (1963) analysed the spinodal decomposition expected in alloys such as these and proposed that if the decomposition occurred at a temperature close to the Curie temperature (where the alloy on cooling first becomes magnetic), then the magnetostatic energy favours the preferred development of 'composition' waves in the magnetic direction. De Vos (1969), however, suggests that the preferred development occurs not during the initial spinodal decomposition when composition waves develop in all three ⟨100⟩ directions, but during the subsequent annealing when the microstructure coarsens to reduce the interfacial area. Both of these rival hypotheses make use of the same physical idea: that magnetic energy favours the formation of magnetic rods aligned parallel to the magnetic field. For optimum magnetic properties the Alnico alloys are cast to give columnar grains with a ⟨100⟩ fibre texture parallel to the columnar growth direction, and then heat-treated with a magnetic field parallel to the ⟨100⟩ axis. This produces the maximum amount of elongated magnetic $\alpha'$ precipitates which owe their resistance to demagnetisation to their elongated shape and to the fact that their size causes them to be single domain particles.

## 6.5.2 Phase transformations in a magnetic field

Satyanarayan and Miodownik (1969) reviewed the work that had been carried out on the influence of a magnetic field on the martensite reaction in steels where the non-magnetic austenite (fcc) changes into the magnetic ferrite (bcc) phase. This type of reaction occurs when the free energy difference between the two phases is sufficient to overcome the elastic strain energy barrier to the transformation (see, for example, Christian (1965) and Wayman (1983)). The influence of the strongest magnetic fields on the martensite reaction in steels is to increase the martensite start temperature $M_s$ by a few degrees (5–8 K). This change is due to the decrease in magnetic free energy $\Delta G_m$ given by

$$\Delta G_m = H\Delta J \tag{6.16}$$

where $H$ is the magnetic field and $\Delta J$ is the difference in magnetisation between the phases (mainly due to the ferromagnetic $\alpha$ phase). The difference in $M_s$ was successfully predicted using the relationship

$$M_s' - M_s = (T_0 - M_s)\,\Delta G_m / \Delta G_e \tag{6.17}$$

where $T_0$ is the temperature at which both phases (austenite and ferrite) have the same free energy and $\Delta G_s$ is the energy that must be provided to overcome interfacial and elastic restraints. The smallness of the change is, of course, due to the small values of $\Delta G_m$ compared to the free energy changes characteristic of metallic phase changes.

Equivalent changes have been reported by Peters and Miodownik (1973) for the non-martensitic diffusional reaction between the bcc ($\alpha$) and fcc ($\gamma$) phases in Fe–Co alloys. The two-phase boundaries

$$\alpha/(\alpha+\gamma)$$

and

$$\gamma/(\alpha+\gamma)$$

were found both theoretically and experimentally to increase to higher temperatures under the influence of strong magnetic fields. The increase is due to the lower free energy of the magnetic $\alpha$ phase in a magnetic field. It was also found, as might be expected, that the rate of formation of the $\alpha$ phase on cooling was increased and its dissolution on heating slowed down.

In the same way Mullins (1956) has shown that grain boundaries in a magnetic field with a suitable *magneto-crystalline anisotropy* will migrate when annealed in a magnetic field. This is due to the preferred growth of suitably oriented grains with a lower magnetic free energy at the expense of grains with other orientations and higher energies.

## 6.6 Deformation

All forms of plastic deformation result in large and very important changes in the microstructure of metals with respect to dislocation structure and to point defect concentrations. These changes are a central topic in materials science but cover so much information that, although aspects of this have been discussed in chapter 4, they cannot be further discussed here without distorting the content of this book. In this section only aspects of deformation and the microstructural stability of second-phase particles will be discussed.

### 6.6.1 Effect of plastic deformation and fatigue loading on metastable precipitates

Finely spaced coherent precipitates are readily sheared by dislocations as this usually requires a smaller applied stress than the alternative bypassing mechanisms (see, for example, Martin (1980)). Such dislocation motion can easily shear small precipitates into two or more even smaller ones. This process would be expected, if carried far enough, to reduce precipitates to below the critical size (§2.2.2) and so

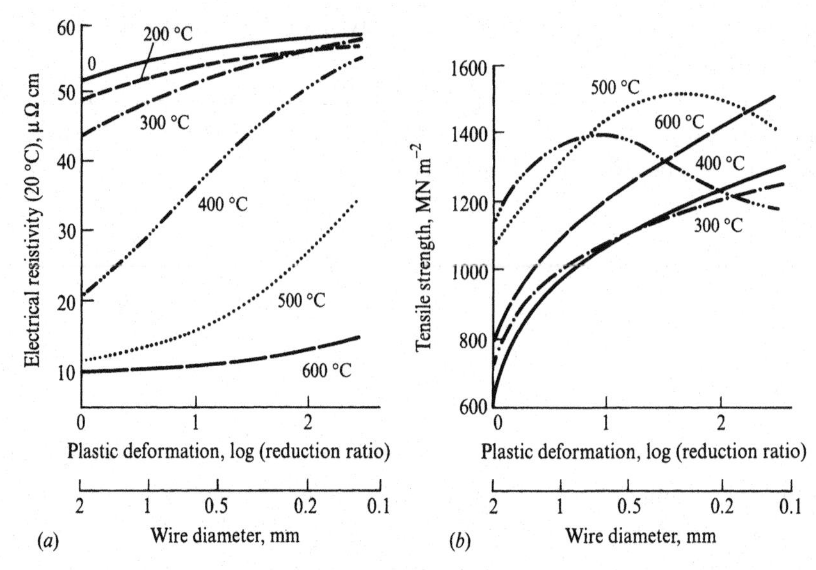


**Figure 6.13** (*a*) Increase of the electrical resistivity of Cu–4 wt% Ti as a function of wire drawing strain applied at room temperature in samples previously aged at the temperatures indicated. (*b*) The change of tensile strength – note the strain softening at 400 and 500 °C that correlates with the large rise in resistivity indicating precipitate dissolution. (After Fargette 1979. Courtesy of Metals Technology.)

make them liable to redissolve. Fargette (1979) and Doherty (1983) briefly reviewed the evidence for deformation-induced dissolution of shearable precipitates in copper alloys. Fig. 6.13(*a*) shows the increase of electrical resistivity for a series of precipitated Cu – 4 wt% Ti samples previously aged at the temperatures indicated. Alloys aged at 400 °C and above were almost fully precipitated as indicated by the low value of the resistivity in the undeformed samples. The low value of resistivity corresponds to low amounts of Ti in solution. On deformation there is, for all the samples, an increase in resistivity that will be due in part to the increase in dislocation density – the samples all exhibited work-hardening, fig. 6.13(*b*). The alloy with the finest sized precipitates, that aged at 400 °C, however, showed a very large increase of resistivity – an increase from values indicating almost no solute in solution to values corresponding to unprecipitated samples, those pre-aged at low temperatures. The 400 °C sample showed, fig. 6.13(*b*), a marked work softening at wire drawing strains above 1. This indicated that softening due to the loss of precipitation hardening could offset the additional strain hardening. The sample pre-aged at 500 °C behaved similarly with respect to resistivity and strength. Fargette's observations clearly demonstrate that fine precipitates can be unstable against dissolution under large plastic strains. Calorimetric evidence for the deformation-induced dissolution of coherent $\delta'$ precipitates in Al–Li alloys has been presented by Lendvai, Gundladt and Gerold (1988).

Early evidence for precipitate resolution resulting from cold-rolling of the coherent precipitates in aged Cu–Be was found by Gruhl and Gruhl (1954), who reported the loss of precipitate X-ray diffraction effects and an apparent loss of precipitation strengthening during cold-rolling. The precipitates reappeared on further ageing, supporting the resolution hypothesis. Rabkin and Buinov (1961) also found that after dislocation shear of both GP zones and metastable $q'$ precipitates in Al–Cu alloys, some of the precipitates redissolved. Other examples of this type of resolution caused at room temperature by unidirectional deformation have been discussed by McEvily, Clark, Utly and Herrnstein (1963), and it appears that this deformation-induced instability is well established. It is to be expected that during strain-induced dissolution, accelerated diffusion might occur due to deformation-induced point defects (§2.2.4).

As mentioned in §2.2.1, the instability of precipitates during fatigue deformation is a more intensively studied phenomenon. This is a major problem, particularly in aluminium alloys where heat-treatment to give high tensile strength produces a microstructure with a low *fatigue ratio*, the fatigue strength at $10^6$ cycles to the UTS, see Martin (1980).

In cyclic loading, it is well recognised that strain localisation occurs in 'persistent slip bands' (PSBs). Precipitation hardened alloys differ in their fatigue behaviour from single-phase alloys in so far as the additional interaction between dislocations and precipitate particles plays a dominant role. The changes in microstructure brought about by fatigue deformation have been reviewed by Doherty (1983), Mughrabi (1983) and Gerold and Meier (1987).

There have been several mechanisms proposed for the origin of precipitate damage leading to softening and strain localisation. These include:

### (a) Overageing

Johnson and Johnson (1965) discussed the evidence for a high concentration of point defects in the slip bands during fatigue, and found that vacancy concentrations as high as $5 \times 10^{-4}$ could be detected in copper at 4.2 K. Such vacancy concentrations were not found after room temperature fatigue in the pure metal. However, in alloys high vacancy concentrations may possibly be achieved by fatigue at 300 K, as is suggested by the continued enhanced diffusion rates in quenched aluminium alloys, the stability of the high vacancy concentration being maintained by solute–vacancy interaction. It has been suggested by Broom, Mazza and Whittaker (1957) that the combined action of fatigue-induced dislocations and vacancies may produce overageing. This is the replacement of the metastable precipitates by more stable and more widely spaced precipitates. There has been, however, little unambiguous evidence for this process.

Fatigue at elevated temperature may be expected to lead to accelerated overageing and this has been reported by Stubbington and Forsyth (1966) in

Al–Zn–Mg alloys that were also shown to display reversion (see below) at room temperature.

### (b) Reversion by complete local resolution

It is suggested that cyclic straining causes the dislocations to pass backwards and forwards through the coherent precipitates, so reducing them to below the critical size for stability. This must require that the two slip modes occur on different planes otherwise the damage would be self healing. As for unidirectional deformation, a high concentration of vacancies would greatly accelerate the diffusion of solute away from the subcritical precipitates. Mughrabi (1983) has expressed a simple criterion for the complete softening by the random chopping-up process of particles. This is that, on average, the root mean square irreversible displacement, $(\langle x\rangle^2)^{1/2}$, between adjacent planes should exceed the particle diameter, $d$, measured in the glide plane in the direction of the Burgers vector $b$. For random to-and-fro slip, and allowing for dislocations to be arranged in groups of $n$ like dislocations (this would correspond to planar slip), the Mughrabi condition is:

$$(\langle x\rangle^2)^{1/2}=(4N\gamma_{loc}pnba)^{1/2}>d \quad (6.18)$$

Here $N$ is the number of cycles, $\gamma_{loc}$ is the local shear strain amplitude, $a$ the slip plane spacing and $p$ is the *slip reversibility* (the fraction of the shear strain that is reversible).

Mughrabi observed, with $p=1$ and $n=1$ (corresponding to wavy slip), for $p\gamma_{loc}=0.25$ and for a particle diameter of 10 nm that $N\approx 2000$ cycles would be required for complete softening. The important point to note is that, depending on the precipitate size and the local irreversible shear strain amplitude, very different numbers of cycles may be required for softening.

In precipitation-hardened alloys the PSBs are very narrow – their width being below 100 nm, compared with 1–2 $\mu$m for single-phase materials. Fig. 6.14 shows a transmission electron micrograph of a PSB in an Al–Ag alloy containing spherical GP zones. The primary dislocations are out of contrast while the narrow PSB can be seen because of the disappearance of the zones that have been sheared many times by the heavy to-and-fro dislocation movement in the PSB. The strain concentration can amount to an average shear amplitude of 30–60% (Lee and Laird, 1983).

### (c) Precipitate disordering

Many age-hardening systems which contain precipitates of an ordered phase (e.g. $Ni_3Al$ precipitates in nickel based superalloys) gain a large part of their strength from the order within the precipitates. Plastic deformation can disorder the precipitate if the flow is not fully reversible. Localised softening and PSB

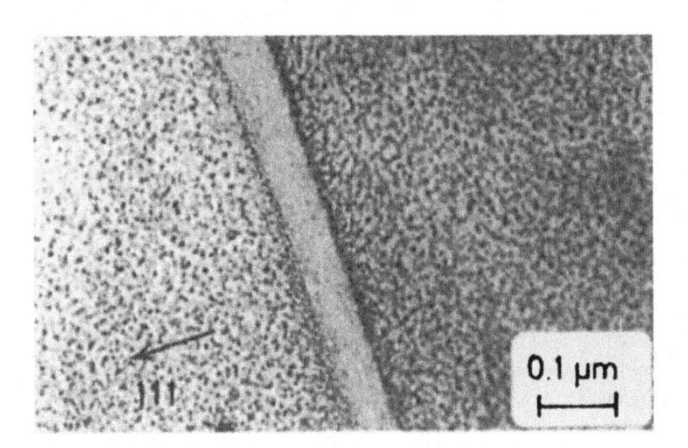


**Figure 6.14** PSB in fatigued Al–Ag containing spherical GP zones. (Kohler Steiner and Gerold 1983. Courtesy of Riso National Laboratories.)

damage can thus be brought about by dislocation-induced disordering as proposed by Calabrese and Laird (1974). Significant softening can then occur leading to strain localisation without precipitate dissolution being necessary. Dissolution may occur *subsequently*, however, as a consequence of the disordering, since the disordered precipitates will have a higher free energy and thus a higher solubility than the more stable ordered precipitates.

### (d) Dissolution due to the Gibbs–Thomson effect

Sargent and Purdy (1974) pointed out that it would not be necessary for the radius of a precipitate in a PSB to be cut to below the critical radius for nucleation for dissolution to occur. Any reduction in precipitate radius would render the precipitate liable to dissolution since the adjacent precipitates outside the PSB retain their full radius. This means that there will be a higher solubility, by virtue of the Gibbs–Thomson effect, see §5.5.2, of the small precipitates in the PSB compared to the adjacent precipitate. This will lead to loss of the solute out of the PSB by diffusion over distances of the order of the PSB width. As a result not only will the precipitates dissolve in the PSB but the PSB will become almost solute-free and softening will arise from loss of both precipitate and solute hardening. (Similar effects will also follow the effects of (*b*) full reversion and (*c*) disordering). Brett and Doherty (1978) demonstrated the existence of a solute depleted layer in the fatigue fracture surface of an underaged Al–Cu alloy by means of a glancing angle microanalysis technique. Fig. 6.15(*a*), shows the evidence for the copper depletion in the higher ratio of Al to Cu X-ray counts from the fracture facet of the failed {111} PSB compared to a mechanically polished surface. Brett and Doherty were able to show a similar change in the X-ray count rates by vapour deposition of 25 nm of pure Al onto a mechanically polished piece of the alloy. The Al – 4.2 wt% Cu alloy had been aged at 130 °C for 18 h and had fractured after $1.5\times10^5$ cycles of tension-zero loading. Confirmation of the solute loss from the fracture facet was subsequently provided by Auger electron spectroscopy by Lea, Brett and Doherty (1979) as shown in fig. 6.15(*b*).

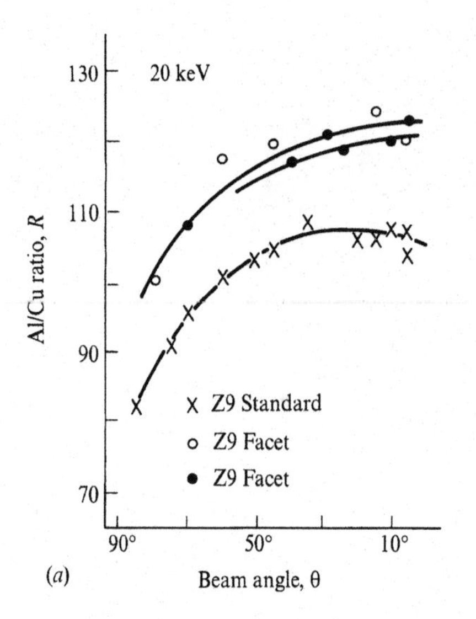


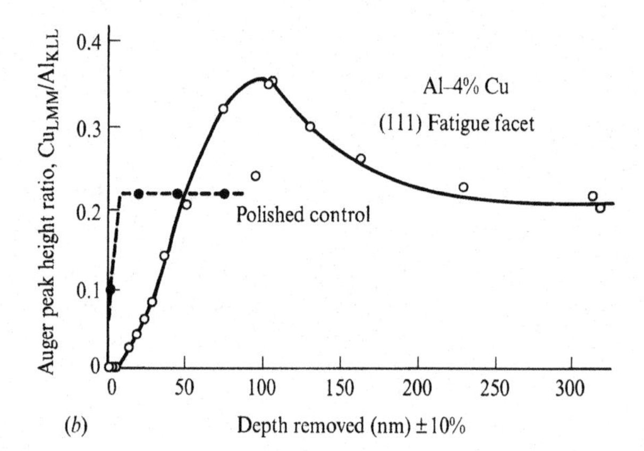


**Figure 6.15** Al – 4.2 wt% Cu alloy, underaged at 130 °C for 18 h, fatigue fractured along {111} PSB in the coarse grained sample. (*a*) Al/Cu $K_\alpha$ X-ray count rate ratio determined from energy dispersive X-ray analysis as a function of electron beam incidence angle, $\theta$. ($\theta$=90° is normal incidence). Two opposite {111} fatigue fracture facets of an underaged Al–4 wt% Cu alloy (Z9) are shown together with a mechanically polished sample of Z9. (Brett and Doherty (1978). Courtesy of *Materials Science and Engineering*.) (*b*) Same samples studied by Auger electron spectroscopy with argon ion beam depth profiling. Solute loss from the PSB (50 nm partial thickness) into the subsurface region. (Lea *et al.* (1979). Courtesy of *Scripta Met.*)

### (e) Mechanical weakening of spinodally decomposed alloys

Copper based alloys also show the phenomenon of low fatigue ratio in the precipitation-hardened condition, as has been reported for example by Ham, Kirkaldy and Plewes (1967) when the precipitates are coherent as in Cu–Co and in Cu–Be alloys. In the same study the fatigue ratio of spinodally decomposed Cu–Ni–Fe alloys was investigated and found to be much higher than in the microstructures produced by the nucleation and growth mechanism. The suggested explanation for this difference was the very short critical wavelength (equivalent to the critical radius of the nuclei) of the spinodally decomposed solid solutions.

Quin and Schwartz (1980) and Sinning (1982) discussed the role of plastic and elastic strain on the stability of spinodal microstructures. They have shown that both types of strain will lead to instability in alloys where the precipitates have a significant misfit, see eq. (2.13). Quin and Schwartz pointed out that the mechanical demodulation of the modulated alloys that they observed directly by diffraction analysis was likely to have been caused by slip irreversibility – so accounting for the observed cylic softening.

### 6.6.2 Enhancement of dislocation recovery by fatigue strain

Another example where fatigue can modify an initial microstructure is found in work-hardened metals. The dislocation substructure that is produced by unidirectional strain can be extensively modified by subsequent fatigue strain. Early work by Broom and Ham (1957) showed that work-hardened copper could be softened by fatigue if this was carried out at a temperature at which vacancies were mobile. This suggests that dislocation climb, leading to polygonisation, was involved. More extensive investigations by, for example, Coffin and Tavernelli (1959) demonstrated that high-cycle fatigue strain gave the same properties and hence structure in both annealed and unidirectionally work-hardened metals.

### 6.6.3 Stress annealing

As for the other constraints applied to metals at elevated temperatures where atomic mobility is high, it might be expected that elastic distortion will modify microstructural morphology. Directional order of atoms has been induced, for example, in Fe–Al alloys by this treatment by Birkenbeil and Cahn (1962) and, as with such directional order produced by magnetic annealing, large uniaxial magnetic anisotropy was caused by this structural change. Tien and Copley (1971) studied the effect of a uniaxial elastic distortion (both tensile and compressive) on a nickel-based superalloy, Udimet 700, containing a large volume fraction of the $\gamma'$ coherent $Ni_3Al$ based precipitate. They found that this heat-treatment caused a change of precipitate shape. Fig. 6.16 illustrates the changes seen for both tensile and compressive loading along different crystal directions. The original shape was the normal cubic form with the cube faces parallel to the cube faces of the fcc crystals. The analysis of the precipitate shape change presented by Tien and Copley showed that a change in the shape of a coherent precipitate will cause a change in shape of the host crystal, the form of this distortion depending on the sign of two system parameters $A$ and $\phi$.

$A = a' - a$ (the difference in lattice parameters between precipitate and matrix)

$\phi = E' - E$ (the difference in the tensile elastic modulus)

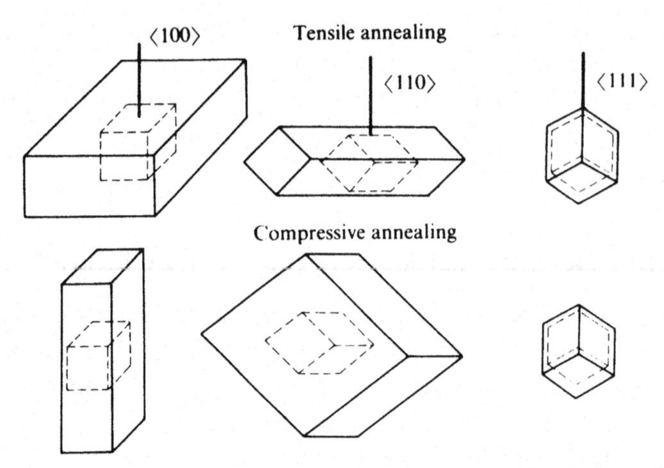


**Figure 6.16** The change in precipitate shape from the initial cube shape (shown dotted) to the final shape (full line) as a function of stress direction and sense. Only the ⟨111⟩ strain caused no change of shape. (After Tien and Copley 1972. Courtesy of *Met. Trans.*)

Since the precipitate shape change will distort the sample in response to an imposed elastic strain on the matrix, the elastic free energy of the system will be reduced by that precipitate change of shape which relaxes the elastic strain. Tien and Copley applied their analysis to the nickel based alloy containing $\gamma'$ precipitates that they had studied, and successfully accounted for the shape changes that they found. (The parameter $A$ was positive and $\phi$ was negative.)

Although the evolution of precipitate morphology can be understood qualitatively using an energy analysis, dealing with the kinetics of the process is a more formidable matter. Johnson (1987) has used the thermodynamics of non-hydrostatically stressed crystalline solids and interfaces to examine the influence of an externally applied stress field on equilibrium interfacial concentrations and the kinetics of precipitate morphological evolution. Johnson showed that the local state of stress at a precipitate–matrix interface depends on the sign and magnitude of the precipitate misfit, the applied stress and the elastic constants on the precipitate and the matrix. Interfacial concentrations may be either enhanced or diminished along various directions, and so the presence of an external stress field alters the local growth rate of a precipitate and results in the precipitate changing shape. The predicted shape changes from a sphere are shown schematically in fig. 6.17, where the forms of the resulting ellipsoids of revolution depend on the product $\epsilon\tau(1-\delta)$, where $\delta$ is the ratio of precipitate to matrix shear modulus, $\epsilon$ the coherency strain and $\tau$ the non-dimensional applied stress. When this product is greater than zero the sphere will evolve toward an oblate spheroid; when it is less than zero the sphere will evolve towards a prolate spheroid. When $\epsilon$ and t are of the same

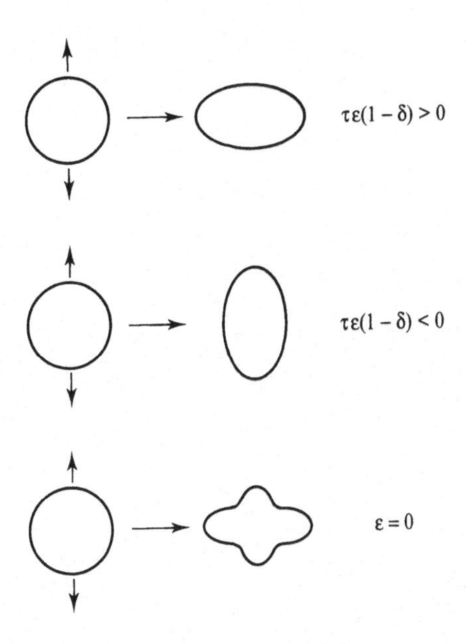


**Figure 6.17** Schematic representation of the predicted precipitate shape changes from the sphere. When $|\epsilon| \gg |\tau|$ the shape changes can be approximated as an ellipsoid of revolution, the orientation depending on the product $\epsilon\tau(q-\delta)$. The evolution of precipitate shape may be toward a metastable state. (Johnson 1987.)

order, the precipitate tends to a more complex shape than an ellipsoid of revolution.

Johnson, Berkenpas and Laughlin (1988) examined precipitate shape transitions that may occur in cubic materials during growth or coarsening of a precipitate in the presence of a constant external stress field and also shape transitions that result from changing the external stress at a constant precipitate size. Their approach shows the combination of material parameters that influence the shape transition and whether or not the transition will be continuous or discontinuous.

## 6.6.4 Changes in precipitate distribution during diffusion creep

Metals with a fine grain size can be deformed, albeit at a low strain rate at high temperatures, by the processes of vacancy (diffusion) creep. This can be by lattice diffusion of vacancies first discussed by Herring (1950) and denoted as Nabarro–Herring creep, or by grain boundary diffusion of vacancies as described by Coble (1963) and known as Coble creep. In either case a higher than normal concentration of vacancies occurs at grain boundaries transverse to a tensile stress, and a lower than normal concentration at longitudinal boundaries. This concentration gradient leads to a flow of vacancies from the transverse to the longitudinal boundaries causing an opposite flow of atoms that creates the tensile strain in the direction of the stress. Under such conditions precipitates should take no part in the process and merely act as inert markers like those in the Kirkendahl experiment. Boundaries on which atoms are deposited will then

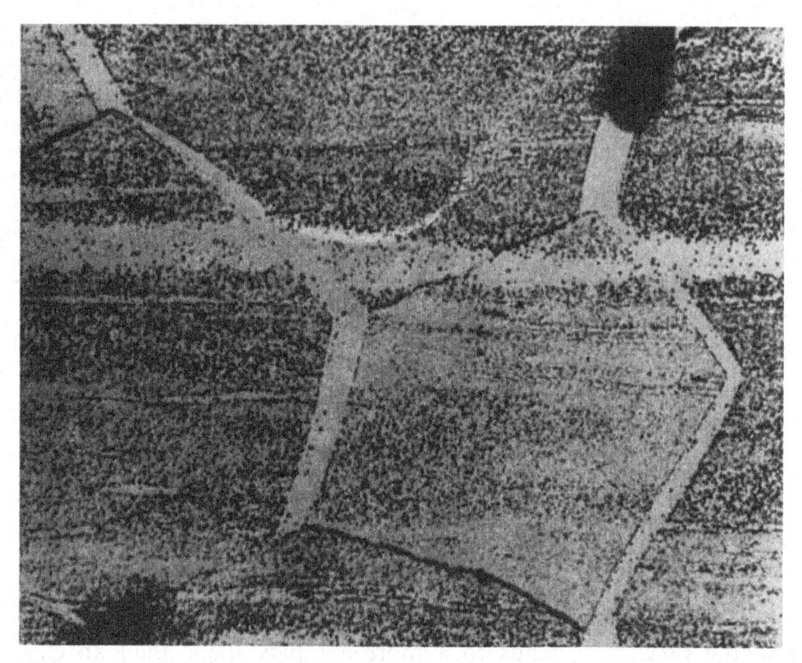

**Figure 6.18** Creep tested Mg–Zn alloy after 11 206 h at 723 K showing precipitate-free zones along the transverse grain boundary and dense concentrations of precipitates on longitudinal boundaries. (After Vickers and Greenfield 1967. Courtesy of *J. Nucl. Mat.*)

show, after some creep strain, a precipitate-free zone of the deposited matrix, while boundaries from which matrix atoms are removed will become rich in the precipitate phase (see fig. 6.18). Development of such microstructural inhomogeneities has been frequently reported, particularly in magnesium based alloys used in low temperature nuclear reactors, but also in other metals (see, for example, Harris and Jones (1963), Squires, Weiner and Phillips (1963), Pickles (1967), Vickers and Greenfield (1967), Harris, Jones, Greenwood and Ward (1968), Jones (1969), Burton (1971)).

Many of the theories which describe the inhibition of diffusion creep by precipitates postulate that a precipitate–matrix interface cannot absorb or emit vacancies. However, Clegg and Martin (1983) have shown that inhibition is not due to this effect. The Cu–$SiO_2$ interface *can* absorb vacancies (Lewis and Martin 1963), yet a dispersion of silica particles in copper inhibits diffusion creep, with creep rates lower than those predicted by the Coble creep equation in a manner similar to that observed by other workers. Clegg and Martin suggested that inhibition arises because, as the grains change their shape by creep, the boundaries move and as they do so they drag the particles with them (Harris 1973), see also §5.8.5. If a particle is to move, then it is necessary for particle atoms to move into the vacant lattice sites formed by the removal of matrix atoms (Ashby 1980). The rate at which particle migration occurs is then controlled by the slower of the two processes – either the diffusion of matrix atoms or of particle atoms. As many of the particles have high melting points, it is likely that the latter may become rate controlling. This is consistent with the observation that more refractory particles are, the more effectively they inhibit diffusion creep (Burton 1971).

## 6.7 Precipitate stability under irradiation

Nelson, Hudson and Mazey (1972) have shown both theoretically and by actual observation that an established precipitate microstructure can be made to redissolve by heavy radiation. Two mechanisms were proposed by which the precipitate would go back into solid solution at temperatures well below the equilibrium solvus. These were:

(1) *Recoil dissolution* Here atoms are removed from the precipitate into the matrix by means of the kinetic energy given to them by collision with the incoming radiation. Matrix atoms coming the other way may also aid dissolution, since they will cause the precipitate composition to move towards that of the matrix.

(2) *Disordering dissolution* Precipitates such as $\gamma'$ in nickel based alloys can be disordered by irradiation, and since the precipitate stability depends on their atomic order, irradiation can lead to the disordered precipitate becoming soluble in the matrix and redissolving. In the model of this process there is competition between that diffusion which reorders the precipitates and that which leads to resolution. Irradiation is therefore assumed to involve the dissolution of only the outer layer of precipitate.

Irradiation will also enhance the matrix diffusion coefficient from $D$ to $D^*$ by increasing the vacancy concentration. By consideration of all the effects of irradiation Nelson *et al.* (1972) derived the rate of change of precipitate size for conditions of disordering dissolution:

$$dr/dt = -\psi K + (3D^* C/4\pi\rho r) - D^* r^2 n \tag{6.19}$$

where $K$ is the damage rate (number of displacements per atom per second), $\psi$ is the dissolution parameter defined as the number of dissolved atoms per displacement, $C$ the total solute content of the alloy, $\rho$ the atom fraction of solute in the precipitate and $n$ the density of precipitates.

At low temperatures, where $D^*$ is negligible, all precipitates will redissolve, but at higher temperatures, where $D^*$ is sufficiently large to encourage some return to equilibrium, the form of eq. (6.19) indicates that *small* precipitates will grow but larger ones will shrink – the opposite of the usual case of Ostwald ripening. Under these conditions a *small* steady state precipitate radius will be produced. Experimental observation of both $\gamma'$ and $M_{23}C_6$ precipitates in a nickel based alloy showed that the precipitates were redissolved by charged particle irradiation at room temperature, but at higher temperatures the precipitate size was significantly *reduced*, just as predicted.

# References

## Chapter 1


Aaronson, H.I. (1962). *Decomposition of austenite by diffusional processes*, Interscience, New York, p. 387.

Aaronson, H.I., Laird, C. and Kinsman, K.R. (1970). *Phase transformations*, A.S.M., Metals Park, Ohio, and Chapman & Hall, London, p. 313.

Atkinson, C. (1981). *Proc. R. Soc.*, **A378**, 351.

Atkinson, C. (1982). *Proc. R. Soc.*, **A384**, 167.

Bainbridge, B.G. (1972). D.Phil. thesis, University of Sussex, UK.

Balluffi, R.W. (1982). *Met. Trans.*, **13B**, 527.

Barrett, C.S. (1952). *Structure of metals*, 2nd edn, McGraw-Hill, New York, p. 544.

Brown, A.M. and Ashby, M.F. (1982). *Act. Met.*, **28**, 1085.

Cahn, J.W. (1968). *Trans. Met. Soc. AIME*, **242**, 167.

Cahn, J.E. and Hilliard, J.E. (1958). *J. Chem. Phys.*, **28**, 258.

Cahn, J.E. and Hilliard, J.E. (1959). *J. Chem. Phys.*, **31**, 688.

Corti, C.W. and Cotterill, P. (1972). *Scripta Met.*, **6**, 1047.

DeHoff, R.T. and Rhines, F.N. (1968). *Quantitative microscopy*, McGraw-Hill, New York.

Doherty, R.D. (1982). *Metal Sci.*, **16**, 1.

Doherty, R.D. (1983) in *Physical metallurgy* (eds. Cahn, R.W. and Haasen, P.) Elsevier, Amsterdam, p. 934.

Doherty, R.D. and Cantor, B. (1982) in *Proc. Int. Conf. on Solid–Solid Phase Transformations*, AIME, Warrendale PA., p. 547.

Doherty, R.D. and Cantor, B. (1988). *Scripta Met.*, **22**, 301.

Doherty, R.D. and Feest, A. (1968). *Solidification of metals*, ISI Publication 110, p. 102.

Doherty, R.D. and Rajab, K.E. (1989). *Acta Met.*, **37**, 2723.

Enomoto, M. (1987). *Acta Met.*, **35**, 935.

Fairs, S. (1978). D.Phil thesis, University of Sussex.

Ferrante, M. and Doherty, R.D. (1979). *Acta Met.*, **27**, 1603.

Gaskell, D.R. (1983) in *Physical metallurgy* (eds. Cahn, R.W. and Haasen, P.) Elsevier, Amsterdam, p. 272.

Gibbs, J.W. (1878) in *The scientific papers of J. Willard Gibbs*, Volume 1, *Thermodynamics*, Dover Publications Inc., New York (1961).

Gleiter, H. and Chalmers, B. (1972). *Prog. Mat. Sci.*, **16**, 1.

Haessner, F. and Hofmann, S. (1978) in *Recrystallization of metallic materials* (ed. Haessner, F.) Dr Riederer Verlag GmbH, Stuttgart, p. 63.

Hilliard, J.E. (1966). *Recrystallization, grain growth and textures*, ASM, Metals Park, Ohio, and Chapman & Hall, London, p. 267.

Hilliard, J.E. (1969). *Quantitative relation between properties and microstructure* (eds. Brandon, D.G. and Rosen A.), Israel University Press, p. 3.

Howe, J.M., Aaronson, H.I. and Gronsky, R. (1985). *Acta Met.*, **33**, 639 and 649.

Howe, J.M., Dahmen, U. and Gronsky, R. (1987). *Phil. Mag.*, **56A**, 31.

Jackson, K.A. (1958) in *Liquid metals and solidification*, ASM, Metals Park, Ohio, p. 174.

Jackson, K.A. and Hunt, J.D. (1965). *Acta Met.*, **13**, 1212.

Jones, D.R.H. and Chadwick, G.A. (1971). *Phil. Mag.*, **24**, 995.

Rajab, K.E. and Doherty, R.D. (1989). *Acta Metall.*, **37**, 2709.

Reed-Hill, R.E. and Abbaschian, R. (1992). *Physical metallurgy principles*, 3rd edn, PWS-Kent Publishing Co., Boston, p. 538.

Servi, I.S. and Turnbull, D. (1966). *Acta Met.*, **14**, 161.

Shewmon, P.G. (1989). *Diffusion in solids*, 2nd edn. Reprinted by TMS, Warrendale, PA 15086.

Singh, R.P. and Doherty, R.D. (1992). *Met. Trans.*, **23A**, 307.

Stobbs, W.M. (1991). *Met. Trans.*, **22A**, 1139.

Underwood, E.E. (1970). *Quantitative stereology*, Addison-Wesley, New York.

Vander Voort, G.F. (1984). *Metallography – principles and practice*, McGraw-Hill, New York.

Weatherly, G.C. (1971). *Acta Met.*, **19**, 181.

Woodruff, D.P. (1973). *The solid–liquid interface*, Cambridge University Press, Cambridge, p. 151.

## Chapter 2

Aaron, H.D. and Kotler, G.R. (1971). *Met. Trans.*, **2**, 393.

Aaron, H.D., Fainstein, D. and Kotler, G.R. (1970). *J. Appl. Phys.*, **41**, 4404.

Aaronson, H.I., Aaron, H.B. and Kinsman, K.R. (1971). *Metallography*, **4**, 1.

Aaronson, H.I., Laird, C. and Kinsman, K.R. (1970). *Phase transformations*, ASM, Metals Park, Ohio, and Chapman & Hall, London, p. 313.

Abbott, K. and Haworth, C.W. (1973). *Acta Met.*, **21**, 951.

Ahearn, P.J. and Quigley, F.C. (1966). *JISI*, **204**, 16.

Annavarapu, S. and Doherty, R.D. (1993). *Int. J. Powder Metal.*, **29**, 331.

Annavarapu, S. and Doherty, R.D. (1995). *Acta Metall. et Mater.*, **43**, 3207.

Annavarapu, S., Liu, J. and Doherty, R.D. (1992) in *Grain growth in polycrystalline materials* (eds. Abbruzzese, G. and Brosso, P.) Trans. Tech. Publications, Zurich, Switzerland, p. 649.

Baker, R.G., Brandon, D.G. and Nutting, J. (1959). *Phil. Mag.*, **4**, 1339.

Bastien, P.G. (1957). *JISI*, **187**, 281.

Baty, D.L., Tanzilli, R.A. and Heckel, R.W. (1970). *Met. Trans.*, **1**, 1651.

Becker, R. and Döring, W. (1935). *Ann. Phys.*, **24**, 719.

Biloni, H. (1984) in *Physical metallurgy*, 3rd edn (eds. Cahn, R.W. and Haasen, P.), North-Holland Publishing, p. 477.

Bonar, L.G. (1962). PhD thesis, University of Cambridge.

Brooks, H. (1952). *Metal interfaces*, ASM, Metals Park, Ohio, p. 20.

Cahn, J.W. (1956). *Acta Met.*, **4**, 449.

Cahn, J.W. (1957). *Acta Met.*, **5**, 168.

Cahn, J.W. (1961). *Acta Met.*, **9**, 795.

Cahn, J.W. (1962). *Acta Met.*, **10**, 179.

Cahn, J.W. (1964). *Acta Met.*, **12**, 1457.

Cahn, J.W. (1968). *Trans. Met. Soc. AIME*, **242**, 166.

Cahn, J.W. and Hilliard, J.E. (1958). *J. Chem. Phys.*, **28**, 258.

Cahn, J.W. and Hilliard, J.E. (1959). *J. Chem. Phys.*, **31**, 688.

Cerezo, A., Hyde, J.M., Miller, M.K., Petts, S.C., Setna, R.P. and Smith, G.D.W. (1992). *Phil. Trans. R. Soc. London*, **A341**, 313.

Chalmers, B. (1964). *Principles of solidification*, John Wiley, New York.

Christian, J.W. (1975). *The theory of transformations in metals and alloys*, 2nd edn, Pergamon, Oxford.

Ciach, R., Dukiet-Zawadzka, B. and Ciach, T.D. (1978). *J. Mater. Sci.*, **13**, 2676.

Cole, G.S. (1971). *Met. Trans.*, **2**, 357.

Daniel, V. and Lipson, H. (1943). *Proc. Roy. Soc.*, **A181**, 368.

Daniel, V. and Lipson, H. (1944). *Proc.. Roy. Soc.*, **A182**, 378.

Darken, L.S. (1948). *Trans. Met. Soc. AIME*, **174**, 184.

da Silva, L.C. and Mehl, R.F. (1951). *Trans. AIME*, **191**, 155.

Doherty, R.D. and Melford, D.A. (1966). *JISI*, **204**, 1131.

Doherty, R.D. (1979a). *J. Mater. Sci.*, **14**, 2276.

Doherty, R.D. (1979b) in *Crystal Growth*, 2nd edn (ed. Pamplin, B.) Pergamon Press, Oxford.

Doherty, R.D. (1993). Research at Drexel University.

Doherty, R.D., Feest, A. and Lee, Ho-In (1984). *Mater. Sci. and Eng.*, **65**, 181.

Dubé, C.A. (1948). PhD thesis, Carnegie Institute of Technology.

Eshelby, J.D. (1957). *Proc. Roy. Soc.*, **A241**, 376.

Feest, E.A. and Doherty, R.D. (1973). *Met. Trans.*, **4**, 125.

Gaskell, D.R. (1983) in *Physical metallurgy* (eds. Cahn, R.W. and Haasen, P.) Elsevier, Amsterdam, p. 272.

Gibbs, J.W. (1876). *Trans. Conn. Acad. Sci.*, **3**, 228.

Gleiter, H. (1967a). *Acta Met.*, **15**, 1213.

Gleiter, H. (1967b). *Acta Met.*, **15**, 1223.

Gleiter, H. (1967c). *Z. f. Naturforschung*, **58**, 101.

Grange, R.A. (1971). *Met. Trans.*, **2**, 417.

Gündos, M. and Hunt, J.D. (1985). *Acta Met.*, **33**, 1651.

Hall, M.G. and Haworth, C.W. (1970). *Acta Met.*, **18**, 331.

Ham, R.K., Kirkaldy, J.S. and Plewes, J.W. (1967). *Acta Met.*, **15**, 861.

Hannerz, N.E. (1968). *Met. Sci. J.*, **2**, 148.

Heckel, R.W. and Balasubramanian (1971). *Met. Trans.*, **2**, 379.

Hewitt, P. and Butler, E.P. (1986). *Acta Met.*, **34**, 1163.

Hillert, M. (1957). *Jernkorntorets Ann.*, **141**, 757.

Hillert, M. (1961). *Acta Met.*, **9**, 525.

Hilliard, J.E. (1970). *Phase transformations*, ASM, Metals Park, Ohio, and Chapman & Hall, London, p. 497.

Hone, M. and Purdy, G.R. (1969). Published in Purdy and Kirkaldy (1971).

Hornbogen, E. (1967). *Aluminium*, **43**, 115.

Hu, H. and Smith, C.S. (1956). *Acta Met.*, **4**, 638.

Humphreys, F.J. (1968). *Acta Met.*, **16**, 1069.

Kahlweit, M. (1968). *Scripta Met.*, **2**, 251.

Kattamis, T.Z., Coughlin, J.C. and Flemings, M.C. (1967). *Trans. Met. Soc. AIME*, **239**, 1504.

Kinsman, K.R. and Aaronson, H.I. (1967). *Transformation and hardenability in steels*, Climax Molybdenum C., Ann Arbor, Michigan, p. 39.

Kirkaldy, J.S., Destinon-Forstmann, J. von and Brigham, R.J. (1962). *Can. Met. Quart.*, **1**, 59.

Kohn, A. and Doumerc, J. (1955). *Mem. Sci. Rev. Met.*, **53**, 249.

Kohn, A. and Philibert, J. (1960). *Metal Treatment*, **27**, 351.

Lewis, M.H. and Martin, J.W. (1964). *J. Inst. Metals*, **92**, 126.

Lorimer, G.W. and Nicholson, R.B. (1969). *The mechanism of phase transformation in crystalline solids*, Institute of Metals, London, p. 36.

Matyja, H., Giessen, B.C. and Grant, N.J. (1968). *J. Inst. Metals*, **96**, 30.

Nemoto, M. (1974). *Acta Met.*, **22**, 847.

Nicholson, R.B. (1970). *Phase transformations*, ASM, Metals Park, Ohio, and Chapman & Hall, London, p. 269.

Phillips, V.A. and Livingstone, J.D. (1962). *Phil. Mag.*, **7**, 969.

Polmear, I.J. (1966). *J Australian Inst. Metals*, **11**, 246.

Proud, L.W. and Wynne, E.J. (1968). *ISI Sp. Rpt*, **108**, 157.

Purdy, G.R. (1971). *Met. Sci. J.*, **5**, 81.

Purdy, G.R. and Kirkaldy, J.S. (1971). *Met. Trans.*, **2**, 371.

Rundman, K.B. and Hilliard, J.E. (1967). *Acta Met.*, **15**, 1025.

Servi, I.S. and Turnbull, D. (1966). *Acta Met.*, **14**, 161.

Shewmon, P.G. (1965). *Trans. Met. Soc. AIME*, **233**, 736.

Shewmon, P.G. (1966). *Recrystallization, grain growth and textures*, ASM, Metals Park, Ohio, and Chapman and Hall, London, p. 165.

Shewmon, P.G. (1989). *Diffusion in solids*, 2nd edn, TMS-AIME, Warrendale, PA.

Smigelskas, A. and Kirkendall, E. (1947). *Trans. Met. Soc. AIME*, **171**, 130.

Soffa, W.A. and Laughlin, D.E. (1983) in *Solid state phase transformations, the Pittsburgh Conference* (ed. Aaronson H.I.) Met. Soc. AIME, Warrendale, PA, p. 159.

Townsend, R.D. and Kirkaldy, J.S. (1968). *Trans. Amer. Soc. Metals*, **61**, 605.

Vietz, J.T. and Polmear, I.J. (1966). *J. Inst. Metals*, **94**, 410.

Vogel, A., Cantor, B. and Doherty, R.D. (1979). *Sheffield conference on solidification*, The Metals Society (now the Institute of Materials), London, p. 518.

Volmer, M. and Weber, A. (1925). *Z. Phys. Chem.*, **119**, 277.

Ward, R.G. (1965). *JISI*, **203**, 930.

Weatherly, G.C. (1968). *Phil. Mag.*, **17**, 791.

Weatherly, G.C. and Nicholson, R.B. (1968). *Phil. Mag.*, **17**, 801.

Weinberg, F. and Buhr, R.K. (1969). *JISI*, **207**, 1114.

Whelan, M.J. (1969). *Met. Sci. J.*, **3**, 95.

Zener, C. (1949). *J. Appl. Phys.*, **20**, 950.

## Chapter 3

Ahmadzadeh, M. and Cantor, B. (1981). *J. Non-Cryst. Solids*, **43**, 189.

Anantharaman, T.R. (1988). *Current Sci.*, **57**, 578.

Atzmon, M., Veerhoven, J.R., Gibson, E.R. and Johnson, W.L. (1984). *Appl. Phys. Lett.*, **45**, 1052.

Bak, P. (1986). *Scripta Met.*, **20**, 1199.

Bardenheuer, P. and Bleckmann, R. (1939). *Mitt KW Inst Eisenforschung*, **21**, 201.

Battezzati, L. (1990). *Phil. Mag.*, **B61**, 511.

Benjamin, J.S. (1976). *Sci. Amer.*, May, 40.

Bernal, J.D. (1960). *Nature*, **188**, 910.

van den Beukel, A. (1988). *Scripta Met.*, **22**, 877.

van den Beukel, A. and Radelaar, S. (1983). *Acta Met.*, **31**, 419.

Bevk, J. (1983). *Annual reviews of material science* (eds Huggins, R.A., Bube, R.H. and Vermilyea, D.A.) Annual Reviews Inc, Palo Alto, vol 13.

Bhatia, A.B. and Thornton, D.E. (1970). *Phys Rev.*, **B2**, 3004.

Bhatti, A.R. and Cantor, B. (1994). *J. Mater. Sci.*, **29**, 816.

Bloch, J. (1962). *J. Nucl. Mater.*, **6**, 203.

Boettinger, W.J. and Coriell, S.R. (1986) in *Science and technology of the undercooled melt* (eds Sahm, P.R., Jones, H. and Adam, C.M.) Martinus Nijhoff, Dordrecht, p. 81.

Boettinger, W.J. and Perepezko, J.H. (1993) in *Rapidly solidified alloys* (ed. Liebermann, H.H.) Dekker, New York, p. 17.

Boettinger, W.J., Shechtman, D., Schaefer, R.J. and Biancaniello, F.S. (1984). *Metall. Trans.*, **A15**, 55.

Bøttiger, J., Greer, A.L. and Karpe, N. (1994). *Mater, Sci. Eng.*, **A178**, 65.

Breinan, E.M., Kear, B.H. and Banas, C.M. (1976). *Physics Today*, Nov., 44.

Brenner, A. (1963). *Electrodeposition of alloys: Principles and practice*, Academic Press, New York.

Brenner, A., Couch, D.E. and Williams, E.K. (1950). *J. Res. Nat. Bur. Stand.*, **44**, 109.

Brown, A.M. and Ashby, M.F. (1980). *Acta Met.*, **28**, 1085.

Buckel, W. and Hilsch, R. (1952). *Z. Phys.*, **131**, 420.

Buckel, W. and Hilsch, R. (1954). *Z. Phys.*, **138**, 109.

Buckel, W. and Hilsch, R. (1956). *Z. Phys.*, **146**, 27.

Bursill, L. and Lin, J. (1985). *Nature*, **316**, 50.

Cahn, R.W. (1991) in *Glasses and amorphous materials* (ed. Zarzycki, J.) vol. 9 of *Materials science and technology* (eds. Cahn, R.W., Haasen, P. and Kramer, E.J.) VCH, Weinheim, p. 493.

Calka, A., Matyja, H., Polk, D.E., Giessen, B.C., van der Sande, J. and Madhava, M. (1977). *Scripta Met.*, **11**, 65.

Cantor, B. (1985) in *Rapidly quenched metals V* (eds Steeb, S. and Warlimont, H.) North Holland, Amsterdam, vol. 1, p. 595.

Cantor, B. (1986). *Science and technology of the undercooled melt* (eds. Sahm, P.R., Jones, H. and Adam, C.M.) Martinus Nijhoff, Dordrecht, p. 3.

Cantor, B. and Cahn, R.W. (1976a). *Acta Met.*, **24**, 845.

Cantor, B. and Cahn, R.W. (1976b). *J. Mater. Sci.*, **11**, 1066.

Cantor, B. and Cahn, R.W. (1983) in *Amorphous metallic alloys* (ed. Luborsky, F.E.) Butterworths, London, p. 487.

Cantor, B., Kim, W.T., Bewlay, B.P. and Gillen, A.G. (1991). *J. Mater. Sci.*, **26**, 1266.

Cargill, G.S. III (1969). *Acta Crystall.*, **A25**, 595.

Cargill, G.S. III (1970). *J. Appl. Phys.*, **41**, 12.

Carslaw, H.S. and Jaeger, J.C. (1959). *Conduction of heat in solids*, OUP, Oxford.

Chen, H.S. (1983) in *Amorphous metallic alloys* (ed. Luborsky, F.E.) Butterworths, London, p. 169.

Christian, J.W. (1975). *The theory of transformations in metals and alloys*, Pergamon, Oxford.

Cochrane, R.F., Schumacher, P. and Greer, A.L. (1991). *Mater. Sci. Eng.*, **A133**, 367.

Cohen, M., Kear, B.H. and Mehrabian, R. (1980) in *Rapid solidification processing: Principles and technologies II* (eds. Mehrabian, R., Kear, B.H. and Cohen, M.) Claitors, Baton Rouge, p. 1.

Dahlgren, S.D. (1978) in *Rapidly quenched metals III* (ed. Cantor, B.) Metals Society, London, p. 36.

Davies, H.A. (1983) in *Amorphous metallic alloys* (ed. Luborsky, F.E.) Butterworths, London, p. 8.

de Bruijn, N.G. (1981a). *Math. Proc.*, **A84**, 39.

de Bruijn, N.G. (1981b). *Ned. Akad. Wet. Proc. Ser.*, **A43**, 39.

Dong, C., Chattopadhyay, K. and Kuo, K.H. (1987). *Scripta Met.*, **21**, 1307.

Drehman, A.J. and Greer, A.L. (1984). *Acta Met.*, **32**, 323.

Duine, P.A., Wonnell, S.K. and Sietsma, J. (1994). *Mater. Sci. Eng.*, **A179/180**, 270.

Duine, P.A., Sietsma, J. and van den Beukel, A. (1993). *Phys. Rev.*, **B37**, 6957.

Duwez, P., Willens, R.H. and Klement Jr, W. (1960). *J. Appl. Phys.*, **31**, 36.

Egami, T. (1978). *Mater. Res. Bull.*, **13**, 557.

Egami, T. (1984). *Rep. Prog. Phys.*, **47**, 1601.

Egami, T. (1993) in *Rapidly solidified alloys* (ed. Liebermann, H.H.) Dekker, New York, p. 231.

Elser, V. (1985). *Phys. Rev.*, **B34**, 4892.

Elser, V. (1986). *Acta Crystall.*, **A42**, 36.

Evetts, J.E. (1985) in *Rapidly quenched metals V* (eds. Steeb, S. and Warlimont, H.) North Holland, Amsterdam, vol. 1, p. 607.

Faupel, F., Hüppe, P.W. and Rätzke, K. (1990). *Phys. Rev. Lett.*, **65**, 1219.

Field, R.D. and Fraser, H. (1985). *Mater. Sci. Eng.*, **68**, L17.

Finney, J.L. (1983) in *Amorphous metallic alloys* (ed. Luborsky, F.E.) Butterworths, London, p. 42.

Finney, J.L. and Wallace, J. (1981). *J. Non-Cryst. Solids.*, **43**, 165.

Frank, W., Horner, A., Schwarwaechter, P. and Kronmuller, H. (1994). *Mater. Sci. Eng.*, **A179/180** 36.

Fukamichi, K., Kikuchi, M., Masumoto, T. and Matsura, M. (1979). *Phys. Lett.*, **A73**, 436.

Gao, W. and Cantor, B. (1989). *Acta Met.*, **37**, 3409.

Gaskell, P.H. (1978) in *Rapidly quenched metals III* (ed. Cantor, B.) Metals Society, London, vol. 2, p. 277.

Gaskell, P.H. (1982) in *Rapidly quenched metals IV* (eds Masumoto, T. and Suzuki, K.) Japan Institute of Metals, Sendai, vol. 1, p. 267.

Gibbs, M.R.J. and Evetts, J.E. (1983). *J. Mater. Sci.*, **18**, 278.

Giessen, B.C. (1969) in *Advances in X-ray analysis* (ed. Barrett, C.S.) Plenum, New York, vol. 12, p. 23.

Gillen, A.G. and Cantor, B. (1985). *Acta Met.*, **33**, 1813.

Grant, W.A., Ali, A., Chadderton, L.T., Grundy, P.J. and Johnson, E. (1978) in *Rapidly quenched metals III* (ed. Cantor, B.) Metals Society, London, p. 63.

Greer, A.L. (1991). *Mater. Sci. Eng.*, **A133**, 16.

Greer, A.L. (1993) in *Rapidly solidified alloys* (ed. Liebermann, H.H.) Dekker, New York, p. 269.

Greer, A.L. (1994). *Mater. Sci. Eng.*, **A179**, 41.

Greer, A.L. and Leake, J.A. (1978) in *Rapidly quenched metals III* (ed. Cantor, B.) Metals Society, London, vol. 2, p. 299.

Hafner, J. (1980). *Phys. Rev.*, **B21**, 406.

Hafner, J. (1983). *Phys. Rev.*, **B28**, 1734.

Hafner, J. (1985) in *Rapidly quenched metals V* (eds. Steeb, S. and Warlimont, H.) North-Holland, Amsterdam, vol. 1, p. 421.

Hagiwara, M., Inoue, A. and Masumoto, T. (1981). *Sci. Rep. Tohoku Univ.*, **29**, 351.

Hasegawa, R. and Ray, R. (1978). *J. Appl. Phys.*, **49**, 4174.

Hayzelden, C., Rayment, J.J. and Cantor, B. (1983). *Acta Met.*, **31**, 379.

Herlach, D.M., Cochrane, R.F., Egry, I., Fecht, H.J. and Greer, A.L. (1993). *Int. Met. Rev.*, **38**, 273.

Höfler, H.J., Averbach, R.S., Rummel, G. and Mehrer, H. (1992). *Phil. Mag. Lett.*, **66**, 301.

Holland, J.R., Mansur, L.K. and Potter, D.I. (eds.) (1981). *Phase stability during irradiation*, TMS-AIME, Warrendale.

Hughes, M.A., Bhatti, A.R., Gao, W. and Cantor, B. (1988). *Mater. Forum*, **11**, 21.

Hunt, A.J., Cantor, B. and Kijek, M. (1985) in *Rapidly solidified materials* (eds. Lee, P.W. and Carbonara, R.S.) ASM, Metals Park, p. 169.

Inoue, A. and Masumoto, T. (1991). *Mater. Sci. Eng.*, **A133**, 6.

Jackson, K.A. and Hunt, J.D. (1966). *Trans. Met. Soc. AIME*, **236**, 1129.

Johnson, W.L. (1986). *Prog. Mater. Sci.*, **30**, 81.

Jones, H. (1973). *Rep. Prog. Phys.*, **36**, 1425.

Jones, H. (1982). *Rapid solidification of metals and alloys*, Institution of Metallurgists, London.

Jones, H. (1991). *Mater. Sci. Eng.*, **A133**, 33.

Jones, H. (1994). *Mater. Sci. Eng.*, **A179/180**, 1.

Kauzmann, W. (1948). *Chem. Rev.*, **43**, 219.

Kelton, K.F. (1993). *Int. Mater. Rev.*, **38**, 105.

Kijek, M.M., Akhtar, D., Cantor, B. and Cahn, R.W. (1982) in *Rapidly quenched metals IV* (eds. Masumoto, T. and Suzuki, K.) Japan Institute of Metals, Sendai, vol. 1, p. 573.

Kim, D.H. and Cantor, B. (1990) in *Quasicrystals and incommensurate structures in condensed matter* (eds. Yacaman, M.J., Romeu, D., Castano, V. and Gomez, A.) World Scientific, Singapore, p. 273.

Kim, D.H. and Cantor, B. (1994). *Phil. Mag.*, **69**, 45.

Kim, D.H., Hutchison, J. and Cantor, B. (1990). *Phil. Mag.*, **61**, 167.

Kim, W.T. and Cantor, B. (1991). *J. Mater. Sci.*, **26**, 2868.

Kim, W.T., Cantor, B., Griffth, W.D. and Jolly, M.R. (1992). *Int. J. Rapid Solidification*, **7**, 245.

Kim, W.T., Cantor, B. and Kim, T.H. (1990). *Int. J. Rapid Solidification*, **5**, 251.

Kim, W.T., Clay, K., Small, C. and Cantor, B. (1991). *J. Non-Cryst. Solids*, **127**, 273.

Klar, E. and Shaefer, W.M. (1972) in *Powder metallurgy for high performance applications* (ed. Burke, J. and Weiss, V.) Syracuse University Press, Syracuse, p. 57.

Klement Jr, W., Willens, R.H. and Duwez, P. (1960). *Nature*, **187**, 869.

Knowles, K.M., Greer, A.L., Saxton, W.O. and Stobbs, W.M. (1985). *Phil. Mag.*, **B52**, L31.

Koch, C.C. (1991) in *Processing of metals and alloys* (ed. Cahn, R.W.) vol. 15 of *Materials science and technology* (eds. Cahn, R.W., Haasen, P. and Kramer, E.J.) VCH, Weinheim, p. 193.

Köster, U. and Herold, U. (1982) in *Rapidly quenched metals IV* (eds. Masumoto, T. and Suzuki, K.) Japan Institute of Metals, Sendai, vol. 1, p. 719.

Köster, U. and Schünemann, U. (1993) in *Rapidly solidified alloys* (ed. Liebermann, H.H.) Dekker, New York, p. 303.

Kurz, W. and Gilgien, P. (1994). *Mater. Sci. Eng.*, **A178**, 171.

Kurz, W. and Trivedi, R. (1991). *Metall. Trans.*, **22**, 3051.

Kurz, W., Giovanola, B. and Trivedi, R. (1986). *Acta Met.*, **34**, 823.

Lavernia, E.J., Ayers, J.D. and Srivatsan, T. (1992). *Int. Mater. Rev.*, **37**, 1.

Lee, M.H., Chang, I.T.H., Dobson, P.J. and Cantor, B. (1994). *Mater. Sci. Eng.*, **A179/180**, 545.

Liebermann, H.H. (1983) in *Amorphous metallic alloys* (ed. Luborsky, F.E.) Butterworths, London, p. 26.

Luborsky, F.E. and Walter, J.L. (1977). *IEEE Trans. Magn.*, **MAG-13**, 953, 1635.

Lux, B. and Hiller, W. (1977). *Praktische Metallog*, **8**, 218.

Mackay, A.L. (1976). *Phys. Bull.*, Nov., 495.

Mackay, A.L. (1981). *Sov. Phys. Crystall.*, **26**, 517.

Mader, S., Nowick, A.S. and Widmer, H. (1967). *Acta Met.*, **15**, 203.

Mingard, K.P. and Cantor, B. (1993). *J. Mater. Res.*, **8**, 274.

Mobley, C.E., Clauer, A.H. and Wilcox, B.A. (1972). *J. Inst. Metals.*, **100**, 142.

Nagel, S.R. and Tauc, J. (1977). *Solid State Comm.*, **22**, 129.

Narasimhan, M.C. (1979). US Pat 4142571.

Ojha, S.N., Ramachandrarao, P. and Anantharaman, T.R. (1983). *Trans. Ind. Inst. Met.*, **36**, 51.

Pauling, L. (1985). *Nature*, **317**, 512.

Pauling, L. (1987). *Phys. Rev. Lett.*, **58**, 365.

Penrose, R. (1974a). *Bull. Inst. Math. Appl.*, **10**, 266.

Penrose, R. (1974b). *Math. Intelligence*, **2**, 32.

Perepezko, J.H. (1980). *Rapid solidification processing: Principles and technologies II* (eds. Mehrabian, R., Kear, B.H. and Cohen, M.) Claitors, Baton Rouge, p. 56.

Picraux, S.T. and Choyke, W.J. (eds.) (1982). *Metastable materials formed by ion implantation*, North-Holland, Amsterdam.

Polk, D.E. (1970). *Scripta Met.*, **4**, 117.

Pond, R.B. (1958). US Pat 2825108.

Pond Jr, R. and Maddin, R. (1969). *Trans. AIME*, **245**, 2475.

Ramachandrarao, P. (1980). *Z. Metallk.*, **71**, 172.

Rätzke, K., Hüppe, P.W. and Faupel, F. (1992). *Phys. Rev. Lett.*, **68**, 2347.

Ray, R., Giessen, B.C. and Grant, N.J. (1968). *Scripta Met.*, **2**, 357.

Ruhl, R.C., Giessen, B.C., Cohen, M. and Grant, N.J. (1967). *Acta Met.*, **15**, 1693.

Saunders, N. and Miodownik, A.P. (1986). *J. Mater. Res.*, **1**, 38.

Saunders, N. and Miodownik, A.P. (1988). *Mater. Sci. Tech.*, **4**, 768.

Schechtman, D. and Blech, I. (1985). *Metall. Trans.*, **A16**, 1005.

Schechtman, D., Blech, I., Gratias, D. and Cahn, J.W. (1984). *Phys. Rev. Lett.*, **53**, 1951.

Schwarz, R.B. and Johnson, W.L. (1983). *Phys. Rev. Lett.*, **51**, 415.

Scott, M.G. (1983) in *Amorphous metallic alloys* (ed. Luborsky, F.E.) Butterworths, London, p. 144.

Scott, M.G., Gregan, G. and Dong, Y.D. (1982b) in *Rapidly quenched metals IV* (eds. Masumoto, T. and Suzuki, K.) Japan Institute of Metals, Sendai, vol. 1, p. 671.

Scott, M.G., Cahn, R.W., Kursumovic, A., Girt, E. and Njuhovic, N.B. (1982a) in *Rapidly quenched metals IV* (eds. Masumoto, T. and Suzuki, K.) Japan Institute of Metals, Sendai, vol. 1, p. 469.

Singer, A.R.E. (1970). *Metals and Materials*, **4**, 246.

Spaepen, F. and Cargill, G.S. III. (1985) in *Rapidly quenched metals V* (eds. Steeb, S. and Warlimont, H.) North Holland, Amsterdam, vol. 1, p. 581.

Spaepen, F. and Taub, A.I. (1983) in *Amorphous metallic alloys* (ed. Luborsky, F.E.) Butterworths, London, p. 231.

Steinhardt, P.J. and Ostlund, S. (1987). *The physics of quasicrystals*, World Scientific, Singapore, p. 29.

Steurer, W. (1995) in *Physical Metallurgy* (ed. Cahn, R.W. and Haasen, P.) North Holland, Amsterdam, in press.

Strange, E.A. and Pim, C.H. (1908). US Pat 905758.

Suryanarayana, C. (1991) in *Processing of metals and alloys* (ed. Cahn, R.W.) vol. 15 of *Materials science and technology* (eds. Cahn, R.W., Haasen, P. and Kramer, E.J.) VCH, Weinheim, p. 57.

Suryanarayana, C., Froes, F.H. and Rowe, R.G. (1991). *Int. Mater. Rev.* **36**, 85.

Suzuki, K. (1983) in *Amorphous metallic alloys* (ed. Luborsky, F.E.) Butterworths, London, p. 74.

Takahara, Y. and Matsuda, H. (1994). *Mater. Sci. Eng.*, **A179/180**, 279.

Takayama, S. and Oi, T. (1969). *J. Appl. Phys.*, **50**, 4962.

Thompson, C.V., Greer, A.L. and Spaepen, F. (1983). *Acta Met.*, **31**, 1883.

Trivedi, R. (1994). *Mater. Sci. Eng.*, **A178**, 129.

Tsao, S.S. and Spaepen, F. (1985). *Acta Met.*, **33**, 881.

Tsaur, B.Y., Lau, S.S. and Mayer, J.W. (1980). *Appl. Phys. Lett.*, **36**, 823.

Turnbull, D. (1969). *Contemp. Phys.*, **10**, 473.

Uhlman, D.R. (1972). *J. Non-Cryst. Solids*, **7**, 337.

Vecchio, K.S. and Williams, D.B. (1988). *Metall. Trans.*, **A19**, 2875.

Vonnegut, B. (1948). *J. Colloid Sci.*, **3**, 563.

Wagner, A.V. and Spaepen, F. (1994). *Mater. Sci. Eng.*, **A1179/180**, 265.

Wagner, C.N.J. (1983) in *Amorphous metallic alloys* (ed. Luborsky, F.E.) Butterworths, London, p. 58.

Watanabe, T. and Scott, M.G. (1980). *J. Mater. Sci.*, **15**, 1131.

Yeh, X.L., Samwer, K. and Johnson, W.L. (1983). *Appl. Phys. Lett.*, **42**, 242.

Yermakov, A.Y., Yurchikov, Y.Y. and Barinov, V.A. (1981). *Phys. Met. Metall.*, **52**, 50.

Zielinski, P.G., Ostalek, J., Kijek, M. and Matyja, H. (1978) in *Rapidly quenched metals III* (ed. Cantor, B.) Metals Society, London, vol. 2, p. 337.

## Chapter 4


Aaronson, H.I. (1962). *Decomposition of austenite by diffusional processes*, Interscience, New York, p. 387.

Aaronson, H.I., Laird, C. and Kinsman, K.R. (1970). *Phase transformations*, ASM, Metals Park, Ohio, and Chapman & Hall, London, p. 313.

Ahlborn, H., Hornbogen, E. and Köster, U. (1969). *J. Mat. Sci.*, **4**, 944.

Anderson, W.A. and Mehl, R.F. (1945). *Trans. Met. Soc. AIME*, **161**, 140.

Ashby, M.F. (1971). *Strengthening mechanisms in solids* (eds. Kelly, A. and Nicholson, R.B.) Elsevier, Amsterdam, p. 137.

Åström, H.U. (1955). *Arkiv fur Fysik*, **10**, 197; *Acta Met.*, **3**, 508.

Aust, K.T. and Rutter, J.W. (1959). *Trans. Met. Soc. AIME*, **215**, 119, 820.

Aust. K.T. and Rutter, J.W. (1963) in *Recovery and recrystallization of metals* (ed. Himmel, L.) Interscience, New York, p. 131.

Averbach, B.L., Bever, M.D., Comerford, M.F. and Leach, J.L. (1956). *Acta Met.*, **4**, 477.

Bailey, J.E. and Hirsch, P.B. (1960). *Phil. Mag.*, **5**, 485.

Bailey, J.E. and Hirsch, P.B. (1962). *Proc. Roy. Soc.*, **A267**, 11.

Baker, I. and Martin, J.W. (1983a). *Metal Science*, **17**, 459.

Baker, I. and Martin, J.W. (1983b). *Metal Science*, **17**, 469.

Balluffi, R.W., Koehler, J.S. and Simmons, R.O. (1963) in *Recovery and recrystallization of metals* (ed. Himmel, L.) Interscience, New York, p. 1.

Barrett, C.S. (1945). *Trans. Met. Soc. AIME*, **161**, 15.

Barrett, C.S. and Massalski, T.D. (1980). *Structure of metals*, 3rd (revised) edn, McGraw-Hill, New York.

Basinski, S.J. and Basinski, Z.S. (1966) in *Recrystallization, grain growth and textures*, ASM, Metals Park, Ohio, and Chapman & Hall, London, p. 1.

Bay, B. (1970). *J. Mat. Sci.*, **5**, 617.

Beck, P.A. and Hu, H. (1966) in *Recrystallization, grain growth and textures*, ASM, Metals Park, Ohio, and Chapman & Hall, London, p. 393.

Beck, P.A. and Sperry, P.R. (1950). *J. Appl. Phys.*, **21**, 150.

Bellier, S.P. (1971). D.Ph. thesis, University of Sussex.

Bellier, S.P. and Doherty, R.D. (1973). *Acta Met.*, **25**, 521.

Bever, M.B. (1957). *Creep and recovery*, ASM, Metals Park, Ohio, p. 14.

Bever, M.B., Holt, D.L. and Titchener, A.L. (1973). *Prog. Mat. Sci.*, **17**, 1.

Bishop, G.H. and Chalmers, B. (1971). *Phil. Mag.*, **24**, 515.

Brimhall, J.L. and Huggins, R.A. (1965). *Trans. Met. Soc. AIME*, **233**, 1076.

Brimhall, J.L., Klein, M.J. and Huggins, R.A. (1966). *Acta Met.*, **14**, 459.

Burke, J.E. and Turnbull, D. (1952). *Prog. Met. Phys.*, **3**, 220.

Byrne, J.G. (1965). *Recovery, recrystallization and grain growth*, Macmillan, London.

Cahn, J.W. (1962). *Acta Met.*, **10**, 789.

Cahn, R.W. (1949). *J. Inst. Metals*, **76**, 121.

Cahn, R.W. (1966) in *Recrystallization, grain growth and textures*, ASM, Metals Park, Ohio, and Chapman & Hall, London, p. 99.

Cahn, R.W. (1971). *Recrystallization of metallic materials* (ed. Haessner, F.) Riederer, Stuttgart, p. 43.

Cahn, R.W. (1980) in *Recrystallization and grain growth of multi-phase and particle-containing alloys* (eds. Hansen, N. *et al.*) Risø National Laboratory, Denmark, p. 77.

Cahn, R.W. (1983). *Physical metallurgy*, 3rd edn (eds. Cahn, R.W. and Haasen, P.), North-Holland, Amsterdam, p. 1595. Also (1996), 4th edn, in press.

Chalmers, B. (1964). *Principles of solidification*, John Wiley, New York.

Clarebrough, L.M. (1950). *Austr. J. Sci. Res.*, **A1**, 70.

Clarebrough, L.M., Hargreaves, M.E. and Loretto, M.H. (1963) in *Recovery and recrystallization of metals* (ed. Himmel, L.) Interscience, New York, p. 62.

Clarebrough, L.M., Hargreaves, M.E. and West, G.W. (1955). *Proc. Roy. Soc.*, **A232**, 252.

Clarebrough, L.M., Humble, P. and Loretto, M.H. (1967). *Acta Met.*, **15**, 1007.

Clarebrough, L.M., Hargreaves, M.E., Mitchell, D. and West, G.W. (1952). *Proc. Roy. Soc.*, **A215**, 507.

Clarebrough, L.M., Segall, R.L., Loretto, M.H. and Hargreaves, M.E. (1964). *Phil. Mag.*, **9**, 377.

Cotterill, R.M.J. (1961). *Phil. Mag.*, **6**, 1351.

Cottrell, A.H. (1953). *Dislocations and plastic flow in crystals*, Oxford University Press, p. 39.

Dillamore, I.L., Smith, C.J.E. and Watson, T.W. (1967). *Met. Sci. J.*, **1**, 49.

Dimitrov, O., Fromageau, R. and Dimitrov, C. (1978) in *Recrystallization of metallic materials* (ed. Haessner, F.) Riederer, Stuttgart, p. 137.

Doherty, R.D. (1974). *Met. Sci. J.*, **8**, 132.

Doherty, R.D. (1978) in *Recrystallization of metallic materials* (ed. Haessner, F.) Riederer, Stuttgart, p. 23.

Doherty, R.D. and Cahn, R.W. (1972). *J. Less-Common Metals*, **28**, 279.

Doherty, R.D. and Martin, J.W. (1962/3). *J. Inst. Metals*, **91**, 332.

Doherty, R.D. and Martin, J.W. (1964). *Trans. Amer. Soc. Metals*, **57**, 874.

Farren, W.S. and Taylor, G.I. (1925). *Proc. Roy. Soc.*, **A107**, 422.

Feltner, C.E. and Laughhunn, D.J. (1962). *Acta Met.*, **8**, 563.

Fredriksson, H. (1990). *Mat. Sci. Tech.*, **6**, 811.

Frois, C. and Dimitrov, O. (1961). *C.R. Acad. Sc.*, **253**, 2532.

Frois, C. and Dimitrov, O. (1966). *Ann. Chim. Paris*, **1**, 113.

Fujita, M. (1961). *J. Phys. Soc. Japan*, **16**, 397.

Furu, T., Marthinsen, K. and Nes, E. (1990). *Materials Science and Technology*, **6**, 1093.

Gilman, J.J. (1955). *Acta Met.*, **3**, 277.

Gladman, T., McIvor, I.D. and Pickering, F.B. (1971). *JISI*, **209**, 380.

Gordon, P. (1955). *Trans. Met. Soc. AIME*, **203**, 1043.

Gordon, P. and Vandermeer, R.A. (1966) in *Recrystallization, grain growth and textures*, ASM, Metals Park, Ohio, and Chapman & Hall, London, p. 205.

Grabski, M.W. and Korski, R. (1970). *Phil. Mag.*, **22**, 707.

Greenfield, P. and Bever, M.B. (1957). *Acta Met.*, **5**, 125.

Grünwald, W. and Haessner, F. (1970). *Acta Met.*, **18**, 217.

Haessner, F., Hornbogen, E. and Mukherjee, N. (1966). *Z. f. Metallk.*, **57**, 171, 270.

Hansen, N. (1990). *Mat. Sci. and Tech.*, **6**, 1039.

Hatherly, M. and Malin, A.S. (1979). *Metals Tech.*, **6**, 308.

Hibbard, W.R., Jr and Dunn, C.G. (1956). *Acta Met.*, **4**, 306.

Hibbard, W.R., Jr and Dunn, C.G. (1957). *Creep and recovery*, ASM, Metals Park, Ohio, p. 52.

Honeycombe, R.W.K. (1951). *J. Inst. Metals*, **80**, 49.

Honeycombe, R.W.K. (1968). *The plastic deformation of metals*, Edward Arnold, London.

Honeycombe, R.W.K. and Boas, W. (1948). *Austr. J. Sci. Res.*, **A1**, 70.

Hornbogen, E. (1970). *Praktische Metallographie*, **9**, 349.

Hornbogen, E. and Köster, U. (1978) in *Recrystallization of metallic materials* (ed. Haessner, F.) Riederer, Stuttgart, p. 159.

Hu, H. (1962). *Trans. Met. Soc. AIME*, **224**, 75.

Hu, H. (1969). *Textures in research and practice* (eds. Grewen, J. and Wasserman, G.) Springer, Hamburg, p. 200.

Hu, H. (1981) in *Metallurgical treatises* (eds. Tien, J. and Elliott, J.F.) Met. Soc. AIME, p. 385.

Humphreys, F.J. (1977). *Acta Met.*, **25**, 1323.

Humphreys, F.J. (1979). *Metal Science*, **13**, 136.

Humphreys, F.J. (1983). *Micromechanisms of plasticity and fracture* (eds. Lewis, M.H. and Taplin, D.M.R.) Trinity College Dublin, Parsons Press, 1983.

Humphreys, F.J. and Kalu, P.N. (1990). *Acta Metall. et Mater.*, **38**, 917.

Humphreys, F.J. and Martin, J.W. (1966). *Acta Met.*, **14**, 775.

Humphreys, F.J. and Martin, J.W. (1967). *Phil. Mag.*, **16**, 927.

Humphreys, F.J. and Martin, J.W. (1968). *Phil. Mag.*, **17**, 365.

Humphreys, F.J. and Stewart, A.T. (1972). *Surface Sci.*, **31**, 389.

Jackson, K.A. and Hunt, J.D. (1965). *Acta Met.*, **13**, 1212.

Johnson, W.A. and Mehl, R.F. (1939). *Trans. Met. Soc. AIME*, **135**, 416.

Kamma, C. and Hornbogen, E. (1976). *J. Mat. Sci.*, **11**, 2340.

Kelly, A. and Nicholson, R.B. (1963). *Prog. Mat. Sci.*, **10**, 151.

Kimura, H. and Hasiguti, R.R. (1962). *J. Phys. Soc. Japan*, **17**, 1724.

Kimura, H. and Maddin, R. (1971). *Quench hardening in metals*, North-Holland, Amsterdam.

Klein, M.J. and Huggins, R.A. (1962). *Acta Met.*, **10**, 55.

Köster, U. (1971) in *Recrystallization of metallic materials* (ed. Haessner, F.) Riederer, Stuttgart, p. 215.

Köster, U. and Hornbogen, E. (1968). *Z. f. Metallkunde*, **59**, 792.

Kreye, H. and Hornbogen, E. (1970). *Praktische Metallographie*, **9**, 349.

Kuhlmann, D. (1948). *Z. Physik*, **124**, 468.

Kuhlmann-Wilsdorf, D. (1965). *Acta Met.*, **13**, 257.

Leslie, W.C., Michalek, J. and Aul, F.W. (1963). *Iron and its dilute solid solutions* (eds. Spencer, C.W. and Werner, F.E.) Interscience, New York, p. 119.

Lewis, M.H. and Martin, J.W. (1963). *Acta Met.*, **11**, 1207.

Li, J.C.M. (1960). *Acta Met.*, **8**, 563.

Li, J.C.M. (1962). *J. Appl. Phys.*, **33**, 2958.

Li, J.C.M. (1966) in *Recrystallization, grain growth and texture*, ASM, Metals Park, Ohio, and Chapman & Hall, London, p. 45.

Lücke, K. and Detert, K. (1957). *Acta Met.*, **5**, 628.

Lücke, K. and Stüwe, H.P. (1963) in *Recovery and recrystallization of metals* (ed. Himmel, L.) Interscience, New York, p. 171.

McGeary, R.K. and Lustman, D. (1953). *Trans. Met. Soc. AIME*, **197**, 284.

Mäder, K. and Hornbogen, E. (1974). *Scripta Met.*, **8**, 979.

Martin, J.W. (1957). *Metallurgia*, **56**, 161.

Martin, J.W. (1980). *Micromechanisms in particle-hardened alloys*, Cambridge University Press, Cambridge.

Meijering, J.L. and Druyvesteyn, M.J. (1947). *Philips Res. Rep.*, **2**, 81, 260.

Meshii, M. and Kauffman, J.-W. (1960). *Phil. Mag.*, **5**, 687, 939.

Michell, D. (1956). *Phil. Mag.*, **1**, 584.

Michell, D. and Haig, F.D. (1957). *Phil. Mag.*, **2**, 15.

Mori, T. and Meshii, M. (1963). *J. Metals*, **15**, 80.

Mould, P.R. and Cotterill, P. (1967). *J. Mat. Sci.*, **2**, 241.

Nakada, Y. (1965). *Phil. Mag.*, **11**, 251.

Nes, E. (1976a). *Acta Met.*, **24**, 391.

Nes, E. (1976b). *Scripta Met.*, **10**, 1025.

Preston, O. and Grant, N.J. (1961). *Trans. Met. Soc. AIME*, **221**, 164.

Price, W.L. and Washburn, J. (1963). *J. Austral. Inst. Metals.*, **8**, 1.

Ralph, B., Barlow, C., Cooke, B. and Porter, A. (1980) in *Proc Risø international symposium on recrystallization and grain growth of multi-phase and particle containing materials* (eds. Hansen, N., Jones, A.R. and Leffers, T.) Risø National Laboratory, Denmark, p. 229.

Ringer, S.P., Li, W.B. and Easterling, K.E. (1992). *Acta Met. et Mat.*, **40**, 275.

Rollason, T.C. and Martin, J.W. (1970). *Acta Met.* **18**, 1267.

Sato, S. (1931). *Sci. Rep. Tôhoku Univ.*, **20**, 140.

Scharf, G. and Gruhl, W. (1969). *Z.f.Metallk.*, **60**, 413.

Schober, T. and Balluffi, R.W. (1969). *Phil. Mag.*, **20**, 511.

Schober, T. and Balluffi, R.W. (1970). *Phil. Mag.*, **21**, 109.

Shewmon, P. (1966) in *Recrystallization, grain growth and textures*, ASM, Metals Park, Ohio, and Chapman & Hall, London, p. 165.

Shin, P.W. and Meshii, M. (1963). *J. Metals*, **15**, 80.

Sinha, P.P. and Beck, P.A. (1961). *J. Appl. Phys.*, **32**, 1222.

Stark, J.P. and Upthegrove, W.R. (1966). *Trans. Amer. Soc. Metals*, **59**, 479.

Stibitz, G.R. (1936). *Phys. Rev.*, **49**, 862.

Talbot, J. (1963) in *Recovery and recrystallization of metals* (ed. Himmel, L.) Interscience, New York, p. 269.

Thompson-Russell, K.C. (1974). *Planseeberichte für Pulvermetallurgie*, **22**, 155.

Titchener, A.L. and Bever, M.B. (1958). *Prog. Met. Phys.*, **7**, 247.

Tizhnova, N.V. (1946). *J. Tech. Phys. (Moscow)*, **16**, 1, 389.

Turnbull, D. and Hoffman, R.E. (1954). *Acta Met.*, **2**, 419.

van Arkel, A.E. and Burgers, W.G. (1930). *Z. Physik*, **48**, 690.

van Drunen, G. and Saimoto, S. (1971). *Acta Met.*, **19**, 213.

Vasudevan, A.K., Petrovic, J.J. and Robertson, J.A. (1974). *Scripta Met.*, **8**, 861.

Warrington, D.H. (1961). *Proc. Eur. Reg. Conf. on Elec. Mic.*, De Nederlandse Vereniging Electronemicroscopie, Delft, p. 354.

Williams, R.O. (1962). *Trans. Met. Soc. AIME*, **224**, 719.

Williams, R.O. (1964). *Acta Met.*, **12**, 744.

Yoshida, S., Kiritani, M., Shimomura, Y. and Yoshinaka, A. (1965). *J. Phys. Soc. Japan*, **20**, 628.

Zener, C. quoted by C.S. Smith (1948). *Trans. Met. Soc. AIME*, **175**, 345.

## Chapter 5

Aaronson, H.I. (1962). *The decomposition of austenite by diffusional processes*. Interscience.

Aaronson, H.I., Laird, C. and Kinsman, K.R. (1970). *Phase transformations*, ASM, Metals Park, Ohio, and Chapman & Hall, London, p. 313.

Abbruzzese, G. and Lucke, K. (1986). *Acta Met.*, **34**, 905.

Aboav, D.A. (1972). *Metallography*, **5**, 251.

Adachi, M. and Grant, N.J. (1960). *Trans. Met. Soc. AIME.*, **218**, 881.

Anand, L. and Gurland, J. (1975). *Metall. Trans.*, **6A**, 928.

Anderson, M.P., Srolovitz, D.J., Grest, G.S. and Sahni, P.S. (1984). *Acta Met.*, **32**, 783.

Anderson, M.P., Grest, G.S. and Srolovitz, D.J. (1985). *Scripta Met.*, **19**, 225.

Anderson, M.P., Grest, G.S. and Srolovitz, D.J. (1989a). *Phil. Mag.*, **59B**, 293.

Anderson, M.P., Li, K., Doherty, R.D., Grest, G.S. and Srolovitz, D.J. (1989b). *Scripta Met.*, **23**, 753.

Andrade, E.N. da C. and Aboav, D.A. (1966). *Proc. Roy. Soc.*, **A291**, 18.

Annavarapu, S. and Doherty, R.D. (1993). *Int. J. of Powder Met.*, **29**, 331.

Annavarapu, S. and Doherty, R.D. (1995). *Acta Metall. et Mater.*, **43**, 3207.

Annavarapu, S., Liu, J. and Doherty, R.D. (1992). *Grain growth in polycrystalline materials* (eds. Abbruzzese, G. and Brosso, P.) *Materials Science Forum*, Trans Tech Publications, Zurich, **94–96**, p. 649.

Ardell, A.J. (1967). *Acta Met.*, **15**, 1772.

Ardell, A.J. (1968). *Acta Met.*, **16**, 511.

Ardell, A.J. (1969). *Mechanism of phase transformations in crystalline solids*, Institute of Metals, London, p. 111.

Ardell, A.J. (1970). *Met. Trans.*, **1**, 525.

Ardell, A.J. (1972a). *Acta Met.*, **20**, 61.

Ardell, A.J. (1972b). *Acta Met.*, **20**, 601.

Ardell, A.J. (1972c). *Metallography*, **5**, 285.

Ardell, A.J. (1972d). *Met. Trans.*, **3**, 1395.

Ardell, A.J. (1988) in *Phase transformations* (ed. Lorimer, G.W.) The Institute of Metals, London, p. 485.

Ardell, A.J. (1990). *Scripta Metal. et Mater.*, **24**, 343.

Ardell, A.J. and Nicholson, R.B. (1966a). *Acta Met.*, **14**, 1295.

Ardell, A.J. and Nicholson, R.B. (1966b). *J. Phys. Chem. Solids*, **27**, 1793.

Ardell, A.J. and Lee, S.S. (1986). *Acta Met.*, **34**, 2411.

Ardell, A.J. and Przystrupa, M.A. (1984). *Mech. Mater.*, **3**, 319.

Argon, A.S. (1965). *Phys. Stat. Sol.*, **12**, 121.

Ashby, M.F. (1972). *Acta Met.*, **20**, 887.

Ashby, M.F. (1980) in *Recrystallization and grain growth in multi-phase and particle containing materials* (eds. Hansen, N., Jones, A.R. and Leffers, T.) *1st Risø Symposium on Metallurgy and Materials Science*, Røskilde, Denmark, p. 325.

Ashby, M.F. and Centamore, R.M.A. (1968). *Acta Met.*, **16**, 1081.

Ashby, M.F., Harper, J. and Lewis, J. (1969). *Trans. AIME*, **245**, 413.

Ashcroft, T.B. and Faulkner, R.G. (1972). *Metal Sci. J.*, **6**, 224.

Asimov, R. (1963). *Acta Met.*, **11**, 72.

Atkinson, H.V. (1988). *Acta Met.*, **36**, 469.

Aubauer, H.P. and Warlimont, H. (1974). *Z. f. Metallkunde*, **65**, 297.

Bainbridge, B.G. (1972). D.Phil. thesis, University of Sussex.

Bannyh, O., Modin, H. and Modin, S. (1962). *Jernkontorets Ann.*, **146**, 774.

Bardeen, J. and Herring, C. (1952) in *Imperfections in nearly perfect crystals*, (eds, Shockley, W. *et al.*) Wiley, New York, p. 261.

Bayles, B.J., Ford, J.A. and Salkind, M.T. (1967). *Trans. Met. Soc. AIME*, **239**, 844.

Beck, P.A., Holzworth, M.L. and Sperry, P.R. (1949). *Trans. Met. Soc. AIME*, **180**, 163.

Beck, P.A., Kramar, J.C., Demer, L.J. and Holzworth, M.L. (1948). *Trans. Met. Soc. AIME*, **175**, 372.

Beck, P.A. and Sperry, P.R. (1949). *Trans. Met. Soc. AIME*, **185**, 240.

Bender, W. and Ratke, L. (1993). *Scripta Metal. et Mater.*, **28**, 737.

Benedek, R.A. (1974). Ph.D. thesis, University of Sussex, UK.

Benedek, R.A. and Doherty, R.D. (1974). *Scripta Met.*, **8**, 675.

Berry, F.G., Davenport, A.T. and Honeycombe, R.W.K. (1969). *The mechanism of phase transformation in crystalline solids*, Institute of Metals, London, p. 288.

Bhattacharya, S.K. and Russell, K.C. (1972). *Metall. Trans.*, **3**, 2195.

Bhattacharya, S.K. and Russell, K.C. (1976). *Metall. Trans.*, **7A**, 453.

Bikerman, J.J. (1965). *Phys. Stat. Sol.*, **10**, 3.

Bishop, G.H., Hartt, W.H. and Bruggeman, G.A. (1971). *Acta Met.*, **19**, 37.

Bjorklund, S. and Hillert, M. (1974). Private communication.

Bjorklund. S., Donaghey, L.F. and Hillert, M. (1972). *Acta Met.*, **20**, 7.

Blakely, J.M. (1973). *Introduction to the properties of crystal surfaces*, Pergamon Press, Oxford, p. 53.

Blakely, J.M. and Mykura, H. (1962). *Acta Met.*, **10**, 565.

Bolling, G.F. and Winegard, W.C. (1958). *Acta Met.*, **6**, 283.

Bonis, L.J. and Grant, N.J. (1962). *Trans. Met. Soc. AIME*, **224**, 308.

Bower, E.N. and Whiteman, J.A. (1969). *The mechanism of phase transformation in crystalline solids*, Institute of Metals, London, p. 119.

Boyd, J.D. and Nicholson, R.B. (1971a). *Acta Met.*, **19**, 1101.

Boyd, J.D. and Nicholson, R.B. (1971b). *Acta Met.*, **19**, 1379.

Brailsford, A.D. and Wynblatt, P. (1979). *Acta Met.*, **27**, 484.

Brown, L.C. (1989). *Acta Metal.*, **37**, 71.

Brown, L.C. (1990a). *Scripta Metal. et Mater.*, **24**, 963.

Brown, L.C. (1990b). *Scripta Metal. et Mater.*, **24**, 2231.

Brown, L.C. (1992a). *Scripta Metal. et Mater.*, **26**, 1939.

Brown, L.C. (1992b). *Scripta Metal. et Mater.*, **26**, 1945.

Brown, L.M., Cook, R.H., Ham, R.K. and Purdy, G.R. (1973). *Scripta Met.*, **7**, 815.

Buffington, F.S., Hirano, K. and Cohen, M. (1961). *Acta Met.*, **9**, 434.

Burke, J.E. (1949). *Trans. Met. Soc. AIME*, **180**, 73.

Burke, J.E. and Turnbull, D. (1952). *Prog. Met. Phys.*, **3**, 220.

Butcher, B.R., Weatherley, G.C. and Petit, H.R. (1969). *Met. Sci. J.*, **3**, 7.

Byrne, J.G. (1965). *Recovery recrystallization and grain growth*, Macmillan, New York.

Cahn, J.W. (1962). *Acta Met.*, **10**, 789.

Cahn, J.W. and Hilliard, J.E. (1958). *J. Chem. Phys.*, **28**, 258.

Cahn, J.W. and Pedawar, G.E. (1965). *Acta Met.*, **13**, 1091.

Calvet, J. and Renon, C. (1960). *Mem. Sci. Rev. Met.*, **57**, 345.

Chadwick, G. (1963). *J. Inst. Metals.*, **93**, 18.

Chalmers, B. (1959). *Physical metallurgy*, John Wiley, New York, p. 134.

Chalmers, B., King, R. and Shuttleworth, R. (1948). *Proc. Roy. Soc.*, **A193**, 465.

Chellman, D.J. and Ardell, A.J. (1974). *Acta Met.*, **22**, 577.

Chen, Y.C., Fine, M.E. and Weertman, J.R. (1990). *Acta Metall. et Mater.*, **38**, 771.

Chen, Y.C., Fine, M.E., Weertman, J.R. and Lewis, R. (1987). *Scripta Met.*, **21**, 1003.

Chojnowski, E.A. and McTegart, W.J. (1968). *Met. Sci. J.*, **2**, 14.

Christian, J.W. (1975). *The theory of transformations in metals and alloys*, Part 1, Pergamon Press, Oxford, pp. 448–54.

Cline, H.E. (1971). *Acta Met.*, **19**, 481.

Coble, R.L. and Burke, J.E. (1963). *Progr. Ceramic Sci.*, **3**, 197.

Corti, C.W., Cotterill, P. and Fitzpatrick, G.A. (1974). *Intern. Metall. Revs.*, **19**, 77.

Cratchley, D. (1965). *Met. Revs.*, **10**, 133.

Darken, L.S. (1948). *Trans. Met. Soc. AIME*, **175**, 184.

Das, S.K., Biswas, A. and Ghosh, R.N. (1993). *Acta Metall. et Mater.*, **41**, 777.

Davenport, A.T., Berry, F.G. and Honeycombe, R.W.K. (1968). *Met. Sci. J.*, **2**, 104.

Davies, C.K.L., Nash, P. and Stevens, R.N. (1980). *Acta Met.*, **28**, 179.

Davies, R.G. and Johnston, T.L. (1970). *Ordered alloys*, Claitors Publishing Division, Baton Rouge, p. 447.

DeHoff, R.T. and Rhines, F.N. (1968). *Quantitative microscopy*, McGraw-Hill, New York, chapter 10.

Dillamore, I.L. (1978) in *Recrystallization of Metallic Materials* (ed. Haessner, F.) Dr. Rieder Verlag GMBH Stuttgart, p. 236.

Doherty, R.D. (1975). *Met. Trans.*, **6**, 588.

Doherty, R.D. (1978) in *Recrystallization of metallic materials* (ed. Haessner, F.), Rieder, Stuttgart, p. 23.

Doherty, R.D. (1982). *Metal Sci.*, **16**, 1.

Doherty, R.D. (1984). *J. Mater. Educ.*, **6**, 841.

Doherty, R.D. and Cantor, B. (1982) in *Solid–solid phase transformations* (eds. Aaronson, H.T. *et al.*) AIME, Warrendal, PA, p. 547.

Doherty, R.D. and Cantor, B. (1988). *Scripta Met.*, **22**, 301.

Doherty, R.D. and Martin, J.W. (1962/3). *J. Inst. Metals*, **91**, 332.

Doherty, R.D. and Martin, J.W. (1964). *Trans. Amer. Soc. Metals*, **57**, 874.

Doherty, R.D. and Purdy, G.R. (1982) in *Solid–solid phase transformations* (eds. Aaronson, H.I. *et al.*) AIME, Warrendal, PA, p. 561.

Doherty, R.D. and Rajab, K.E. (1989). *Acta Met.*, **37**, 2723.

Doherty, R.D., Srolovitz, D.J., Rollett, A.D. and Anderson, M.P. (1987). *Scripta Met.*, **21**, 675.

Doherty, R.D., Li, K., Kashyap, K., Rollett, A.D. and Anderson, M.P. (1989). *Materials architecture* (eds. Bilde-Sorensen, J.B., Hansen, N., Juul-Jensen, D., Leffers, T., Lilholt, H. and Pedersen, O.B.) *10th Riso Symposium*, Roskilde, Denmark, p. 31.

Doherty, R.D., Li, K., Anderson, M.P., Rollett, A.D. and Srolovitz, D.J. (1990) in *Recrystallization '90* (ed. Chandra, T.) TMS-AIME, Warrendale, PA, p. 129.

Drolet, J.P. and Galibois, A. (1971). *Met. Trans.*, **2**, 53.

Dromsky, A., Lenel, F.V. and Ansell, G.S. (1962). *Trans. Met. Soc. AIME*, **224**, 236.

Dunn, C.G. and Walter, J.L. (1966). *Recrystallization, grain growth and textures*, ASM, Metals Park, Ohio, and Chapman & Hall, London, p. 461.

Eichelkraut, H., Abbruzzese, G. and Lucke, K. (1988). *Acta Met.*, **36**, 55.

Enomoto, M. (1987). *Acta Met.*, **35**, 935.

Enomoto, Y., Tokuyama, M. and Kawasaki, K. (1986). *Acta Met.*, **34**, 2119.

Eshelby, J.D. (1961). *Prog. Solid Mech.*, **2**, 89.

Fang, Z. and Patterson, B.R. (1993). *Acta Metall. et Mater.*, **41**, 1993.

Faulkner, R.G. and Ralph, B. (1972). *Acta Met.*, **20**, 703.

Feltham, P. (1957). *Acta Met.*, **5**, 97.

Ferran, G., Cizeron, G. and Aust, K.T. (1967). *Mem. Sci. Rev. Met.*, **64**, 1069.

Ferrante, M. and Doherty, R.D. (1979). *Acta Met.*, **27**, 1603.

Fillingham, P.J., Leamy, H.J. and Tanner, L.E. (1972). *Electron microscopy and structure of materials*, University of California Press, Berkeley, p. 163.

Fiset, M., Braunovic, M. and Galibois, A. (1971). *Scripta Met.*, **5**, 325.

Flemings, M.C. (1974). *Solidification processing*, McGraw-Hill, New York.

de Fontaine, D. (1973). *Scipta Met.*, **7**, 463.

Footner, P.K. and Alcock, C.B. (1972). *Met. Trans.*, **3**, 2633.

Fradkov, V., Shvindlerman, L. and Udler, D. (1985). *Scripta Met.*, **19**, 1285.

Frank, F.C. (1958) in *Growth and perfection of crystals*, (eds. Doremus, R.H. *et al.*) Wiley, New York, p. 411.

Frank, F.C. (1963). *Metal surfaces*, ASM, Metals Park, Ohio, p. 1.

Frost, H.J. and Ashby, M.F. (1982). *Deformation – mechanism maps*, Pergamon Press, Oxford UK.

Frost, H.J. and Thompson, C.V. (1986). *Computer simulation of microstructural evolution* (ed. Srolovitz, D.J.) TMS-AIME, Warrendale, PA 15086, p. 33.

Frost, H.J., Thompson, C.V. and Walton, D.T. (1990). *Acta Metall. et Mater.*, **38**, 1455.

Gaskell, D.R. (1984) in *Physical metallurgy*, 3rd edn (eds. Cahn, R.W. and Haasen, P.) North-Holland, Amsterdam, p. 310.

Gaudig, W. and Warlimont, H. (1969). *Z. f. Metallkunde*, **60**, 488.

Gawne, D.T. and Higgins, G.T. (1971). *JISI*, **209**, 562.

Gibbs, J.W. (1878). *Trans. Conn. Acad.*, **3**, 102, 345; reprinted in *The collected works of J. Willard Gibbs*, vol. 1, Dover Publications, New York, 1961.

Gjostein, N. and Rhines, F.N. (1959). *Acta Met.*, **7**, 319.

Gladman, T. (1966). *Proc. Roy. Soc.*, **A294**, 298.

Glazier, J.A., Gross, S.P. and Stavans, J. (1987). *Phys. Rev.* **36A**, 306.

Gleiter, H. and Hornbogen, E. (1967). *Z. f. Metallkunde*, **58**, 157.

Glicksman, M.E., Rajan, K., Nordberg, J., Palmer, M., Marsh, S.P. and Pande, C.P. (1992) in *Grain growth in polycrystalline materials* (eds. Abbruzzese, G. and Brosso, P.) *Mater. Sci. Forum*, **94–96**, 909.

Godwin, A.W. (1966) quoted by Grey and Higgins (1973).

Gordon, P. and El-Bassyouni, T.A. (1965). *Trans. Met. Soc. AIME*, **233**, 391.

Gordon, P. and Vandermeer, R.A. (1966). *Recrystallization, grain growth and textures*, ASM, Metals Park, Ohio, and Chapman & Hall, London, p. 205.

Graham, L.D. and Kraft, R.W. (1966). *Trans. Met. Soc. AIME.*, **236**, 94.

Grant, N.J. (1970). *Fizika*, **2**, 16.

Green, D.J. (1982). *J. Am. Ceram. Soc.*, **65**, 610.

Greenwood, G.W. (1956). *Acta Met.*, **4**, 243.

Greenwood, G.W. (1969). *The mechanism of phase transformations in crystalline solids*, Institute of Metals, London, p. 103.

Gregory, E. and Grant, N.J. (1954). *Trans. Met. Soc. AIME*, **200**, 247.

Grest, G.S., Anderson, M.P. and Srolovitz, D.J. (1988). *Phys. Rev.*, **38B**, 4752.

Grey, E.A. and Higgins, G.T. (1972). *Scripta Met.*, **6**, 253.

Grey, E.A. and Higgins, G.T. (1973). *Acta Met.*, **21**, 309.

Grohlich, M., Haasen, P. and Frommeyer, G. (1982). *Scripta Met.*, **16**, 367.

Gunduz, M. and Hunt, J.D. (1985). *Acta Met.*, **33**, 1651.

Gyeguzin, Ya. Ye. and Krivoglaz, M.A. (1971). *The movement of macroscopic inclusions in solids*, Metallurgizat, Moscow. (English translation from Consultant's Bureau, New York.)

Haasen, P. and Wagner, R. (1992). *Met. Trans.*, **23A**, 1901.

Hardy, S.C. and Voorhees, P.W. (1988). *Metall. Trans.*, **19A**, 2713.

Haroun, N.A. and Budworth, D.W. (1968). *J. Mater. Sci.*, **3**, 326.

Harper, J.G. and Dorn, J.E. (1957). *Acta Met.*, **5**, 887.

Hayward, E.R. and Greenough, A.P. (1960). *J. Inst. Metals*, **88**, 217.

Hazzledine, P.M., Hirsch, H.B. and Louat, N. (1980) in *Recrystallization and grain growth in multi-phase and particle containing materials.* (eds. Hansen, N., Jones, A.R. and Leffers, T.) 1st Riso Symposium on Metallurgy and Materials Science, Roskilde, Denmark, p. 165.

von Heimendahl, M. and Thomas, G. (1964). *Met. Trans.*, **230**, 1520.

Hellman, P. and Hillert, M. (1975). *Scand. J. Metall.*, **4**, 211.

Herring, C. (1949). *Symposium on the physics of powder metallurgy*, McGraw-Hill, New York.

Hertzberg, R.W. (1965). *Fibre composite materials*, ASM, Metals Park, Ohio, p. 77.

Higgins, G.T. (1974). *Met. Sci. J.*, **8**, 143.

Higgins, G.T., Wiryolukito, S. and Nash, P. (1992). *Grain growth in polycrystalline materials* (eds. Abbruzzese, G. and Brosso, P.) *Materials Science Forum*, Trans Tech Publications, Zurich, **94–96**, 671.

Higgins, J. and Wilkes, P. (1972). *Phil. Mag.*, **25**, 599.

Hillert, M. (1965). *Acta Met.*, **13**, 227.

Hillert, M. (1988). *Acta Met.*, **36**, 3177.

Hillert, M., Hunderi, O. and Ryum, N. (1992a). *Scripta Metall. et Mater.*, **26**, 1933.

Hillert, M., Hunderi, O. and Ryum, N. (1992b). *Scripta Metall. et Mater.*, **26**, 1943.

Hillert, M., Hunderi, O., Ryum, N. and Saetre, T.O. (1989). *Scripta Met.*, **23**, 1979.

Hirata, T. and Kirkwood, D.H. (1977). *Acta Met.*, **25**, 1425.

Hogan, L.M., Kraft, R.W. and Lemkey, F.D. (1971). *Adv. Mater. Res.*, **5**, 83.

Hondros, E.D. (1965). *Proc. Roy. Soc.*, **A285**, 479.

Hondros, E.D. (1968). *Acta Met.*, **16**, 1377.

Hondros, E.D. (1969) in *Interfaces conference* (ed. Gifkins, R.C.) Butterworths, London.

Hopper, R.W. and Uhlmann, D.R. (1972). *Scripta Met.*, **6**, 327.

Hornbogen, E. and Roth, M. (1967). *Z. f. Metallkunde*, **58**, 842.

Howe, J.M., Aaronson, H.I. and Gronsky, R. (1985). *Acta Met.*, **33**, 649.

Hoyt, J.J. (1990). *Scripta Met.*, **24**, 163.

Hsu, C.Y. (1984). Ph.D. thesis, Massachusetts Institute of Technology.

Hull, D. and Bacon, D.J. (1984). *Introduction to dislocations*, 3rd edn, Pergamon Press Oxford.

Humphreys, F.J. (1977). *Acta Met.*, **25**, 1323.

Humphreys, F.J. (1992). *Mater. Sci. and Techn.*, **8**, 135.

Hunderi, O. (1979). *Acta Met.*, **27**, 167.

Hunderi, O., Ryum, N. and Westengen, H. (1979). *Acta Metall.*, **27**, 161.

Hunt, J.D. (1968). *Crystal growth*, North-Holland, Amsterdam, p. 82.

Hutchinson, W.B. and Duggan, B.J. (1978). *Metal Sci.*, **12**, 372.

Inman, M.C. and Tipler, H.R. (1963). *Met. Revs.*, **10**, 3.

Jaffrey, D. and Chadwick, G. (1970). *Metal. Trans.*, **1**, 3389.

James, D.W. and Leak, G.M. (1965). *Phil. Mag.*, **12**, 491.

Jayanth, C.S. and Nash, P. (1990). *Mat. Sci. and Tech.*, **6**, 405.

Johnson, W.C. (1987a). *Metal. Trans.*, **18A**, 233.

Johnson, W.C. (1987b). *Metal. Trans.*, **18A**, 1093.

Johnson, W.C. and Voorhees, P.W. (1985). *Metall. Trans.*, **16A**, 337.

Johnson, W.C. and Voorhees, P.W. (1987). *Metal. Trans.*, **18A**, 1213.

Jones, H. (1969). *Mat. Sci. Eng.*, **5**, 1.

Jones, H. (1971). *Met. Sci. J.*, **5**, 15.

Jones, H. and Leak, G.M. (1967). *Met. Sci. J.*, **1**, 211.

Kahlweit, M. (1968). *Scripta Met.*, **2**, 251.

Kampmann, L. and Kahlweit, M. (1967). *Ber. Bunsen, Gesell. Phys. Chem.*, **71**, 78.

Kampmann, L. and Kahlweit, M. (1970). *Ber. Bunsen, Gesell, Phys. Chem.*, **74**, 456.

Kattamis, T.Z. and Flemings, M.C. (1966). *Trans. Met. Soc. AIME*, **236**, 1525.

Kattamis, T.Z., Coughlin, J.M. and Flemings, M.C. (1967). *Trans. Met. Soc. AIME*, **239**, 1504.

Kattamis, T.Z., Holmberg, H.T. and Flemings, M.C. (1967). *J. Inst. Metals*, **95**, 343.

Kelly, A. (1966). *Strong solids*, Oxford University Press, Oxford.

Kelly, A. and Davies, G.J. (1965). *Met. Revs.*, **10**, 1.

Kelly, A. and Nicholson, R.B. (1971). *Strengthening mechanisms in crystals*, Elsevier, Amsterdam.

Kelvin, Lord (1887). *Phil. Mag.*, **24**, 53.

Kirchner, H.P.K. (1971). *Met. Trans.*, **2**, 2861.

Kneller, E. and Wolff, M. (1966). *J. Appl. Phys.*, **37**, 1350, 1838.

Kreye, H. (1970). *Z. f. Metallkunde*, **61**, 103.

Kurtz, S.K. and Carpay, F.M.A. (1980). *J. Appl. Phys.*, **51**, 5725, 5745.

Kurz, W. and Fisher, D.J. (1989). *Fundamentals of solidification*, 3rd edn, Trans. Tech. Publications, Switzerland, pp. 93 and 261.

Lange, F.F. and Hirlinger, M.M. (1984). *J. Am. Ceram. Soc.*, **67**, 164.

Lee, H.M. (1990). *Scripta Metall. et Mater.*, **24**, 2443.

Lee, H.M. and Allen, S.M. (1991). *Metal Trans.*, **22A**, 2877.

Lee, H.M., Allen, S.M. and Grunjicic, M. (1991a). *Metal. Trans.*, **22A**, 2863.

Lee, H.M., Allen, S.M. and Grunjicic, M. (1991b). *Metal Trans.*, **22A**, 2869.

Li, C., Blakely, J.M. and Feingold, A.H. (1966). *Acta Met.*, **14**, 1397.

Li, K. (1992). Ph.D. Thesis, Drexel University.

Li, K., Anderson, M.P. and Doherty, R.D. (1994). Research in progress.

Lifshitz, I.M. and Slyozov, V.V. (1961). *J. Phys. Chem. Solids*, **19**, 35.

Lin, P., Lee, S.S. and Ardell, A.J. (1989). *Acta Met.*, **37**, 739.

Ling, S., Anderson, M.P., Grest, G.S. and Glazier, J.A. (1992). *Grain growth in polycrystalline materials*, (eds. Abbruzzese, G. and Brosso, P.) *Materials Science Forum*, Trans Tech Publications, Zurich, **94–96**, p. 39.

Liu, J. (1987). Ph.D. thesis, Drexel University, Philadelphia, PA 19104.

Livingston, J.D. (1959). *Trans. Met. Soc. AIME*, **215**, 566.

Livingston, J.D. and Cahn, J.W. (1974). *Acta Met.*, **22**, 495.

Louat, N.P. (1974). *Acta Met.*, **22**, 721.

Louat, N.P. and Imam, M.A. (1990) in *Recrystallization '90* (ed. Chandra, T.) TMS-AIME Warrendale, PA 15086, p. 387.

Mackenzie, J.K., Moore, A.J.W. and Nicholas, J.F. (1962). *J. Phys. Chem. Solids*, **23**, 185.

McLean, M. (1971). *Acta Metall.*, **19**, 387.

McLean, M. and Gale, B. (1969). *Phil. Mag.*, **20**, 1033.

Mahalingam, K., Liedl, G.L. and Sanders, T.H. Jnr (1987). *Acta Met.*, **35**, 483.

Malcolm, J.A. and Purdy, G.R. (1967). *Trans. Met. Soc. AIME*, **239**, 1391.

Marich, S. (1971). *Met. Trans.*, **1**, 2953.

Marquese, J.A. and Ross, J. (1984). *J. Chem. Phys.*, **80**, 536.

Martin, J.W. (1980). *Micromechanisms in particle hardened alloys.* Cambridge University Press, Cambridge, p. 51.

Martin, J.W. and Humphreys, F.J. (1974). *Scripta Met.*, **8**, 679.

May, J.E. and Turnbull, D. (1958). *Trans. Met. Soc. AIME*, **212**, 769.

Merle, P. and Doherty, R.D. (1982). *Scripta Met.*, **16**, 357.

Merle, P. and Merlin, J. (1981). *Acta Met.*, **29**, 1929.

Mills, B., McLeall, M. and Hondros, E.D. (1973). *Phil. Mag.*, **27**, 369.

Mirkin, I.I. and Kancheev, O.D. (1967). *Met. Sciences and Heat Treatment*, **10** (quoted by Davies, R.G. and Johnston, T.L. (1970). *Ordered alloys*, Claitors Publishing Division, Baton Rouge, p. 447).

Moore, A.J.W. (1963). *Metal surfaces*, ASM, Metals Park, Ohio, p. 155.

Moore, A. and Elliot, R. (1968). *The solidification of metals*, ISI, London. Publication No. 167.

Morral, J.E. and Ashby, M.F. (1974). *Acta Met.*, **22**, 567.

Morral, J.E. and Louat, N.P. (1974). *Scripta Met.*, **8**, 91.

Mukherjee, T. and Sellars, C.M. (1969). *The mechanism of phase transformations in crystalline solids*, Institute of Metals, London, p. 122.

Mukherjee, T., Stumpf, W.E., Sellars, C.M. and McTegart, W.T. (1969). *JISI*, **207**, 621.

Mullins, W.W. (1958). *Acta Met.*, **6**, 414.

Mullins, W.W. (1959). *J. Appl. Phys.*, **30**, 77.

Mullins, W.W. (1963). *Metal surfaces*, ASM, Metals Park, Ohio, p. 17.

Mullins, W.W. (1989). *Acta Met.*, **37**, 2979.

Mullins, W.W. and Vinals, J. (1989). *Acta Met.*, **37**, 991.

Mykura, H. (1961). *Acta Met.*, **9**, 570.

Mykura, H. (1966). *Solid surfaces and interfaces*, Routledge & Kegan Paul, London and Dover, New York.

Nakagawa, Y.G. and Weatherly, G.C. (1972). *Acta Met.*, **20**, 345.

Nelson, R.S., Mazey, D.J. and Barnes, R.S. (1965). *Phil. Mag.*, **11**, 91.

Nes, E., Ryum, N. and Hunderi, O. (1985). *Acta Met.*, **33**, 11.

Nicholson, R.B. (1969). *Interfaces conference* (ed. Gifkins, R.C.) Butterworths, London, p. 139.

Niemi, A.N. and Courtney, T.H. (1981). *Metal. Trans.*, **12A**, 1987.

Novikov, V. Yu. (1978). *Acta Met.*, **26**, 1739.

Olgaard, D.L. and Evans, B. (1986). *J. Am. Ceram. Soc.*, **61**, C272.

Oriani, R.A. (1964). *Acta Met.*, **12**, 1399.

Oriani, R.A. (1966). *Acta Met.*, **14**, 84.

Ostwald, W. (1900). *Z. Phys. Chem.*, **34**, 495.

Palmer, M.A., Fradkov, V.E., Glicksman, M.E. and Rajan, K. (1994). *Scripta Metall. et Mater.*, **30**, 633.

Pande, C.S. (1987). *Acta Met.*, **35**, 671.

Parratt, N. (1966). *Chem. Eng. Prog.*, **62**, 61.

Petrasak, P. and Weston, J. (1964). *Trans. Met. Soc. AIME*, **230**, 977.

Petrovic, J.J. and Ebert, L.J. (1972a). *Met. Trans.*, **3**, 1123.

Petrovic, J.J. and Ebert, L.J. (1972b). *Met. Trans.*, **3**, 1131.

Purdy, G.R. (1971). *Met. Sci. J.*, **5**, 81.

Rajab, K.E. and Doherty, R.D. (1989). *Acta Met.*, **37**, 2709.

Randle, V. and Ralph, B. (1986). *Acta Met.*, **34**, 891.

Rastogi, P.K. and Ardell, A.J. (1971). *Acta Met.*, **19**, 321.

Rath, B.B. and Hu, H. (1969). *Trans. Met. Soc. AIME*, **245**, 1577.

Rayleigh, Lord (1879). *London Math. Soc. Proc.*, **10**, 4.

Rhines, F.N. (1967). *Proceedings of the second international congress for stereology.* Springer, New York, p. 234.

Rhines, F.N. and Craig, K.R. (1974). *Met. Trans.*, **5**, 413.

Rhines, F.N. and Patterson, B.R. (1982). *Met. Trans.*, **13A**, 985.

Rollett, A.D., Srolovitz, D.J. and Anderson, M.P. (1989). *Acta Met.*, **37**, 1227.

Rollett, A.D. and Anderson, M.P. (1990). *Simulation and theory of evolving microstructures*, TMS-AIME, Warrendale, PA 15086.

Rollett, A.D., Srolovitz, D.J., Anderson, M.P. and Doherty, R.D. (1992). *Acta Metall. et Mater.*, **40**, 3475.

Salkind, S.M., George, F. and Tice, W. (1969). *Trans. Met. Soc. AIME*, **245**, 2339.

Salkind, S.M., Leverant, G. and George, F. (1967). *J. Inst. Metals*, **95**, 349.

Schwartz, D.M. and Ralph, B. (1969a). *Phil. Mag.*, **19**, 1069.

Schwartz, D.M. and Ralph, B. (1969b). *Met. Sci. J.*, **3**, 216.

Servi, I. S. and Turnbull, D. (1966). *Acta Met.*, **14**, 161.

Shewmon, P.G. (1963). *Diffusion in solids* McGraw-Hill, New York, p. 171.

Shewmon, P.G. (1966). *Recrystallization grain growth and textures*, ASM, Metals Park, Ohio, and Chapman and Hall, London, p. 165.

Shewmon, P.G. and Robertson, W.M. (1963). *Metal surfaces* ASM, Metals Park, Ohio, and Chapman and Hall, London, p. 67.

Simpson, C.J., Aust, K.T. and Winegard, W.C. (1971). *Met. Trans.*, **2**, 987.

Skapski, A.S. (1948). *J. Chem. Phys.*, **16**, 389.

Smartt, H.M., Tu, L.K. and Courtney, T. (1971). *Met. Trans.*, **2**, 2717.

Smith, A. (1967). *Acta Met.*, **15**, 1867.

Smith, A.F. (1965). *J. Less-Common Metals*, **9**, 233.

Smith, C.S. (1952). *Metal interfaces*, ASM, Metals Park, Ohio, p. 65.

Smith, C.S. (1964). *Met. Revs.*, **9**, 1.

Soutiere, B. and Kerr, A.W. (1969). *Trans. Met. Soc. AIME*, **245**, 2595.

Speich, G.R. and Oriani, R.A. (1965). *Trans. Met. Soc. AIME*, **233**, 623.

Speight, M.V. (1968). *Acta Met.*, **16**, 133.

Srolovitz, D.J., Grest, G.R. and Anderson, M.P. (1985). *Acta Met.*, **33**, 2233.

Srolovitz, D.J., Anderson, M.P., Sahni, P.S. and Grest, G.S. (1984a). *Acta Met.*, **32**, 793.

Srolovitz, D.J., Anderson, M.P., Grest, G.S. and Sahni, P.S. (1984b). *Acta Met.*, **32**, 1429.

Stumpf, W.F. and Sellars, C.M. (1969). *The mechanism of phase transformation in crystalline solids*, Institute of Metals, London, p. 120.

Sundquist (1964). *Acta Met.*, **12**, 67.

Swalin, R.A. (1972). *Thermodynamics of solids*, John Wiley, New York.

Tanner, L.E. (1966). *Phil. Mag.*, **14**, 111.

Thompson, C.V., Frost, H.J. and Spaepen, F. (1987). *Acta Met.*, **35**, 887.

Thursfield, G., Jones, I.J., Burden, M.H., Dibling, R.C. and Stowell, M.J. (1970). *Fizika*, **2**, 19.

Towner, R.J. (1958). *Metal Progress*, **73**, 70.

Turnbull, D. (1950). *J. Appl. Phys.*, **21**, 1022.

Tweed, C.J., Hansen, N. and Ralph, B. (1983). *Metal. Trans.*, **14A**, 2235.

Udin, H., Shaler, A.J. and Wulff, J. (1949). *Trans. Met. Soc. AIME*, **185**, 186.

Umantsev, A. and Olson, G.B. (1993). *Scripta Metall. et Mater.*, **29**, 1135.

Vandermeer, R.A. (1992). *Acta Metall. et Mater.*, **40**, 1159.

Voorhees, P.W. and Glicksman, M.E. (1984a). *Metal Trans.*, **15A**, 1081.

Voorhees, P.W. and Glicksman, M.E. (1984b). *Acta Met.*, **32**, 2001.

Voorhees, P.W. and Glicksman, M.E. (1984c). *Acta Met.*, **32**, 2013.

Voorhees, P. and Johnson, W.C. (1986a). *Metal Trans.*, **16A**, 337.

Voorhees, P. and Johnson, W.C. (1986b). *J. Chem. Phys.*, **84**, 5108.

Voorhees, P.W. and Schaeffer, R.J. (1987). *Acta Met.*, **35**, 327.

Wagner, C. (1961). *Z. Elektrochem.*, **65**, 581.

Warlimont, H. and Thomas, G. (1970). *Met. Sci. J.*, **4**, 47.

Weare, D. and Glazier, J.A. (1992). *Grain growth in polycrystalline materials* (eds. Abbruzzese, G. and Brosso, P.) *Materials Science Forum*, Trans Tech Publications, Zurich, **94–96**, p. 27.

Weare, D. and Kermode, J.P. (1983). *Phil. Mag.*, **48B**, 245.

Weatherly, G.C. (1971). *Acta Met.*, **19**, 181.

Weatherly, G.C. and Nakagawa, Y.G. (1971). *Scripta Met.*, **5**, 777.

Weatherly, G.C. and Nicholson, R.B. (1968). *Phil. Mag.*, **17**, 801.

Wendt, H. and Haasen, P. (1983). *Acta Met.*, **31**, 1649.

Weyl, H. (1957). *Symmetry*, Princeton University Press.

Whelan, E.P. and Howarth, C.W. (1964). *J. Inst. Metals*, **94**, 402.

Williams, W.M. and Smith, C.S. (1952). *Trans. Met. Soc. AIME*, **194**, 755.

Wulff, G. (1901). *Z. Krist.*, **53**, 440.

Xiao, S.Q. and Haasen, P. (1991). *Acta Metall. et Mater.*, **39**, 651.

Yang, S.C., Higgins, G.T. and Nash, P. (1992). *Mater. Sci. and Tech.*, **8**, 10.

Youle, A., Schwartz, D.M. and Ralph, B. (1971). *Met. Sci. J.*, **5**, 131.

Zener, C. (1948). Private communication to C.S. Smith, *Trans, Met. Soc. AIME*, **175**, 15.

## Chapter 6

Anthony, T.R. (1970). *Acta Met.*, **18**, 877.

Anthony, T.R. and Cline, H.E. (1970). *Phil. Mag.*, **22**, 893.

Anthony, T.R. and Cline, H.E. (1971). *J. Appl. Phys.*, **42**, 1823.

Anthony, T.R. and Cline, H.E. (1972). *Acta Met.*, **20**, 247.

Anthony, T.R. and Cline, H.E. (1973). *Acta Met.*, **21**, 117.

Anthony, T.R. and Sigsbee, R.A. (1971). *Acta Met.*, **19**, 1029.

Anthony, T.R. and Turnbull, D. (1966). *Phys. Rev.*, **151**, 495.

Ashby, M.F. (1980) in *Recrystallization and grain growth in multi-phase and particle containing materials* (eds. Hansen, N., Jones, A.R. and Leffers, T.) 1st Riso Symposium on Metallurgy and Materials Science, Roskilde, Denmark, p. 325.

Bardsley, W., Boulton, J.S. and Hurle, D.T.J. (1962). *Solid State Electronics*, **5**, 135.

Barnes, R.S. (1964). *J. Nucl. Mat.*, **11**, 135.

Barnes, R.S. and Mazey, D.J. (1963). *Proc. Roy. Soc.*, **A275**, 47.

Barnes, R.S. and Mazey, D.J. (1964). *Proc. Roy. Soc.*, **A275**, 47.

Bellamy, R.G. (1962). *Proc. Symposium on Uranium and Graphite*, Institute of Metals Monograph no. 27, p. 53.

Birkenbeil, H.J. and Cahn, R.W. (1962). *Proc. Phys. Soc.*, **79**, 831.

Bleach, I.R. and Meieran, E.S. (1967). *Appl. Phys. Letters*, **11**, 263.

Bleach, I.R. and Meieran, E.S. (1969). *J. Appl. Phys.*, **40**, 485.

Brett, S.J. and Doherty, R.D. (1978). *Mater. Sci. and Eng.*, **32**, 255.

Broom, T. and Ham, R.K. (1957). *Proc. Roy. Soc.*, **A242**, 17.

Broom, T., Mazza, J.A. and Whittaker, V.N. (1957). *J. Inst. Metals*, **86**, 617.

Burton, B. (1971). *Met. Sci. J.*, **5**, 11.

Cahn, J.W. (1963). *J. Appl. Phys.*, **34**, 3581.

Calaberese, C. and Laird, C. (1974). *Mater. Sci. and Eng.*, **13**, 141.

Chikazumi, S. (1950). *J. Phys. Soc. Japan*, **5**, 327.

Christian, J.W. (1965). *The theory of transformation in metals and alloys*, 1st edn, Pergamon Press, Oxford, p. 910.

Clegg, W.J. and Martin, J.W. (1983) in *Deformation of multi-phase and particle containing materials*, (eds. Bilde-Sorensen, J.B., Hansen, N., Horsewell, A., Leffers, T. and Lilholt, H.) Riso National Laboratory, Denmark, p. 199.

Cline, H.E. and Anthony, T.R. (1971a). *Acta Met.*, **19**, 175.

Cline, H.E. and Anthony, T.R. (1971b). *Acta Met.*, **19**, 491.

Cline, H.E. and Anthony, T.R. (1972). *J. Crystal Growth*, **13**, 790.

Coble, R.L. (1963). *J. Appl. Phys.*, **34**, 1679.

Coffin, L.F. and Tavernelli, J.F. (1959). *Trans. Met. Soc. AIME*, **215**, 794.

Cullity, B.D. (1972). *Introduction to magnetic materials*, Addison-Wesley, London, pp. 357, 565.

Denbigh, K.H. (1951). *The thermodynamics of the steady state*, Methuen, London, p. 69.

de Vos, K.J. (1969). *Magnetism and metallurgy* (eds. Berkowitz, A.E. and Kneller, E.) Academic Press, London, vol. 1, p. 473.

de Vries, G., Van Geest, D.W., Gersdorf, R. and Rathenau, G.W. (1959). *Physics*, **25**, 1131.

Doherty, R.D. (1983) in *Micromechanisms of plasticity and fracture*, (eds Lewis, M.H. and Taplin, D.M.R.) Parsons Press, Trinity College, Dublin, Ireland, p. 303.

Doherty, R.D. and Strutt, T. (1976). *J. Mat. Sci.*, **11**, 2169.

Embury, J.D. and Nicholson, R.B. (1965). *Acta Met.*, **13**, 403.

Fargette, B. (1979). *Metals Techn.*, **6**, 343.

Flemings, M.C. (1974). *Solidification processing*, McGraw-Hill, New York.

Gerold, V. and Meier, B. (1987). *Fatigue '87*, vol. III, (eds. Ritchie, R.O. and Starke, E.A. Jr) EMAS, Warley, UK, p. 1517.

Ghate, P.B. (1967). *Appl. Phys. Letters*, **11**, 14.

Graham, C.D. (1959). *Magnetic properties of metals and alloys*, ASM, Metals Park, Ohio, p. 288.

Greenwood, G.W. (1970). *Scripta Met.*, **4**, 171.

Grey, E.A. and Higgins, G.T. (1972). *Scripta Met.*, **6**, 256.

Gruber, E.E. (1967). *J. Appl. Phys.*, **38**, 243.

Gruhl, W. and Gruhl, U. (1954). *Z. f. Metallkunde*, **8**, 20.

Haessner, F., Hofmann, S. and Seekel, H. (1974). *Scripta Met.*, **8**, 299.

Ham, R.K. (1966). *Can. Met. Quart.*, **5**, 161.

Ham, R.K., Kirkaldy, J.S. and Plewes, J.T. (1967). *Acta Met.*, **15**, 861.

Harris, J.E. (1973). *Metal Sci.* **7**, 1.

Harris, J.E. and Jones, R.B. (1963). *J. Nucl. Mat.*, **10**, 360.

Harris, J.E., Jones, R.B., Greenwood, G.W. and Ward, M.J. (1968). *J. Austral. Inst. Metals.*, **14**, 154.

Herring, C. (1950). *J. Appl. Phys.*, **21**, 437.

Hillig, W.B. (1958). *Growth and perfection of crystals*. John Wiley, London, p. 350.

Ho, P.S. (1966). *J. Phys. Chem. Solids*, **27**, 1331.

Ho, P.S. (1970). *J. Appl. Phys.*, **41**, 64.

Ho, P.S. and Huntingdon, H.B. (1966). *J. Phys. Chem. Solids*, **27**, 1319.

Jackson, K. (1958) in *Liquid metals and solidification*, ASM, Metals Park, Ohio, p. 174.

Jackson, K., Uhlmann, D.R. and Hunt, J.D. (1967). *J. Crystal Growth*, **1**, 1.

Johnson, E.W. and Johnson, H.H. (1965). *Trans. Met. Soc. AIME*, **233**, 1333.

Johnson, W.C. (1987). *Met. Trans.*, **18A**, 233.

Johnson, W.C., Berkenpas, M.B. and Laughlin, D.E. (1988). *Acta Met.*, **36**, 3149.

Jones, D.R.H. (1974a). *Mat. Sci. Eng.*, **15**, 203.

Jones, D.R.H. (1974b). *Met. Sci. J.*, **8**, 37.

Jones, D.R.H. and Chadwick, G.A. (1971). *Phil. Mag.*, **24**, 1327.

Jones, D.R.H. and May, G.J. (1975). *Acta Met.*, **23**, 29.

Jones, R.B. (1969). *Quantitative relation between properties and microstructure* (eds. Brandon, D.G. and Rosen, A.) Israel University Press, p. 343.

Kelly, A. and Nicholson, R.E. (1963). *Prog. in Mat. Sci.*, **10**, 151.

Kohler, E., Steiner, D. and Gerold, V. (1983) in *Deformation of multi-phase and particle containing materials* (eds. Bilde-Sorensen, J.B., Hansen, N., Horsewell, A., Leffers, T. and Lilholt, H.) Riso National Laboratory, Denmark, p. 345.

Kramer, J.J. and Tiller, W.A. (1963). *J. Chem. Phys.*, **37**, 841.

Kramer, J.J. and Tiller, W.A. (1965). *J. Chem. Phys.*, **42**, 257.

Kurz, W. and Fisher, D.J. (1989). *Fundamentals of solidification*, 3rd edn, Trans. Tech. Publications, Switzerland.

Lea, C., Brett, S.J. and Doherty, R.D. (1979). *Scripta Met.*, **13**, 45.

Lee, J.J. and Laird, C. (1983). *Phil. Mag.*, **47A**, 579.

Lendvai, J., Gudladt, H.J. and Gerold, V. (1988). *Scripta Met.*, **22**, 1755.

Lewis, M.H. and Martin, J.W. (1963). *Acta Met.*, **11**, 1207.

Lormand, G., Rouais, J.-C. and Eyraud, C. (1965). *C.R. Acad. Sci.*, **261**, 1291.

Lormand, G., Rouais, J.-C. and Eyraud, C. (1974). *Acta Met.*, **22**, 793.

Luborsky, F.E., Livingston, J.D. and Chin, G.Y. (1983) in *Physical metallurgy*, (eds. Cahn, R.W. and Haasen, P.) North-Holland Physics Publishing, Amsterdam, p. 1673.

Martin, J.W. (1980). *Micromechanisms in particle hardened alloys*, Cambridge University Press, Cambridge, pp. 51–64.

McEvily, A.J., Jr, Clark, J.B., Utley, E.C. and Herrnstein III, W.H. (1963). *Trans. Met. Soc. AIME*, **227**, 1093.

McGrath, J.T. and Bratina, W.J. (1967). *Acta Met.*, **15**, 329.

McLean, M. (1975). *Scripta Met.*, **9**, 439.

McLean, M. and Loveday, M.S. (1974). *J. Mat. Sci.* **9**, 1104.

Mughrabi, H. (1983) in *Deformation of multi-phase and particle containing materials* (eds. Bilde-Sorensen, J.B., Hansen, N., Horsewell, A., Leffers, T. and Lilholt, H.) Riso National Laboratory, Denmark, p. 65.

Mullins, W.W. (1956). *Acta Met.*, **4**, 421.

Nabarro, F.R.N. (1948). *Report of conference on strength of solids*, Physical Society, London, p. 74.

Nelson, R.S., Hudson, J.A. and Mazey, D.J. (1972). *J. Nucl. Mat.*, **44**, 318.

Nichols, F.A. (1969). *J. Nucl. Mat.*, **30**, 143.

Nichols, F.A. (1972). *Acta Met.*, **20**, 207.

Oriani, R.A. (1959). *Thermodynamics and transport properties of gases, liquids and solids*, McGraw-Hill, London, p. 123.

Peters, C.T. and Miodownik, A.P. (1973). *Scripta Metall.*, **7**, 955.

Pfann, W. (1955). *Trans. Met. Soc. AIME*, **203**, 961.

Pfann, W. (1966). *Zone melting*, 2nd edn, John Wiley, New York, chapter 10.

Pickles, B.W. (1967). *J. Inst. Metals*, **95**, 333.

Quin, M.P. and Schwartz, L.H. (1980). *Mater. Sci. and Eng.*, **46**, 249.

Rabkin, V.G. and Buinov, N.N. (1961). *Physics of metals and metallography*, **11**, 61.

Rigney, D.A. and Blakely, J.M. (1966). *Acta Met.*, **14**, 1375.

Sargent, C.M. and Purdy, G.R. (1974). *Scripta Met.*, **8**, 569.

Satyanarayan, K.R. and Miodownik, A.P. (1969). *The mechanism of phase transformation in crystalline solids*, Institute of Metals, London, p. 162.

Sens, P.F. (1972). *J. Nucl. Mat.*, **43**, 293.

Shewmon, P.G. (1964). *Trans. Met. Soc. AIME*, **230**, 1134.

Sinning, H.R. (1982). *Acta Met.*, **30**, 1019.

Squires, R.L., Weiner, R.T. and Phillips, M. (1963). *J. Nucl. Mat.*, **8**, 77.

Stubbington, C.A. and Forsyth, P.J.E. (1966). *Acta Met.*, **14**, 5.

Swalin, R.A. and Yin, C.A. (1967). *Acta Met.*, **15**, 245.

Tien, J.K. and Copley, S.M. (1971). *Met. Trans.*, **2**, 215, 543.

Tiller, W.A. (1963). *J. Appl. Phys.*, **34**, 2757, 2763.

Tiller, W.A. (1966). *Acta Met.*, **14**, 1383.

Tiller, W.A. (1969). *J. Crystal Growth*, **6**, 77.

Vickers, W. and Greenfield, P. (1967). *J. Nucl. Mat.*, **24**, 249.

Vogel, D.L. and Rieck, G.D. (1971). *Acta Met.*, **19**, 233.

Wayman, C.M. (1983) in *Physical metallurgy*, (eds. Cahn, R.W. and Haasen, P.) North-Holland Physics Publishing, Amsterdam, p. 1031.

Wernick, J.H. (1956). *J. Chem. Phys.*, **25**, 47.

Wernick, J.H. (1957). *Trans. Met. Soc. AIME*, **209**, 1169.

Whitman, W.D. (1926). *Amer. J. Sci. Ser.* **5**XI, 126.

Willertz, L.E. and Shewmon, P.G. (1970). *Met. Trans.*, **1**, 2217.

Woodruff, D.P. (1973). *The solid–liquid interface*, Cambridge University Press, Cambridge, chapter 3.

# Index

Printed in the United States
By Bookmasters